THEORY OF MACHINES

Kinematics and Dynamics

(Second Edition)

THEORY OF MACHINES
Kinematics and Dynamics
(Second Edition)

B.V.R. GUPTA

Formerly Professor & Dean
Faculty of Engineering
Andhra University
Visakhapatnam, A.P.

I.K. International Pvt. Ltd.

NEW DELHI

Published by
I.K. International Pvt. Ltd.
4435-36/7, Ansari Road, Daryaganj
New Delhi-110 002 (India)
E-mail: info@ikinternational.com
Website: www.ikbooks.com

ISBN: 978-93-90620-82-1

© 2023 I.K. International Pvt. Ltd.

All rights reserved. No part of this publication may be reproduced, stored in a retrieval system, or transmitted in any form or any means: electronic, mechanical, photocopying, recording, or otherwise, without the prior written permission from the publisher.

Published by Krishan Makhijani for I.K. International Pvt. Ltd., 4435-36/7, Ansari Road, Daryaganj, New Delhi-110 002 and Printed by Rekha Printers Pvt. Ltd., Okhla Industrial Area, Phase II, New Delhi-110 020.

Dedication

I dedicate this book on **THEORY OF MACHINES** to my beloved brother.

Sri B.P. Gupta

He is responsible for what I am today. He wrote books on mathematics for engineering students in late fifties when I was very young. His motivation and encouragement helped me in all my endeavours including writing books. I present this book and my previous book on ENGINEERING DRAWING to my beloved student community of past, present and future.

Preface to the Second Edition

Life has become mechanical nowadays. No mechanical no mobility or motion. All components are designed, manufactured and maintained by mechanicals only. Theory of machines is an important subject not only to mechanical engineering discipline but also to other engineering disciplines. The book covers both kinematics (motion) and dynamics (motion and forces) in the machines. This book is useful to students of degree level or diploma level or those preparing for AMIE.

In this revised edition few more points have been added with pictures here for better understanding of the concepts like in gyroscopes and in other chapters. Few printing mistakes observed in figures or in expressions or in write up etc have been corrected. Thanks to all those friends who have given good suggestions.

B.V.R. GUPTA

Preface to the First Edition

Theory of machines is an important subject not only to mechanical engineering discipline, but also to other engineering disciplines. All products whether they are mechanical, electrical, civil, computers or electronics, are designed, manufactured and maintained by mechanicals only. This is one of the interesting and useful subject which deals with the mechanisms and machines. The book covers both kinematics and dynamics of machines. The book will be useful not only to degree level students of mechanical engineering but also to those preparing for AMIE and various other competitive examinations.

Chapter 1 introduces the concepts of mechanisms and machines. In Chapter 2 various mechanisms consisting of lower pairs are discussed. Chapter 3 deals with the analysis of their motions such as velocities and accelerations of various mechanisms using different methods. Chapter 4 deals with the study of kinetics, the time-varying forces in machines such as torque and power. Chapter 5 gives turning moment diagram and design of flywheels. Chapter 6 gives about friction and its effect on mechanical efficiency. The friction effects on bearings and its uses in power transmission. In Chapter 7, types of governors for speed regulation are presented. Chapter 8 deals with transmission of power using belts, ropes and chains. Chapter 9 deals with gyroscopic effects on the vehicles such as two wheelers, four wheelers, aeroplanes and ship. Chapter 10 gives about different types of cams and their design. Chapter 11 deals with gears and their types. Chapter 12 gives the application of gears used for transmission of power. Chapter 13 and Chapter 14 deal with balancing of rotating masses and reciprocating masses. Chapter 15 and Chapter 16 are on the study of longitudinal, transverse and torsional vibrations.

Generally in many books, several examples will be given but in this book a few common examples are shown for different methods. This will enable students or new faculty to understand the relative merits and demerits of the methods. This will help them in choosing the right method for the given problem and the time it requires. For example, in the velocity and accelerations, whether to use instantaneous centre method or relative velocity method and in the governors, whether to use analytical or graphical method, etc. are given so that the students will not try in the examination time or new faculty in the classrooms while teaching.

Another highlight of this book is unlike other books where different problems are shown for different parameters whereas in this book, the same problem has been shown by changing each time one parameter. This will give a clear idea to students the effect of that particular parameter. For example in cams, the effect of follower, motion, whether it is offset or radial cam on its profile. In governors, Porter or Proell, unequal or equal arms, pivoted on the spindle or away from the spindle, etc. Students should not attempt by trial and error way especially in the examinations time.

The author has written this book by his forty years of teaching experience. This book will help not only to students but also to new faculty members for preparing well in all the above aspects with confidence. The international System of Units (SI) has been adopted throughout the book. The author thanks all those who influenced him in writing this book.

B.V.R. GUPTA
'HARINIVAS'
D.No. 4-50-6, Lawsons Bay Colony,
Visakhapatnam–530 017.

Acknowledgements

I thank my teachers who taught me this subject and inspired me. The doubts I got during the student days and later as a faculty member while teaching this subject has made me to write this book. Many books are available but they lack proper information from the student's point of view and junior faculty members. I must thank one most important person, **Prof. S. Tiruvengalum**, Professor of Electrical Engineering (Retd.), JNTU Engineering College, Anantapur for his blessings and encouragement right from my student days till now. I also must thank my teacher and colleague **Prof. N. L. N. Rao**, Professor of Mechanical Engineering (Retd.), Andhra University College of Engineering Visakhapatnam for his wonderful guidance throughout my career. My thanks to my mentors, **Prof. N. Ganesan** and **Prof. S. Narayanan**, Mechanical Department, I.I.T., Madras under whom I did my Ph.D. in Dynamics.

I thank the Hon'ble Vice-Chancellors of Andhra University, JNTU, Kakinada and Anantapur for their encouragement in writing the book.

I also thank Head of the Department of Mechanical Engineering, Head of the Department of Marine Engineering, Andhra University and Head of the Department of Mechanical Engineering, Govt. Polytechnic, Visakhapatnam for their advices and good support.

I must thank my sons **Mr. Yoganand** and **Mr. Hari Kishan**, Engineers for their guidance in computer usage in preparing the manuscript. My thanks to **Mr. G. Mahesh**, son of my classmate **Mr. G. Mohan Krishna, E.E. (R&B), Retd.** for his guidance in AutoCAD.

I must thank my wife, daughter-in-laws and grandchildren, **Chy. Samanvitha, Chy. Akhil** and **Chy. Gayatri** for providing peaceful atmosphere throughout the preparation of this book.

Lastly, I must thank **Mr. Krishan Makhijani**, Managing Director, **I.K. International Publishing House Pvt. Ltd., New Delhi** for his support and coming forward to publish my book.

B.V.R. GUPTA
'HARINIVAS'
D.No. 4-50-6, Lawsons Bay Colony,
Visakhapatnam–530 017.

Machine Design Section
Department of Mechanical Engineering
Indian Institute of Technology (Madras) Chennai–600036

Foreword

Theory of Machines and Mechanisms also known as Kinematics and Dynamics of Machinery is one of the most important core subjects taught to Mechanical Engineering undergraduate students all over the world. This subject is a natural sequel to the study of the basic courses in Engineering Mechanics, Statics and Dynamics. This subject introduces the student to different types of mechanisms, their motions and their analysis and synthesis and forms an essential pre-requisite for leading the student to the concepts and processes of Machine Design. It is also important to other engineering disciplines like Aerospace Engineering, Applied Mechanics and Engineering Design. With the advent of Micro Electronic Mechanical Systems (MEMS), Mechatronic devices and Robotics in various applications, the importance of the subject extends to other multidisciplinary areas.

Though there are many textbooks written by a number of authors giving emphasis to different aspects of the subject, the present book **"Theory of Machines" Kinematics and Dynamics** written by Prof. B.V.R. Gupta, Principal of Simhadhri Educational Society Group of Institutions, Andhra University College of Engineering, Visakhapatnam brings to fore his rich and long years of experience of teaching this course to several batches of Mechanical Engineering students in his college and elsewhere. Another feature of this book is that when different methods of solution are available the same example problem is considered to illustrate the working of the different methods and bring out the similarity and differences between them.

I know Prof. B.V.R. Gupta for more than three decades now. He was my doctoral student at IIT Madras. He is known for his excellent teaching and profound knowledge of the subject. This book is the result of his careful and meticulous attention to details and concepts and hard work. This book is written to all Mechanical Engineering undergraduate students in the country and the curriculum of all Universities and Autonomous Institutions in mind. I am certain that this book will be an important and useful addition to the number of books on this subject. I recommend this book for adoption by the students and the institutions.

                                              Dr. S. Narayanan
                                              Professor Emeritus,
                                              (Formerly Dean Academic Research),
                                              Department of Mechanical Engineering,
                                              IIT Madras,
                                              Chennai–600 036.

13th March 2010

G.M.Rao
Group Chairman
GMR Group
IBC Knowledge Park, Phase 2, D Block, 11th Floor
4/1. Bannerghatta Road,
Bangalore.

Message

Dear Dr. Gupta garu,

I am very happy to know that you have written yet another book titled Theory of Machines. Please accept my heartiest best wishes on this occasion and I am confident, this book would be of great help to the engineering students and also a good reference work for the faculty. This is my message for your great book launch.

Warm regards,
G M Rao

N. Sambasiva Rao, I.P.S.
Commissioner of Police
Visakhapatnam City

✆ 0891-2562709, 2562763 (O)

Date: 18/3/10

Message

I am glad that my professor Dr. B.V.R. Gupta, Retired from Andhra University, Engineering College, Visakhapatnam has written a book on **"THEORY OF MACHINES"**. I have gone through the book and it has reminded me my student days. He has presented the subject in the same manner he taught us. I did my project work also under his guidance. The concepts are well presented in simple words along with figures wherever possible. It is easy for any student to follow on his own. The book covers both **KINEMATICS AND DYNAMICS OF MACHINES**. The book gives clarifications to all doubts that the students get. The application of different methods and which method the student has to choose during the examination time has been well explained.

This is his second book. The first book he wrote is on **"ENGINEERING DRAWING"** which is an important subject to all branches of students. In a short period the book has become so popular and it has prescribed in most of the universities in Andhra Pradesh. He is a well known teacher for these two subjects.

I am sure that these two books will be made use by all students of both B.Tech/B.E and Diploma in all the colleges in the country.

(N. SAMBASIVA RAO)

Prof B.S.K. Sundara Siva Rao
PhD, FIE,CE
Head of the Department &
Chairman Board of Examiners
Tel. (O) 0891-2844804

ANDHRA UNIVERSITY
Department of Mechanical Engineering
Visakhapatnam 530003
Email: hodmechau@gmail.com
Fax (O)   0891-2747969
Cell   9848186121

Message

It gives me great pleasure to introduce this textbook **'THEORY OF MACHINES'** by Prof. B. V. R. Gupta. Prof. Gupta is my teacher and senior colleague in the Department of Mechanical Engineering, Andhra University. The author has been teaching this subject for the last forty years to both U.G. and P.G. students. The students used to express their appreciation for the way he deals with the subject.

In addition to the basic engineering subjects like Engineering Drawing, Engineering Mechanics etc., the subject Theory of Machines is an important subject for Mechanical Engineers.

This book covers the syllabus usually prescribed for the Diploma courses conducted by various boards, Undergraduate courses in Universities and various Institutions of Engineers.

This book is written in a simple, lucid way with the average student in mind. The pictures provided in the book gives students a proper understanding and feel for practical applications of components.

The multiple choice questions, short answer questions given at the end of each chapter are useful for the students appearing for their university examinations and competitive examinations.

My Best Wishes to author Prof. B.V.R. Gupta and students who are going to make use of this Textbook.

28-03-2010
Visakhapatnam

(Dr. B.S.K. Sundara Siva Rao)

ANDHRA UNIVERSITY: COLLEGE OF ENGINEERING
DEPARTMENT OF MARINE ENGINEERING

Prof. T.V.K. Bhanuprakash April 4, 2010
B.E.(AU), M.Tech (IIT-Kgp), Ph.D(U Miami, USA)
Dept. of Marine Engineering
College of Engineering
Andhra University
Visakhapatnam - 530 003.

Message

I am very happy to know that my Prof. B.V.R. Gupta has brought out a new book in **Theory of Machines**. I believed that only a good teacher can write a good book and the present book is a classic example of that. Though there are many books on Theory of Machines, this particular book is written specially keeping the student in mind. Theory has been presented clearly and worked out examples are solved with many different techniques and the student can now choose the technique that suits him best. Figures are clear and the presentation is simple. End of the book exercises, short questions and multiple-choice questions help the student to better understand the subject.

I am sure that this beautiful book will enhance the understanding of the subject of Theory of Machines and will remain as a book with long standing patronage. I once again congratulate the author Prof. B.V.R. Gupta for this excellent work and wish him all the best in his future endeavours.

(T.V.K. Bhanuprakash)

24-4-7, Harbour Road, Visakhapatnam - 530 001.

Message

I am very glad to know that Prof. B.V.R. Gupta, my teacher, when I was a student (1970-75) of Mechanical Engineering with Marine as an elective subject from College of Engineering - Andhra University, has recently written another important book on **THEORY OF MACHINES**. He taught us this subject and also **ENGINEERING DRAWING** (his first book) in our first year. I was one of the top few students to obtain good percentage of marks in both these subjects. Prof. Gupta has presented in such a way that each student not only understands clearly the subject but also faces the examinations with ease. He clearly explained the mistakes which students are often prone to commit in examinations. These two books are the outcome of his keen interest on these subjects. I congratulate him for his efforts in bringing out this present book with illustrious pictures for good comprehension even by an average student and I am sure all the engineering students will get immense benefit from these two books.

K. Bimdhu Mohan, BE(Mech-Marine), FIMarE, FIE., C. Engr,
Chief Engineer,
Chairman & Managing Director
KVA Rao Marine Technocrats (P) Ltd., Vizag.

CONTENTS

Preface to the Second Edition — *vii*
Preface to the First Edition — *ix*
Acknowledgements — *xi*
Foreword — *xiii*
Units and their Values — *xxxv*
Abbreviations, Notations and Symbols — *xxxvii*

1. Simple Mechanisms — 1

1.1 Introduction — 3
1.2 Kinematic Link or Element — 3
1.3 Kinematic Pair — 5
    1.3.1 Nature of Relative Motion between the Elements — 5
    1.3.2 Nature of Contact between the Elements — 5
    1.3.3 Nature of the Mechanical Arrangement for Complete or Successful Constraint between the Elements — 6
1.4 Kinematic Chain — 7
    1.4.1 First Equation Using Pairs — 8
    1.4.2 Second Equation Using Joints — 9
    1.4.3 According to the Type of Closure between Elements — 10
    1.4.4 Degrees of Freedom — 11
1.5 Mechanism — 11
1.6 Inversion — 12
    1.6.1 Single Slider Crank Chain — 12
    1.6.2 Double Slider Crank Chain — 15
    1.6.3 Four-Bar Mechanisms — 17
1.7 Exercise — 20
    1.7.1 Short Answer Questions — 20
    1.7.2 Problems — 22
    1.7.3 Multiple Choice Questions — 22

2. Mechanisms with Lower Pairs 25

    2.1   Introduction   27
    2.2   Pantograph   27
    2.3   Mechanisms for Straight Line Motions   28
            2.3.1   Peaucellier Mechanism   29
            2.3.2   Hart Mechanism   29
            2.3.3   Scott-Russell Mechanism   30
    2.4   Approximate Straight Line Mechanism   31
            2.4.1   Watt Mechanism   31
            2.4.2   Grasshopper Mechanism   32
            2.4.3   Tchebicheff Straight Line Motion   32
            2.4.4   Roberts Mechanism   33
    2.5   Steering Gear Mechanism   33
            2.5.1   Davis Steering Gear (Exact)   34
            2.5.2   Ackermann Steering Gear (Approximate)   35
    2.6   Hooke's Joint (or) Universal Joint   36
    2.7   Double Hooke's Joint   41
    2.8   Exercise   43
            2.8.1   Short Answer Questions   43
            2.8.2   Problems   45
            2.8.3   Multiple Choice Questions   46

3. Velocities and Accelerations in Mechanisms 49

    3.1   Introduction   51
    3.2   Motion   51
            3.2.1   Translatory Motion   51
            3.2.2   Rotary Motion   51
            3.2.3   Speed   51
            3.2.4   Angular Displacement (θ)   52
            3.2.5   Radian   52
            3.2.6   Angular Velocity (ω)   52
            3.2.7   Relation between Linear Velocity and Angular Velocity   53
    3.3   Instantaneous Centre Method   53
            3.3.1   Properties of Instantaneous Centres   54
            3.3.2   Number of Instantaneous Centres in a Mechanism   55
            3.3.3   Types of Instantaneous Centres   55

3.3.4	Location of Instantaneous Centres	55
3.3.5	Kennedy's Theorem or Three-centres-in-line Theorem	56
3.3.6	Application of Instantaneous Centre to Any Mechanism	57
3.3.7	Steps in Determining the Unknown Instantaneous Centres	57

3.4 Relative Velocity Method — 64
3.5 Acceleration in Mechanisms — 70
    3.5.1 Introduction — 70
    3.5.2 Angular Acceleration — 70
    3.5.3 Vector form between Linear and Angular Acceleration — 70
    3.5.4 Various Steps to be Followed in the Acceleration Analysis — 71
3.6 Coriolis Component of Acceleration — 78
3.7 Exercise — 83
    3.7.1 Short Answer Questions — 83
    3.7.2 Problems — 83
    3.7.3 Multiple Choice Questions — 88

4. Inertia Forces in Reciprocating Parts — 91

4.1 Introduction — 93
    4.1.1 Terms Used in Static — 93
    4.1.2 D'Alembert's Principle — 94
4.2 Analytical Method for Reciprocating Mechanism — 95
    4.2.1 Displacement of Piston (X_p) — 96
    4.2.2 Velocity of Piston (v_p) — 97
    4.2.3 Acceleration of Piston (a_p) — 97
    4.2.4 Angular Velocity of Connecting Rod (ω_c) — 98
    4.2.5 Angular Acceleration (α_c) — 98
4.3 Klein's Construction for Reciprocating Mechanisms — 100
    4.3.1 Klein's Velocity Diagram — 100
    4.3.2 Klein's Acceleration Diagram — 101
4.4 Forces on the Reciprocating Parts of an Engine — 104
    4.4.1 Neglecting the Weight of the Connecting Rod — 104
    4.4.2 Considering the Weight of the Connecting Rod — 109
4.5 Equivalent Dynamical System — 110
    4.5.1 Dynamically Equivalent System — 110
    4.5.2 Determination of Dynamically Equivalent System of Two Masses Placed Arbitrarily (Analytically) — 111

		4.5.3	Determination of Dynamically Equivalent System of Two Masses Placed Arbitrarily (Graphically)	112

4.6	Inertia Forces in a Reciprocating Engine	113
	4.6.1 Graphical Method	113
	4.6.2 Analytical Method	114
4.7	Exercise	120
	4.7.1 Short Answer Questions	120
	4.7.2 Problems	120
	4.7.3 Multiple Choice Questions	122

5. Turning Moment Diagrams and Design of Flywheel — 125

5.1	Introduction	127
5.2	Single-Cylinder Double-Acting Steam Engine	127
5.3	Four-Stroke Cycle Internal Combustion Engine	128
	5.3.1 Fluctuation of Energy	129
5.4	Flywheel	130
	5.4.1 Coefficient of Fluctuation of Speed	131
	5.4.2 Energy Stored in the Flywheel (E)	131
	5.4.3 Design of Flywheel	132
5.5	Typical Worked Examples	133
5.6	Flywheel in Punching Press	141
5.7	Exercise	144
	5.7.1 Short Answer Questions	144
	5.7.2 Problems	144
	5.7.3 Multiple Choice Questions	145

6. Friction — 147

6.1	Introduction	149
6.2	Laws of Friction	150
	6.2.1 Friction between Dry Surfaces	151
	6.2.2 Friction between Rough Surfaces	151
	6.2.3 Friction is Self Adjusting	151
	6.2.4 Angle of Friction (ϕ)	151
	6.2.5 Rolling Friction	152
6.3	Equilibrium of Body on a Rough Inclined Plane	153
	6.3.1 Motion Up the Plane	154

	6.3.2	Motion Down the Plane	154
	6.3.3	Maximum Efficiency	155
6.4	Screw Friction		156
	6.4.1	Square Thread	156
	6.4.2	Relation Between Effort and Weight Lifted by a Screw Jack	157
	6.4.3	V-Thread	158
	6.4.4	Mechanical Advantage	158
6.5	Pivot and Collar Friction		159
	6.5.1	Uniform Intensity of Pressure	161
	6.5.2	Uniform Rate of Wear	162
6.6	Clutches		164
	6.6.1	Single-plate Clutch	165
	6.6.2	Multi-plate Clutch	165
	6.6.3	Cone Clutch	166
6.7	Brakes and Dynamometers		168
	6.7.1	Introduction	168
	6.7.2	Types of Brakes	168
	6.7.3	Dynamometers	176
	6.7.4	Types of Frictions	178
6.8	Exercise		181
	6.8.1	Short Answer Questions	181
	6.8.2	Problems	182
	6.8.3	Multiple Choice Questions	185

7. Governors   187

7.1	Introduction		189
7.2	Centrifugal Governors		189
7.3	Various Parts and Terms Used in Governors		191
	7.3.1	Height of the Governor (h)	191
	7.3.2	Equilibrium Speed	191
	7.3.3	Sleeve Lift	191
7.4	Simple Watt Governor		191
	7.4.1	Analytical Method	192
	7.4.2	Graphical Method	193
7.5	Porter Governor		194
	7.5.1	Analytical Method	195
	7.5.2	Graphical Method	197

7.6	Proell Governor		197
	7.6.1	Analytical Method	198
	7.6.2	Graphical Method	199
	7.6.3	Comparison between Flywheel and Governor	209
7.7	Hartnell Governor		209
7.8	Hartung Governor		213
7.9	Definitions		218
	7.9.1	Sensitiveness	218
	7.9.2	Stable and Unstable	218
	7.9.3	Isochronous/Isochronism	218
	7.9.4	Hunting	218
	7.9.5	Effort	218
	7.9.6	Power	218
	7.9.7	Controlling Force	218
	7.9.8	Coefficient of Insensitiveness	219
7.10	Wilson-Hartnell Governor		219
7.11	Exercise		221
	7.11.1	Short Answer Questions	221
	7.11.2	Problems	221
	7.11.3	Multiple Choice Questions	223

8. Belt, Rope and Chain Drives 225

8.1	Introduction		227
8.2	Types of Belts		227
	8.2.1	Flat Belt	228
	8.2.2	V-belt	228
	8.2.3	Circular Belt or Rope	229
8.3	Types of Belt Drives		230
	8.3.1	Compound Belt Drives	231
	8.3.2	Stepped or Cone Pulley	232
8.4	Speed Ratio or Velocity Ratio of a Belt Drive		232
	8.4.1	Velocity Ratio of a Compound Belt Drive	233
	8.4.2	Slip of the Belt	234
	8.4.3	Effect of Creep on Velocity Ratio	235
8.5	Length of an Open Belt		235
8.6	Length of a Crossed Belt		237
8.7	Ratio of Tensions		239

	8.7.1	Power Transmitted by a Belt	241
	8.7.2	Effect of Centrifugal Tension T_C on Power Transmitted	241
	8.7.3	Condition for Maximum Power	242
	8.7.4	Effect of Initial Tension (T_0)	243
8.8	Rope Drive		243
	8.8.1	Ratio of Tensions	243
8.9	Chain Drives		245
	8.9.1	Types of Chains	246
8.10	Exercise		248
	8.10.1	Short Answer Questions	248
	8.10.2	Problems	249
	8.10.3	Multiple Choice Questions	251

9. Gyroscope   255

9.1	Introduction	257
9.2	Gyroscopic Couple and its Effect	258
9.3	Effect of Gyroscopic Couple on an Aeroplane	259
9.4	Special Terms Used in Ships	263
	9.4.1 Effect of Gyroscopic Couple on the Ship During Steering	264
	9.4.2 Effect of Gyroscopic Couple on the Ship During Pitching	265
	9.4.3 Effect of Gyroscopic Couple on the Ship During Rolling	266
9.5	Stability of Four-wheeler	268
	9.5.1 Effect of the Gyroscopic Couple	269
	9.5.2 Effects of the Centrifugal Couple	270
9.6	Stability of a Two-wheeler	273
	9.6.1 Effect of the Gyroscopic Couple	274
	9.6.2 Effects of the Centrifugal Couple	275
9.7	Exercise	278
	9.7.1 Short Answer Questions	278
	9.7.2 Problems	279
	9.7.3 Multiple Choice Questions	280

10. Cams   283

10.1	Introduction	285
10.2	Classification of Followers	286
	10.2.1 Based on the Surface in Contact	287

10.2.2	Based on the Type of Movement of the Follower	287
10.2.3	Based on the Line of Motion of Follower	287
10.2.4	Based on the Desired Mathematical Motions	288
10.3	Types of Cams	288
10.3.1	Based on Follower Motion	288
10.3.2	Based on the Shape of the Cam	288
10.4	Terminology	289
10.4.1	Cam Profile	289
10.4.2	Base Circle	289
10.4.3	Trace Point	289
10.4.4	Pitch Curve	290
10.4.5	Prime Circle	290
10.4.6	Pressure Angle	290
10.4.7	Cam Angle	290
10.4.8	Pitch Point	290
10.4.9	Lift or Stroke(s)	290
10.4.10	Pitch Circle	291
10.5	Analysis of Motion of the Follower	291
10.5.1	Uniform Velocity	292
10.5.2	Simple Harmonic Motion (SHM)	294
10.5.3	Uniform Acceleration and Retardation	296
10.5.4	Cycloidal Motion	297
10.6	Construction of Displacement Diagrams	299
10.6.1	Displacement Diagram for Uniform Velocity	300
10.6.2	Displacement Diagram for Simple Harmonic Motion (SHM)	300
10.6.3	Displacement Diagram for Uniform Acceleration and Retardation (UAR)	301
10.6.4	Displacement Diagram for Cycloidal Motion	302
10.7	Construction of Cam Profiles	303
10.7.1	Cam Profile with Radial Knife Edge Follower Having Outward Cycloidal Motion and Return Uniform Velocity Motion	303
10.7.2	Cam Profile with a Radial Knife Edge Follower Having Outward SHM and Return Uniform Acceleration and Retardation (UAR)	305
10.7.3	Cam Profile with an Offset Knife Edge Follower Having Outward SHM and Return UAR	305
10.7.4	Cam Profile with the Radial Roller Follower with Outward Cycloidal Motion and Return Uniform Velocity	306
10.7.5	Cam Profile with an Offset Roller Follower with Outward Cycloidal Motion and Return with Uniform Velocity	308

	10.7.6 Cam Profile for Radial Flat Faced Radial Follower with Outward Cycloidal Motion and Return Uniform Velocity	309
10.8	Cams with Specified Contours	312
	10.8.1 Circular Arc Cam with Flat-faced Reciprocating Follower	312
	10.8.2 Tangent Cam with Reciprocating Roller Follower	315
10.9	Exercise	318
	10.9.1 Short Answer Questions	318
	10.9.2 Problems	319
	10.9.3 Multiple Choice Questions	320

11. Toothed Gearing 323

11.1	Introduction	325
11.2	Classification of Toothed Gearing	325
	11.2.1 According to Axes	325
	11.2.2 According to the Range of Peripheral Velocity	326
	11.2.3 According to Position of Teeth on the Gear Surface	326
	11.2.4 According to Type of Gearing	327
	11.2.5 According to Materials Used for Gears	328
11.3	Terminology Used in Gears	328
	11.3.1 Pitch Circle	328
	11.3.2 Addendum (a)	329
	11.3.3 Addendum Circle	329
	11.3.4 Dedendum (d)	330
	11.3.5 Dedendum Circle	330
	11.3.6 Clearance	330
	11.3.7 Face	330
	11.3.8 Flank	330
	11.3.9 Face Width	331
	11.3.10 Top Land	331
	11.3.11 Tooth Profile	331
	11.3.12 Circular Pitch (P_c)	331
	11.3.13 Pitch Point (P)	331
	11.3.14 Diametral Pitch (P_d)	331
	11.3.15 Module (m)	331
	11.3.16 Pressure Angle or Obliquity (ψ)	332
	11.3.17 Path of Contact	332
	11.3.18 Length of Path of Contact	332

	11.3.19 Arc of Contact	332
11.4	Condition for Constant Velocity Ratio or Law of Gearing	332
11.5	Length of the Arc of Contact	335
11.6	Minimum Number of Teeth on the Pinion to Avoid Interference	340
11.7	Interference in Involute Gears	344
11.8	Methods of Avoiding Interference	344
11.9	Forms of Teeth	344
	11.9.1 Cycloidal Teeth	345
	11.9.2 Involute Tooth	346
11.10	Helical Gears	346
11.11	Bevel Gears	347
11.12	Spiral Gears	348
11.13	Exercise	348
	11.13.1 Short Answer Questions	348
	11.13.2 Problems	349
	11.13.3 Multiple Choice Questions	351

12. Gear Trains353

12.1	Introduction	355
12.2	Simple Gear Train or Simple Gear Drive	355
	12.2.1 Speed Value or Speed Ratio or Velocity Ratio (VR)	356
	12.2.2 Train Value	356
	12.2.3 Power Transmitted by a Simple Gear Train	357
12.3	Compound Gear Train	358
12.4	Reverted Gear Train	359
12.5	Epicyclic Gear Train	362
12.6	Torque in Epicyclic Gear Trains	369
12.7	Compound Epicyclic Gear Train	371
12.8	Epicyclic Gear Trains with Bevel Gears	375
12.9	Exercise	380
	12.9.1 Short Answer Questions	380
	12.9.2 Problems	382
	12.9.3 Multiple Choice Questions	386

13. Balancing of Rotating Masses — 389

    13.1 Introduction — 391
    13.2 Checking of a Rotating Element — 391
    13.3 Types of Balancing of Rotating Elements — 392
        13.3.1 Balancing of a Single Unbalanced Rotating Mass — 392
        13.3.2 Balancing of Several Unbalanced Rotating Masses — 392
    13.4 Balancing of a Single Unbalanced Rotating Mass — 393
        13.4.1 By a Single Balancing Mass Rotating in the Same Plane — 393
        13.4.2 By Two Balancing Masses in Two Different Planes — 394
    13.5 Balancing of Several Unbalanced Masses Rotating in the Same Plane — 399
        13.5.1 Analytical Method — 400
        13.5.2 Graphical Method — 400
    13.6 Balancing of Several Unbalanced Masses Rotating in Several Planes — 402
    13.7 Exercise — 410
        13.7.1 Short Answer Questions — 410
        13.7.2 Problems — 410
        13.7.3 Multiple Choice Questions — 412

14. Balancing of Reciprocating Masses — 413

    14.1 Introduction — 415
    14.2 Partial Balancing — 416
    14.3 Effect of Partial Balancing in Two-Cylinder Locomotives — 417
        14.3.1 Tractive Force (F_T) — 418
        14.3.2 Swaying Couple — 419
        14.3.3 Hammer Blow — 419
        14.3.4 Types of Locomotives — 420
    14.4 Multi-cylinder In-line Engines — 428
    14.5 Radial Engines — 434
        14.5.1 Direct and Reverse Crank Method — 435
        14.5.2 Analytical Method — 436
    14.6 V-Engines — 439
        14.6.1 Analytical Method — 440
        14.6.2 Direct and Reverse Crank Method — 441
    14.7 Exercise — 443
        14.7.1 Short Answer Questions — 443
        14.7.2 Problems — 443
        14.7.3 Multiple Choice Questions — 445

15. Longitudinal and Transverse Vibrations 447

    15.1 Introduction 449
    15.2 Basic Elements of Any Vibratory System 449
        15.2.1 Inertial Element or Mass 449
        15.2.2 Restoring Element or Spring 449
        15.2.3 Damping Elements or Damper 450
    15.3 Various Terms Used in Vibration and their Meanings 450
        15.3.1 Period 450
        15.3.2 Cycle 450
        15.3.3 Frequency 450
        15.3.4 Resonance 450
    15.4 Types of Vibrations 450
        15.4.1 Free or Natural Vibrations 450
        15.4.2 Forced Vibrations 450
        15.4.3 Damped Vibrations 450
    15.5 Types of Vibrations Based on the Deflection 451
        15.5.1 Longitudinal Vibrations 451
        15.5.2 Transverse Vibrations 451
        15.5.3 Torsional Vibrations 451
    15.6 Natural Frequency of Free Longitudinal Vibrations 451
        15.6.1 Equilibrium Method 452
        15.6.2 Energy Method 453
        15.6.3 Rayleigh's Method 454
    15.7 Natural Frequency of Free Transverse Vibrations 459
        15.7.1 Energy (Rayleigh's) Method of a Shaft Subjected to Number of Point Loads 464
        15.7.2 Dunkerley's Method for a Shaft Subjected to a Number of Point Loads 464
    15.8 Critical Speed or Whirling Speed of a Shaft 468
    15.9 Frequency of Free Damped Vibrations (Viscous Damping) 471
        15.9.1 When the Roots are Real (Overdamping or Large Damping) 473
        15.9.2 When the Roots are Equal (Critical Damping) 473
        15.9.3 When the Roots are Complex Conjugate (Underdamping or Small Damping) 473
        15.9.4 Logarithmic Decrement 474
    15.10 Frequency of Forced Damped Vibration 476
        15.10.1 Magnification Factor or Dynamic Magnifier (D) 478
    15.11 Exercise 480
        15.11.1 Short Answer Questions 480
        15.11.2 Problems 481
        15.11.3 Multiple Choice Questions 483

16. Torsional Vibrations — 485

   16.1 Introduction — 487
   16.2 Natural Frequency of Free Torsional Vibrations — 487
   16.3 Torsional Vibrations of a Shaft with Number of Rotors — 488
       16.3.1 Free Torsional Vibrations of a Single Rotor System — 488
       16.3.2 Free Torsional Vibrations of a Two-Rotor System — 490
       16.3.3 Free Torsional Vibrations of a Three Rotor System — 493
   16.4 Torsionally Equivalent Shaft — 499
   16.5 Free Torsional Vibrations of a Geared System — 505
   16.6 Exercise — 509
       16.6.1 Short Answer Questions — 509
       16.6.2 Problems — 509
       16.6.3 Multiple Choice Questions — 510

Bibliography — 513

Index — 515

Units and their Values

Standard SI Prefixes

I. Prefixes used in basic units

Factor by which the unit is multiplied	Standard form	Prefix	Abbreviation
1 000 000 000 000	10^{12}	tera	T
1 000 000 000	10^{9}	giga	G
1 000 000	10^{6}	mega	M
1 000	10^{3}	kilo	k
100	10^{2}	hecto	h
10	10^{1}	deca	da
0.1	10^{-1}	deci	d
0.01	10^{-2}	centi	c
0.001	10^{-3}	milli	m
0.000 001	10^{-6}	micro	μ
0.000 000 001	10^{-9}	nano	n
0.000 000 000 001	10^{-12}	pico	p

II. The following derived units will be used in this book

Density (or Mass density)	kg/m^3
Force (in Newtons)	N
Pressure	N/mm^2 or N/m^2
Work done (in Joules)	J = N-m
Power (in Watts)	W = J/s

III. Fundamental units in S.I. system and their standard abbreviations, which are internationally recognised, are as given below:

m: for metre or metres; **km:** for kilometre or kilometres
kg: for kilogram or kilograms; **t:** for tonne or tonnes
s: for second or seconds; **min:** for minute or minutes
N: for newton or newtons; **N-m:** for newton metre (i.e., work done)
kN-m: for kilonewton × metres; **rad:** for radian or radians

IV. The units and their values to represent the units as per recommendations of ISO and BSO as follows:

7500	not	7 500	or	7,500
86 589 000	not	86589000	or	8,65,89,000
0.015 22	not	0.01522	or	0.01522
40×10^6	not	4,00,00,000	or	4×10^7

Abbreviations, Notations and Symbols

a–area, acceleration, addendum, distance between pivots, track
acw–anticlockwise
A–area, amplitude
b–breadth or width
B–breadth
c–damping coefficient, connecting rod
cw–clockwise
ccw–counterclockwise
C–couple, coefficient
CG–centre of gravity
D, d–diameter, dedendum
e–eccentricity, offset
E–Young's modulus, energy
Eq–equivalent
Ex–example
f–force, frequency
F–force, friction
g–acceleration due to gravity
G–geat ratio, modulus of rigidity, centre of mass
h–height, lift
H–height
I–mass moment of inertia, instantaneous centre
IDC–inner dead centre
J–polar moment of inertia, number of joints
k–radius of gyration
K–coefficient of fluctuation of speed, ratio
KE–kinetic energy
l–length
L–length, lead, links
m–mass, metre
M–mass
MA–mechanical advantage
Max, Min–maximum, minimum
n–number, $\left(\dfrac{L}{r}\right)$ ratio, degrees of freedom, number of links
N–rotational speed in rpm, Newtons

ODC–outer dead centre
p–pitch, pressure, circular pitch, pairs
P–power, pressure, diametral pitch, number of pairs, piston
PE–potential energy
Pic–picture
q–torsional stiffness
r–radius
r.p.m.–revolutions per minute
R–reaction, radius
Ref–reference plane
s–displacement, stiffness
S–slip, length of lever
SE–strain energy
SHM–simple harmonic function
t–thickness, time
T–teeth, torque, tension, Table
v–velocity
V–velocity, volume
VR–velocity ratio
w–width, weight, wheel base
W–weight
W E–worked example
x–displacement
X–displacement, amplitude
y–deflection
x, y, z–Cartesian coordinates
θ–angle
α–angular acceleration, angle
β–angle
ψ–pitch angle
δ–angle, increment, logarithmic decrement
μ–coefficient of friction
ω–angular velocity (rad/sec)
ρ–mass density
λ–angle
ϵ–transmissibility
ζ–damping factor
ϕ–angle, pressure angle, angle of friction
π–3.1416
η–efficiency
τ, σ–hoop stress
Σ–sum of quantities
Δ–displacement

Subscripts:

b–balancing
cor–coriolis
c–circular pitch, connecting rod, centrifugal
d–diametral pitch
i–input, inertia
min, max–minimum and maximum
n–nose, normal
o–output
p–pair
r–radial
t–tangential

Superscripts:

c–centripetal
t–tangential
s–sliding

Simple Mechanisms

1

M–Wheel;   a–Crank;   b–Connecting rod;   c–Pedal

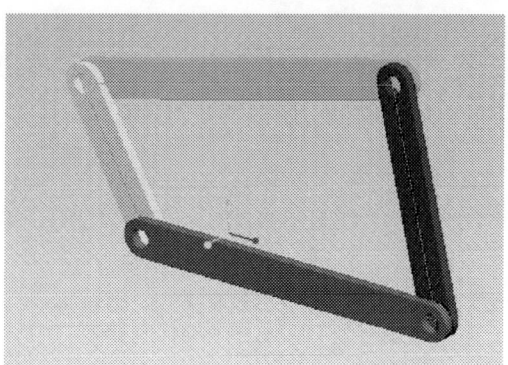

Simple Mechanisms

1.1 INTRODUCTION

This chapter deals with machines and mechanisms. A **machine** is an apparatus which receives energy in some available form and uses it to do some particular kind of work. A **mechanism** is a combination of different parts assembled in such a way that the motion of one causes predictable motion to others. So a machine is a mechanism or a combination of mechanisms which apart from imparting definite motions to the parts, also transmits and modifies the available mechanical energy into some kind of desired work.

Examples: Punching machine, typewriters, clocks, watches, toys made of springs, petrol engine, lathe machine, etc.

The theory of machines is an applied science which is used to understand the relationships between the geometry and motions of the parts of a machine or mechanism and the forces which produce these motions. Hence, it comprises the study of relative motion between the various parts of a machine and the study of the forces that act on those parts.

The study of relative motion between the parts is called **kinematics of a machine** while the study of the forces which act on the parts is called **dynamics of a machine**.

Statics deals with the analysis of stationary systems, i.e., those in which time is not a factor, and dynamics deals with systems which change with time. Kinetics deals with the parts in motion due to the application of forces. Therefore, the subject '**Theory of Machines**' comprises the study of both kinematics and dynamics of machines.

1.2 KINEMATIC LINK OR ELEMENT

The word link is used to designate a machine part or a component of a mechanism. A machine consists of several resistant bodies or parts which are said to be rigid. A body is said to be rigid when it does not suffer any distortion or in other words the distance between any two points does not change. Each rigid body of a machine which has motion relative to some other rigid body is termed **kinematic link** or **element** or simply a link.

Kinematic link or element may consist of several parts which are manufactured as separate units and assembled. There will not be any relative motion between the parts of the same link or element. The picture of the reciprocating engine is shown in Pic. 1.1 and the various parts of the reciprocating engine are shown in the line diagram in Fig. 1.1.

1. Piston, piston rod and cross head are rigidly fastened with no relative motion between them. All the three parts together is called a **link** or **element**.
2. Similarly, connecting rod with big and small ends together is called a link or element.
3. Crankshaft, crankpin and flywheel together is called a link or element.
4. Cylinder head, bed plate and main bearings of the crankshaft together is called a link or element.

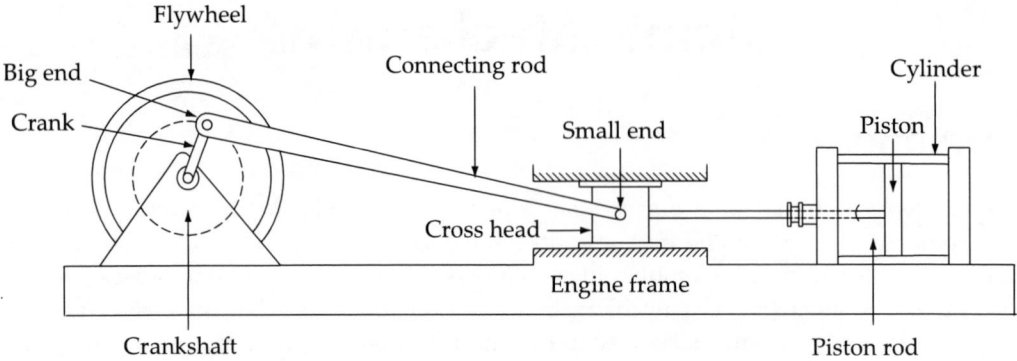

Fig. 1.1 Reciprocating engine.

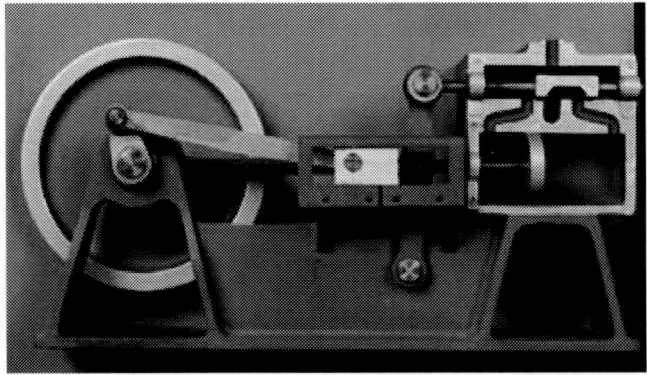

Pic. 1.1

Link need not necessarily be a rigid body, but it must be a resistant body, i.e., it must be capable of transmitting the required motion and force with negligible deformation.

For example:

(a) Chain or belt or rope, which are resistant to tensile forces called semi-rigid bodies used for transmitting motion and force.
(b) Fluids, which are resistant to compressive forces are also used as links in hydraulic presses, hydraulic brakes and hydraulic jacks, etc.

Links can be classified into binary, ternary and quarternary, etc. depending upon its ends on which revolving or turning pairs can be placed. They are rigid links as shown in Fig. 1.1(a). There will be no relative motion between the joints within the link, i.e. the lengths between the joints remain constant in (i) between two points 1 and 2, in (ii) between the three points 1, 2 and 3, in (iii) between the four points 1, 2, 3 and 4.

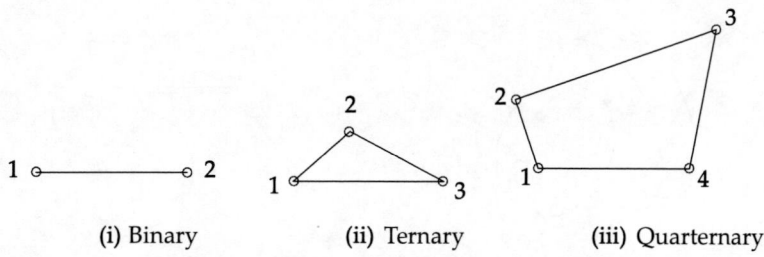

Fig. 1.1(a) Types of links.

1.3 KINEMATIC PAIR

The links of a mechanism must be connected together in such a manner that these transmit motion from the driver link or input link to the follower link or output link. Two elements or links which are connected or joined together in such a way that their relative motion is completely constrained forms a **kinematic pair**. Constraint means definite predictable relative motion between the two links. Accordingly, the kinematic pairs are classified on the following considerations:

(a) Nature of relative motion between the elements.
(b) Nature of contact between the elements.
(c) Nature of the mechanical arrangement for complete or successful constraint between the elements.

1.3.1 Nature of Relative Motion between the Elements

There are five types of kinematic pairs depending upon the relative motion between the elements and they are named accordingly as given below:

1. Turning pair called pin joint – Crankshaft in a bearing.
2. Sliding pair   – Piston in a cylinder.
3. Rolling pair   – Gear wheel and pinion wheel.
4. Spherical pair   – Ball in a socket.
5. *Screw pair or helical pair – Bolt and nut.

*This pair has only one degree of freedom because the sliding and rotational motions are related by the helix angle of the thread. If the helix angle is made zero, the screw pair becomes a turning pair and if it is made 90°, the screw pair becomes a sliding pair.

1.3.2 Nature of Contact between the Elements

There are two types of kinematic pairs according to the nature of contact between the elements as given below:

1. **Higher pairs:** A pair of links having a point or line contact between the members is called a higher pair. For example, contact between cam and follower, contact between two mating gears, a wheel rolling on a rail and a ball rolling on a flat surface, etc. Figure 1.2(a) shown below are two discs having line contact used to transmit motion using friction. Figure 1.2(b) shown below is a cam and follower with line contact also used to transmit motion.

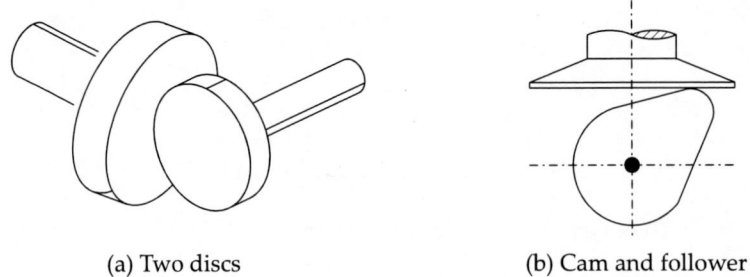

(a) Two discs   (b) Cam and follower

Fig. 1.2 Higher pair.

2. **Lower pairs:** A pair of links having a surface or area contact between the members. For example, shaft turning in a bearing, piston sliding in a cylinder, etc., various examples are shown in Fig. 1.3(a, b, c, d) having surface contacts.

1.3.3 Nature of the Mechanical Arrangement for Complete or Successful Constraint between the Elements

The relative motion between the two elements of a pair having a definite motion irrespective of the direction of the forces applied on them is called constrained motion. There are three types of constraints as follows:

1. **Completely constrained:** The motion between the two elements may be only either sliding or rotary motion.

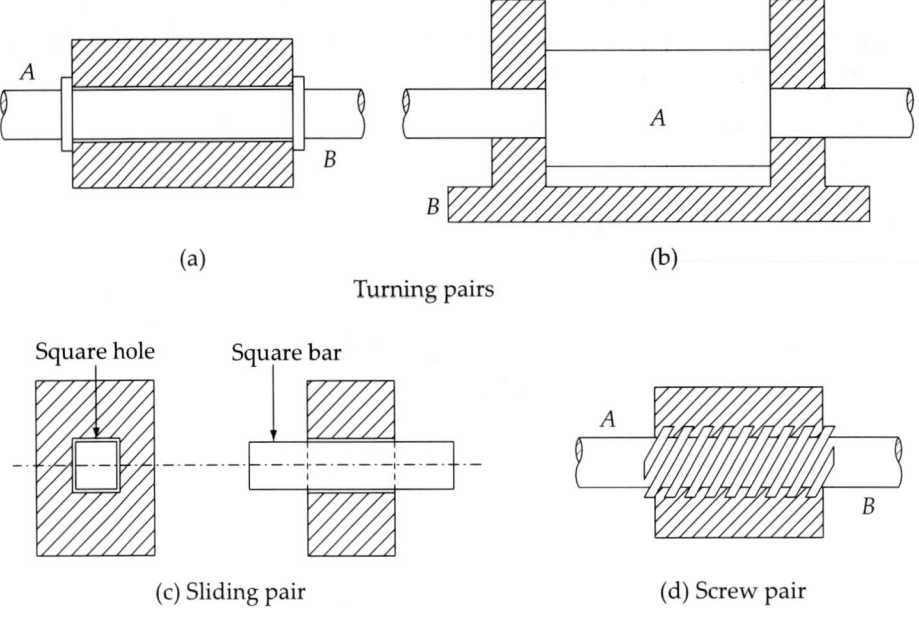

Fig. 1.3(a, b, c, d) Completely constrained.

In Fig. 1.3(a) and (b), the motion between the two elements—A shaft and B bearing is purely a rotation. In Fig. 1.3(c), the motion between the two elements—square bar in a square bearing is purely a sliding motion. In Fig. 1.3(d), the motion between the two elements is purely a screw motion.

2. **Incompletely constrained motion:** The motion between the two links is not having a definite motion as it is possible to have two types of motions, i.e. sliding as well as rotary as shown in Fig. 1.3(e) below a shaft in a bearing.

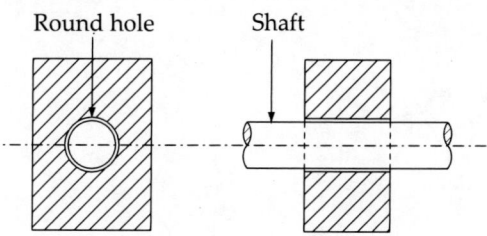

Fig. 1.3(e) Incompletely constrained.

3. **Successfully constrained:** The examples for successfully constrained pairs are cam and follower and the piston in a cylinder as shown in Fig. 1.3(f) and (g).

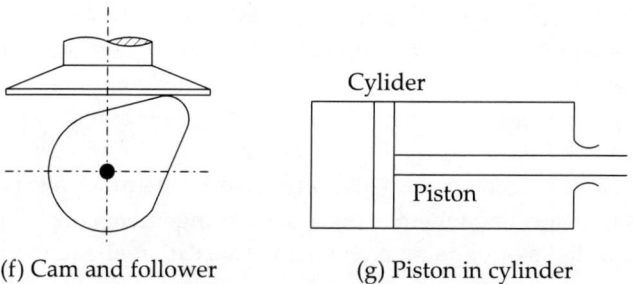

(f) Cam and follower   (g) Piston in cylinder

Fig. 1.3(f, g) Successfully constrained.

1.4 KINEMATIC CHAIN

When several links are movably connected together by joints, they are said to form a kinematic chain. A kinematic chain is a combination of kinematic pairs in which each element or link forms part of two pairs and the relative motion is completely constrained.

The various combinations of turning pairs are shown in Fig. 1.4. Each element forms part of two turning pairs as per the definition of a kinematic chain in all the three arrangements.

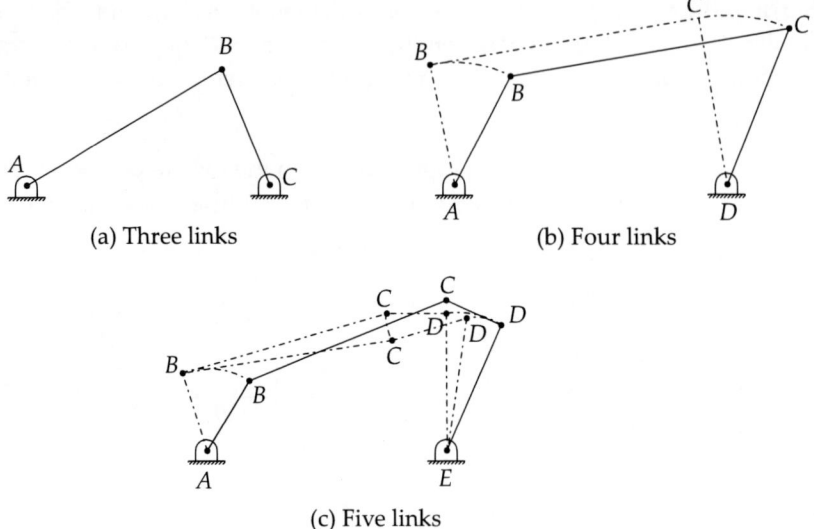

(a) Three links (b) Four links (c) Five links

Fig. 1.4 Kinematic chains with turning pairs.

In Fig. 1.4(a), with three pin-jointed links forms a rigid frame with no relative motion between them. In Fig. 1.4(b), with four pin-jointed links, the relative motion is completely constrained by fixing the link AD. Definite rotation given to AB, the resulting displacements of the remaining two links BC and CD are perfectly definite and predictable as shown by the dotted lines. With five pin-jointed links as shown at Fig. 1.4(c), the relative motion is not completely constrained. By fixing AE and when AB is displaced, the resulting displacements of the remaining links BC, CD and DE are not predictable as shown by the dotted lines. The links BC, CD and DE can take two different positions. Hence, only the arrangement shown in Fig. 1.4(b) constitutes a **kinematic chain**.

There are two ways to determine whether the given arrangement constitutes a kinematic chain or not from the number of links or pairs given by using two different equations as follows.

1.4.1 First Equation Using Pairs

The equation $L = 2P - 4$ where L represents number of links and P represents number of pairs. If the equation is satisfied, then that arrangement can be called a kinematic chain.

W E 1.1: Let us apply this equation to three different kinematic chains shown in Fig. 1.4(a),(b) and (c).

In Fig. 1.4(a): Links $L = 3$ and pairs $P = 3$, therefore, applying equation $3 = 2 \times 3 - 4$ gives 2. The equation is not satisfied. Hence, it is not a kinematic chain. It is called a structure as there is no relative motion between links.

In Fig. 1.4(b): Links $L = 4$ and pairs $P = 4$, therefore, $4 = 2 \times 4 - 4$ gives 4. The equation is satisfied. Hence, it is a kinematic chain.

In Fig. 1.4(c): Links $L = 5$ and pairs $P = 5$, therefore, $5 = 2 \times 5 - 4$ gives 6. The equation is not satisfied. Hence, it is not a kinematic chain.

1.4.2 Second Equation Using Joints

The equation $L = \dfrac{2(J+2)}{3}$ where L represents the number of links and J represents the number of joints. If the equation is satisfied, then that arrangement can be called a kinematic chain.

W E 1.2: Let us apply this equation again to three different kinematic chains shown in Fig. 1.4(a), (b) and (c).

In Fig. 1.4(a): Links $L = 3$ and joints $J = 3$, therefore, $3 = \dfrac{2(3+2)}{3}$ gives $\dfrac{10}{3}$. The equation is not satisfied. Hence, it is not a kinematic chain.

In Fig. 1.4(b): Links $L = 4$ and joints $J = 4$, therefore, $4 = \dfrac{2(4+2)}{3}$ gives 4. The equation is satisfied. Hence, it is a kinematic chain.

In Fig. 1.4(c): Links $L = 5$ and joints $J = 5$, therefore, $5 = \dfrac{2(5+2)}{3}$ gives $\dfrac{14}{3}$. The equation is not satisfied. Hence, it is not a kinematic chain.

The chains which satisfy the above two equations are called the constrained kinematic chains.

The above two equations only apply to kinematic chains in which the lower pairs are used. In case the above two equations are applied to kinematic chains in which the higher pairs are used, then each higher pair must be taken as equivalent to two lower pairs plus an additional link or element. Thus, one higher pair is equivalent to two lower pairs plus an additional link.

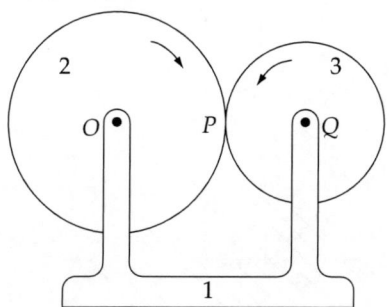

Fig. 1.5 Kinematic chain with three links.

For example, there are only three links in the kinematic chain, i.e. two wheels as shown in Fig. 1.5. This kinematic chain consists of two lower pairs (turning pairs) having surface contacts between links 1-2 and between links 1-3. There is one higher pair (rolling pair) having a line contact between links 2-3 which is equivalent to two lower pairs plus one additional link. Therefore, the total number of pairs (P) in the kinematic chain is now four and the total number of links (L) is four.

Therefore, applying the first equation, $L = 2P - 4$; $4 = 2 \times 4 - 4$ gives 4. The equation is satisfied and hence it is a kinematic chain.

1.4.3 According to the Type of Closure between Elements

There are two types of pairs as given below:

(i) Self-closed pair: Example: Cam and follower mechanism and a reciprocating mechanism with a roller shown in Fig. 1.6(a) are the examples for self-closed pair. These are also examples for mechanisms having one higher pair. Between 2 and 3 in one figure and 1 and 4 in the other figure as the contact is a line contact.

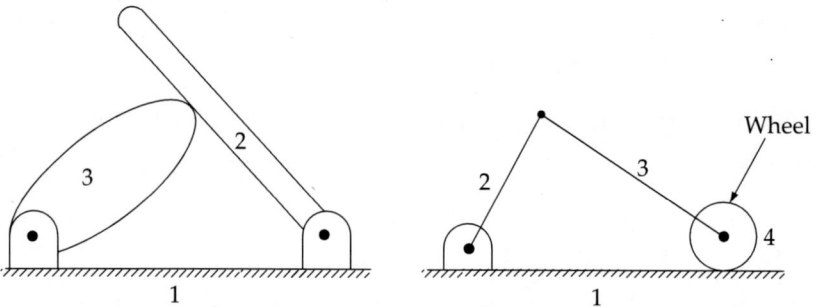

Fig. 1.6(a) Self-closed pairs.

(ii) Forced closed pair: Here the elements are kept closed by the forces exerted due to spring force or gravitational forces. Example: The foot step bearing shown in Fig. 1.6(b) uses gravitational force.

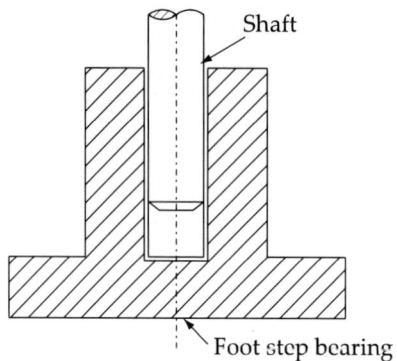

Fig. 1.6(b) Forced closed pair.

The examples of kinematic chains having three lower pairs are shown in Fig. 1.7. In both the mechanisms, there are only three links. The links A and B are sliders and link C is the frame as shown in the mechanism in Fig. 1.7(a). The three links form three sliding pairs with A-B, B-C and C-A. The three links A, B and C shown in a pressing machine in Fig. 1.7(b) have three different lower pairs. The three links A is the slider, B is the screw and C is the frame and A-B forms a turning, B-C forms a screw pair and C-A forms a sliding pair.

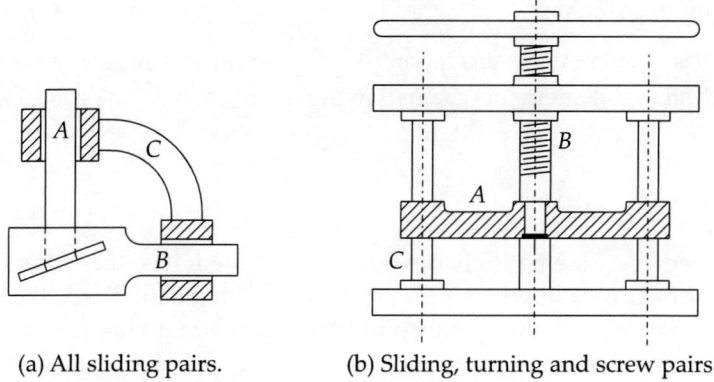

(a) All sliding pairs.   (b) Sliding, turning and screw pairs.

Fig. 1.7 Kinematic chain with three lower pairs.

1.4.4 Degrees of Freedom

One of the first concerns in either design or the analysis of a mechanism is the number of degrees of freedom, also called the mobility of the device. The mobility of a mechanism is the number of input parameters which must be controlled independently in order to bring the device into a particular position. The mobility of the mechanism can be determined from the number of links and the number of joints which it includes.

An unconstrained rigid body can describe translational motion and rotational motions about three mutually perpendicular axes. Therefore, the degrees of freedom means the number of independent relative motions a pair can have. The equation to determine the degrees of freedom of a given mechanism is $[3(L-1) - 2L_p - H_p]$ where L is the number of links, L_p is number of lower pairs and H_p is the number of higher pairs.

W E 1.3: Let us apply this to four-bar mechanism shown at Fig. 1.4(b) where $L = 4$; $L_p = 4$; $H_p = 0$. Therefore, applying the equation $[3(4-1) - 2 \times 4 - 0]$ gives 1 indicating single degree of freedom.

Let us apply this equation to the mechanism given in Fig. 1.4(c) where $L = 5$; $L_p = 5$; $H_p = 0$. Therefore, the number of degrees of freedom of the mechanism $[3(5-1) - 2 \times 5 - 0]$ gives 2 indicating two degrees of freedom. Thus, the link D can have two positions for every position of link B.

1.5 MECHANISM

The purpose of any mechanism is only the transformation of motion. There are several mechanisms such as indexing mechanism, oscillating mechanism and reciprocating mechanisms, etc. In many applications, mechanisms are used to perform repetitive operations such as pushing parts along an assembly line, clamping parts together while they are welded and folding cardboard boxes in an automated packaging machine, etc.

When one element or link of a kinematic chain is fixed, then that chain may be used for transmitting or transforming motion from one element to another. It is then termed a '**Mechanism**'.

There are two types of mechanisms:

1. *Simple mechanisms:* Kinematic chains having 4 links are called simple mechanisms.
2. *Compound mechanisms:* Kinematic chains having more than 4 links are called compound mechanisms.

1.6 INVERSION

A mechanism is defined as a kinematic chain with one link fixed. The fixed link is also called the frame. When different links are chosen as a frame in a given kinematic chain, the relative motions between the various links are not altered much but their absolute motions (those measured with respect to the frame link) may change drastically. The process of choosing different links of a chain as a frame is known as kinematic inversion or simply inversion.

A kinematic chain with L number of links gives L different mechanisms by choosing each time one link as frame (fixed). They are called inversions of the given kinematic chain. As an example, the inversions of three different mechanisms having four links are explained below.

1.6.1 Single Slider Crank Chain

This mechanism consists of four links with three turning pairs and one sliding pair. The various inversions are shown in Fig. 1.8(a) to (e). The corresponding pairs are indicated by the same letter in each inversion, so that there should be no difficulty in recognising each mechanism based on the same kinematic chain.

(a) Reciprocating Mechanism

This is the most usual form in which the single slider crank mechanism appears which is called reciprocating engine mechanism where the frame and cylinder represented by OQ is fixed as shown in Fig. 1.8(a). As crank OC rotates, the piston P reciprocates in the stationary cylinder.

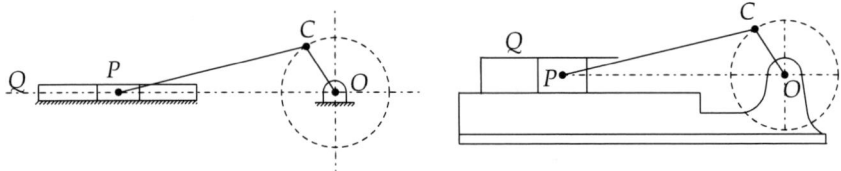

Fig. 1.8(a) Reciprocating engine mechanism.

(b) Oscillating Cylinder Mechanism

This inversion is derived from the slider crank chain by fixing the connecting rod CP as shown in Fig. 1.8 (b). As the crank OC rotates about an axis through C, the slotted link (cylinder) is pivoted to the fixed link at P. The actual form of the mechanism is used in the oscillating cylinder engine shown on the right side. The cylinder is pivoted at P, and as the piston slides inside the cylinder when the crank revolves and the cylinder oscillates about the axis of the pivot. The picture of the oscillating cylinder is shown in Pic. 1.2.

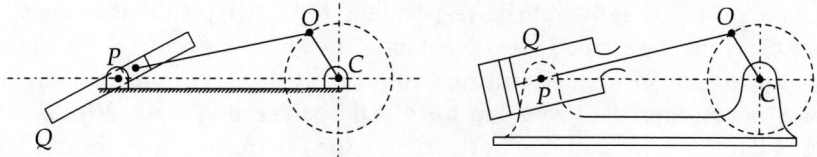

Fig. 1.8(b) Oscillating cylinder engine.

Pic. 1.2 Oscillating cylinder engine.

(c) Crank and Slotted Lever Quick-Return Mechanism

This inversion is obtained by fixing the link (crank) OC. This inversion is known as Whitworth quick-return motion and is used in slotting and shaping machines. The driving crank CP rotates at uniform velocity, the die block attached to the crankpin P, slides along the slotted link OQ and causes this link to oscillate about O with a variable angular velocity. From the pin Q on the slotted link, a connecting rod passes to a pin R on the ram which carries the tool box, and R reciprocates along a line of stroke normal to OC as shown in Fig. 1.8(c).

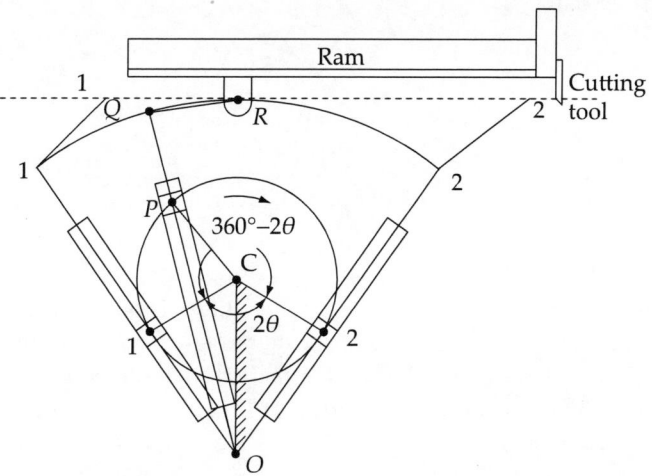

Fig. 1.8(c) Crank and slotted lever quick-return motion.

The two extreme positions of the ram corresponding to the two positions of the slotted link are indicated by 1 and 2, while the crank CP revolves clockwise about the centre C. The ratio of the time taken by the ram during advance called **cutting stroke** or advance, i.e. from 1 to 2 and return stroke called **idle stroke** (not cutting), i.e. from 2 to 1 is given by the ratio $(360° - 2\theta)/2\theta$ or $(180° - \theta)/\theta$ as shown in Fig. 1.8 (c). The ratio of the time taken by the ram to complete the cutting and return strokes is given by the ratio of the angles $(180° - \theta)/\theta$.

This ratio can also be found by knowing the crank length CP or C1 or C2 and the distance between O and C. Either by drawing or by calculations, the ratio can be determined. Then angle θ can be determined as $\angle O1C$ or $\angle O2C$ is 90°. Once θ is known, the ratio can be calculated.

The quick-return mechanism with higher ratio is desirable for such repetitive operations than one in which this ratio is lower. The mechanisms for which the ratio of advance to return is greater than unity are called **quick-return mechanisms**. Another example is the Whitworth mechanism also called crank-shaper mechanism. It is the kinematic property of the mechanism found from the geometry of the device.

Another example is the rotary internal combustion engine used in aircraft where a single crank is connected to different pistons through different connecting rods as shown in Fig. 1.8(d). OC the crank is fixed. The complete assembly of cylinders and crankcase rotate about the centre O and the pistons reciprocate along their respective cylinders.

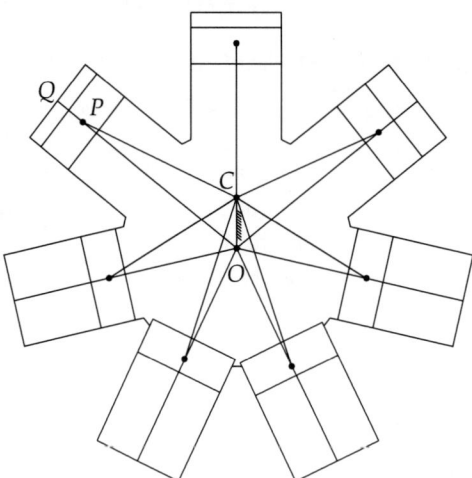

Fig. 1.8(d) Rotary I.C. engine.

(d) Bull Engine or Pendulum Engine

A fourth inversion of the slider crank chain shown in Fig. 1.8(e) is obtained by fixing the die block. It is then possible for the slotted link OQ to reciprocate along a vertical straight line. At the same time OC rotates while CP oscillates about the pin P attached to the fixed block. This is not of much practical use. The mechanism of the pendulum pump or Bull engine is based on this. This is shown diagrammatically on the right side.

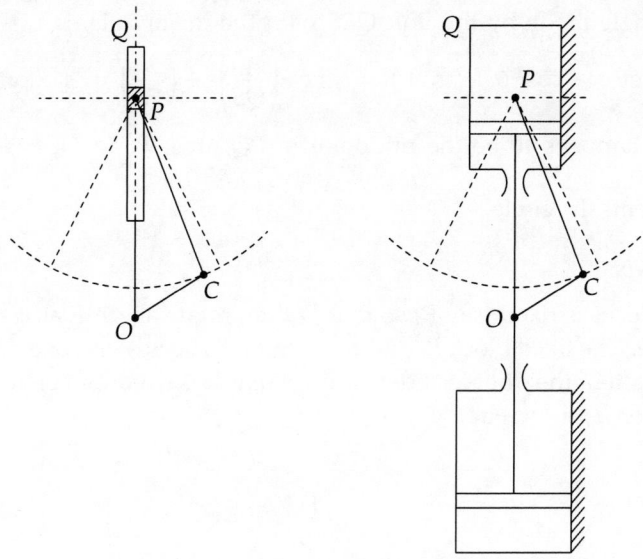

Fig. 1.8(e) Bull engine or pendulum engine.

1.6.2 Double Slider Crank Chain

This kinematic chain consists of two turning and two sliding pairs. Two die blocks slide along two slots at 90° in a frame and the pins P and Q on the die blocks are connected by the link PQ. Each of the die block forms a sliding pair with the frame and a turning pair with the link PQ. There are three inversions as shown below in Fig. 1.9(a, b, c).

(a) Ellipse Trammels

Sliding connectors are used when one slider (the input) to drive another slider (output). The two sliders operate in the same plane but in different directions. As shown in the Fig. 1.9(a), the slotted frame is fixed. Any point S on the link PQ except the midpoint will trace an ellipse as the blocks P and Q slide along their respective slots. The QS and PS represent the semi-major and semi-minor axes of the ellipse. This inversion is known as the ellipse trammels.

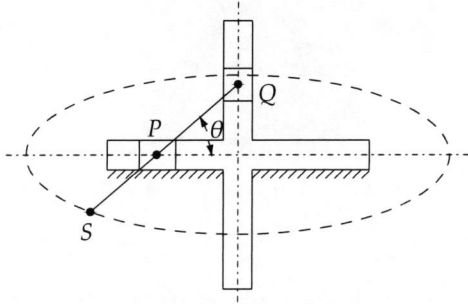

Fig. 1.9(a) Ellipse trammels.

Suppose θ is the angle made by the link QPS with the major axis, i.e. x-axis. $x = QS\cos\theta$ and $y = PS\sin\theta$. Therefore, $(\cos^2\theta + \sin^2\theta) = 1$, i.e. $\left(\dfrac{x}{QS}\right)^2 + \left(\dfrac{y}{PS}\right)^2 = 1$ represents the equation for ellipse. In case, S happens to be the midpoint of PQ, then $PS = QS = \dfrac{PQ}{2}$. The equation $x^2 + y^2 = \left(\dfrac{PQ}{2}\right)^2$ represents the circle.

(b) Scotch Yoke Mechanism

Here, one of the two blocks is fixed (say P) so that PQ can rotate about P as centre and thus cause the frame to reciprocate. The fixed block P guides the frame. The Scotch yoke mechanism shown in Fig. 1.9(b) is the inversion of the double slider crank chain. It is used for converting rotary motion into reciprocating motion as shown in Pic. 1.3.

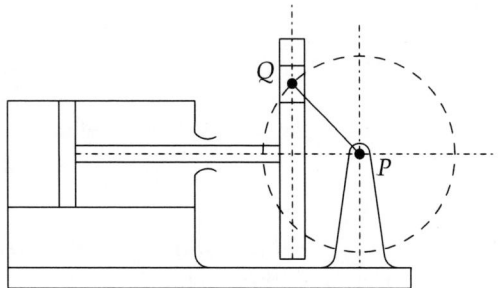

Fig. 1.9(b) Scotch yoke mechanism.

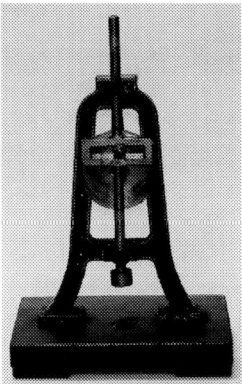

Pic. 1.3 Scotch yoke mechanism.

(c) Oldham Coupling

Couplings are used to transmit motion between two parallel shafts. The third inversion is obtained by fixing the link PQ. Each of the two die blocks may then turn about the pins P and Q. If one block is turned through a definite angle, the frame and the block turn through the same angle.

As rotation takes place, the frame will slide relative to each of the two blocks. The Oldham shaft coupling is an example of this inversion shown in Fig. 1.9(c). A circular disc D, with a tongue passing diametrically across the face and two tongues set at right angles to each other, is placed between the two half couplings, so that each tongue fits into its corresponding grooves in one of the half couplings. The tongues are a sliding fit in their grooves. This is used to transmit power between two parallel shafts.

The midpoint of the disc D describes a circle with distance between the axes as radius 'd'. The maximum sliding velocity of each tongue in the slot will be the peripheral velocity of the midpoint of the disc along the circular path.

Maximum sliding velocity = peripheral velocity along the circular path = angular velocity of shaft × distance between the shafts.

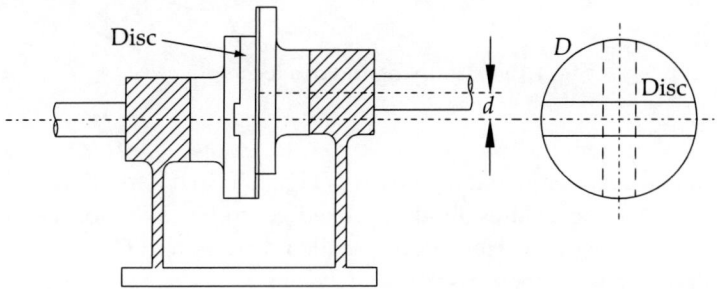

Fig. 1.9(c) Oldham coupling.

W E 1.4: The distance between two parallel shafts is 20 mm and they are connected by an Oldham's coupling. The driving shaft revolves at 150 r.p.m. What will be the maximum speed of sliding of the tongue (disc) of the intermediate piece along its groove?

Given:

Distance between the shafts (d) = 20 mm = 0.02 m

$$N = 150 \text{ r.p.m.}$$

and hence $\omega = \dfrac{2\pi N}{60}$

$$= \dfrac{2\pi 150}{60} = 15.72 \text{ rad/s.}$$

Maximum velocity of sliding $= \omega \times d$
$= 15.72 \times 20$
$= 31.42$ mm/s or 0.03142 m/s.

1.6.3 Four-Bar Mechanisms

Figure 1.4(b) is called the four-bar mechanism. All the four pairs are turning pairs. All the four inversions also become four different mechanisms with links having different lengths. The practical application of this mechanism depends solely upon the relative lengths of the four elements or links. Some of the applications are shown in Fig. 1.10.

(a) Coupling Rod of a Locomotive

This is as shown in Fig. 1.10(a). In this inversion, the opposite links are equal in length. Links AB and CD can rotate and so they are called cranks. Hence, this inversion is also called double cranks or crank-crank or rotary-rotary or double rotary mechanism.

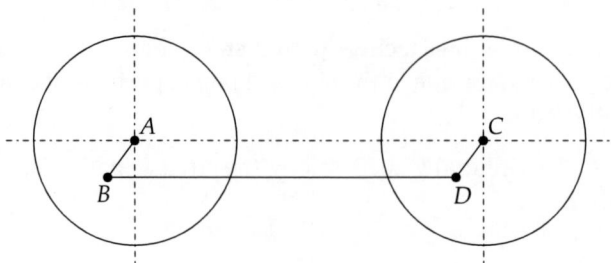

Fig. 1.10(a) Coupling rod of a locomotive.

(b) Beam Engine

Part of the mechanism of a beam engine is as shown in Fig. 1.10(b) where AB rotates about the fixed centre A, while the beam CDE oscillates about the fixed centre D. Here in this inversion only AB can rotate and CD can only oscillate. Hence, AB is called a crank and CD as a rocker. Hence, this inversion is called a crank-rocker mechanism.

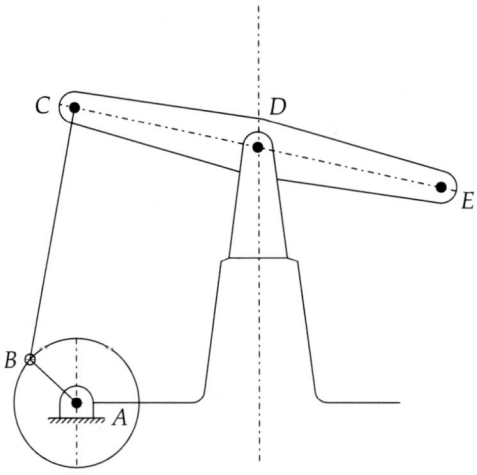

Fig. 1.10(b) Beam engine.

(c) Ackermann Steering Gear

The mechanism of the Ackermann steering gear of an automobile is shown in Fig. 1.10(c and d). In this inversion, the two short links are equal in length, while the longer links are unequal in length. When the vehicle is moving along a straight path, the long links AD and AE are parallel as shown at Fig. 1.10(c). When the vehicle moves along a curved path, the mechanism takes up the position shown in Fig. 1.10(d). The proportion of the links are such that the axes of all four wheels intersect at the same point *I*. This ensures that the relative motion between the tyres and the road surface shall be of pure rolling. Here *AB* and *CD* links only oscillate and this inversion is also called rocker-rocker mechanism or double rocker mechanism.

Note: The Pic. 1.4 and Pic 1.5 show the Ackermann steering gear and pressing machine. The Pic. 1.6 and Pic. 1.7 are the foot operated air pump and weight lifting jack made of simple mechanisms.

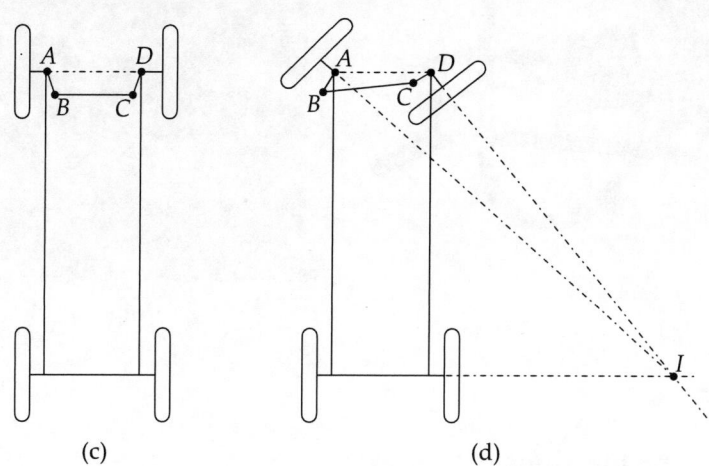

(c) (d)

Fig. 1.10(c, d) Ackermann steering gear.

Pic. 1.4   Pic. 1.5

Pic. 1.6   Pic. 1.7

1.7 EXERCISE

1.7.1 Short Answer Questions

1. Define the following terms and illustrate with sketches: element or link, mechanism, machine and inversion.
2. Explain complete, incomplete and successful constraint.
3. What is a kinematic link?
4. What is a machine and a structure? Differentiate between a machine and a structure, giving examples.
5. Differentiate between mechanism and machine, lower and higher pairs, turning and rolling pairs, closed and unclosed pairs.
6. How are the kinematic pairs classified? Explain with examples.
7. Explain kinematic pair, lower pair and higher pair and give one example of each one.
8. (a) Explain degrees of freedom of a kinematic chain and what is the expression for it?
   (b) Find the degrees of freedom of a planar mechanism having 4 links, 3 lower pairs and one higher pair.
9. What is meant by inversion and how do you find the number of inversions?
10. (a) Describe three kinds of lower pairs giving a sketch for each kind with an example.
    (b) What is the difference between the Whitworth quick-return motion and crank and slotted quick-return motion? Explain with sketches.

11. (a) Draw any six-bar mechanism and show when it becomes a structure.
    (b) What are the inversions of a slider crank chain? Explain with neat sketches.
12. What is double slider crank chain? Explain all the inversions of a double slider crank chain with neat sketches.
13. (a) Define kinematic chain and mechanism.
    (b) Sketch any one inversion of quadric chain.
14. What are the different types of constrained motions? Explain each type with an example.
15. Sketch and explain the inversions of four-bar mechanism.
16. How do you check the given chain, whether it is a kinematic chain or not?
17. What are the different types of constrained motions? Explain each type with an example.
18. Describe the construction of Oldham coupling. In what way the Oldham's coupling is useful in connecting two parallel shafts.
19. What is meant by an equivalent mechanism?
20. Name some inversions of double slider crank mechanism.
21. Sketch and describe Ackermann steering gear mechanism.
22. Explain the term (i) degrees of freedom of a mechanism and (ii) inversion of a kinematic chain.
23. How are the mechanisms classified?
24. Distinguish between complete, incomplete and successful constraint of the relative motion between two links.
25. Give diagrammatic sketches of the following mechanisms and state on which kinematic chain each is based (i) Ellipse trammel (ii) Whitworth quick-return motion (iii) Oscillating cylinder engine and (iv) Oldham shaft coupling.
26. Distinguish between: (i) Machine and Mechanism (ii) Chain, Structure and Mechanism.
27. How are the kinematic pairs classified? Explain with examples.
28. What is a kinematic pair? Classify kinematic pairs according to nature of relative motion.
29. Sketch and explain the mechanism, which is used to connect two parallel non-collinear shafts.
30. Explain various methods of classifying kinematic pairs, giving at least two examples of each category.
31. Explain with suitable sketches inversions of slider crank chain.
32. What is the difference between an element and a kinematic link of a mechanism? How do you classify links of a mechanism?
33. What do you mean by degrees of freedom of a kinematic pair? How the pairs are classified? Give examples.
34. Describe elliptical trammel. Show that it can describe a true ellipse.
35. Distinguish between a structure and a machine.

22   Theory of Machines

36. Enumerate the inversions of the single slider crank chain. Explain each of them with their applications.
37. Define 'Machine' and 'Mechanism'. How are these different from each other?
38. Explain completely, partially and incompletely constrained motions of a kinematic pair with examples.
39. Explain different types of kinematic pairs as classified by relative motion.
40. Name the steering gear mechanisms and indicate any one by a neat sketch.

1.7.2 Problems

1. In a quick-return motion mechanism of crank and slotted lever type, the ratio of maximum velocities is 2. If the length of stroke is 25 cm, find (i) the length of the slotted lever, (ii) the ratio of cutting and return strokes, and (iii) the maximum cutting velocity, if the crank rotates at 30 r.p.m.
2. In a crank and slotted lever quick-return mechanism, the distance between the fixed centres is 15 cm and the driven crank is 10 cm long. Find the ratio of the time taken during the cutting and return strokes.
3. The distance between the axes of parallel shafts connected by Oldham's coupling is 25 mm. The speed of rotation of the shafts is 250 r.p.m. Determine the maximum velocity of sliding of each tongue in its slot.
4. The distance between the axes of two parallel shafts (r) connected by Oldham's coupling is 1.5 cm. Determine the kinetic energy in the intermediate piece when the shafts rotate at 300 r.p.m. Mass (m) of the intermediate disc is 4 kg and its radius of gyration (k) is 15 cm.

$$\text{Hint: } KE = \left(\frac{I\omega^2}{2}\right) + \left(\frac{m\omega^2 r^2}{2}\right) \text{ where } I = mr^2$$

1.7.3 Multiple Choice Questions

1. The piston in a cylinder forms
   (a) completely constrained   (b) successfully constrained motion
   (c) incompletely constrained motion   (d) none of them.   [*Ans.* (b)]

2. The cam and follower forms
   (a) lower pair   (b) higher pair   (c) screw pair   (d) none of them.   [*Ans.* (b)]

3. The bolt and nut form
   (a) sliding pair   (b) rolling pair   (c) screw pair   (d) turning pair.   [*Ans.* (c)]

4. The relation between the number of links and number of pairs is
   (a) $L = 2p - 2$ (b) $L = 2p - 4$ (c) $L = 3 - 2p$ (d) $L = 2p - 3$.   [*Ans.* (b)]

5. In a kinematic pair when the elements have surface contact while in motion, it is a
   (a) higher pair   (b) closed pair   (c) lower pair   (d) unclosed pair.   [*Ans.* (c)]

6. The transmission angle is maximum when the crank angle with the fixed link is
   (a) 0°   (b) 90°   (c) 180°   (d) 270°.   [*Ans.* (c)]

7. The transmission angle is minimum when the crank angle with the fixed link is
   (a) 0°   (b) 90°   (c) 180°   (d) 270°.   [*Ans.* (a)]

8. Which of the following is an inversion of single-slider chain?
   (a) Elliptical trammel   (b) Handpump
   (c) Scotch yoke   (d) Oldham's coupling.   [*Ans.* (b)]

9. How many links must be there in a simple mechanism?
   (a) 3   (b) 4   (c) 5   (d) 6   [*Ans.* (b)]

10. How many links must be there in a compound mechanism?
    (a) 3   (b) 4   (c) 5   (d) 6   [*Ans.* (d)]

11. Which of the following is an inversion of double-slider crank chain?
    (a) Whitworth quick return mechanism   (b) Reciprocating
    (c) Scotch yoke   (d) Rotary engine.   [*Ans.* (c)]

12. Oldham's coupling is used to connect two shafts which are
    (a) intersecting   (b) parallel   (c) perpendicular   (d) none.   [*Ans.* (b)]

Mechanisms with Lower Pairs

2

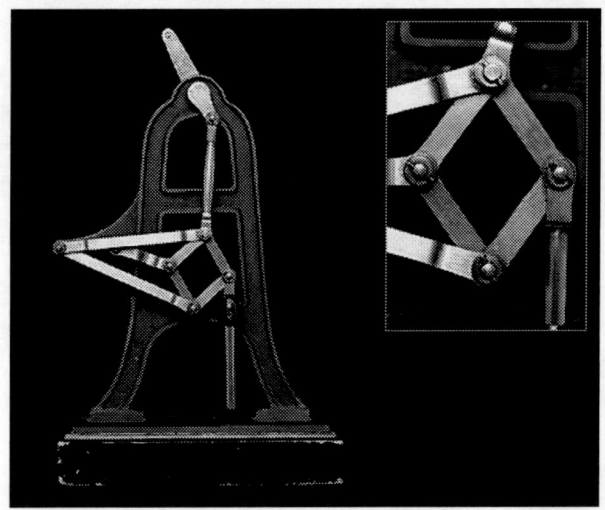

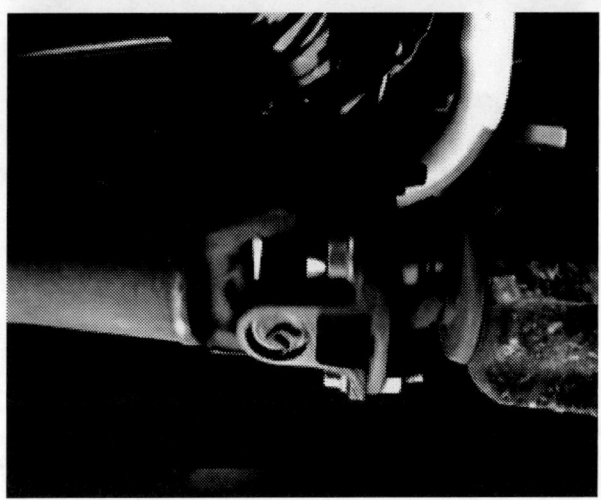

Mechanisms with Lower Pairs

2.1 INTRODUCTION

In the previous chapter, the mechanisms and their inversions have been discussed in general. This chapter deals with the mechanisms constituted by lower pairs, i.e. turning or sliding pairs, developed to copy the available graphs or curves to a reduced scale or to an enlarged scale and also to generate the exact or approximate straight line motions.

2.2 PANTOGRAPH

The pantograph shown in Fig. 2.1 is used to trace figures of larger or smaller size. It comprises lower pairs only, and is used to reproduce drawings to an enlarged or a reduced scale as exactly as possible. This can be used for guiding cutting tools or for welding two parts along the path described by a given point.

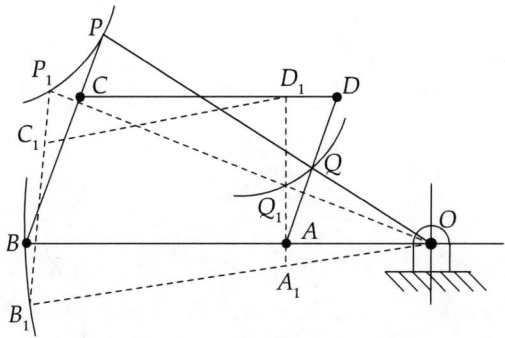

Fig. 2.1 Pantograph.

The pantograph shown at Fig. 2.1 has four links pin-jointed at A, B, C and D. Link AB is parallel to DC and AD is parallel to BC. The link AB is extended up to the fixed pin O. The three points P on the extension of BC, Q on AD and O are all on a single straight line. The point P reproduces the motion of Q to an enlarged scale or the point Q reproduces the motion of P to a reduced scale. Thus, the points P and Q move parallel and similar to each other over any path, straight or curved. Their motions will be proportional to their distances from the fixed point O.

Proof: The mathematical proof of the mechanism is as follows:

The four links are arranged in such a way that a parallelogram $ABCD$ is formed. Thus, $AB = DC$ and $BC = AD$. When the mechanism is in the position shown by full lines, the two triangles OAQ, OBP are similar, therefore, $\dfrac{OQ}{OP} = \dfrac{OA}{OB}$.

Let P be moved to the new position P_1 as shown by the dotted lines and accordingly Q moves to Q_1, then $A_1B_1C_1D_1$ is again a parallelogram. The two triangles OA_1Q_1 and OB_1P_1 are again similar triangles.

Therefore, $\dfrac{OQ_1}{OP_1} = \dfrac{OA_1}{OB_1} = \dfrac{OA}{OB}$ as $OB_1 = OB$ and $OA_1 = OA$.

Hence, the two triangles OPP_1 and OQQ_1 are similar and PP_1 and QQ_1 are parallel. The displacement of Q must therefore be parallel to the displacement of P in the proportion $OA : OB$.

Therefore, the points P and Q must trace out similar paths. The path of P will be a reproduction of the path of Q to an enlarged scale or the path of Q will be a reproduction of the path of P to a reduced scale.

2.3 MECHANISMS FOR STRAIGHT LINE MOTIONS

The principle involved in the exact straight line motions is explained below. In Fig. 2.2, the line OP oscillates about O as centre while point Q on it moves along a circular arc of diameter OR, then the path traced by the point P will be a straight line perpendicular to the diameter OR.

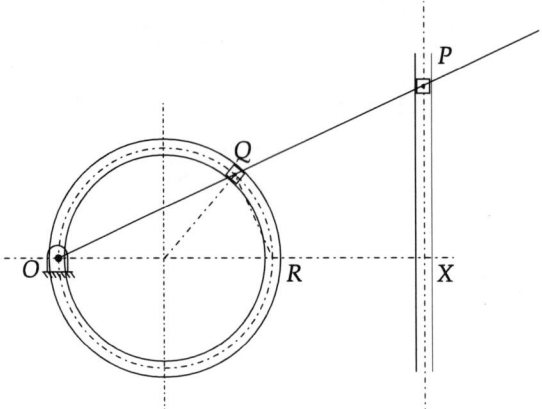

Fig. 2.2 Condition for straight line motion.

The condition to be satisfied is that the position of P must be such that $OP \times OQ$ is constant.

Proof:
The two right-angled triangles OQR, OXP are similar. Hence, $\dfrac{OQ}{OR} = \dfrac{OX}{OP}$ and therefore $OX = \dfrac{OP.OQ}{OR}$.

Since the diameter OR is constant, OX will be constant if the product $OP.OQ$ is constant. Hence, the point P moves along the straight path XP perpendicular to OR.

Several mechanisms are there for satisfying the above condition. Two such mechanisms with all turning pairs and one mechanism with both turning and sliding pairs are given below.

2.3.1 Peaucellier Mechanism

This mechanism consists of total eight links. The pin Q is constrained to move along the circumference of a circle of diameter OR by means of the link QA as shown in Fig. 2.3. The crank QA and the fixed link OA are equal. The pins P and Q are at opposite corners of a four-bar chain with all the four links being equal. The pins B and C are connected by links OB and OC of equal length to the fixed pin O.

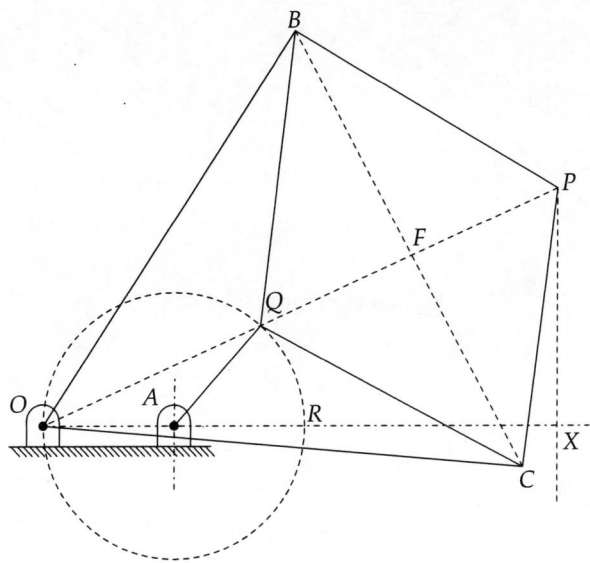

Fig. 2.3 Peaucellier mechanism.

As the point Q moves along the circle of diameter OR, the product $OQ \cdot OP$ remains constant. Join BC to bisect PQ at F.

Proof:
From the two right-angled triangles OBF, BPF;

$$OB^2 = OF^2 + FB^2 \text{ and } BP^2 = BF^2 + FP^2$$

then,
$$OB^2 - BP^2 = OF^2 - FP^2 = (OF + FP)(OF - FP) = OP \cdot OQ$$

Since the lengths of OB and BP are constant, the product $OP \cdot OQ$ also remains constant. Therefore, the point P traces out a straight path PX perpendicular to OR.

2.3.2 Hart Mechanism

This mechanism consists of total six links. The four links BC, CD, DE and EB form a crossed parallelogram as shown in Fig. 2.4 with $BC = DE$ and $CD = EB$. The lines joining B to D and C to E will be parallel for all positions of the links. If the three points O, Q and P lie on a straight line parallel to BD or CE for one position of the mechanism, then they will lie for all other positions too. It is proved that the product $OP \cdot OQ$ remains constant for all positions as below.

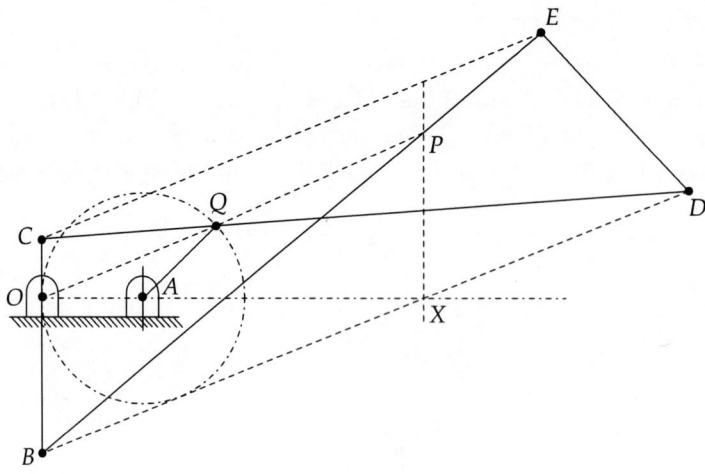

Fig. 2.4 Hart mechanism.

Proof:
In the triangle EDB, $BE^2 = BD^2 + DE^2 - 2BD \cdot DE \cos \angle EDB$ where $\cos \angle EDB = \dfrac{BE - CE}{2DE}$.

Therefore, $BE^2 = BD^2 + DE^2 - \dfrac{2BD \cdot DE(BD - CE)}{2DE} = BD^2 + DE^2 - BD^2 + BD \cdot CE$

$$BE^2 = DE^2 + BD \cdot CE \text{ or } BD \cdot CE = BE^2 - DE^2 = \text{constant} \qquad (1)$$

as the lengths of the links BE and DE are constant.

The triangles BCD, OCQ are similar and therefore $\dfrac{BD}{BC} = \dfrac{OQ}{OC}$ or $BD = \dfrac{OQ \cdot BC}{OC}$.

The triangles CBE, OBP are similar and therefore $\dfrac{CE}{CB} = \dfrac{OP}{OB}$ or $CE = \dfrac{CB \cdot OP}{OB}$.

On substituting in (1), for $BD \cdot CE = \left(\dfrac{BC \cdot OQ}{OC}\right)\left(\dfrac{CB \cdot OP}{OB}\right) = \dfrac{BC^2 \cdot OQ \cdot OP}{OB \cdot OC} = \text{constant}$.

Because BC, OB and OC lengths are constants, hence the product $OQ \cdot OP$ will be constant.

The path traced by P will be a straight line normal to OX, the extension of the diameter OA, while Q moves along the circular path through O. This mechanism has six links as compared to the previous mechanism which has eight links. The disadvantage of this mechanism is, it occupies a large amount of space.

2.3.3 Scott-Russell Mechanism

Figure 2.5 shows an exact straight line mechanism using a slider. Here the straight line is not generated but it merely copies the straight line motion. This mechanism is just similar to that of the reciprocating engine mechanism. The crank OC and the connecting rod CP are equal in length. The connecting rod is extended upto Q as shown in Fig. 2.5 such that CQ and CP are equal.

As OC rotates, the point Q follows a straight path as P is constrained to move along the straight path between guides.

Disadvantage: It makes use of sliding pair, the friction and wear of which are always higher than the turning pairs. *P* cannot move in a straight line motion due to wear.

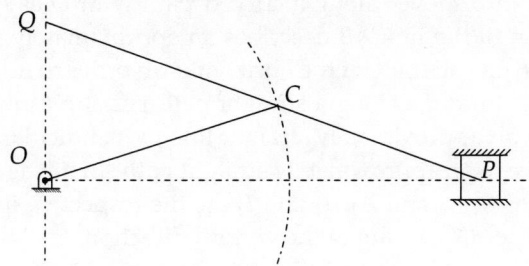

Fig. 2.5 Scott-Russell mechanism.

2.4 APPROXIMATE STRAIGHT LINE MECHANISM

Most of these mechanisms are derived from the four-bar chain and because of simplicity they are very popular even today.

2.4.1 Watt Mechanism

The links *OA* and *DC* oscillate about the fixed centres *O* and *C* and the point *Q* on the connecting rod *AD*, describes an approximate straight line path as shown in Fig. 2.6. The best position for point *Q* on *AD* is fixed using the instantaneous centre *I* as shown. Then for small oscillations of *OA* and *CD*, *Q* will trace an approximate straight line.

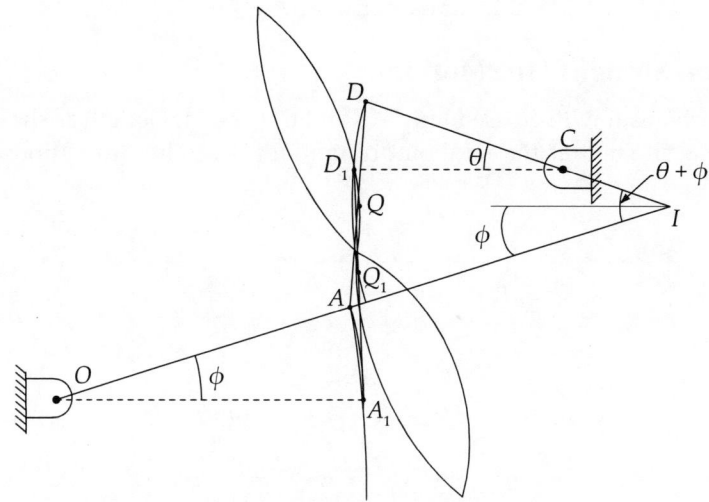

Fig. 2.6 Watt mechanism.

2.4.2 Grasshopper Mechanism

In Fig. 2.7 the centres O and Q are fixed. Pin A moves along a curved path with Q as centre and QA as radius whereas the pin B moves along a curved path with O as centre and OB as radius. The point P on the extension of the link AB describes an approximately straight path for smaller angular displacements of OB on each side of the horizontal. In order to make P to describe an exact straight line, the point A has to move along a straight path passing through O. If the length QA is large enough, A moves in an approximately straight line perpendicular to QA for small angular movements, then P will move in an approximately straight path. In the figure, this construction has been shown for two positions of A_1 and A_2. I_1 and I_2 are the respective instantaneous centres and P_1 and P_2 are the respective points moving in the vertical direction.

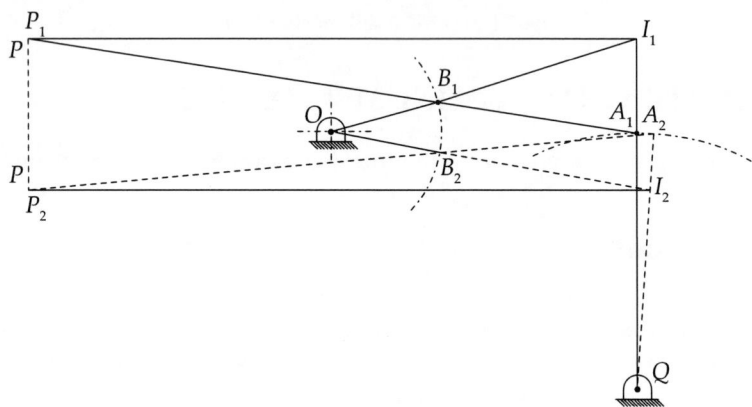

Fig. 2.7 Grasshopper mechanism.

2.4.3 Tchebicheff Straight Line Motion

This is a four-bar mechanism with crossed links AB and CD of equal length as shown in Fig. 2.8(a). The tracing point P is situated at the midpoint of the link BC. The proportions of the links are

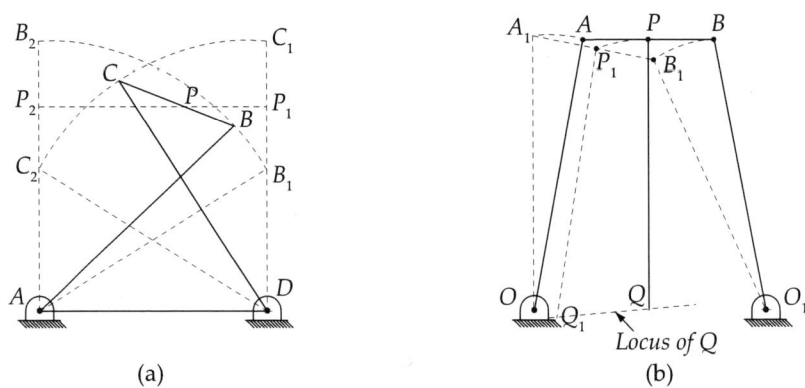

Fig. 2.8 (a) Tchebicheff straight line mechanism.   (b) Roberts mechanism.

usually such that P is directly above A or D in the extreme positions of the mechanisms as shown when CB lies along AB or when CB lies along CD. It can easily be shown that in these circumstances the tracing point P will lie on a straight line parallel to AD in the two extreme positions and in the mid position if the length of the links CB, AD and AB are in 3 : 4 : 5.

2.4.4 Roberts Mechanism

This is also a four-bar mechanism, which in its mean position has the form of a trapezium. The links OA and O_1B are of equal length and OO_1 is fixed. Bar PQ is rigidly attached to the link AB at its middle point P.

A little consideration will show that if the mechanism is displaced as shown by the dotted lines as in Fig. 2.8(b) the point Q will trace an approximately straight line.

2.5 STEERING GEAR MECHANISM

The steering gear is a mechanism used for changing the direction of wheel axels with reference to the chassis, so as to move the vehicle in any desired path. Usually, the two rear wheels will have a common axis fixed in direction with respect to the chassis and the steering is done by means of the front wheels.

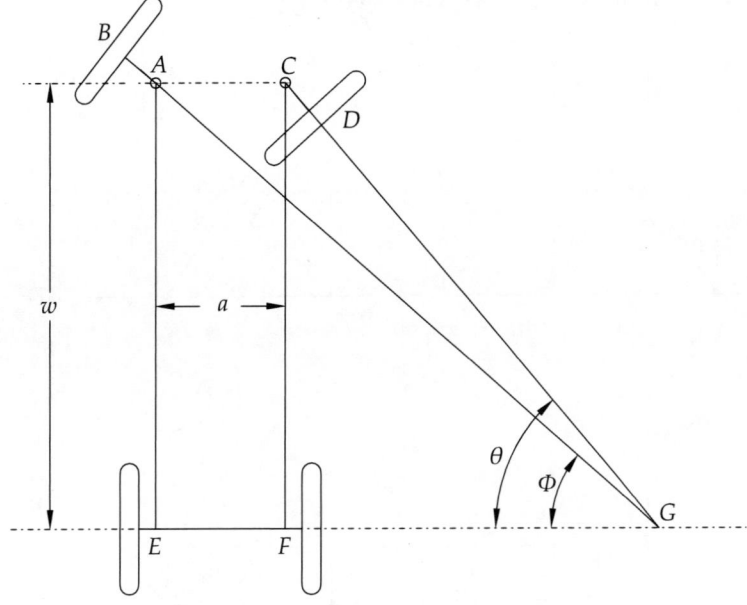

Fig. 2.9 Ackermann steering gear.

Requirements

1. The relative motion between the wheels of a vehicle and the road surface should be of pure rolling.
2. The paths of the point of contact of each wheel with the ground must be concentric circular arcs.

Steering is effected by turning the axes of rotation of the two front wheels relative to the chassis or body of the vehicle. To satisfy the above conditions, the axis of the wheel on the inside of the curve (CD) must be turned through a larger angle than the axis of the wheel on the outside of the curve (AB) as shown in Fig. 2.9. Hence, the front wheels are mounted on short separate axles which are pivoted to the chassis at A and C.

w–wheel base; a–distance between pivots or track; ϕ–angle turned by outer; θ–angle turned by inner wheels.

AB and CD are two axles with pivots at A and C. Point G is the instantaneous centre where the axes AB and CD intersect on the common axis EF of the rear wheels as per requirement (2). $AC = EF = EG - FG = AE \cot\phi - CF \cot\theta$, since $AE = CF$, $\cot\phi - \cot\theta = \dfrac{AC}{AE}$ or $\cot\phi - \cot\theta = \dfrac{a}{w}$.

The front axes must therefore be operated by the steering gear in such a way that this equation is always satisfied for any radius of curvature of the path followed by the vehicle. There are two types of steering mechanisms as given below.

2.5.1 Davis Steering Gear (Exact)

Figure 2.10 shows diagrammatically the front wheels. The arms AK and CL are fixed to the axles so as to form bell-crank levers and the angles BAK and DCL are equal to α. M and N are the die blocks pivoted to lever MN which slides in guides while steering.

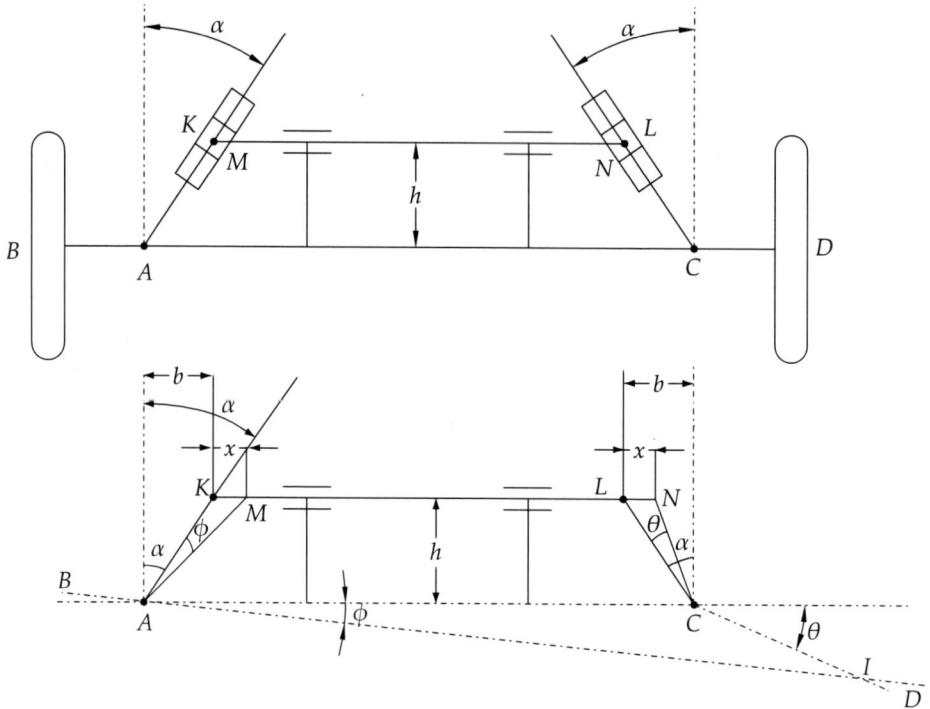

Fig. 2.10 Davis steering.

Steering is effected by sliding MN either to the right or to the left. Let MN be moved by an amount say x towards right for right turn. Then bell crank levers BAK and DCL move to new positions. BA and CD intersect at a point when produced.

The initial position in Fig. 2.10(a) indicates the position when the vehicle is going straight and from that position, $\tan \alpha = \dfrac{b}{h}$;

From the displaced position, when link MN slides say towards right for right turn by an amount say x, then the bell crank lever BAK rotates by an angle ϕ and DCL by an angle θ as shown, hence $\tan(\alpha+\phi) = \dfrac{(b+x)}{h}$ and $\tan(\alpha - \theta) = \dfrac{(b-x)}{h}$.

Then by expanding $\tan(\alpha + \phi)$ and $\tan(\alpha - \theta)$ and substituting for $\tan \alpha$ in them gives after simplifying, $\tan \phi = \dfrac{xh}{(h^2 + b^2 + bx)}$ and $\tan \theta = \dfrac{xh}{(h^2 + b^2 - bx)}$.

Therefore, $\cot \phi - \cot \theta = \left[\dfrac{(h^2 + b^2 + bx)}{xh}\right] - \left[\dfrac{(h^2 + b^2 - bx)}{xh}\right] = \dfrac{2bx}{xh} = \dfrac{2b}{h} = 2 \tan \alpha$, since $\dfrac{b}{h} = \tan \alpha$.

But for correct steering $\cot \phi - \cot \theta = \dfrac{a}{w}$ where w is the wheel base; a is the track width.

Hence, $\dfrac{2b}{h}$ or $2 \tan \alpha$ must be equal to $\dfrac{a}{w}$ or $\tan \alpha = \dfrac{a}{2w}$.

Thus, the Davis steering gear gives correct steering. But the disadvantage of this steering mechanism is due to the sliding pairs which cause wear due to friction effecting the correct steering.

2.5.2 Ackermann Steering Gear (Approximate)

This is simpler than the previous one and consists of only turning pairs. This works based on four-bar chain as mentioned in earlier chapter where one set of parallel links are unequal and other set is equal.

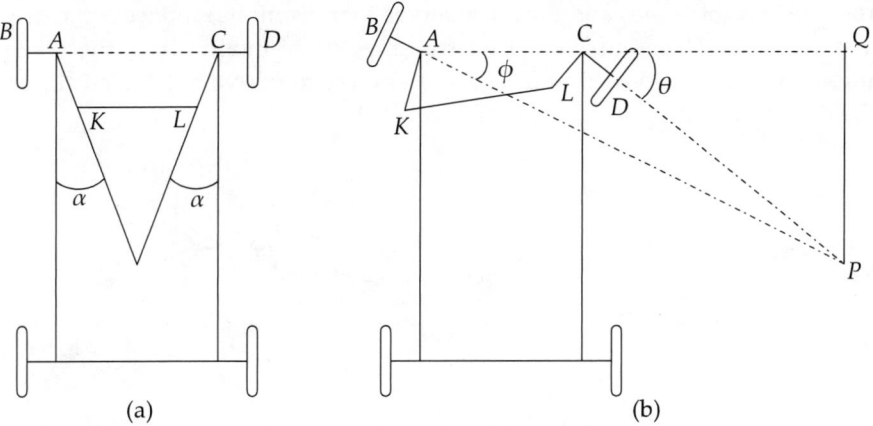

Fig. 2.11 Ackermann steering gear.

In Fig. 2.11 the mechanism is behind the front wheels whereas in the Davis steering gear the mechanism is on the front side of the front wheels. This mechanism gives correct steering only for three positions. They are while moving straight, one correct angle while turning towards right and one correct angle while turning towards left.

From Fig. 2.11(b) for correct steering $\cot \phi - \cot \theta = \dfrac{(AQ - CQ)}{QP} = \dfrac{AC}{QP} = \dfrac{a}{w}$. In the mid position, when the car is moving in straight path as shown in Fig. 2.11(a), the links AC and KL are parallel and other links AK and CL are inclined at α to the longitudinal axis.

The value of ϕ for a given value of θ depends on the ratio of $\dfrac{AK}{AC}$ and the angle α.

For the given value of AK, AC and α, the value of ϕ for different values of θ can be found.

For cars with $a = 0.4w$, for smaller values of θ, ϕ will be greater than the values for correct steering and vice versa. As seen from the table, the difference between $\cot \phi$ and $\cot \theta$ will increase first slowly and later more rapidly as θ increases.

Note: Front two wheels are mounted on short axles known as stub axles. a–distance between the pivots of front axle; w–wheel track.

For example: Let $AK = 1$; $AC = 8.5$; $\alpha = 18°$

θ	10°	20°	30°	40°	50°
ϕ_a (actual)	9° 25′	17° 43′	24° 49′	30° 34′	34° 43′
$\cot \phi_a - \cot \theta$	0.356	0.383	0.431	0.501	0.604
ϕ_c For correct steering	9° 21′	17° 38′	25° 8′	32° 8′	28° 54′

Negotiable

2.6 HOOKE'S JOINT (OR) UNIVERSAL JOINT

This joint is used to connect two non-parallel, intersecting shafts. The end of each shaft is forked and each fork provides two bearings for the arms of a cross as shown in Fig. 2.12 and Pic. 2.1. The arms of the cross are at right angles and the cross serves to transmit motion from the driving shaft (1) to the driven shaft (2). This joint is used for transmission of motion from the gearbox to the back axle in automobiles and also used in the transmission of the drive to the spindles of multiple drilling machines. It is customary to use two for parallel shafts.

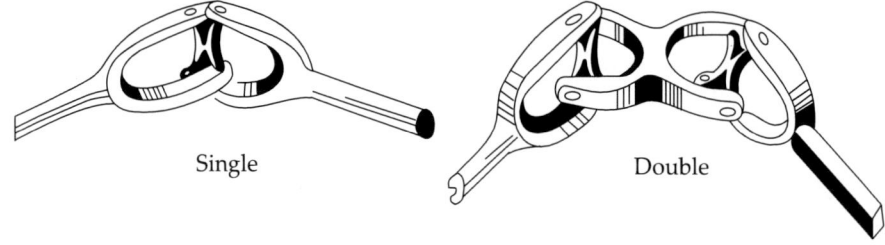

Fig. 2.12 Hooke's joint.

Pic. 2.1 Hooke's joint.

The elevation (front view) and plan (top view) looking along the axis of the driving shaft (1) are shown by line diagrams in Fig. 2.13.

In the top view, the planes of rotation of the two arms of the cross are represented by AA_1OB_1B and C_1CODD_1. In the front view, the plane of rotation of the arms attached to the driving shaft (1) is represented as circle by the plane of the paper.

Let the driving shaft be turned through an angle θ, so that the arms AB and CD turn to A_1B_1 and C_1D_1 by the same angle θ in their respective planes of rotation. But the plane of rotation of CD is inclined to the plane of rotation of AB by α. So the actual angle of rotation of OC_1' is given by OC_2', i.e., the actual angle through which the driven shaft (2) turns is given by ϕ.

From the front view of the figure, $\tan\phi = \dfrac{O'N'}{N'C_2'}$ and $\tan\theta = \dfrac{O'M'}{M'C_1'} = \dfrac{O'M'}{N'C_2'}$ since $M'C_1' = N'C_2'$

Therefore,
$$\frac{\tan\phi}{\tan\theta} = \frac{O'N'}{N'C_2'} \times \frac{N'C_2'}{O'M'} = \frac{O'N'}{O'M'}$$

But $O'M' = OM = ON_1 \cos\alpha = ON \cos\alpha = O'N' \cos\alpha$ from the top view.

Therefore,
$$\frac{\tan\phi}{\tan\theta} = \frac{1}{\cos\alpha} \text{ or } \tan\theta = \cos\alpha.\tan\phi \tag{1}$$

Let ω_1 be the angular velocity of the driving shaft (1), i.e. $\left(\dfrac{d\theta}{dt}\right)$ and ω_2 be the angular velocity of the driven shaft (2), i.e. $\left(\dfrac{d\phi}{dt}\right)$.

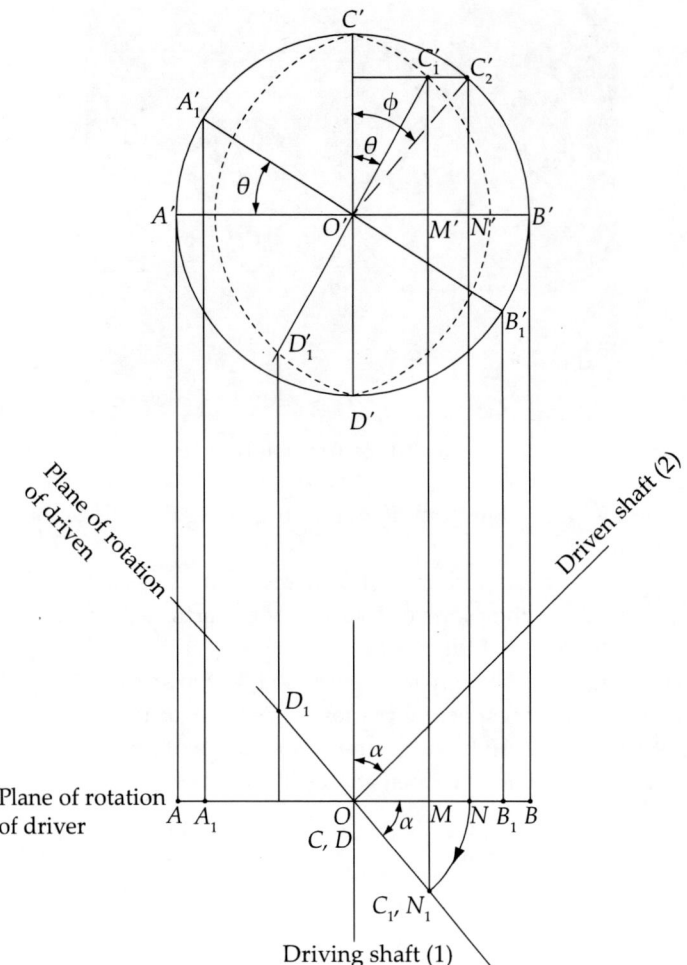

Fig. 2.13 Hooke's joint (front view and top view).

Differentiating both sides of (1) gives

$$\sec^2\theta \frac{d\theta}{dt} = \cos\alpha \sec^2\phi \frac{d\phi}{dt}$$

Therefore,

$$\frac{\omega_1}{\omega_2} = \frac{\cos\alpha \sec^2\phi}{\sec^2\theta} = \cos\alpha \sec^2\phi \cos^2\theta \qquad (2)$$

But $\sec^2\phi = 1 + \tan^2\phi$ since $\tan\phi = \dfrac{\tan\theta}{\cos\alpha}$ from (1)

Hence, $\sec^2\phi = 1 + \tan^2\phi = 1 + \dfrac{\tan^2\theta}{\cos^2\alpha} = 1 + \dfrac{\sin^2\theta}{\cos^2\theta \cos^2\alpha} = \dfrac{\cos^2\theta \cos^2\alpha + \sin^2\theta}{\cos^2\theta \cos^2\alpha}$

Replacing $\cos^2\alpha = (1 - \sin^2\alpha)$ in the numerator gives

$$\sec^2\phi = \frac{(1-\sin^2\alpha)\cos^2\theta + \sin^2\theta}{\cos^2\alpha \cos^2\theta} = \frac{1-\cos^2\theta \sin^2\alpha}{\cos^2\alpha \cos^2\theta}$$

Substituting in (2) gives

$$\frac{\omega_1}{\omega_2} = \cos\alpha \cos^2\theta \left(\frac{1-\cos^2\theta \sin^2\alpha}{\cos^2\theta \cos^2\alpha}\right)$$

$$= \frac{1-\cos^2\theta \sin^2\alpha}{\cos\alpha} = \frac{N_1}{N_2} \qquad (3)$$

The value of this expression is maximum when $\cos\theta = 0$, i.e. when $\theta = \frac{\pi}{2}, \frac{3\pi}{2}$, etc.; and is minimum when $\cos\theta = \pm 1$, i.e. when $\theta = 0, \pi$, etc.

If the speed ω_1 of the driving shaft is constant, the maximum speed of the driven shaft is given by $\omega_{2\max} = \frac{\omega_1}{\cos\alpha}$ and the minimum speed of the driven shaft by $\omega_{2\min} = \omega_1 \cos\alpha$.

Figure 2.14 shows the polar diagram depicting the salient features of the driven shaft speed. The following observations can be made from the diagram for one complete rotation of the driven shaft.

1. At points 1 and 2, the speed of driven is maximum and at 3 and 4 the speed is minimum.
2. Since there are two maximum and two minimum speeds of the driven shaft, therefore, there are four points when the speeds of the driven and driving shaft are same as shown at 5, 6, 7 and 8.
3. Since the angular velocity of the driving shaft is usually constant, it is represented by a circle of radius ω_1. The driven shaft is represented by an ellipse with a semi-major axis of $\frac{\omega_1}{\cos\alpha}$ and semi-minor axis of $\omega_1 \cos\alpha$ as shown.
4. Due to the variation in the speed of the driven shaft, vibrations will occur. In order to control the vibrations, a flywheel having heavy mass is provided on the driven shaft.

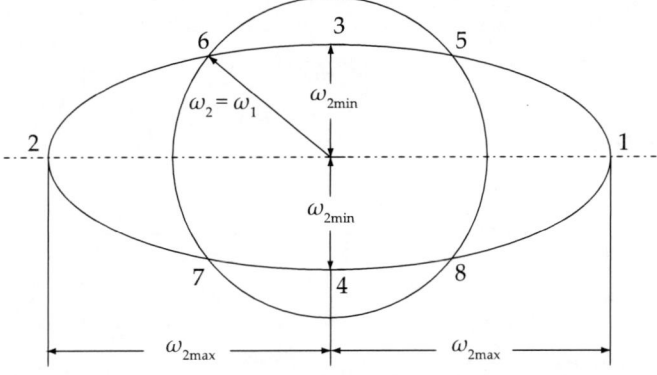

Fig. 2.14 Polar diagram.

Figure 2.14 shows the polar diagram depicting the salient features of the driven shaft speed for one complete revolution as below:

At 5, 6, 7, 8	$\omega_1 = \omega_2$	$\tan\theta = \pm\sqrt{\cos\alpha}$	
At 1 and 2	ω_{2max}	at $\theta = 0$ and $180°$	$\omega_2 = \dfrac{\omega_1}{\cos\alpha}$
At 3 and 4	ω_{2min}	at $\theta = 90$ and $270°$	$\omega_2 = \omega_1 \cos\alpha$

The value of θ for which the speeds of the driving and the driven shafts are equal may be found by equating (3) to unity.

For equal speeds: $\omega_1 = \omega_2$

Then $1 - \cos^2\theta \sin^2\alpha = \cos\alpha$ or $1 - \cos\alpha = \cos^2\theta \sin^2\alpha$;

Therefore,

$$\cos^2\theta = \frac{1-\cos\alpha}{\sin^2\alpha}$$

$$\sin^2\theta = 1 - \cos^2\theta = 1 - \left(\frac{1-\cos\alpha}{\sin^2\alpha}\right) \text{ Replacing } \sin^2\alpha = 1 - \cos^2\alpha$$

$$\sin^2\theta = 1 - \left(\frac{1-\cos\alpha}{1-\cos^2\alpha}\right) = 1 - \left(\frac{1-\cos\alpha}{(1+\cos\alpha)(1-\cos\alpha)}\right)$$

$$= 1 - \left(\frac{1}{1+\cos\alpha}\right) = \frac{1+\cos\alpha-1}{1+\cos\alpha} = \frac{\cos\alpha}{1+\cos\alpha}$$

Therefore,

$$\tan^2\theta = \frac{\sin^2\theta}{\cos^2\theta}$$

$$= \left(\frac{\cos\alpha}{1+\cos\alpha}\right)\left(\frac{\sin^2\alpha}{1-\cos\alpha}\right)$$

$$\tan^2\theta = \frac{\cos\alpha \sin^2\alpha}{1-\cos^2\alpha} = \cos\alpha$$

Hence,

$$\tan\theta = \sqrt{\cos\alpha} \tag{4}$$

The angular acceleration of the driven shaft is given by $\dfrac{d\omega_2}{dt}$, therefore

$$\frac{d\omega_2}{dt} = \frac{d\omega_2}{d\theta} \times \frac{d\theta}{dt} = \omega_1 \frac{d\omega_2}{d\theta} = \frac{-\omega_1^2 \cos\alpha \sin^2\alpha \sin 2\theta}{\left(1-\cos^2\theta \sin^2\alpha\right)^2} \tag{5}$$

The value of θ for which the acceleration is maximum may be found by differentiating with respect to θ and equating to zero gives

$$\cos 2\theta \simeq \frac{2\sin^2\alpha}{\left(2-\sin^2\alpha\right)} \tag{6}$$

Fluctuation of speed $(q) = \omega_1 \left[\left(\dfrac{1}{\cos\alpha}\right) - \cos\alpha\right] = \dfrac{\omega_1 \sin^2\alpha}{\cos\alpha}$ and,

Fluctuation of speed,
$$q = \omega_1 \alpha^2 \tag{7}$$

for smaller values of α, as $\cos\alpha = 1$ and $\sin\alpha = \alpha$.

Max. fluctuation of speed $(q_{max}) = \omega_{2\,max} - \omega_{2\,min}$

$$= \omega_1 \left[\left(\dfrac{1}{\cos\alpha}\right) - \cos\alpha\right]$$

2.7 DOUBLE HOOKE'S JOINT

It has been observed that the velocity of the driven shaft is not constant, but varies from maximum to minimum values. In order to have a constant velocity ratio between the driving and driven shafts, an intermediate shaft with a Hooke's joint at each end is used. This type of joint is known as Double Hooke's Joint as shown in Fig. 2.15.

Note: Fluctuations are confined to intermediate shaft only.

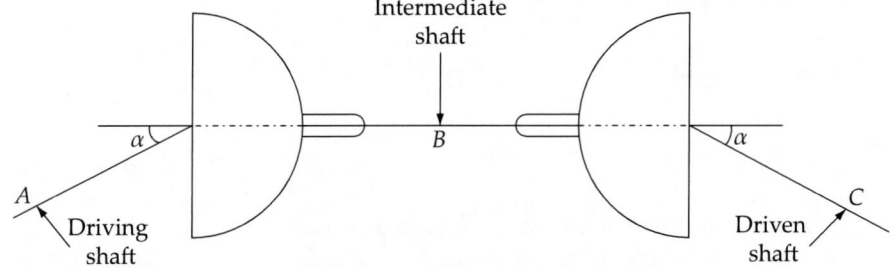

Fig. 2.15 Double Hooke's joint.

From shafts A to B, $\tan\theta = \tan\phi\cos\alpha$ and shafts B to C, $\tan\gamma = \tan\phi\cos\alpha$, therefore, $\theta = \gamma$ or $\omega_A = \omega_C$. Speed of the driving and driven shaft is same and in other words, this joint gives a velocity ratio equal to unity if:

1. The axes of the driving and driven shafts are in the same plane.
2. The driving and driven shafts make equal angles with the intermediate shaft.

Note: Axes of driving and driven should not be in perpendicular planes.

W E 2.1: In a Davis steering gear, the distance between the pivots of the front axle is 1 m and the wheel base is 2.5 m. Find the inclination of the track arm to the longitudinal axis of the car, when it is moving along a straight path.

Given:
Distance between pivots of the front axle $(a) = 1$ m and wheel base $(w) = 2.5$ m.

Let α be the inclination of the track arm to the longitudinal axis of the car, when it is moving along a straight path.

Then $\tan \alpha = \dfrac{a}{2w} = \dfrac{1}{(2 \times 2.5)} = \dfrac{1}{5} = 0.2$ (or)

$\alpha = \tan^{-1}(0.2) = 11.30°$

W E 2.2: A Hooke's joint connects two shafts whose axes intersect at 150°. The driving shaft rotates uniformly at 120 r.p.m. The driven shaft operates against a steady torque of 150 N-m and carries a flywheel whose mass is 45 kg and radius of gyration is 150 mm. Find the maximum torque which will be exerted by the driving shaft.

Given:

$\alpha = 180° - 150° = 30°$ $N_1 = 120$ r.p.m.; $T_1 = 150$ N-m;
$m = 45$ kg; $k = 150$ mm $= 0.15$ m

Angular velocity of driving shaft $\omega_1 = \dfrac{2\pi N_1}{60} = 12.56$ rad/sec;

Mass moment of inertia $I = mk^2 = 45 \times (0.15)^2 = 1.01$ kg-m²

Using the equation (6) for $\cos 2\theta \simeq \dfrac{2 \sin^2 \alpha}{2 - \sin^2 \alpha}$.

For $\alpha = 30°$ gives $2\theta = 73°24'$ or $286°36'$ and $\theta = 36°42'$ or $143°18'$.

Choose $2\theta = 286°36'$ for α_2 to be positive, then $\dfrac{-\omega_1^2 \cos \alpha \sin^2 \alpha \sin 2\theta}{(1 - \cos^2 \theta \sin^2 \alpha)^2}$ giving $\alpha_2 = 44.87$ rad/s².

Maximum torque exerted due to acceleration on the driven shaft

$$T_2 = I\alpha_2$$
$$= 1.01 \times 44.87 = 45.43 \text{ N-m}$$

The maximum total torque exerted by the driving shaft is

$$T_1 + T_2 = 150 + 45.53$$
$$= 195.43 \text{ N-m}$$

W E 2.3: Two shafts are connected by a Hooke's joint. The driving shaft revolves uniformly at 500 r.p.m. If the total permissible variation in speed of a driven shaft is not to exceed ±6% of the mean speed, find the greatest permissible angle between the centre lines of the shafts. Also determine the maximum and minimum speed of the driven shaft.

Given:
$N_1 = 500$ r.p.m.; $\omega_1 = 52.4$ rad/sec;

Let α be the greatest permissible angle between the centre lines of the shafts.

Since the variation in speed of the driven shaft is ±6% of the mean speed (i.e., speed of the driving speed), therefore, total fluctuation of speed of the driven shaft $q = 12\%$ of mean speed $\omega = 0.12\omega$ where $\omega = \dfrac{\omega_1 + \omega_2}{2}$.

But fluctuation of speed $q = \omega_1 \left(\dfrac{1 - \cos^2 \alpha}{\cos \alpha} \right)$ substituting and solving gives

$$\alpha = 19.64°$$
$$N_{max} = 500 + 0.06 \times 500 = 530 \text{ r.p.m.}$$
$$N_{min} = 500 - 0.06 \times 500 = 470 \text{ r.p.m.}$$

W E 2.4: Two inclined shafts are connected by means of a universal joint. The speed of the driving shaft is 1000 r.p.m. If the total fluctuation of speed of the driven shaft is not to exceed 12.5%, what is the maximum possible inclination between the two shafts?

With this angle, what will be the maximum acceleration to which the driven shaft is subjected and when this will occur?

Given:

$N_1 = 1000$ r.p.m.; $\omega_1 = \dfrac{2\pi N_1}{60} = 104.6$ rad/sec

$q = 1.25\omega_1 = \omega_1 \left(\dfrac{1 - \cos^2 \alpha}{\cos \alpha} \right)$ solving gives $\alpha = 20.24°$

Let maximum angular acceleration of the driven shaft be $\dfrac{d\omega_2}{dt}$, and θ be the angle through which the driving shaft turns, $\cos 2\theta = \dfrac{2\sin^2 \alpha}{2 - \sin^2 \alpha}$ giving $\theta = 41.274°$.

Substituting for θ in the expression for

$$\dfrac{d\omega_2}{dt} = \left(\dfrac{d\omega_2}{d\theta} \right) \left(\dfrac{d\theta}{dt} \right) = \omega_1 \cdot \left(\dfrac{d\omega_2}{d\theta} \right)$$
$$= -\dfrac{\omega_1^2 \cos \alpha \, \sin^2 \alpha \, \sin 2\theta}{\left(1 - \cos^2 \theta \, \sin^2 \alpha \right)^2}$$

giving maximum acceleration = 1570 rad/s².

2.8 EXERCISE

2.8.1 Short Answer Questions

1. Give a neat sketch of Peaucellier mechanism and explain.
2. Give a neat sketch of a Hart's mechanism and give the differences between this and the Peaucellier mechanism.
3. What is Hooke's joint?

4. For what purpose is the pantograph used? Sketch one form of pantograph and show that it satisfies the required conditions.
5. Enumerate straight line mechanisms. Why are they classified into exact and approximate straight line mechanisms?
6. What is a Scott-Russel mechanism? What is its limitation?
7. Name the approximate straight line mechanisms and explain any one with a neat sketch.
8. In what way is a Grasshopper mechanism a deviation of the modified Scott-Russel mechanism? $\left[\text{Hint: } \dfrac{PA}{AB} = \dfrac{AB}{OB} \text{ for approximate straight line motion.}\right]$
9. Discuss the geometrical proportions of a pantograph mechanism to give a magnification ratio of three. Make a note on pantograph.
10. Explain with the help of a neat sketch any one mathematically correct straight line mechanism and prove the result.
11. Explain why two Hooke's joints are used to transmit motion from engine to the differential of an automobile.
12. (a) Sketch neatly the Hart's straight line mechanism and explain with a proof how the tracing point describes a straight line path.
    (b) Compare the two main types of steering gears discussing their relative merits and demerits.
13. What is a pantograph and what are its uses? Explain the working of pantograph mechanism.
14. Sketch and explain Peaucellier straight line mechanism.
15. Prove that a point on one of links of a Hart mechanism traces a straight line on the movement of its links.
16. Sketch and explain the following mechanisms: (i) Pantograph (ii) Robert's mechanism (iii) Grasshopper mechanism.
17. Derive the conditions of exact straight line motion.
18. Sketch a pantograph and explain how the mechanism would be used to enlarge a drawing.
19. Explain why two Hooke's joints are used to transmit motion from the engine to the differential of an automobile.
20. Derive the condition of correct steering in the case of an automobile. Draw a neat sketch of Davis steering gear and show that the condition is correct.
21. Derive the condition of correct steering in the case of an automobile.
22. Prove that the variation of speeds in double Hooke's joint is $\dfrac{1}{\cos \alpha}$ to $\cos \alpha$.
23. Sketch the polar velocity diagram of a Hooke's joint and mark its salient features.
24. Differentiate between Davis and Ackermann steering gears.

25. Sketch and describe the Peaucellier and Hart straight line motion mechanisms.
26. What is a Hooke's joint? What are its applications?
27. What conditions must be satisfied by the steering mechanism of a car in order that the wheels may have a pure rolling motion when rounding a curve?

2.8.2 Problems

1. The ratio between the width of the front axle and that of the wheel base of a steering mechanism is 0.44. At the instant when the front inner wheel is turned by 18°, what should be the angle turned by the outer front wheel for perfect steering? [**Ans.** = 15.90°]

2. In a Davis steering gear, the length of the car between axles is 2.4 m, and the steering pivots are 1.35 m apart. Determine the inclination of the track arms to the longitudinal axis of the car when the car moves in a straight path. [**Ans.** 15°42′]

3. Two shafts of a Hooke's coupling have their axes inclined at 20°. The shaft A revolves at a uniform speed of 1000 r.p.m. The shaft B carries a flywheel of mass 30 kg, if the radius of gyration of the flywheel is 100 mm, find the maximum torque in shaft B. [**Ans.** 411 N-m]

4. Two shafts are connected by a Hooke's joint. The driving shaft revolves uniformly at 500 r.p.m. If the total permissible variation in speed of the driven shaft is not to exceed ±6% of the mean speed, find the greatest permissible angle between the centre lines of the shafts.

5. What is an automobile steering gear and what are the various types? What is the fundamental equation of steering gear? Which steering gear fulfils this condition?

6. A Hooke's joint is to connect two shafts whose axes intersect at 150°. The driving shaft rotates uniformly at 120 r.p.m. The driven shaft operates against a steady torque of 170 N-m and carries a flywheel whose mass is 50 kg and radius of gyration 15 cm. What is the maximum value of the torque which must be exerted by the driving shaft.

7. In the Fig. 2.1 pantograph, length $OA = 0.0$ mm; $AB = 100$ mm; $BC = CP = CD = DA = 100$ mm. If Q is at the midpoint of DA, give the scale of reduction. What are the limitations of this mechanism for the given dimensions? [**Ans.** Scale of reduction = 21]

8. Figure 2.6 shows the Watt mechanism with $OA = 50$ mm; $DC = 50$ mm; horizontal distance between fixed centres O and C is 90 mm and vertical distance between O and C is 40 mm. Plot the path of point Q, mark and measure the straight line segment of the P.

9. (a) Derive the expression for the speed ratio of two shafts connected by Hooke's joint.
$$\frac{N_2}{N_1} = \frac{\cos \alpha}{(1 - \cos^2\theta \sin^2\alpha)}$$ where N_2 is the speed of driven; N_1 is the speed of the driver; α is the angle between shafts; θ is the angle turned by driving shaft.

(b) The angle between the axes of two shafts connected by Hooke's joint is 18°. Determine the angle turned through by the driving shaft when the velocity ratio is (i) maximum and (ii) unity.

10. A circle with EQ as diameter has a point Q on its circumference. P is a point on EQ produced such that if Q turns about E, the product of $EQ \times EP$ is constant. Prove that the point P moves in straight line perpendicular to EQ.

11. Two shafts are connected by a Hooke's joint. The driving shaft revolves uniformly at 500 r.p.m. If the total permissible variation in speed of the driven shaft is not to exceed ±6% of the mean speed, find the greatest permissible angle between the centre line of the shafts. Also calculate the maximum and minimum speeds of the driven shaft.

12. Two shafts are to be connected by a Hooke's joint. The driving shaft is rotated at a uniform speed of 500 r.p.m. and the speed of the driven shaft must be 475 and 526 r.p.m. Determine the maximum permissible angle between the shafts.

13. Derive the condition, for equal speeds of the driving and driven shafts of a Hooke's joint.

14. Derive an expression for the ratio of shafts velocities for Hooke's joint and draw the polar diagram depicting the salient features of driven shaft speed.

15. Deduce the relationship connecting the inclinations of the front stub axles to the rear axle, the distance between the pivot centres for the front axles and wheel base of the car.

2.8.3 Multiple Choice Questions

1. A pantograph consists of
   (a) 4 links (b) 6 links (c) 8 links (d) 10 links   [*Ans.* (a)]

2. A Hart mechanism uses
   (a) 4 links (b) 6 links (c) 8 links (d) 10 links   [*Ans.* (b)]

3. A Peaucellier mechanism has
   (a) 4 links (b) 6 links (c) 8 links (d) 10 links   [*Ans.* (c)]

4. Which of these mechanisms gives an approximately straight line?
   (a) Hart (b) Watt (c) Peaucellier (d) Tchebicheff   [*Ans.* (b)]

5. Which of these mechanisms has six links?
   (a) Tchebicheff (b) Hart (c) Peaucellier (d) Watt   [*Ans.* (b)]

6. Which of these mechanisms use two identical mechanisms?
   (a) Hart (b) Watt (c) Peaucellier (d) none   [*Ans.* (d)]

7. The Davis steering gear is not used because

   (a) it has turning pairs (b) it is costly
   (c) it has sliding pair   (d) it does not fulfil the condition of correct gearing   [*Ans.* (b)]

8. The Davis steering gear fulfils the condition of correct steering at

   (a) two positions   (b) three positions   (c) all positions   (d) one position   [*Ans.* (c)]

9. The Ackermann steering gear fulfils the condition of correct steering at

   (a) no position   (b) one position   (c) three positions   (d) all positions   [*Ans.* (c)]

10. A Hooke's joint is used to join two shafts which are

    (a) aligned   (b) intersecting   (c) parallel   [*Ans.* (b)]

11. The maximum velocity of the driven shaft of a Hooke's joint is

    (a) $\omega_1 \cos\alpha$ (b) $\dfrac{\omega_1}{\cos\alpha}$ (c) $\omega_1 \sin\alpha$ (d) $\dfrac{\omega_1}{\sin\alpha}$ [*Ans.* (b)]

12. The maximum velocity of the driven shaft of a Hooke's joint is at θ equal to

    (a) 0° and 180°   (b) 90° and 270°   (c) 90° and 180°   (d) 180° and 270°   [*Ans.* (a)]

Velocities and Accelerations in Mechanisms

3

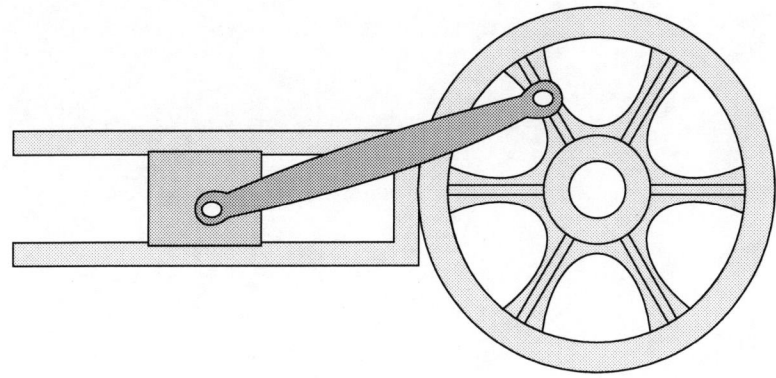

Reciprocation Mechanism

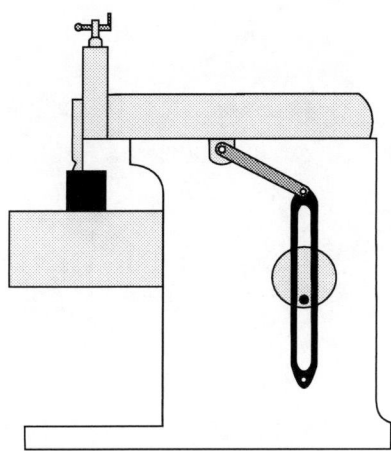

Shaping Machine

Velocities and Accelerations in Mechanisms

3.1 INTRODUCTION

Every particle of every link of a machine must move along a definite constrained path which may be a straight path or curvilinear. Curvilinear may be a circular with a fixed radius or curved in which the path may be an arc with different radii from instant to instant. **Displacement** involves magnitude as well as direction and hence it is a vector quantity.

Velocity of a particle is defined as the rate of change of displacement with respect to its surroundings, in a particular direction. The direction of the velocity at any instant will be same as the direction of the displacement at the same instant. Thus, velocity also has magnitude and direction and so is a vector quantity. Before studying the various methods such as instantaneous centre and relative velocity methods to determine the velocities in mechanisms, the different types of motions and velocities are to be understood.

Acceleration of a particle is defined as the rate of change of velocity with respect to its surroundings, in a particular direction. The acceleration also has magnitude and direction and so is a vector quantity.

3.2 MOTION

A rigid body is said to be at rest, if it occupies the same position with respect to its surroundings at all moments. The body may undergo translatory motion or rotary motion or a combination of both the motions.

3.2.1 Translatory Motion

If all the particles of a rigid body describe parallel paths so that the lines joining any two points in the body always remain parallel to their initial position, then such a motion is called the translatory motion.

In translatory motion, all the particles of a rigid body will have same linear displacements and linear velocities.

3.2.2 Rotary Motion

If all the particles of a rigid body describe circular paths around a point, called the axis of rotation, then such a motion is called the rotary motion. In rotary motion all the particles of a rigid body will have same angular displacements and angular velocities. Let us know some of the important physical quantities associated with rotary motion.

3.2.3 Speed

The speed of a body may be defined as its rate of change of displacement with respect to its surroundings. The speed of the body is irrespective of its direction and is thus a scalar quantity.

3.2.4 Angular Displacement (θ)

The angle through which the radius vector representing the position of a particle rotates is called angular displacement (or) the change in position of a particle in a circular path with respect to its centre is called **angular displacement**. Look at the Figs. 3.1 and 3.2.

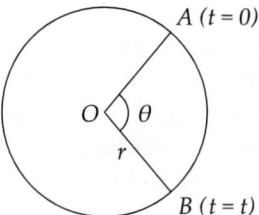

Fig. 3.1

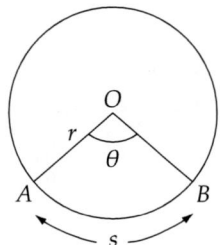

Fig. 3.2

Let a particle rotate around the circumference of the circle whose centre is at 'O'. Let us assume that the initial position of the particle at time $(t) = 0$ is A. The position of the particle at time $(t) = t$ is B. Then, angular displacement of the particle is θ.

The angular displacement can be measured in degrees. But the most convenient unit to measure it is radians. A radian is approximately equal to 60°. But let us define the radian precisely.

3.2.5 Radian

Radian is the SI unit of angular displacement. One radian is defined as the angle subtended at the centre of a circle by an arc which is equal to length of the arc divided by the radius of the circle.

Here in this figure, the angle in radians is defined by the equation $\theta = \dfrac{s}{r}$. where s represents the circumference, then $\theta = \dfrac{2\pi r}{r} = 2\pi = 360°$

Or π radians = 180° or 1 radian = $\dfrac{180°}{\pi}$ = 57.3°

3.2.6 Angular Velocity (ω)

The rate of change of angular displacement of a particle moving on a circle is called the angular velocity represented by ω. Average angular velocity = Angular displacement per unit time or rate of change of angular displacement. The SI unit of angular velocity is radian/sec.

Direction of angular velocity: The popular way of specifying the direction of rotation of any wheel is clockwise or counterclockwise. But this alone will not give us all the information about which the wheel is turning. Then how can we describe the direction of the turning wheel completely? It is specified by the right-hand screw rule as follows:

Right-Hand Screw Rule: Angular velocity is a vector quantity. Its magnitude can be computed from the rate of change of angular displacement, i.e. by $\dfrac{d\theta}{dt}$. Its direction is given by the right-hand screw rule.

The right-hand screw rule tells us that, if a right-hand screw is rotated in the direction of the rotating particle, then the direction of angular velocity of the particle is given by the direction in which the head of the screw moves.

Here is a simple question. Earth turns towards the east. What is the direction of the angular velocity of the earth?

The answer is simple. Use the right-hand screw rule, then you can easily make out that the angular velocity of earth points toward the north, which remains fixed on the earth's axis.

3.2.7 Relation between Linear Velocity and Angular Velocity

Linear velocity of a particle is given by the product of the radius of the circle on which the particle is moving and the angular velocity of the particle.

Linear velocity (v) = Angular velocity (ω) × radius (r). Hence, $v = \omega \times r$.

3.3 INSTANTANEOUS CENTRE METHOD

One of the interesting concepts in kinematics is that of an instantaneous axis for rigid bodies which move relative to one another. In fact, an axis exists which is common to both bodies and about which either body can be regarded as rotating with respect to the other. Each axis is perpendicular to the plane of the motion and they are called instant centres. The rigid body sometimes undergoes combined motion of rotation and translation. Both these motions can be considered taking place one after the other. This combined motion of rotation and translation may be assumed to be a pure rotation about some centre, which also goes on changing. Such a centre which goes on changing from one instant to another is known as instantaneous centre.

Hence, this method is based upon the fact that at any instant the motion of a rigid body is equivalent to a motion of the body as a whole about some centre called "instantaneous centre or instant centre or virtual centre of rotation" indicated by I. It changes its location relative to both bodies as the motion progresses and the paths of the instant centres are called centrodes.

This method is easy to apply for mechanisms having less number of links say 4 to 6 links. But the main disadvantage of this method is that this method cannot be extended to acceleration analysis.

The instantaneous centre of a body is useful in finding the velocity of any point on that body as they will be proportional to their distances from the instantaneous centre and their directions will be perpendicular to the line joining that point and the instantaneous centre. The instantaneous centre of the link AB shown in Fig. 3.3(a) can be obtained from the initial and final positions of the rigid body or link. Let AB and A_1B_1 be the two positions respectively. Join A to A_1 and B to B_1.

Draw perpendicular bisectors to AA_1, BB_1 and extend them to intersect at I representing the centre of rotation or instantaneous centre of AB as shown.

Let ω be the angular velocity of the link AB. Then the linear velocity of A,

$$v_A = \omega \times IA \quad \text{or} \quad \omega = \frac{v_A}{IA} \tag{i}$$

Similarly,

$$\omega = \frac{v_B}{IB} \tag{ii}$$

Equating (i) and (ii), the ratio of the velocities is $\dfrac{v_B}{v_A} = \dfrac{IB}{IA}$

The direction of v_A and v_B will be at right angles to IA and IB respectively as shown in Fig. 3.3(b). The velocities of any other points on the link AB such as X, Y and Z which are rigidly attached to AB are proportional to the product of their distances from I and the angular velocity. Their direction will be at right angles to the lines IX, IY and IZ as shown.

The instant centre is designated by using the numbers of the two links associated with it. Thus, I_{23} identifies the instant centre between links 2 and 3. The same centre can be identified as I_{32} since the order of the numbers has no significance.

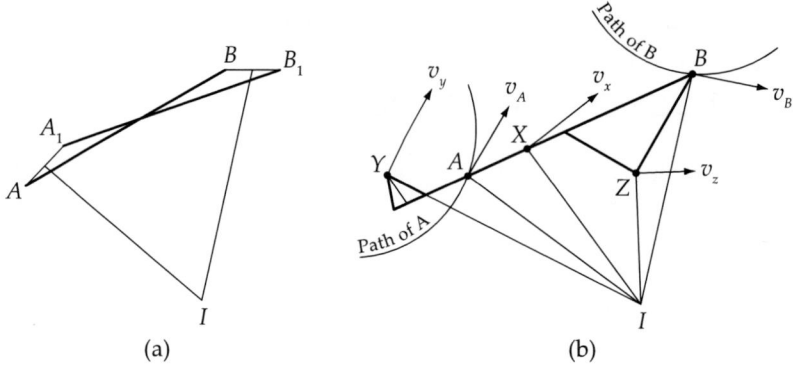

Fig. 3.3 Instantaneous centre

3.3.1 Properties of Instantaneous Centres

There are two important properties:

1. A rigid link rotates instantaneously relative to another link at the instantaneous centre for the given configuration of the mechanism.
2. The two rigid links have no linear velocity relative to each other at the instantaneous centre. Both will have the same linear velocity. The velocity of the instantaneous centre relative to any third rigid link will be same whether the instantaneous centre is regarded as a point on the first or on the second rigid link.

3.3.2 Number of Instantaneous Centres in a Mechanism

A mechanism has as many instantaneous centres as there are ways of pairing the link numbers. Thus, the total number of instantaneous centres N in a mechanism having n links is $N = \dfrac{n(n-1)}{2}$.

For $n = 4$, the number of instantaneous centres, $N = \dfrac{4 \times 3}{2} = 6$.

3.3.3 Types of Instantaneous Centres

There are two types of instantaneous centres. They are primary and secondary instantaneous centres. The primary centres are again of two types: fixed or permanent and secondary centres. The secondary centres are neither fixed nor permanent but they change from instant to instant.

3.3.4 Location of Instantaneous Centres

The following rules are to be applied for locating the instantaneous centres of any given mechanism depending upon the relative motion between the two links (1) and (2) as follows:

(a) When two links (1) and (2) are connected by a pin joint (or pivot joint), the instantaneous centre of them I_{12} lies at the centre of the pin as shown in Fig. 3.4(a).

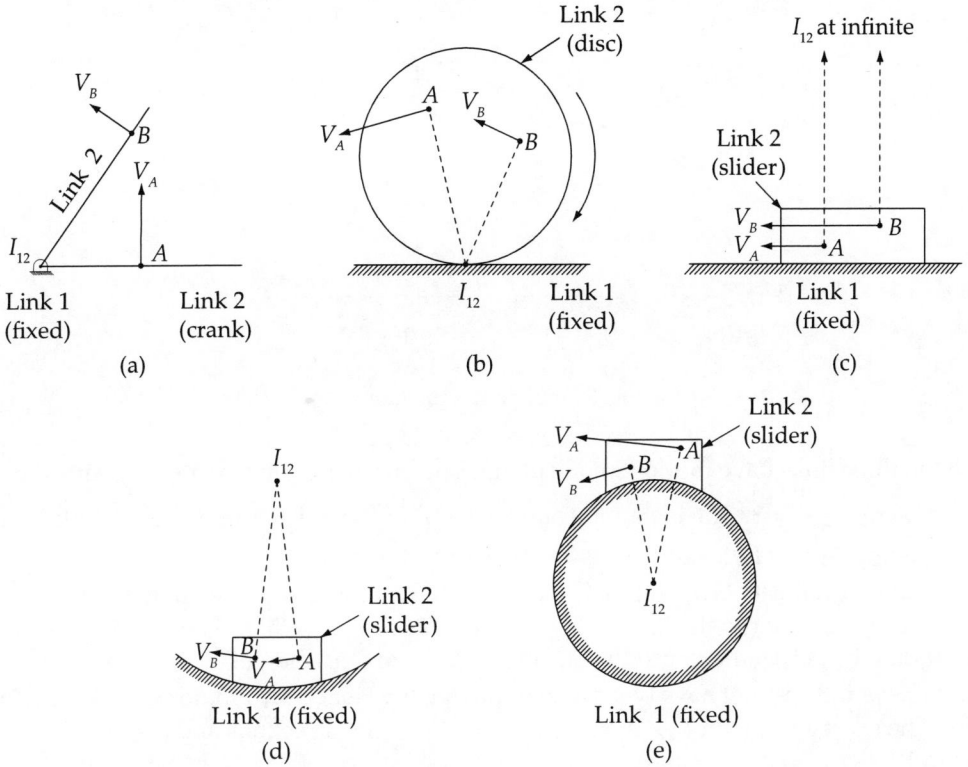

Fig. 3.4 Location of instantaneous centres.

(b) When two links (1) and (2) have a pure rolling contact, the instantaneous centre I_{12} lies at the point of contact as shown in Fig. 3.4(b).
(c) When two links (1) and (2) have a sliding contact, the instantaneous centre I_{12} lies on the common normal drawn as follows in three different cases shown in Fig. 3.4 (c, d, e).

 (i) Link 2 (slider) slides on a fixed link 1 having straight surface, the instantaneous centre I_{12} lies at an infinite distance and each point on the slider having same velocity.
 (ii) Link 2 (slider) slides on a fixed link 1 having curved surface, the instantaneous centre I_{12} lies at the centre of curvature of the curvilinear path at that instant.
 (iii) Link 2 (slider) slides on a fixed link 1 having constant radius of curvature, the instantaneous centre I_{12} lies at the centre of curvature, i.e. at the centre of the circle.

3.3.5 Kennedy's Theorem or Three-centres-in-line Theorem

Consider three kinematic links 1, 2 and 3 as shown in Fig. 3.5 having relative plane motions between them. P is the point of contact between 2 and 3. The three instantaneous centres are I_{31} is 3 relative to 1, I_{21} is 2 relative to 1 and I_{32} is 3 relative to 2. Kennedy's theorem states that these three instantaneous centres must lie on a straight line.

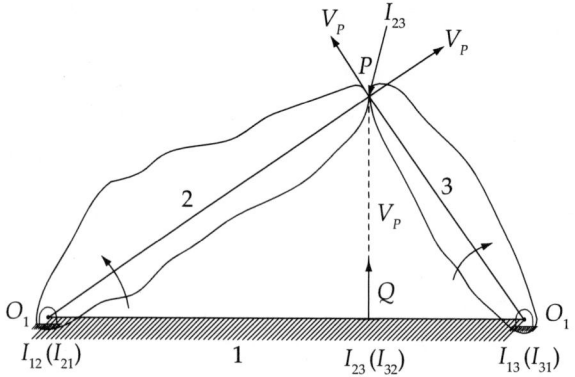

Fig. 3.5 Kennedy's theorem.

Let all the three links have motion in one plane. The number of instantaneous centres for three links are equal to $\dfrac{3 \times 2}{2} = 3$. The instantaneous centre I_{23} of the link 3 relative to the link 2 must lie on the line joining I_{21} and I_{31} according to the Kennedy's theorem.

P is the point of contact between link 2 and link 3. When P is considered a point on link 2, then v_p should be perpendicular to the line joining I_{21} and P as shown. When P is considered a point on 3, then v_p should be perpendicular to the line joining I_{13} and P as shown.

But a single point P cannot have two different directions. Hence, I_{23} should be on the line joining I_{12} and I_{13}. The exact location of I_{23}, i.e. at Q, depends upon the directions and magnitudes of 2 and 3 relative to 1.

Note: $I_{21} = I_{12}$; $I_{13} = I_{31}$; $I_{32} = I_{23}$

3.3.6 Application of Instantaneous Centre to Any Mechanism

The various steps are as given below:

1. Draw the mechanism to a suitable scale and give numbers to all the links.
2. Count the number of links in the given mechanism, say 'n'
3. Calculate the total number of instantaneous centres of the given mechanism using the expression $\left[\dfrac{n(n-1)}{2}\right]$.
4. Mark the known primary instantaneous centres on the mechanism, i.e. fixed and permanent.
5. Find the remaining centres using the three-centres-in-line theorem.
6. Determine the velocities of the required points using instantaneous centres.

Note: (i) Velocities of required links can be determined using some of the instantaneous centres.
(ii) All the instantaneous centres are to be determined only when they are asked. Otherwise determine only those particular centres essential to determine the velocities of required points.

3.3.7 Steps in Determining the Unknown Instantaneous Centres

Follow this method to determine the unknown instantaneous centres or for indicating and checking the instantaneous centres and their names.

1. Draw a circle of some diameter and divide the circumference into number of parts equal to the number of links in the given mechanism.
2. Then identify the known instantaneous centres for a given pair of links and mark on the circle by joining the corresponding points by a full line.
3. Indicate the unknown instantaneous centres, by joining such points by dotted lines.
4. Take one-by-one dotted line representing the unknown centres which should be the common sides of two triangles having other two sides as full lines.
5. The point of intersection of the two lines passing through the known centres represented by the two sides of the triangles with full lines, gives the unknown centre. Then immediately make that dotted line as a full line indicating that the centre is determined.
6. Repeat the same for all other unknown centres.

This is explained through the following examples. The same worked examples are done by other methods also to see the relative merits and demerits of the methods.

W E 3.1: The crank of a slider crank mechanism rotates clockwise at a constant speed of 300 r.p.m. The crank is 150 mm and the connecting rod is 600 mm long. Determine: 1. Linear velocity of the piston A and midpoint of the connecting rod D, and 2. Angular velocity of the connecting rod, at a crank angle of 45° from inner dead centre position using instantaneous centre method.

Given:
Speed of crank OB (2) $N_2 = 300$ r.p.m. or $\omega_2 = \dfrac{2\pi 300}{60} = 31.42$ rad/s;
$OB = 150$ mm $= 0.15$ m; $BA = 600$ mm $= 0.6$ m; Crank angle $\theta = 45°$

Steps:

1. Draw the space diagram choosing a suitable scale as shown at Fig. 3.6 (a). Count the number of links of the given mechanism ($n = 4$) and give numbers to them starting with number 1 to fixed link as shown.

2. Calculate the total number of instantaneous centres using the formula $\dfrac{n(n-1)}{2}$. Here the number of links (n) are 4, so the total number of instantaneous centres are $\dfrac{4(4-1)}{2} = 6$. Their names can be determined by taking the various combinations in the order of 1, 2, 3 and 4 as $I_{12}, I_{13}, I_{14}, I_{23}, I_{24}$ and I_{34}. Mark the known primary instantaneous centres, i.e. fixed and permanent centres such as I_{12}, I_{23}, I_{34} and I_{14} as shown. The other two unknown secondary centres, i.e. neither fixed nor permanent centres I_{13} and I_{24} are to be determined using the Fig. 3.6(b) as follows.

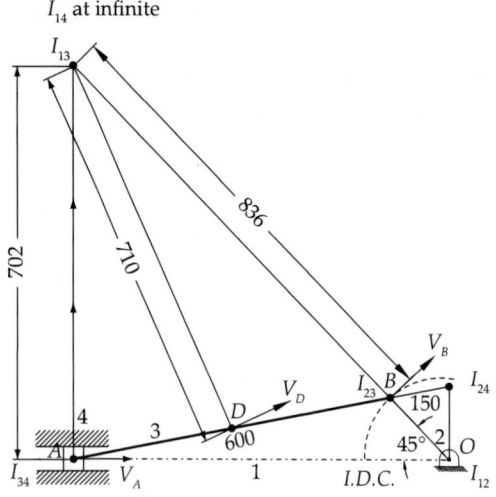

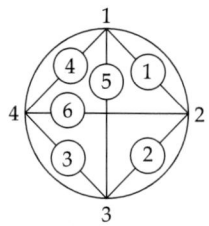

Fig. 3.6(a) Space diagram. **Fig. 3.6(b)** Order to find centres.

3. Draw a circle and mark on it 4 points equal to the number of links of the given mechanism as shown.

4. Join the points 1-2, 2-3, 3-4, 4-1 by full lines representing the known centres I_{12}, I_{23}, I_{34} and I_{14}.

5. Take up the other two unknown centres I_{13} and I_{24} one after the other. Join the points 1-3 and 2-4 with dotted lines first and then apply the three-centres-in-line theorem, one after the other.

6. Take first say 1-3 dotted line which is common to triangles 1-2-3 and 1-4-3. The three centres with the combinations of 1-2-3 are I_{12}, I_{23} and I_{13}. All these three must lie on a single straight line. The other three centres with the combinations of 1-4-3 are I_{14}, I_{43} and I_{13}. All these three must lie on another straight line.

7. Hence, the centre I_{13} must lie at the point of intersection of the above two straight lines, i.e. by extending the line joining (crank OB) I_{12} and I_{23} and the perpendicular line at A, i.e. joining

I_{34} and I_{14} (at infinite) gives the point of intersection I_{13} as shown. After finding the centre I_{13}, join points 1-3 by a full line.

8. Now take dotted line 2-4 common to triangles 1-2-4 and 2-3-4. The centre I_{24}, lies at the point of intersection of the straight lines passing through I_{12}, I_{14} and I_{34}, I_{23} as shown. Now join 2-4 by full line. Thus, all the six centres are to be found.

Note: The serial numbers given on the lines in Fig. 3.6(b), gives the order in which the centres are determined.

9. Measure the distances between the centres, i.e. I_{13}, I_{23} = 836 mm, I_{13}, I_{34} = 702 mm and I_{13}, D = 710 mm.

10. The velocity of point B, considering it as on the crank OB is $v_B = \omega_2.OB = \omega_2.I_{12}I_{23} = 31.42 \times 0.15 = 4.713$ m/s. But the velocity of B, v_B can also be calculated by considering it as on the connecting rod AB using $\omega_3.I_{13}I_{23}$. Knowing the distance between I_{13}, I_{23}, calculate the angular velocity of AB,

$$\omega_3 = \frac{v_B}{I_{13}I_{23}} = \frac{4.713}{0.836} = 5.64 \text{ rad/sec}$$

11. Now knowing ω_3, calculate the velocity of piston A, considering it as on link AB and using $v_A = \omega_3.I_{13}.I_{34} = 5.64 \times 0.702 = 3.96$ m/sec.

12. Obtain the velocity of the midpoint D on AB, (v_D) knowing the distance between I_{13} and D. That is, $v_D = \omega_3.I_{13} D = 5.64 \times 0.710 = 4.01$ m/sec.

Note:

1. While measuring lengths, consider the scale adopted for space diagram.
2. Choose lines which are common to two triangles whose other two sides are known.
3. The accuracy of the results depends upon the scale adopted for space diagram.

W.E 3.2: The dimensions and configuration of the four-bar mechanism shown in Fig. 3.7 are as follows: P_1A = 300 mm; P_2B = 360 mm; AB = 360 mm, and P_1P_2 = 600 mm. The angle $\angle AP_1P_2$ = 60°. The crank P_1A has an angular velocity of 10 rad/sec clockwise. Determine the angular velocities of P_2B, AB and the velocity of the point B using instantaneous centre method.

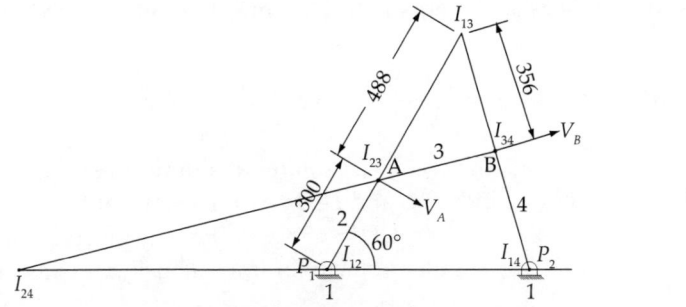
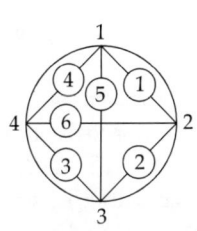

Fig. 3.7(a) Space diagram.   **Fig. 3.7(b)**

Given:
Angular velocity of P_1A, $\omega_2 = 10.0$ rad/s clockwise; $P_1A = 300$ mm $= 0.3$ m;
$BA = 360$ mm $= 0.36$ m; $P_2B = 360$ mm $= 0.36$ m $P_1P_2 = 600$ mm $= 0.6$ m $\angle AP_1P_2 = 60°$.

Steps:

1. Draw the space diagram choosing a suitable scale and give numbers to links as shown at Fig. 3.7(a).

2. Calculate the number of instantaneous centres using the formula $\dfrac{n(n-1)}{2}$ where n is the number of links. Here n is 4, so $\dfrac{4(4-1)}{2} = 6$. Their names can be determined by taking various combinations with 1, 2, 3 and 4 as $I_{12}, I_{13}, I_{14}, I_{23}, I_{24}, I_{34}$. Name the known instantaneous centres on the space diagram, i.e. I_{12}, I_{23}, I_{34} and I_{14} as shown. The other two unknown centres are I_{13} and I_{24}.

3. Draw a circle and mark on it 4 points equal to the number of links as shown at Fig. 3.7(b).

4. Join the points 1-2, 2-3, 3-4, and 4-1 by full lines representing the known centres I_{12}, I_{23}, I_{34} and I_{14}.

5. Take up the unknown centres one-by-one by joining the points 1-3 and 2-4 first with dotted lines.

6. Apply the three centres in line theorem, taking first 1-3 line which is a common side of two triangles 143 and 123. The centre I_{13} lies on the straight lines passing through I_{12}, I_{23} and I_{34}, I_{14} as shown. Extend cranks P_1A and P_2B. The point of intersection gives I_{13} as shown. Now join 1-3 by full line.

7. Next take 2-4 line common to triangles 124 and 234. The centre I_{24} lies on the straight lines passing through I_{12}, I_{14} and I_{34}, I_{23} as shown. Extend AB and P_1P_2. The point of intersection gives I_{24} as shown. Now join 2-4 by full line. Thus, all the six centres are to be found.

8. The velocity of A, considering it as on the crank P_1A is $v_A = \omega_2 \cdot P_1A = 10 \times 0.3 = 3.0$ m/s. But the velocity v_A can also be calculated using $\omega_3 \cdot I_{13}I_{23}$ by considering the point A as on the connecting rod AB. Measure the distance between I_{13} and I_{23} which is equal to 488 mm. Then calculate angular velocity of AB, $\omega_3 = \dfrac{v_A}{(I_{13}I_{23})} = \dfrac{3}{0.488} = 6.1$ rad/s.

9. Now knowing the ω_3 and distance I_{13}, I_{34} equals to 356 mm, the velocity of the point B, $v_B = \omega_3 \cdot I_{13} \cdot I_{34} = 6.1 \times 0.356 = 2.189$ m/s.

10. The angular velocity of P_2B, i.e $\omega_4 = \dfrac{v_B}{P_2B} = \dfrac{2.189}{0.36} = 6.08$ rad/sec.

W E 3.3: In the mechanism shown in Fig. 3.8, the crank OA rotates at a uniform speed of 20 r.p.m. anticlockwise and gives motion to the sliding blocks B and D. The dimensions of the various links are $OA = 300$ mm; $AB = 1200$ mm; $BC = 450$ mm and $CD = 450$ mm. For the given configuration, determine (i) velocities of sliders B and D, (ii) angular velocity of link CD using the instantaneous centre method:

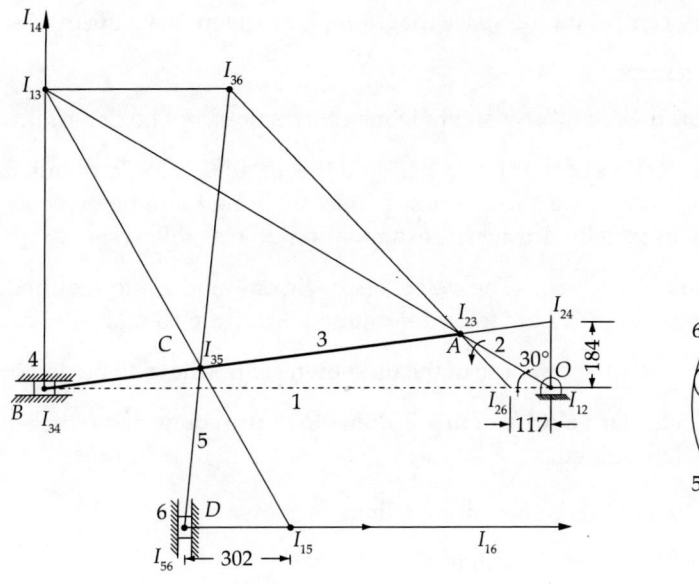
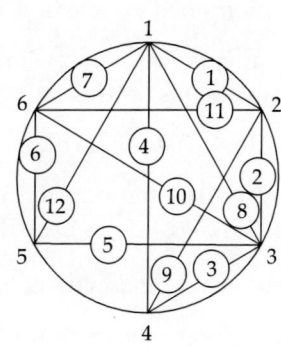

Fig. 3.8(a) Space diagram. **Fig. 3.8(b)**

Given:

Speed of crank OA, $N_{AO} = N_2 = 20$ r.p.m.; $\omega_2 = \dfrac{2\pi 20}{60} = 2.1$ rad/s;

$OA = 300$ mm $= 0.3$ m; $BA = 1200$ mm $= 1.2$ m; $BC = CD = 450$ mm $= 0.45$ m.
Distance between O and slider $D = 1050$ mm $= 1.05$ m.

This mechanism is an extension of the slider crank mechanism seen at W E 3.1. Here there are 6 links with two sliders.

Steps:

1. Draw the space diagram choosing a suitable scale and give numbers to links from 1 to 6 as shown.

2. Calculate the number of instantaneous centres using the formula $\dfrac{n(n-1)}{2}$ where n is the number of links. Since n is 6, the total number of instantaneous centres are $\dfrac{6(6-1)}{2} = 15$. Determine the names of the centres by taking the various combinations as follows.

Centres with number 1	–	I_{12}; I_{13}; I_{14}; I_{15}; I_{16}
Centres with number 2	–	I_{23}; I_{24}; I_{25}; I_{26}
Centres with number 3	–	I_{34}, I_{35}, I_{36}
Centres with number 4	–	I_{45}; I_{46}
Centres with number 5	–	I_{56}

3. Name the 7 known primary centres on the space diagram of the given mechanism as shown.

4. There are still 8 unknown centres.

   Note: In the above given problem, all the instantaneous centres need not be determined.

5. In order to determine the velocities of the sliders B, D and the angular velocity of link CD, i.e. ω_5, only such of those centres necessary from among the remaining 8 unknown centres can be determined to save time especially during the examination time as follows.

6. The angular velocity of link 2 is given. The velocities of links 4 and 6 are required. The instantaneous centres connected with 2 are to be determined. They are I_{24} and I_{26}.

7. In order to determine these two centres, some of the unknown centres have to be determined.

8. Similarly, to determine the angular velocity of link 5, then the centre connected with 5 and the fixed link 1, i.e. I_{15} is to be determined.

9. Draw a circle and mark on it 6 points representing 6 links as shown at Fig. 3.8(b).

10. Join the points of seven known centres by full lines first.

11. Take up the required unknown centres one-by-one, by joining the points first with dotted lines and then apply three-centres-in-line theorem to determine them as follows.

12. To determine I_{24}, I_{26} and I_{15} centres some among the other remaining five have to be determined instead of all the five.

13. First determine I_{13} using I_{12}, I_{23} and I_{14}, I_{34} lines; I_{24} using I_{12}, I_{14} and I_{23}, I_{34} lines; I_{36} using I_{13}, I_{16} and I_{35}, I_{56} lines in the order indicated. Then determine I_{26} using I_{12}, I_{16} and I_{23}, I_{36} lines. Finally, determine I_{15} using I_{13}, I_{35} and I_{16}, I_{56} lines.

    Note: The centres unnecessary for the required results are avoided. They are I_{25}, I_{45} and I_{46}. Definitely some time can be saved.

14. Measure the distances between $I_{12}, I_{24} = 184$ mm and $I_{12}, I_{26} = 117$ mm. Then the velocity of slider B $(v_B) = \omega_2 . I_{12} I_{24} = 2.1 \times 184 = 386$ mm/sec $= 0.386$ m/sec and the velocity of slider D $(v_D) = \omega_2 . I_{12} I_{26} = 2.1 \times 117 = 245.7$ mm/sec $= 0.245$ m/sec.

15. Determine the angular velocity of link 5, by measuring the distance between $I_{15}, I_{56} = 302$ mm $= 0.302$ m. Then $\omega_5 = \dfrac{v_D}{(I_{15} I_{56})} = \dfrac{0.245}{0.302} = 0.811$ rad/sec.

    Note: All the 15 centres need not be determined unless they are asked.

W E 3.4: In the toggle mechanism shown in Fig. 3.9(a), the slider D is constrained to move on a horizontal path. The crank OA is rotating in the counterclockwise direction at a speed of 180 r.p.m. The dimensions of the various links are as follows: OA = 180 mm; CB = 240 mm; AB = 360 mm; and BD = 540 mm. For the given configuration, find velocity of slider D and angular velocity of BD using instantaneous centre method.

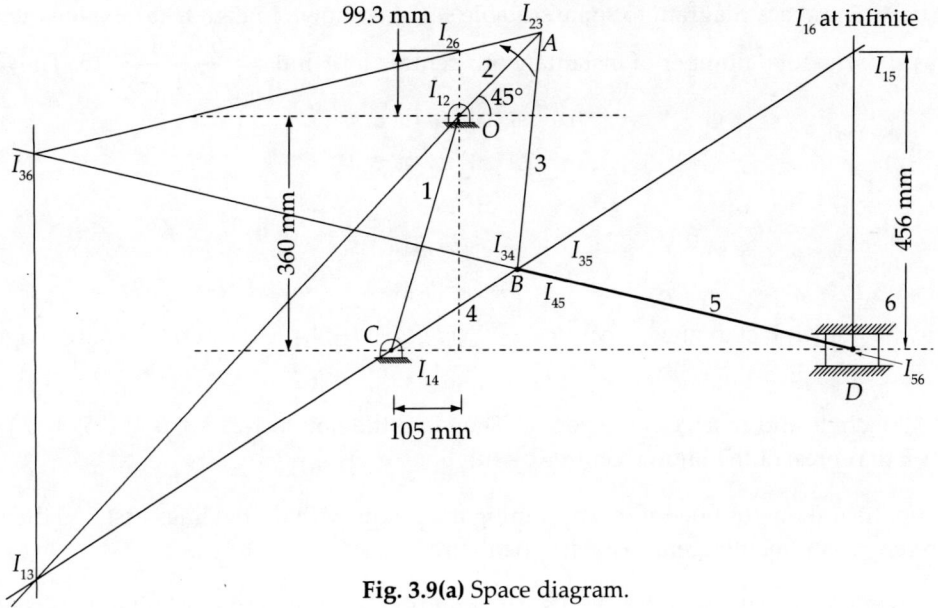

Fig. 3.9(a) Space diagram.

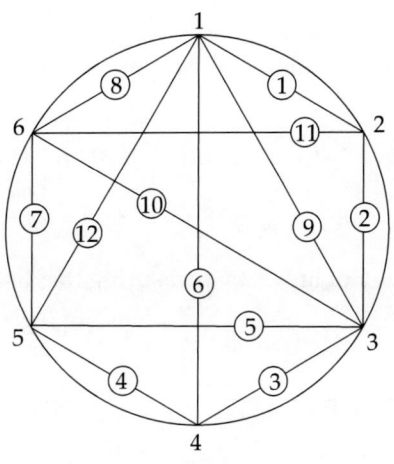

Fig. 3.9(b)

Given:

$N_{AO} = 180$ r.p.m. or $\omega = \dfrac{2\pi 180}{60} = 18.85$ rad/s; $\quad OA = 180$ mm $= 0.18$ m;

$CB = 240$ mm $= 0.24$ m; $\quad AB = 360$ mm $= 0.36$ m; $\quad BD = 540$ mm $= 0.54$ m.

This toggle mechanism is an extension of the four-bar mechanism seen in W E 3.2 above.

Velocity of A with respect to O or velocity of A $(v_A) = \omega_2 \times OA = 18.85 \times 0.18 = 3.4$ m/s.

1. First draw the space diagram to some suitable scale as shown in Fig. 3.9(a). Name the links 1 to 6 as shown. Total number of instantaneous centres for 6 links = $\frac{6(6-1)}{2}$ = 15. They are

$$I_{12}; I_{13}; I_{14}; I_{15}; I_{16}$$
$$I_{23}; I_{24}; I_{25}; I_{26}$$
$$I_{34}, I_{35}, I_{36}$$
$$I_{45}; I_{46}$$
$$I_{56}$$

2. Name the known instantaneous centres $I_{12}; I_{23}; I_{34}; I_{45}; I_{35}; I_{14}; I_{56}; I_{61}$ on the space diagram as shown.

3. Draw the circle and mark 6 points on it. Then join the points 1-2, 2-3; 3-4; 4-5; 3-5; 1-4; 5-6; and 1-6 to represent the known centres by full lines.

4. Take up the unknowns one-by-one by joining the points with dotted lines first and then apply three-centres-in-line theorem to determine them.

5. Determine the velocity of slider D (6), by just finding one unknown instantaneous centres I_{26} which connects link 2. Determine I_{15} for finding the angular velocity of link 5, i.e. ω_5.

6. First determine I_{13} using $I_{12}I_{23}$ and $I_{14}I_{34}$ lines. Determine I_{36} using $I_{13}I_{16}$ and $I_{35}I_{56}$ lines. Then determine I_{26} using $I_{12}I_{16}$ and $I_{23}I_{36}$ lines. Next determine I_{15} using $I_{13}I_{35}$ and $I_{16}I_{56}$ lines as shown at Fig. 3.9 (b).

7. Measure the distances between $I_{12}I_{26}$ = 99.3 mm = 0.099 m. The velocity of slider D,

$$v_D = \omega_2 \cdot I_{12}I_{26} = 18.85 \times 0.099 = 1.866 \text{ m/s}$$

8. Determine the angular velocity of link 5, by measuring the distance between I_{15} and I_{56} equal to 456 mm = 0.456 m. Then $\omega_5 = \dfrac{v_D}{I_{15}I_{56}} = \dfrac{1.866}{0.456} = 4.092$ rad/s.

3.4 RELATIVE VELOCITY METHOD

The principles followed in relative velocity method are based on the relative motions. In the Fig. 3.10(i), A and B are two independent points moving with velocities v_a and v_b respectively. Then the velocity of B relative to A is the velocity with which B appears to be moving to an observer situated at, and moving with, A. Let velocities equal and opposite to v_a be applied to both the points, i.e. $-v_a$ as shown. Then the point A will be coming to rest and the particle B will have a resultant velocity given by the vector sum of v_b and $-v_a$.

This resultant will clearly be the velocity with which, to the observer on A, B appears to be moving. In the figure, i.e. V_b is represented by the vector BC and $-v_a$ by vector BD or the vector CE, so that the velocity of B relative to A is denoted by BE, the vector sum of BC and CE. The velocity of B relative to A is represented by v_{ba}, the vector sum of BC and CE.

In the vector diagram shown at Fig. 3.10(ii), vector ob represents v_b and vector oa represents v_a, then the vector ab represents the difference between the vectors ob and oa represented by v_{ba}.

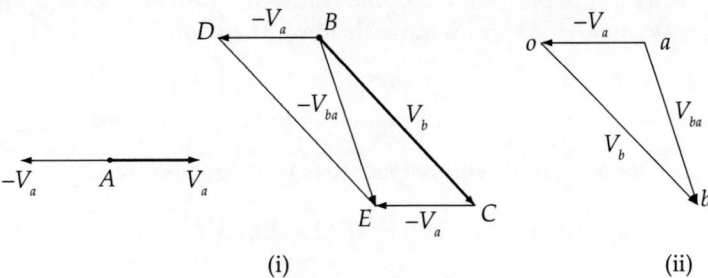

Fig. 3.10 Relative velocity.

Suppose A and B are points on the same link at fixed distance apart as shown in Fig. 3.11, then if the velocity of A is fixed in magnitude and direction then the velocity of B can be fixed only in direction.

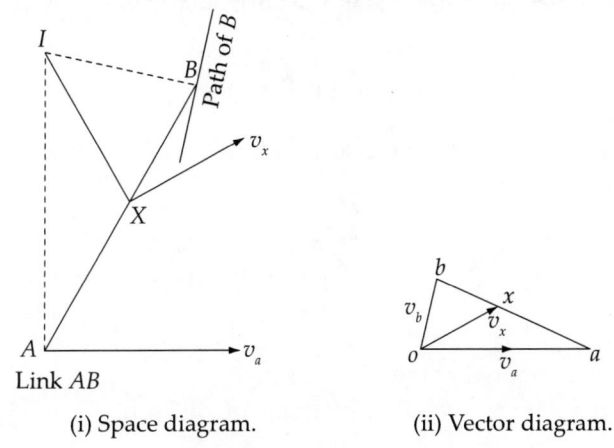

(i) Space diagram.   (ii) Vector diagram.

Fig. 3.11

Draw the vector diagram with oa representing v_a to some scale. Through o, draw ob tangential to the path of B and through a, draw ab perpendicular to AB. The velocity of B relative to A is ab.

The velocity components of v_a and v_b along AB must be equal, otherwise an increase or decrease of the length AB would take place. The vector ab which represents v_{ba} is known as the velocity image of the link AB.

The angular velocity of AB is $\dfrac{ab}{AB}$ radians per second. Obtain the velocity of X on AB by dividing ba in the same proportion as X divides BA and name it as x. Join x to o which represents v_x.

Note: This method is explained by taking the same examples discussed for instantaneous centre method in W E 3.1 to 3.4. This will help in understanding the merits and demerits of the two methods.

W E 3.5: The crank of a slider crank mechanism rotates clockwise at a constant speed of 300 r.p.m. The crank is 150 mm and the connecting rod is 600 mm long. Determine using relative velocity method: 1. Linear velocity of the midpoint D of the connecting rod, and 2. Angular velocity of the connecting rod at a crank angle of 45° from inner dead centre position.

Given:

$N_{BO} = 300$ r.p.m. or $\omega_{BO} = \dfrac{2\pi 300}{60} = 31.42$ rad/s; $\quad OB = 150$ mm $= 0.15$ m;

$BA = 600$ mm $= 0.6$ m. Velocity of B with respect to O or velocity of B, i.e.

$$v_{BO} \text{ or } v_B = \omega_{BO} \times OB = 31.42 \times 0.15 = 4.713 \text{ m/s}$$

1. First draw the space diagram to some suitable scale as shown at Fig. 3.12(a).

2. Take point o to represent the fixed point in the velocity diagram. Draw ob to represent 4.713 m/s perpendicular to OB, the velocity v_B to some suitable scale, say 471 mm. Scale of velocity diagram is 1 mm = 10 mm/sec.

3. Draw ba vector perpendicular to AB link and oa horizontally to represent the velocity of slider A as shown.

4. Both these vectors intersect at a as shown at Fig. 3.12(b).

5. Measure oa and convert according to the scale adopted for velocity diagram. The velocity of piston A, $v_{AO} = oa = 4.02$ m/s as shown.

6. To get the point D, find the midpoint d of vector ba. Then join o to d and measure its length. After converting according to the scale of the velocity diagram, i.e. $v_{DO} = do = 4.05$ m/s.

Note:

1. This method is easier compared to the previous method.
2. The scale for space diagram and velocity diagram are different.
3. The accuracy of the results depends on the scale adopted.

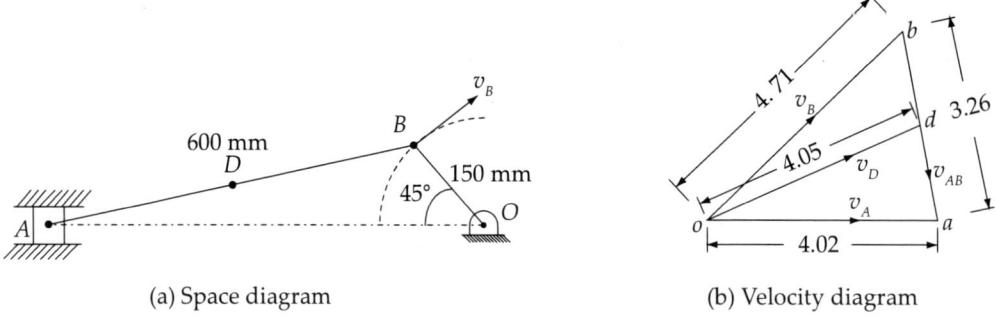

(a) Space diagram (b) Velocity diagram

Fig. 3.12

W E 3.6: The dimensions and configuration of the four-bar mechanism shown in Fig. 3.13 are as follows: $P_1A = 300$ mm; $P_2B = 360$ mm; $AB = 360$ mm and $P_1P_2 = 600$ mm. The angle $AP_1P_2 = 60°$. The crank P_1A has an angular velocity of 10 rad/sec clockwise. Determine the angular velocities of P_2B, and AB and the velocity of the point B by relative velocity method.

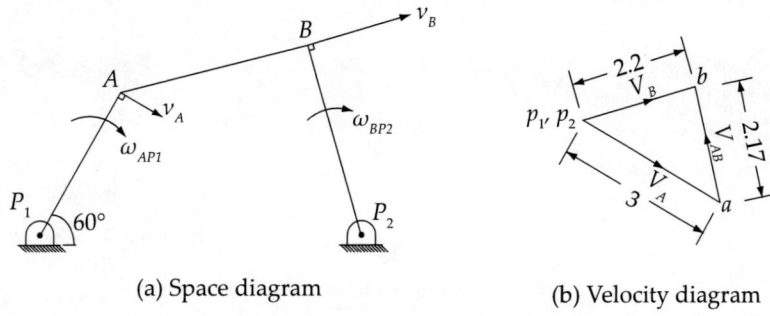

Fig. 3.13 Four-bar mechanism.

Given:
$\omega_{AP1} = 10$ rad/s; $P_1A = 300$ mm $= 0.3$ m; $P_2B = 360$ mm $= 0.36$ m
Velocity of A with respect to P_1 or velocity of A, i.e. v_{AP_1} or $v_A = \omega.P_1A = 10 \times 0.3 = 3$ m/s.

1. First draw the space diagram to some suitable scale as shown at Fig. 3.13(a).

2. Take p_1, p_2 to represent the fixed points in the velocity diagram. Draw p_1a perpendicular to P_1A to represent v_A or v_{AP_1} of 3 m/s to some suitable scale.

3. Draw p_2b and ab vectors perpendicular to P_2B and AB links respectively as shown.

4. Both these vectors intersect at b as shown at Fig. 3.13(b).

5. Measure p_2b and ab and convert according to the scale. $v_{BP_2} = v_B = p_2b = 2.2$ m/s and $v_{BA} = ba = 2.17$ m/s as shown.

6. Angular velocity of P_2B, $\omega_{P_2B} = \dfrac{v_B}{P_2B} = \dfrac{2.2}{0.36} = 6.11$ rad/s (clockwise).

7. Angular velocity of AB, $\omega_{AB} = \dfrac{v_{BA}}{AB} = \dfrac{2.17}{0.36} = 6.028$ rad/s (anticlockwise).

W E 3.7: In the mechanism shown in Fig. 3.14, the crank OA rotates at a uniform speed of 20 r.p.m. anticlockwise and gives motion to the sliding blocks B and D. The dimensions of the various links are $OA = 300$ mm; $AB = 1200$ mm; $BC = 450$ mm and $CD = 450$ mm. For the given configuration, determine using relative velocity method:

(a) velocities of sliding at B and D
(b) angular velocity of CD.

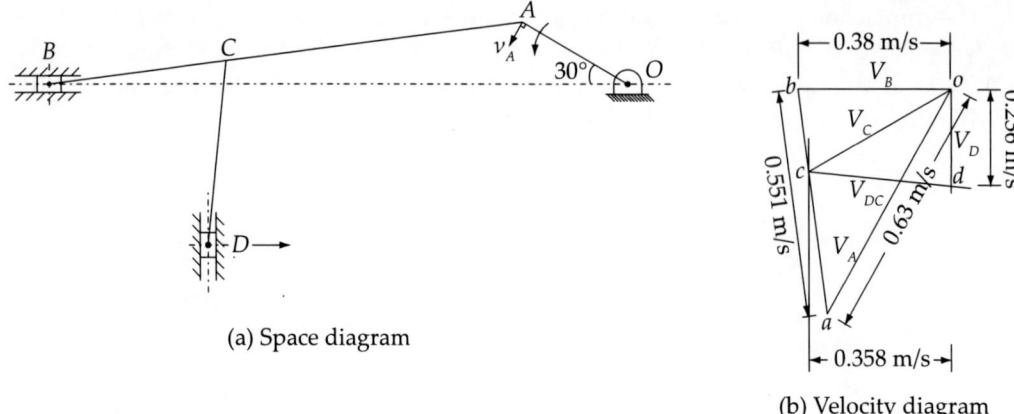

(a) Space diagram

(b) Velocity diagram

Fig. 3.14

Given:

$N_{AO} = 20$ r.p.m.; $\omega_{AO} = \dfrac{2\pi 20}{60} = 2.1$ rad/s; $OA = 300$ mm $= 0.3$ m; $BA = 1200$ mm $= 1.2$ m;

$BC = CD = 450$ mm $= 0.45$ m.

Steps:

1. Draw the space diagram choosing a suitable scale as shown at Fig. 3.14(a).

2. The velocity of A with respect to O or velocity of A, $v_{AO} = v_A = \omega_{AO} \times OA = 2.1 \times 0.3 = 0.63$ m/s.

3. Draw the velocity diagram shown at (b) as follows taking point o to represent the fixed point O.

4. Draw vector oa perpendicular to OA from o, to represent the velocity of v_A to some suitable scale based on its value.

5. From point a, draw vector ab perpendicular to AB to represent the velocity of v_{BA}. From point o, draw a vector ob parallel to path of slider B to represent the velocity of slider B. These two vectors ab and ob intersect at b.

6. Divide vector ab at c in the same ratio as C divides AB. In other words, $\dfrac{BC}{CA} = \dfrac{bc}{ca}$.

7. Now from point c, draw vector cd perpendicular to CD to represent the velocity of D with respect to C, i.e. v_{DC}. From point o, draw vector od parallel to the path of slider D (which is along the vertical direction) to represent the velocity of D.

8. The point of intersection gives d. By measurement, the velocity of slider B, $v_B = $ vector $ob = 0.38$ m/s and velocity of slider D, $v_D = $ vector $o_d = 0.236$ m/s as shown.

9. Obtain the angular velocity of CD by measuring the velocity of D with respect to C, $v_{DC} = $ vector $cd = 0.358$ m/s as shown. Hence, the angular velocity of CD, $\omega_{CD} = \dfrac{v_{DC}}{CD} = \dfrac{0.358}{0.45} = 0.79$ rad/s (anticlockwise).

W E 3.8: In the toggle mechanism shown in Fig. 3.15, the slider D is constrained to move on a horizontal path. The crank OA rotates in the counterclockwise direction at a speed of 180 r.p.m. The dimensions of the various links are as follows: OA = 180 mm; CB = 240 mm; AB = 360 mm; and BD = 540 mm.

For the given configuration, find velocity of slider D and angular velocity of BD using relative velocity method.

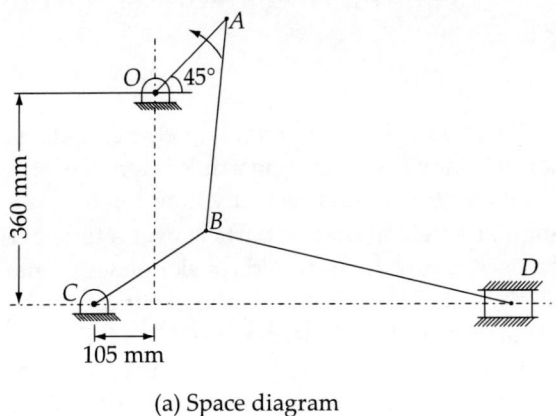

(a) Space diagram (b) Vector diagram

Fig. 3.15

Given: N_{AO} = 180 r.p.m. or $\omega_{AO} = \dfrac{2\pi 180}{60}$ = 18.85 rad/s; OA = 180 mm = 0.18 m; CB = 240 mm = 0.24 m; AB = 360 mm = 0.36 m; BD = 540 mm = 0.54 m

Velocity of A with respect to O or velocity of A = $v_{AO} = v_A = \omega_{AO} \times OA$ = 18.85 × 0.18 = 3.4 m/s.

1. First draw the space diagram to some suitable scale as shown in Fig. 3.15(a) and velocity diagram shown at (b) as follows:

2. Since O and C are fixed points, so they are o and c at one place in the velocity diagram. Draw oa perpendicular to OA to some suitable scale, to represent the velocity of A with respect to O or velocity of A, i.e. v_{AO} or v_A, such that vector $oa = v_{AO} = v_A$ = 3.4 m/s.

3. Draw ab perpendicular to AB to represent the velocity of B with respect to A, i.e. v_{BA}. Draw vector cb perpendicular to CB to represent the velocity of B with respect to C, i.e. v_{BC}. The vectors intersect at b.

4. From point b, draw vector bd perpendicular to BD to represent the velocity of D with respect to B, i.e. v_{DB}. From point c or o, draw vector cd or od in the direction of motion of the slider D, to represent the velocity of D, i.e. v_D. By measurement, velocity of D with respect to B, v_{DB} = vector bd = 2.38 m/s; and velocity of slider D, v_D = vector cd = 2.012 m/s.

5. Angular velocity of BD, $\omega_{BD} = \dfrac{v_{DB}}{BD} = \dfrac{2.38}{0.54}$ = 4.408 rad/s.

3.5 ACCELERATION IN MECHANISMS

3.5.1 Introduction

The rate of change of velocity with respect to time is acceleration. This is also a vector quantity, but the direction of the acceleration vector will be different from that of the velocity. To determine the acceleration of a particle moving along a circular path, the angular displacement, velocity and angular acceleration must be known first. The angular displacement and angular velocity has already been discussed in the previous chapter.

3.5.2 Angular Acceleration

Angular acceleration of a particle is defined as the rate of change of angular velocity. Let a particle have an initial angular velocity of ω_1 and a final angular velocity of ω_2 in a time interval of t seconds. Then the average angular acceleration (α) is the change in angular velocity/time, i.e. $(\omega_1 - \omega_2)/t$.

The above expression gives the average angular acceleration of a particle over a time period of t sec. But to find out the instantaneous angular acceleration, one should consider a very small time interval (δt) so that it tends to zero. Consequently, instantaneous angular acceleration α is $d\omega/dt$.

Angular acceleration is a vector quantity. That means, it has both magnitude and direction. The magnitude can be found out by the above expressions. Just as we assigned the direction to angular velocity by the right-hand screw rule, the direction of the angular acceleration is also taken along the direction of the angular velocity. When the angular velocity increases, acceleration increases and the direction of angular acceleration will be in the direction of the angular velocity. When the angular velocity decreases, the direction of angular acceleration will be in the opposite of angular velocity. In other words, the sense of the angular acceleration depends whether ω is increasing or decreasing. If ω is increasing, a will be in the same direction. If ω is decreasing, a will be in the opposite direction. Linear acceleration 'a' or 'a' of a particle is the product of the radius 'r' of the circle on which the particle is moving and the angular acceleration 'α' of that particle, i.e. $a = \alpha.r$.

3.5.3 Vector form between Linear and Angular Acceleration

The linear acceleration consists of two components. They are **tangential components** 'a^t' equal to $\alpha.r$ and **radial component** 'a^r' equal to $\omega.v$ or $\omega^2 r$. The radial component is directed along the radius towards the centre of rotation. Therefore, it is called the **radial acceleration** or **centripetal**

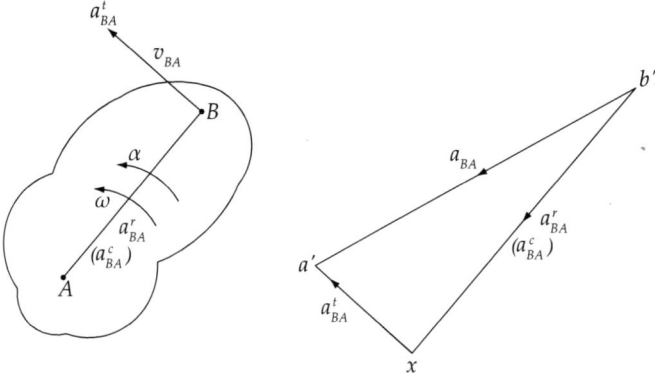

Fig. 3.16

acceleration. The total linear acceleration 'a' of the particle can be obtained by combining the two mutually perpendicular accelerations a^t and a^r in accordance with the parallelogram of vector addition as shown in Fig. 3.16.

3.5.4 Various Steps to be Followed in the Acceleration Analysis

1. Draw the space diagram of the given mechanism by choosing a proper scale based on the dimensions.
2. Calculate the velocity of the driving link.
3. Choose the proper scale for the velocity diagram.
4. Draw the velocity diagram and determine the required angular velocities and linear velocities of various points on various links as in the relative method.
5. Calculate the radial components of accelerations, i.e. $\omega^2 \cdot r$ knowing the angular velocities.
6. Choose the scale for acceleration diagram based on the above calculated values.
7. Draw the acceleration diagram and determine the required unknown accelerations (a) and also angular accelerations (α).

The same mechanisms discussed for instantaneous and relative velocity methods are taken up for acceleration analysis.

Note: This is an extension of the velocity analysis. So the accelerations will be correct if the velocities are correct.

W E 3.9: The crank of a slider crank mechanism rotates clockwise at a constant speed of 300 r.p.m. The crank is 150 mm and the connecting rod is 600 mm long. Determine: 1. The acceleration of the midpoint D of the connecting rod, and 2. Angular acceleration of the connecting rod, at a crank angle of 45° from inner dead centre position.

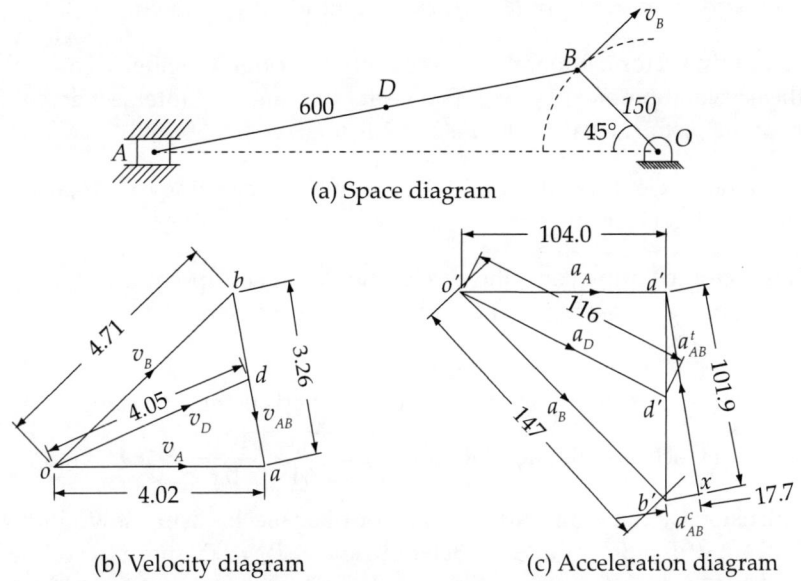

Fig. 3.17 Reciprocating mechanism.

Here to find accelerations one has to determine the necessary velocities. It is not possible to determine directly. Hence, repeat the velocity analysis first as shown at W E 3.5 and then extend as explained below.

The various steps involved for acceleration analysis are as follows:

1. Draw the space diagram and velocity diagram as explained in the W E 3.5 and determine the velocities and angular velocities as they are required to determine the accelerations.

2. Calculate the radial component of acceleration B with respect to O or simply the acceleration of B, is $a^r_{BO} = a_B = \dfrac{v_B^2}{OB} = \dfrac{(ob)^2}{OB} = \dfrac{(4.71)^2}{0.15} = 147$ m/s².

3. The radial component of the acceleration of A with respect to B, is $a^r_{AB} = \dfrac{v_{AB}^2}{BA} = \dfrac{(ab)^2}{AB} = \dfrac{(3.26)^2}{0.6}$ = 17.7 m/s².

4. Now draw the acceleration diagram with vector $o'b'$ parallel to BO, to some suitable scale, to represent the radial component of the acceleration of B with respect to O or simply acceleration of B, i.e. a^r_{BO} or $a_B = 147$ m/s².

   Note: There won't be a tangential component as the crank is rotating at constant speed.

5. The acceleration of A with respect to B has the following two components:
   (a) The radial component of the acceleration of A with respect to B, i.e. a^r_{BA}, and
   (b) The tangential component of the acceleration of A with respect to B, i.e. a^t_{BA}.

   These two components are mutually perpendicular.

6. Therefore, draw vector $b'x$ parallel to AB to represent $a^r_{BA} = 19.3$ m/s² and from point x, draw vector xa', perpendicular to vector $b'x$ whose magnitude is unknown.

7. Now from o', draw vector $o'a'$ parallel to the path of motion of slider A, a horizontal line to represent the acceleration of A, i.e. a_A. The vectors xa' and $o'a'$ intersect at a'. Join $a'b'$. The acceleration of the piston A, a_A = vector $o'a'$ = 104 m/s².

8. In order to find the acceleration of the midpoint D of the connecting rod AB, take the midpoint of vector $a'b'$ and name it as d'.

9. Join $o'd'$. The vector $o'd'$ represents the acceleration of the midpoint D of the connecting rod, i.e. a_D. By measuring according to the scale gives, a_D = vector $o'd'$ = 116 m/s² as shown in Fig. 3.17 (c).

10. Find the angular acceleration of the connecting rod using the tangential component, $a^t_{AB} = 101.9$ m/s². Hence, angular acceleration $a_{AB} = \dfrac{a^t_{AB}}{BA} = \dfrac{101.9}{0.6} = 169.83$ rad/s².

W E 3.10: The dimensions and configuration of the four-bar mechanism shown in Fig. 3.18 are as follows: $P_1A = 300$ mm; $P_2B = 360$ mm; $AB = 360$ mm; and $P_1P_2 = 600$ mm.

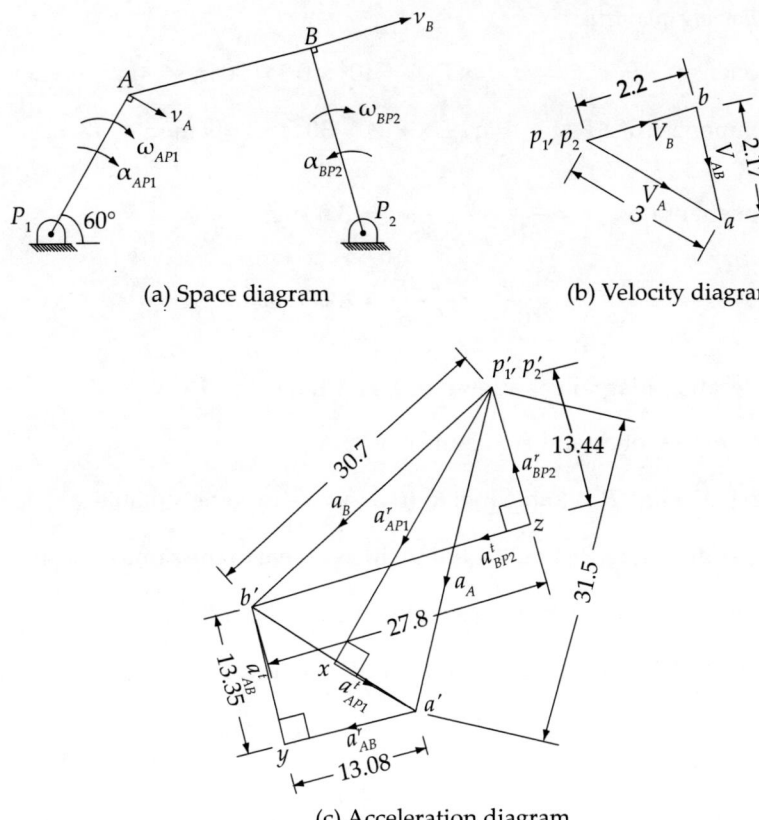

(a) Space diagram

(b) Velocity diagram

(c) Acceleration diagram

Fig. 3.18 Four-bar mechanism.

The angle $\angle AP_1P_2 = 60°$. The crank P_1A has an angular velocity of 10 rad/sec and an angular acceleration of 30 rad/sec², both clockwise. Determine the angular velocities and angular accelerations of P_2B, AB and the velocity and acceleration of the joint B.

Given:
$\omega_{AP_1} = 10$ rad/s; $\alpha_{AP_1} = 30$ rad/s²; $P_1A = 300$ mm $= 0.3$ m; $P_2B = 360$ mm $= 0.36$ m;
velocity of A with respect to $P_1 = v_{AP1} = v_A = \omega_{AP_1} \times P_1A = 10 \times 0.3 = 3$ m/s.

I. First draw the space diagram choosing proper scale as shown at Fig. 3.18(a).

II. Velocity of B and angular velocities of P_2B and AB: Proceed in the same manner as explained in the W E 3.6 by drawing the velocity diagram to determine $v_B, v_{AB}, \omega_{P_2B}$ and ω_{AB}. They are 2.2 m/s; 2.17 m/s; $\omega_{P_2B} = 6.11$ rad/s (CW); $\omega_{AB} = 6.028$ rad/s (ACW).

III. Acceleration of B and angular accelerations of P_2B and AB:

Calculate first the following quantities:

(i) Radial component of AP_1 $a^r_{AP_1} = \omega^2_{AP_1} \times P_1A = 10^2 \times 0.3 = 30$ m/s² $- p'_1 x$

(ii) Tangential component of AP_1 $a^t_{AP_1} = \alpha_{AP_1} \times P_1A = 30 \times 0.3 = 9$ m/s² $- xa'$

(iii) Radial component of BA $a^r_{BA} = \dfrac{v^2_{AB}}{AB} = \dfrac{(2.17)^2}{0.36} = 13.08$ m/s² $- a'y$

(iv) Radial component of BP_2 $a^r_{BP_2} = \dfrac{v^2_{BP_2}}{BP_2} = \dfrac{(2.2)^2}{0.36} = 13.44$ m/s² $- p'_2 z$

Steps to draw acceleration diagram as shown in Fig. 3.18(c)

1. Take a point to represent P_1 and P_2. Name it as p'_1, p'_2.

2. Draw $p'_1 x$ parallel to link P_1A and equal to (i) as above to some suitable scale.

3. Draw xa' perpendicular to P_1A and equal to (ii) as above to the same scale and get point a'.

4. Measure $p'_1 a'$ and convert as per the scale giving $a_A = 31.5$ m/s².

5. Draw $a'y$ parallel to AB and equal to (iii) as above to the same scale and draw perpendicular to it.

6. Draw $p'_2 z$ parallel to P_2B and equal to (iv) as above to the same scale and draw perpendicular to it.

7. Both the perpendiculars intersect at b'. Join p'_2 to b' and measure. It gives $a_B = 30.7$ m/s².

8. Measure $yb' = a^t_{BA} = 13.35$ m/s² and $zb' = a^t_{BP_2} = 27.8$ m/s².

9. Angular acceleration of P_2B equal to $\alpha_{P_2B} = \dfrac{a^t_{BP_2}}{P_2B} = \dfrac{27.8}{0.36} = 77.22$ rad/s². (ACW)

10. Angular acceleration of AB equal to $\alpha_{AB} = \dfrac{a^t_{AB}}{AB} = \dfrac{13.08}{0.36} = 36.33$ rad/s². (ACW)

W E 3.11: In the mechanism, as shown in Fig. 3.19, the crank OA rotates at a uniform speed of 20 r.p.m. anticlockwise and gives motion to the sliding blocks B and D. The dimensions of the various links are $OA = 300$ mm; $AB = 1200$ mm; $BC = 450$ mm and $CD = 450$ mm. For the given configuration. Determine: 1. angular velocity of CD 2. linear acceleration of D, B and angular acceleration of CD.

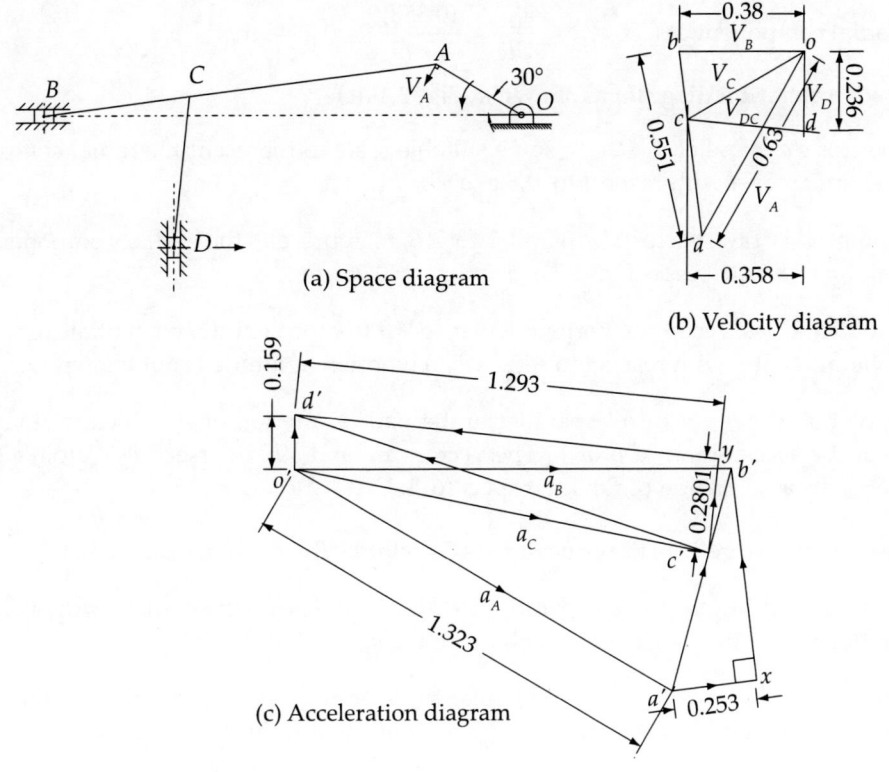

Fig. 3.19

Given:

$N_{AO} = 20$ r.p.m.; $\omega_{AO} = \dfrac{2\pi 20}{60} = 2.1$ rad/s; $OA = 300$ mm $= 0.3$ m;

$BA = 1200$ mm $= 1.2$ m; $BC = CD = 450$ mm $= 0.45$ m.

I. First draw the space diagram choosing proper scale as shown at Fig. 3.19(a).

II. Determination of velocities and angular velocities:
   Proceed in the same manner as explained in the W E 3.7 by drawing the velocity diagram and determine velocities of sliding at B and D.
   They are $v_B = 0.38$ m/s, $v_D = 0.236$ m/s, $v_{DC} = 0.358$ m/s and $\omega_{CD} = 0.789$ rad/s.

III. Determination of accelerations:

   Calculate first the following quantities:

   i. Radial component of AO, $a^r_{AO} = \omega^2_{AO} \times OA = 2.1^2 \times 0.3 = 1.323$ m/s^2 – $o'a'$

   ii. Radial component of BA, $a^r_{BA} = \dfrac{v^2_{AB}}{AB} = \dfrac{(0.551)^2}{0.12} = 0.253$ m/s^2 – $a'x$

iii. Radial component of CD, $a^r_{DC} = \dfrac{v^2_{DC}}{CD} = \dfrac{(0.358)^2}{0.45} = 0.2848$ m/s² – $c'y$

Steps to draw acceleration diagram as shown at Fig. 3.19(c)

1. Draw vector $o'a'$ parallel to OA, to some suitable scale to represent the radial component of the acceleration of A with respect to O, i.e. $o'a' = a^r_{AO} = a_A = 1.323$ m/s².

2. From point a', draw vector $a'x$ parallel to AB to represent the radial component of the acceleration of B with respect to A, i.e. $a'x = a^r_{BA} = 0.253$ m/s².

3. From point x, draw vector xb' perpendicular to AB to represent the tangential component of the acceleration of B with respect to A (i.e., a^t_{BA}) whose magnitude is not known.

4. From point o', draw vector $o'b'$ parallel to the path of motion of B (which is along BO) to represent the acceleration of B (a_B). The vectors xb' and $o'b'$ intersect at b'. Join $a'b'$ which represents the acceleration of B with respect to A.

5. Divide vector $a'b'$ to get c' in the same ratio as C divides AB in the space diagram, i.e. $\dfrac{BC}{CA} = \dfrac{b'c'}{c'a'}$.

6. From point c', draw vector $c'y$ parallel to CD to represent the radial component of the acceleration of D with respect to C, such that $c'y = a^r_{DC} = 0.2801$ m/s².

7. From point y, draw vector yd' perpendicular to CD to represent the tangential component of acceleration of D with respect to C (i.e., a^t_{DC}) whose magnitude is not known.

8. From point o', draw vector $o'd'$ parallel to the path of motion of D (which is along the vertical direction) to represent the acceleration of D (a_D). The vectors yd' and $o'd'$ intersect at d'. By measurement, the linear acceleration of D, a_D = vector $o'd' = 0.159$ m/s².

9. Angular acceleration of CD: From acceleration diagram, the tangential component of the acceleration of D with respect to C, a^t_{DC} = vector $yd' = 1.293$ m/s².

Therefore, angular acceleration of CD is

$$\alpha_{CD} = \dfrac{a^t_{DC}}{CD} = \dfrac{1.293}{0.45} = 2.873 \text{ rad/s}^2$$

W E 3.12: In the toggle mechanism shown in Fig. 3.20, the slider D is constrained to move on a horizontal path. The crank OA is rotating in the counterclockwise direction at a speed of 180 r.p.m. increasing at the rate of 50 rad/sec². The dimensions of the various links are as follows: OA = 180 mm; CB = 240 mm; AB = 360 mm and BD = 540 mm.

For the configuration given, find: 1. Acceleration of the slider D and 2. The angular acceleration of the BD.

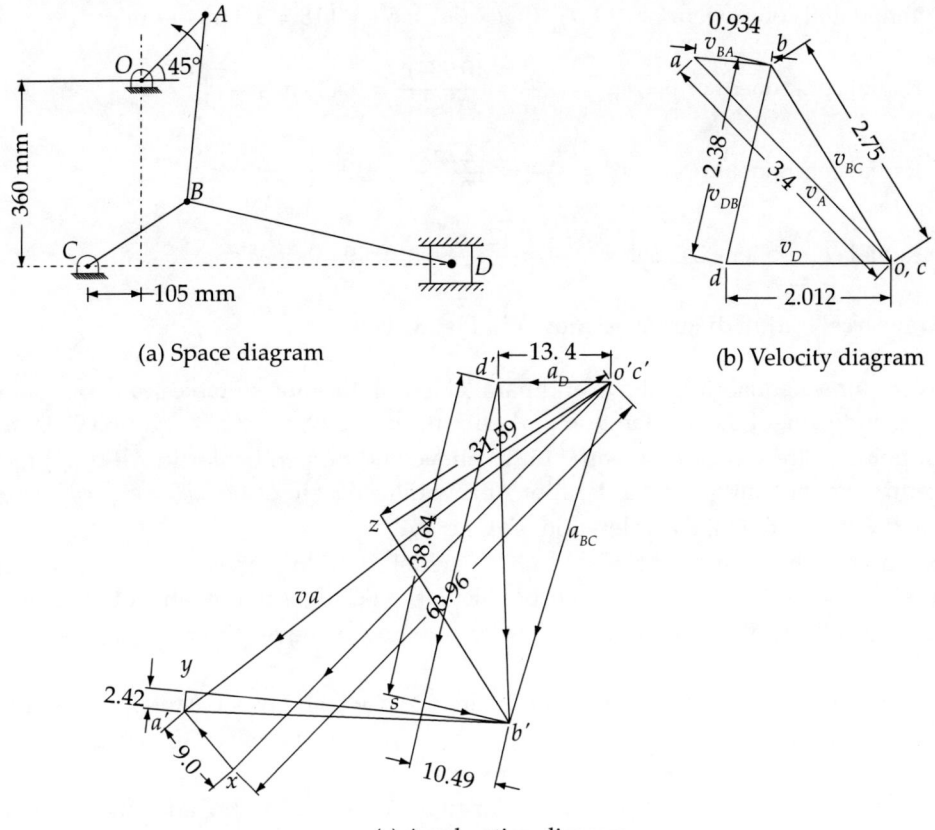

(a) Space diagram
(b) Velocity diagram
(c) Acceleration diagram

Fig. 3.20 Toggle mechanism.

Given:
$N_{AO} = 180$ r.p.m. or $\omega_{AO} = \dfrac{2\pi 180}{60} = 18.85$ rad/s; $\alpha_{AO} = 50$ rad/s^2; $OA = 180$ mm $= 0.18$ m;
$CB = 240$ mm $= 0.24$ m; $AB = 360$ mm $= 0.36$ m; $BD = 540$ mm $= 0.54$ m.

Velocity of A with respect to O or velocity of $A = v_{AO} = v_A = \omega_{AO} \times OA = 18.85 \times 0.18 = 3.4$ m/s

I. First draw the space diagram choosing proper scale as shown Fig. 3.20(a).
II. Determination of velocities and angular velocities: Proceed in the same manner as explained in the W E 3.8 by drawing the velocity diagram and determine velocities of v_D and v_{BD}. They are $v_D = 2.012$ m/s, $v_{BD} = 2.38$ m/s; $v_{AB} = 0.934$ m/s and $\omega_{BD} = \dfrac{2.83}{0.54} = 4.408$ rad/s.
III. Determination of accelerations:

Calculate first the following quantities:

i. Radial component of AO, $a^r_{AO} = \omega^2_{AO} \times OA = 18.85^2 \times 0.18 = 63.958$ m/s^2 – $o'x$

ii. Tangential component of AO, $a^t_{AO} = a \times OA = 50 \times 0.18 = 9.0$ m/s² – xa'

iii. Radial component of BA, $a^r_{BA} = \dfrac{v^2_{BA}}{AB} = \dfrac{(0.934)^2}{0.36} = 2.423$ m/s² – $a'y$

iv. Radial component of BC, $a^r_{BC} = \dfrac{v^2_{BC}}{BC} = \dfrac{(2.75)^2}{0.24} = 31.59$ m/s² – $c'z$

v. Radial component DB, $a^r_{DB} = \dfrac{v^2_{BD}}{DB} = \dfrac{(2.38)^2}{0.54} = 10.49$ m/s² – $b's$

Steps to draw acceleration diagram as shown in Fig. 3.20(c)

1. Draw vector $o'x$ equal to (i) above and parallel to OA, to some suitable scale to represent the radial component of the acceleration of A with respect to O, i.e. $o'x = a^r_{AO} = 63.958$ m/s².
2. From point x, draw vector xa' equal to (ii) above and perpendicular to AB to represent the tangential component of the acceleration of A with respect to O, i.e. $xa' = a^t_{AO} = 9.0$ m/s².
3. Join o' to a' to get the total acceleration of A, i.e. a_A.
4. From point c', draw vector $c'z$ equal to (iv) parallel to BC to represent the radial component of acceleration of BC. From z draw perpendicular to BC to represent tangential component.
5. From a', draw vector $a'y$ equal to (iii) and parallel to AB above. From y draw perpendicular to AB.
6. The above two perpendiculars meet at b'. Join b' to c' which gives the total acceleration of B.
7. From point b', draw vector $b's$ parallel to BD equal to (v) above to represent the radial component of the acceleration of D with respect to B.
8. From point s, draw vector sd' perpendicular to BD to represent the tangential component of acceleration of D with respect to B (i.e., a^t_{DB}) whose magnitude is not known.
9. From point o', draw vector $o'd'$ parallel to the path of motion of D (which is along the horizontal direction) to represent the acceleration of D (a_D). The vectors sd' and $o'd'$ intersect at d'. By measurement, the linear acceleration of D, a_D = vector $o'd' = 13.8$ m/s².
10. From the acceleration diagram, measure the tangential component of the acceleration of D with respect to B, a^t_{DB} = vector $sb' = 38.64$ m/s².
11. Therefore, angular acceleration of BD, $\alpha_{BD} = \dfrac{a^t_{DB}}{BD} = \dfrac{38.64}{0.54} = 71.55$ rad/s² (CW).

3.6 CORIOLIS COMPONENT OF ACCELERATION

When a point on a link while sliding on another link also rotates along with that link, then the Coriolis component of the acceleration occurs. Examples: Whitworth quick-return motion mechanism. The concept and calculation of this component is explained below. Here there are two links, slider B and rotating link OA as shown in Fig. 3.21. All the points on the slider B while sliding on the link OA also rotate along with OA about O. C is the point of contact of slider B on OA and C_1, B_1 are the corresponding points on OA_1 after a time interval of δt. The link OA rotates by $\delta\theta$ in a time interval of δt and takes a new position OA_1 as shown at Fig. 3.21(a).

Let ω represent the angular velocity of link OA at time t seconds and v represent the velocity of slider B along the link OA at time t seconds. Let ωr be the velocity of slider B perpendicular to OA at C at time t seconds and $(\omega + \delta\omega)(r + \delta r)$ at time $(t + \delta t)$ seconds.

From Fig. 3.21(b), bb_1 is the change in the sliding velocity from v to $v + \delta v$ along OA (in radial direction).

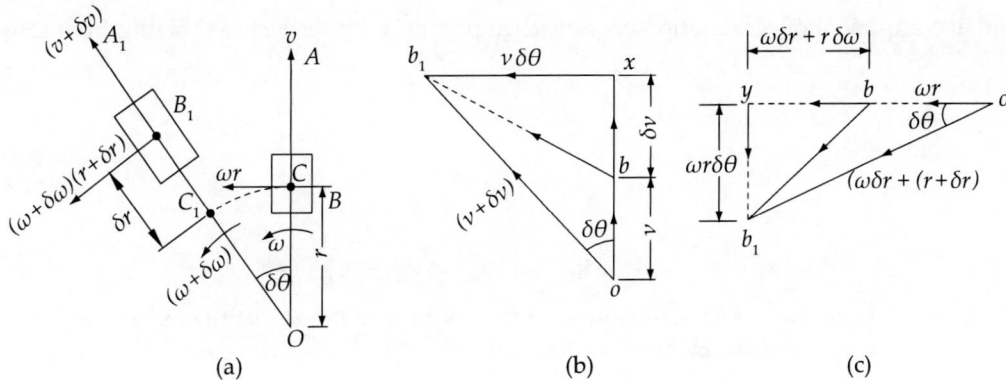

Fig. 3.21 Coriolis component of acceleration.

$$bx \text{ (along } OA\text{)} = ox - ob = (v + \delta v)\cos\delta\theta - v = \delta v \uparrow \qquad \text{(i)}$$

$$xb_1(\text{perpendicular to } OA) = (v + \delta v)\delta\theta = v\delta\theta \leftarrow \qquad \text{(ii)}$$

From Fig. 3.21(c), bb_1 is the change in the tangential velocity from ωr to $(\omega + \delta\omega)(r + \delta r)$ along the perpendicular direction to OA (tangential direction).

$$yb_1(\text{along } OA) = (\omega + \delta\omega)(r + \delta r)\sin\delta\theta = \omega r\delta\theta \downarrow \qquad \text{(iii)}$$

$$by = (\omega + \delta\omega)(r + \delta r)\cos\delta\theta - \omega r = \omega\delta r + r\delta\omega \text{ perpendicular to } OA \qquad \text{(iv)}$$

Combining both the actions at Fig. 3.21(b) and (c)

Total change of velocity along OA radially = (i–iii), i.e.

$$(bx - yb_1) = (\delta v - \omega r\delta\theta) \uparrow$$

Therefore, acceleration along radial direction

$$a^r_{BO} = \frac{(\delta v - \omega r\delta\theta)}{\delta t} = \left[\frac{dv}{dt} - \omega^2 r\right] \qquad \text{(v)}$$

Similarly, total change of velocity tangentially, i.e. perpendicular to OA = (ii + iv), i.e. total acceleration tangential to OA, i.e. perpendicular to OA

$$xb_1 + by = v\delta\theta + (\omega\delta r + r\delta\omega) \leftarrow$$

Therefore,

$$\text{tangential acceleration} = a_{BO}^t = \frac{[v\delta\theta + (\omega\delta r + r\delta\omega)]}{\delta t}$$

$$= \omega v + \omega v + \alpha r = 2v\omega + \alpha r \qquad \text{(vi)}$$

$$\text{since} \frac{dr}{dt} = v$$

Now radial component of acceleration of coincident point C with respect to O acting from C to O,

$$a_{CO}^r = \omega^2 r \downarrow \text{ or } = -\omega^2 r \qquad \text{(vii)}$$

Tangential component

$$a_{CO}^t = \alpha r \uparrow \qquad \text{(viii)}$$

Hence,

$$a_{BC}^r = a_{BO}^r - a_{CO}^r \text{ and } a_{BC}^t = a_{BO}^t - a_{CO}^t$$

This tangential component of acceleration of the slider B with respect to coincident point C is known as "Coriolis Component of Acceleration" and it will be perpendicular to the link on which it is sliding.

The magnitude of the Coriolis component of acceleration is $2\omega v$ and the direction of the Coriolis component for all the four possible cases, is shown in Fig. 3.22, when the directions of ω and v are given.

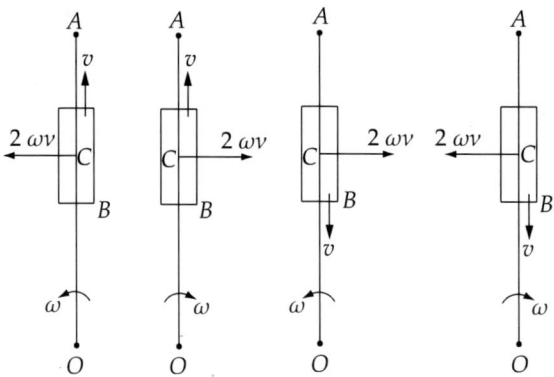

Fig. 3.22 Coriolis component for different cases.

Hint: Rotate vector v in the direction of ω by 90° to get the direction of Coriolis component $2\omega v$.

W E 3.13: A mechanism of a crank and slotted lever quick-return motion is shown in Fig. 3.23. If the crank rotates counterclockwise at 120 r.p.m., determine for the configuration shown, the velocity and acceleration of the ram D. Also determine the angular acceleration of the slotted lever. Crank, $AB = 150$ mm; slotted arm, $OC = 700$ mm; and link $CD = 200$ mm.

Given:

$N_{BA} = 120$ r.p.m.; or $\omega_{BA} = \dfrac{2\pi 120}{60} = 12.57$ rad/s; $AB = 150$ mm $= 0.15$ m;

$OC = 700$ mm $= 0.7$ m; $CD = 200$ mm $= 0.2$ m.

Velocity of B with respect to A, $v_{BA} = \omega_{BA} \times AB = 12.57 \times 0.15 = 1.9$ m/s.

I. First draw the space diagram choosing proper scale as shown in Fig. 3.23(a). Measure $OB' = 516.5$ mm $= 0.5165$ m

II. Determination of velocities and angular velocity:

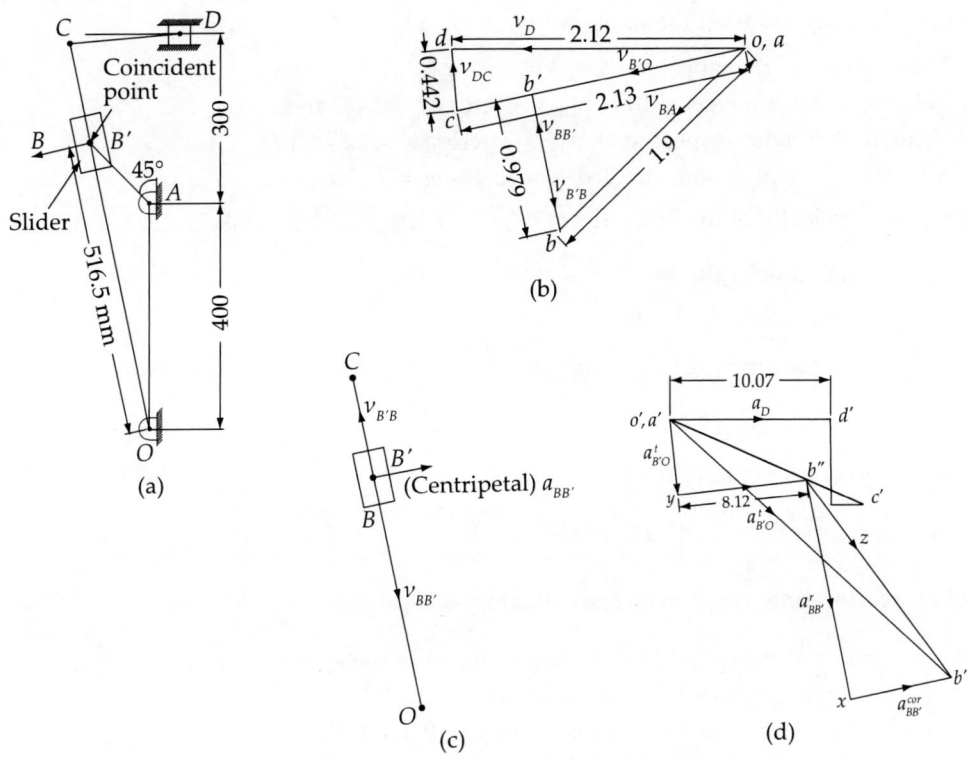

Fig. 3.23 Whitworth quick-return mechanism.

Steps to draw velocity diagram as shown at Fig. 3.23(b)

1. Take a point and name it as o, a to indicate O and A of the space diagram in the velocity diagram.
2. Now draw vector ab from a, perpendicular to AB, using some suitable scale to represent $v_{BA} = 1.9$ m/s.
3. From b, draw vector bb' parallel to OB to represent velocity of B with respect to O, i.e. v_{BO}.
4. From point a, draw vector ac perpendicular to OC to represent the velocity of C.

5. Above two lines intersect at b'. The vector bb' represents the velocity of B with respect to coincident point B', i.e. $v_{BB'}$.
6. Since C lies on OB produced, therefore, c in the velocity diagram lies on the extension of ob_1. Hence, extend in the same ratio as B divides OC in the space diagram such that $\dfrac{oc}{ob'} = \dfrac{OC}{OB}$ or $oc = ob' \times \dfrac{OC}{OB}$.
7. Draw vector cd from point c perpendicular to CD to represent the velocity of D with respect to C, i.e. v_{DC} and from point d, draw vector do parallel to the path of slider D (which is horizontal) to represent the velocity of D, i.e. v_D.
8. By measurement, the velocity of D, v_D = vector od = 2.12 m/s. From the velocity diagram, find also the following required for acceleration diagram:

    (a) Velocity of B with respect to B', $v_{BB'}$ = vector bb' = 0.979 m/s
    (b) Velocity of D with respect to C, v_{DC} = vector cd = 0.442 m/s
    (c) Velocity of B' with respect to O, $v_{B'O}$ = vector ob' = 1.574 m/s
    (d) Velocity of C with respect to O, v_{CO} = vector oc = 2.13 m/s
    (e) Angular velocity of the link OC or OB', $\omega_{CO} = \omega_{B'O} = \dfrac{v_{CO}}{OC}$ = 3.043 rad/s (ACW).

III. Determination of accelerations:

Calculate first the following quantities:

i. Radial component of BA $a^r_{BA} = \omega^2_{BA} \times AB = (12.57)^2 \times 0.15 = 23.7$ m/s² – a'b'

ii. Radial component of B'O, $a^r_{B'O} = \dfrac{v^2_{B'O}}{OB'} = \dfrac{(1.574)^2}{0.5165} = 4.796$ m/s² – a'y or o'y

iii. Coriolis component of BB' $a^{cor}_{BB'} = 2\omega_{CO} \times v_{BB'} = 2 \times 3.043 \times 0.979 = 5.958$ m/s² – xb'

iv. Radial component of DC, $a^r_{DC} = \dfrac{v^2_{DC}}{CD} = \dfrac{(0.442)^2}{0.2} = 0.977$ m/s² – c'z

Steps to draw acceleration diagram as shown at Fig. 3.23(d)

1. Take a point o', a' to represent fixed points O and A in the space diagram.
2. Draw vector a'b' to represent $a^r_{BA'}$ equal to (i) parallel to BA.
3. Draw vector o'y to represent $a^r_{B'O}$ equal to (ii) parallel to OB.
4. Draw vector xb' to represent the Coriolis component of BB' equal to (iii) perpendicular to OB.
5. Draw perpendiculars at y and x, which intersect at b" as shown. Join o' to b".
6. Extend o'b" to c' in the same proportion as OB to C in the space diagram.
7. Draw vector c'z to represent a^r_{DC} equal to (iv) parallel to CD.
8. Draw perpendicular at z to CD and a horizontal from o'. Both intersect at d'.
9. Measure o'd' which gives $a_D = 10.07$ m/s².
10. Angular acceleration of the slotted lever: Measure the length of the tangential component of the coincident point B' with respect to O, $a^t_{B'O}$ = vector yb" = 8.12 m/s². Angular acceleration of the slotted lever, $\alpha_{OC} = \dfrac{8.12}{0.52} = 15.62$ rad/s² (ACW)

3.7 EXERCISE

3.7.1 Short Answer Questions

1. Explain the phenomenon of combined motion of rotation and translation. Give a few examples.
2. (a) Explain the term instantaneous centre. Name three types of instantaneous centres.
   (b) Define Coriolis component of acceleration.
   (c) State Kennedy's theorem.
3. How would you locate the instantaneous centre of a rigid link moving with combined motion of rotation and translate on? Sketch the four-bar mechanism and locate all its instantaneous centres.
4. What do you understand by the term angular velocity and angular acceleration? Do they have any relation between them?
5. (a) How would you find out linear velocity of a rotating body? Obtain an equation between the linear acceleration and angular acceleration of a rotating body.
   (b) Sketch the four-bar mechanism and locate all its instantaneous centres.
6. At a pin joint in a mechanism, the diameter of the pin is 60 mm, the angular velocities of the two links are 20 mm and 30 in rad/s, both being clockwise. Find the rubbing velocity.
7. A rod $AB = 0.3$ m is in plane motion. At an instant when AB is vertical the velocity of A is 10 m/s and of B is 20 m/s. The directions of both velocities are horizontal and to the right. Find the angular velocity of the rod.
8. Name three types of instantaneous centres.
9. Derive an expression for the magnitude and direction of Coriolis component of acceleration.
10. Explain different methods to locate instantaneous centres in a mechanism.
11. State and prove Kennedy's three-centres-in-line theorem.

3.7.2 Problems

1. In a crank and connecting rod mechanism, the crank is 300 mm long and the connecting rod is 1500 mm long. If the crank rotates uniformly at 300 r.p.m., find the velocity of the cross head when the crank is inclined at 30° with the inner dead centre. [*Ans.* 5.39 m/s]

2. The crank and connecting rod of a steam engine are 0.5 m and 2 m respectively. The crank makes 180 r.p.m. in the clockwise direction. When the crank has turned 45° from the inner dead centre, determine (i) velocity of piston and (ii) angular velocity of connecting rod.
[*Ans.* 8.15 m/s; 32.5 r.p.m.]

3. The crank and connection rod of a steam engine are 30 cm and 150 cm in length. The crank rotates at 180 r.p.m. clockwise. Determine the velocity and acceleration of the piston when

the crank is at 40 degrees from the inner dead centre. Also determine the position of the crank for zero acceleration of the piston.

4. The crank OA of a steam engine is 8 cm and the length of the connection rod AB is 24 cm. The centre of gravity of the rod is at G, 8 cm from crankpin. The engine speed is 600 radians per minute. For the position when the rank makes 45 degrees to the horizontal measured from the inner dead centre, find the velocity and acceleration of the piston. Also find the acceleration of the centre of gravity of the connection rod.

5. The crank of a slider crank mechanism rotates clockwise at a constant speed of 400 r.p.m. The crank is 150 mm and the connecting rod is 500 mm long. Determine: 1. Linear velocity and acceleration of the midpoint of the connecting rod, and 2. Angular velocity and angular acceleration of the connecting rod, at a crank angle of 60° from inner dead centre position Check the results by the analytical method.

6. (a) State and prove line-of-three-centres theorem.
   (b) In a four-bar mechanism, the dimensions of the links are as follows: AB (crank) = 200 mm; BC = 200 mm; CD = 300 mm; and AD is the fixed link which is 400 mm long. At the instant when the crank has an angular velocity of 15 rad/sec and angle BAD is 80°. Locate all the instantaneous centres and determine the following:
      i. Velocity of point C and the angular velocity of the links BC and CD.
      ii. Velocity of point F on the link CD when FD = 100 mm.
      iii. Velocity of an offset point E on the link BC if BE = 150 mm, CE = 100 mm and BEC read clockwise.

7. (a) Get an expression to find the magnitude of the Coriolis component of acceleration.
   (b) In the oscillating cylinder mechanism the crank OA is 50 mm long while the piston rod AB is 150 mm only. Determine for 300 r.p.m. of the crank.
      i. The velocity of the piston B relative to circular walls.
      ii. Linear relative velocity of B relative to A on the piston.
      iii. The angular velocity of rod AB and the cylinder.
      iv. The absolute velocity of piston B.
      v. The sliding acceleration of the piston relative to the cylinder walls.
      vi. The linear acceleration of the piston rod AB.

8. In the Fig. 3.24, pinion wheel rotates at 600 r.p.m. (CW) and gear wheel without slipping. Find the velocity difference between points B and A. Radius of gear wheel is 180 mm and pinion is 40 mm. Distance of B from the centre of gear wheel is 160 mm.

9. Two points A and B as in Fig. 3.25, are located along the radius of a wheel, have speeds of 80 and 140 m/s respectively. The distance between points is BA = 300 mm. (a) What is the diameter of the wheel? (b) Find v_{AB}, v_{BA} and angular velocity of the wheel.
[**Ans.** (a) d = 1400 mm, (b) v_{AB} = 60 m/s ω_2 = 200 rad/s]

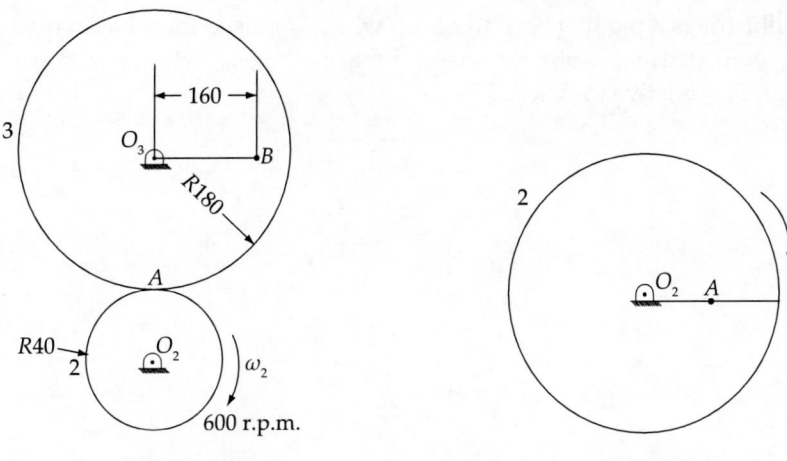

Fig. 3.24   Fig. 3.25

10. The velocity of point B of the linkage shown in the Fig. 3.26 is 40 m/s. Find the velocity of point A and the angular velocity of link C. Determine the acceleration of point A and angular acceleration of link C. AB = 400 mm.

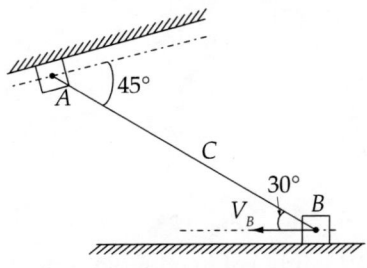

Fig. 3.26

11. The mechanism shown in Fig. 3.27 is driven by link OA at ω_2 = 45 rad/s CCW. Find the angular velocities of links AB and BO_4. O_2A = 100 mm; AB = 250 mm; BO_4 = 300 mm; DC = 75 mm and O_2O_4 = 250 mm. Find the velocities of points B and C also. Find also α_3 and α_4.

[***Ans.*** ω_{AB} = 1.43 rad/s, ω_{BO_4} = 15.4 rad/s CCW]

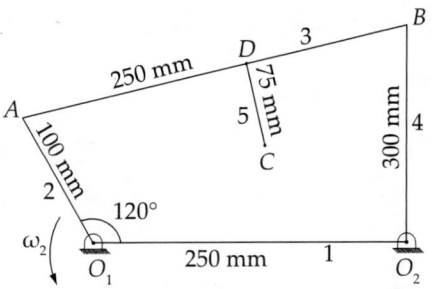

Fig. 3.27

12. Figure 3.28 illustrates a parallel bar linkage in which opposite links have equal lengths. For this linkage, show that ω_{AB} is always zero and that $\omega_{BO} = \omega_{AO}$. How would you describe the motion of link BO_2 relative to link AO_1?

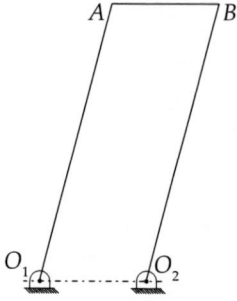

Fig. 3.28

13. The inversion of the slider-crank mechanism shown in Fig. 3.29 is driven by link OA at 60 rad/s CCW. Find the velocity of point B and the angular velocities of link AB and C. $AO_2 = 75$ mm; $BA = 400$ mm; $O_4O_2 = 125$ mm. [**Ans.** $v_B = 4.77$ m/s and $\omega_C = \omega_D = 22$ rad/s CCW]

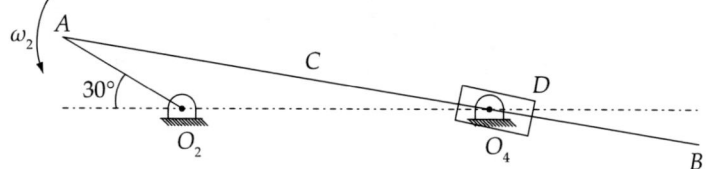

Fig. 3.29

14. Find the velocities of points B, C and D of the double slider mechanism shown in Fig. 3.30. If the crank OA rotates at 42 rad/s CCW, find the acceleration of points B and D. $AO_2 = 50$ mm; $BA = 250$ mm; $CA = 100$ mm; $CB = 175$ mm; $DC = 200$ mm.

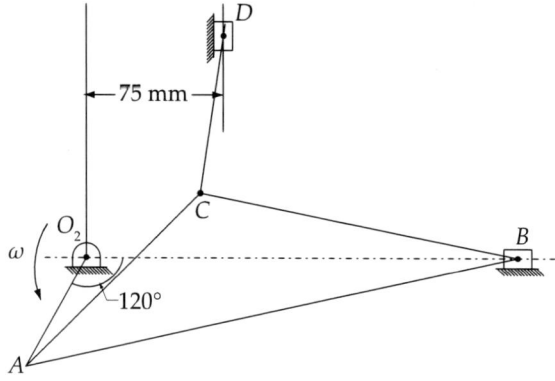

Fig. 3.30

15. Figure 3.31 shows a variation of the Scotch yoke mechanism. It is driven by crank OA at 36 rad/s CCW. Find the velocity of the cross head link B. Determine the acceleration of link B. OA = 250 mm.

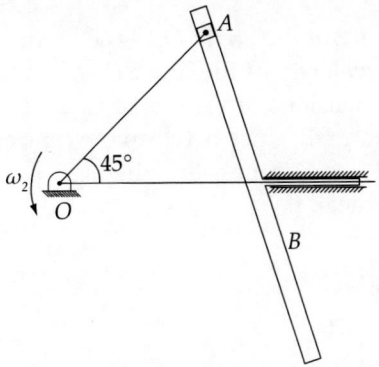

Fig. 3.31

16. A reciprocating engine has connecting rod of length 20 cm and crank 5 cm long. By instantaneous centre and relative velocity methods, determine the velocity of piston when the crank has turned through an angle of 45° from IDC clockwise and is rotating at 240 r.p.m. Also extend relative velocity method and determine the acceleration of piston.

17. In a slider crank mechanism, the lengths of crank OB and connecting rod AB are 100 mm and 400 mm, respectively. If the crank makes an angle of 45° to the horizontal and rotates clockwise with uniform angular velocity of 10 rad/sec. Locate all instantaneous centres of the mechanism and find (i) the velocity of the slider A (ii) angular velocity of the connecting rod AB.

18. A mechanism of a crank and slotted lever quick return mechanism is shown in Fig. 3.32.

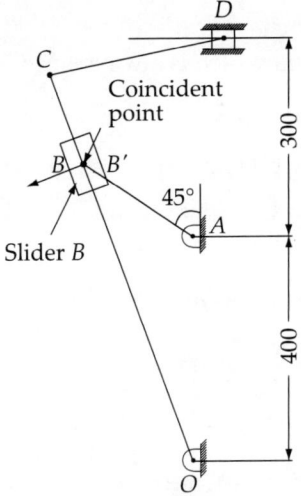

Fig. 3.32

If the crank rotates counterclockwise at 120 r.p.m. Determine for the given configuration, the velocity and acceleration of the ram D. Also determine the angular acceleration of the slotted lever. Crank, $AB = 150$ mm; slotted lever, $OC = 700$ mm and link $CD = 200$ mm.

19. In Fig. 3.33 the angular velocity of the crank OA is 600 mm, determine the linear velocity of the slider D and the angular velocity of the link BD, when the crank is inclined at an angle of 75° to the vertical. The dimensions of various links are $OA = 28$ mm; $AB = 44$ mm; $BC = 49$ mm; $BD = 46$ mm. The centre distance between the centres of rotation O and C is 65 mm. The path of travel of the slider is 11 mm below the fixed point C. The slider moves along a horizontal path and OC is vertical.

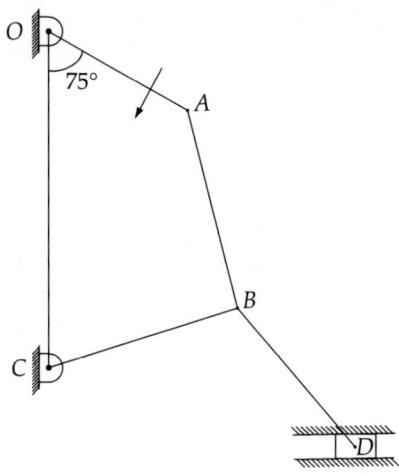

Fig. 3.33

3.7.3 Multiple Choice Questions

1. The instantaneous centre is a point which is always fixed

    (a) Yes (b) No   [*Ans.* (b)]

2. The angular velocity of a rotating body is expressed in terms of

    (a) revolution per minute   (b) radians per second
    (c) any one of the two   (d) none of the two   [*Ans.* (c)]

3. The linear velocity of a rotating body is given by the relation

    (a) $v = r.\omega$; (b) $v = \dfrac{r}{\omega}$; (c) $v = \dfrac{\omega}{r}$; (d) $v = \dfrac{\omega^2}{r}$ [*Ans.* (a)]

4. The linear acceleration of a rotating body is given by the relation

    (a) $a = r.\alpha$; (b) $a = \dfrac{r}{\alpha}$; (c) $a = \dfrac{\alpha}{r}$; (d) $a = \dfrac{\alpha^2}{r}$ [*Ans.* (a)]

5. The relation between linear velocity and angular velocity of a cycle
   (a) exists under all conditions
   (b) does not exist under all conditions
   (c) exists only when it does not slip
   (d) exists only when it moves on horizontal plane
   [Ans. (a)]

6. The velocity of piston in a reciprocating pump mechanism depends upon
   (a) angular velocity of the crank
   (b) radius of the crank
   (c) length of the connecting rod
   (d) both a and b
   (e) all the above
   [Ans. (e)]

7. The linear velocity of a point B on a link rotating at an angular velocity ω relative to another point A on the same link is
   (a) $\omega^2.AB$ (b) $\omega.AB$ (c) $\omega.(AB)^2$ (d) $\dfrac{\omega}{AB}$
   [Ans. (b)]

8. The linear velocity of a point relative to another point on the same link is _____ to the line joining the points.
   (a) perpendicular   (b) parallel   (c) at 45°
   [Ans. (a)]

9. The total number of instantaneous centres of a mechanism having n links is
   (a) $\dfrac{n.(n-1)}{2}$ (b) $\dfrac{(n-1)}{2}$ (c) $\dfrac{n(n+1)}{2}$ (d) $\dfrac{(n+1)}{2}$
   [Ans. (a)]

10. According to Kennedy's theorem the instantaneous centres of three bodies having relative motion lie on a
    (a) curved path   (b) straight line   (c) point
    [Ans. (b)]

11. The instantaneous centres of a slider moving in a linear guide lies at
    (a) pin joints   (b) their point of contact   (c) infinity
    [Ans. (c)]

12. The instantaneous centres of a slider moving in a curved surface lies at
    (a) infinity
    (b) their point of contact
    (c) the centre of curvature
    (d) the pin point
    [Ans. (c)]

13. The fixed instantaneous centre of mechanism
    (a) varies with the configuration   (b) remains at the same place for all configurations
    [Ans. (b)]

14. The instantaneous centre of rotation of a circular disc rolling on a straight path is
    (a) at the centre of the disc
    (b) at their point of contact
    (c) at the centre of gravity of the disc
    (d) at infinity
    [Ans. (b)]

15. The locus of instantaneous centre of a moving body relative to a fixed body is known as the

    (a) space centrode   (b) body centrode
    (c) moving centrode   (d) none of the above   [*Ans.* (a)]

16. The space centrode of a circular disc rolling on a straight path is

    (a) circle   (b) a parabola   (c) a straight line   (d) none of the above   [*Ans.* (c)]

17. The component of the acceleration directed towards the centre of rotation of a revolving body is known as _____ component.

    (a) tangential   (b) centripetal   (c) Coriolis   [*Ans.* (b)]

18. At an instant the link AB of length r has an angular velocity ω and an angular acceleration α. What is the total acceleration of AB?

    (a) $[(w^2.r)^2 + \alpha.r)^2]^{1/2}$ (b) $[(w.r)^2 + \alpha.r)^2]^{1/2}$
    (c) $[(w^2.r)2 + \alpha^2.r)^2]^{1/2}$ (d) $[(w.r)^2 + \alpha^2.r)^2]^{1/2}$ [*Ans.* (a)]

19. At an instant, if the angular velocity of a link is clockwise then the angular acceleration will be

    (a) clockwise   (b) counterclockwise   (c) in any direction   [*Ans.* (c)]

20. Angular acceleration of a link AB is given by

    (a) centripetal acceleration/length   (b) tangential acceleration/length
    (c) total acceleration/length   [*Ans.* (b)]

21. A slider moves with uniform velocity v on a revolving link of length r with angular velocity of ω. The Coriolis acceleration component of a point on the slider relative to a coincident point on the link is equal to

    (a) $2r\omega$ parallel to the link   (b) $2\omega v$ perpendicular to the link
    (c) $2r\omega$ perpendicular to the link   (d) $2\omega v$ parallel to the link   [*Ans.* (b)]

22. The Coriolis acceleration component is taken into account for a _____ mechanism

    (a) double slider crank   (b) four-bar link mechanism
    (c) scotch yoke   (d) quick-return mechanism   [*Ans.* (d)]

23. The Coriolis acceleration component

    (a) lags the sliding velocity by 90°   (b) leads the sliding velocity by 90°
    (c) lags the sliding velocity by 180°   (d) leads the sliding velocity by 180°.   [*Ans.* (b)]

24. In which mechanisms Coriolis component of acceleration is there

    (a) slider crank chain   (b) scotch yoke
    (c) double slider crank   (d) oscillating cylinder mechanism.   [*Ans.* (d)]

Inertia Forces in Reciprocating Parts

4

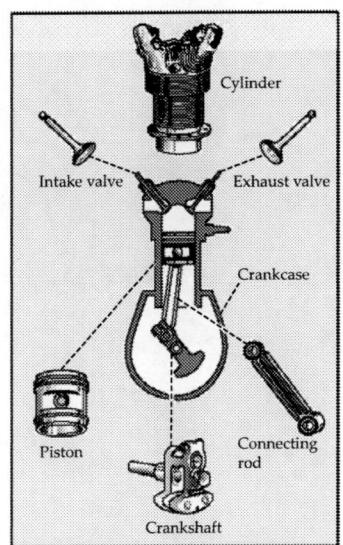

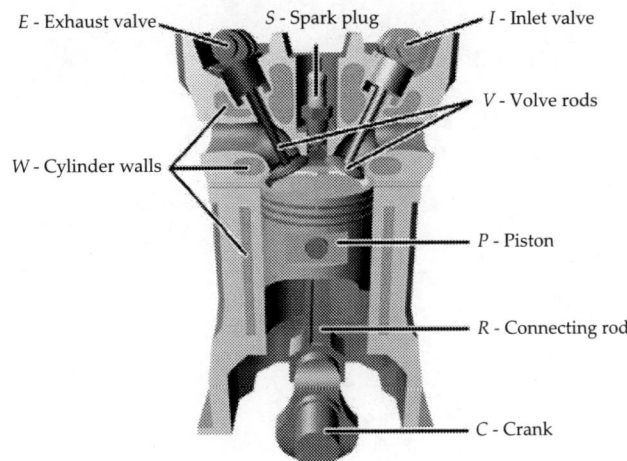

Inertia Forces in Reciprocating Parts

4.1 INTRODUCTION

The fundamental units in kinematic analysis are length and time but in dynamic analysis length, time and force.

Forces are transmitted between machine members through mating surfaces, i.e. gear to gear shaft or from one gear to another gear or from belt to pulley or from cam to follower or from brake drum to brake shoe, etc. Hence, it is necessary to know the magnitudes of these forces for a variety of reasons. In dynamics, only determination of the magnitudes, directions and locations of the forces is studied.

Dynamic forces are associated with accelerating masses. As all machines have some accelerating parts, dynamic forces are always present when machines operate. Depending upon the operating speeds those forces will dominate the external forces. These forces lead to vibrations, wear, noise or even sometimes machine failures.

4.1.1 Terms Used in Static

Some of the terms used are defined as follows:

Force: Force is the action of one body acting on another. The characteristics of the force are magnitude, direction and place of application.

Matter: Matter is any material substance when enclosed is called a body.

Mass: Newton defined mass as the quantity of matter of a body as measured by its volume and density.

Inertia: Inertia is the property of mass that causes it to resist any effort to change its motion.

Weight: Weight is the force which results from gravity acting upon a mass.

Rigid body: All real bodies are either elastic or plastic and will be deformed when acted upon by forces. When the deformation of such a body is small enough to be neglected, such a body is frequently assumed as rigid.

Some of the new terms used in dynamics are defined as follows:

Centroid: The centroid of a system is a point at which a system of distributed forces may be considered concentrated with exactly the same effect.

Centre of mass: The centre of mass means the point at which the mass may be considered concentrated so that the effect is the same. It is indicated by G.

Mass moments and products of inertia: When forces are distributed over an area, then calculating moments about a specified point or axis of rotation and their intensity varies according to its distance from the point or axis of rotation. For rigid bodies, these are constant properties of the body as they are calculated by taking the centre of the mass as the origin of the coordinate system.

4.1.2 D'Alembert's Principle

D'Alembert's principle states that the inertia forces and couples, the external forces and torques on a body together provide statical equilibrium. Simply it states that the resultant force acting on a rigid body together with the inertia force keep the body in equilibrium.

Let F_r be the resultant of the various forces F_1, F_2, F_3, etc., acting on a moving rigid body as shown in Fig. 4.1(a). G indicates the centre of gravity or the mass centre as shown in Fig. 4.1(b). The resultant of the system of forces F_r acts at a distance h from G. The effect of this unbalanced force system is to produce an acceleration of the mass centre of the body since force is mass times acceleration. The moment (M) of the unbalanced force about the mass centre of the body causes angular acceleration (α) since M is equal to inertia times angular acceleration.

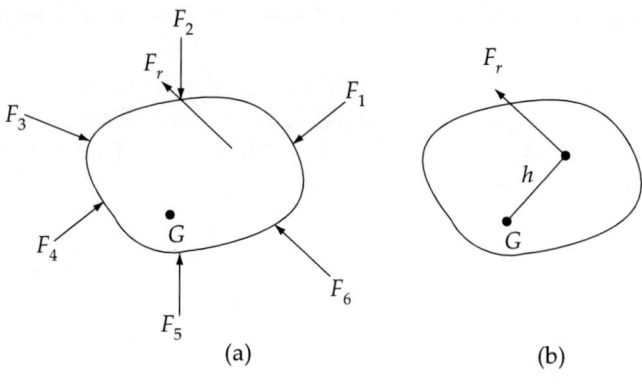

Fig. 4.1

Add one equal and opposite force F_r at G as shown in Fig. 4.2(a). Now the body is subjected to total three forces, two equal and opposite forces at G and one single force at a distance h from G. These three forces can be reduced to (1) a force F_r acting at G and (2) a couple equal to $F_r \cdot h$ as shown at Fig. 4.2(b). Thus, the single resultant force F_r acting away from G can be replaced by the same resultant force F_r acting at centre of gravity G and a couple equal to $F_r \cdot h$ acting on the rigid body.

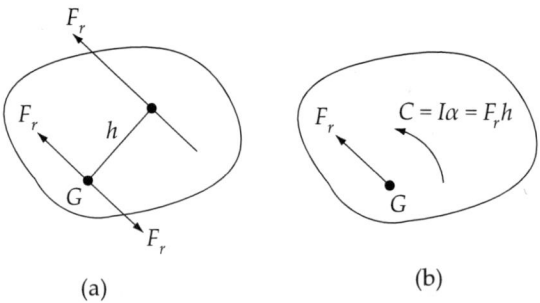

Fig. 4.2

The force F_r causes linear acceleration 'a' equal to $\left(\dfrac{F_r}{m}\right)$ on the body since $F_r = m \cdot a$. The couple C causes angular acceleration 'α' equal to $\left(\dfrac{C}{I}\right)$ on the body since $C = F_r \cdot h = mk^2\alpha = I \cdot \alpha$.

Hence, in order to find the linear acceleration 'a' and angular acceleration 'α' of the rigid body, F_r, m, h and k must be known.

Let F_I be the inertia force of the rigid body, then according to D'Alembert's principle,

$$F_r + F_I = 0$$

Accelerating force F = mass (m) × linear acceleration (a) of the centre of gravity G,

Inertia force F_I = Opposite of the accelerating force or (−Accelerating force) = $-m \cdot a$

Inertia torque T_I = Opposite of the accelerating torque or (−Accelerating torque) = $-I \cdot \alpha$

$\qquad = -mk^2\alpha$ where k is the radius of gyration.

Hence, according to D'Alembert's principle, the several forces and torques acting on the body about G, the vector sum of forces and torques (or couples) have to be zero, i.e. $\Sigma F = 0$; $\Sigma C = 0$. These equations are similar to equations of a body in static equilibrium.

This suggests that first the magnitudes and the directions of inertia forces and couples have to be determined, after which they can be treated just like static loads on the mechanism. Thus, a dynamic analysis problem is reduced to one requiring static analysis.

4.2 ANALYTICAL METHOD FOR RECIPROCATING MECHANISM

This method helps in finding out the velocities and accelerations in reciprocating mechanisms for dynamic analysis without the necessity of drawing space diagrams, velocity diagrams and acceleration diagrams. More time can also be saved and also the correct values can be obtained. Pic. 4.1 shows the reciprocating mechanism.

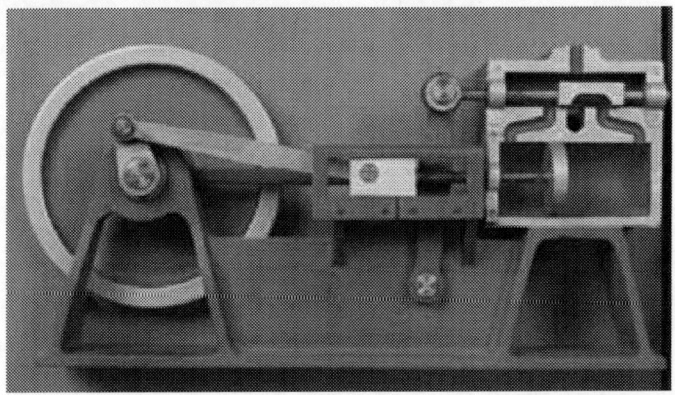

Pic. 4.1

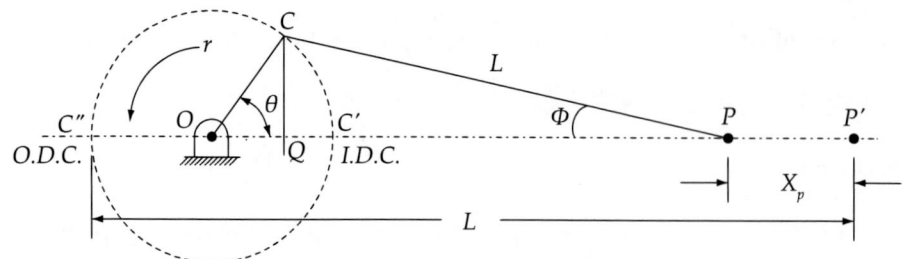

Fig. 4.3 Reciprocating mechanism.

As in Fig. 4.3, let L and r be the lengths of the connecting rod and radius of the crank. Let n be the ratio of $\left(\dfrac{L}{r}\right)$, X_p be the displacement of the piston, θ be the angle made by the crank, ϕ be the angle made by the connecting rod with line of stroke, v_p and a_p be the velocity and acceleration of the piston P.

$$X_p = P'P = OP' - OP = (OC' + C'P') - (OQ + QP)$$
$$= (L + r) - (r\cos\theta + L\cos\phi) = r(1 - \cos\theta) + L(1 - \cos\phi)$$
$$X_p = r\{(1 - \cos\theta) + n(1 - \cos\phi)\} \tag{1}$$

Since $CQ = r\sin\theta = L\sin\phi$ and so

$$\left(\frac{\sin\theta}{\sin\phi}\right) = \left(\frac{L}{r}\right) = n. \text{ (or) } \sin\phi = \frac{\sin\theta}{n} \text{ and so}$$

$$\cos\phi = \sqrt{(1 - \sin^2\phi)} = \left(1 - \frac{\sin^2\theta}{n^2}\right)^{1/2}$$

Expanding this using binomial theorem and neglecting higher terms gives

$$\cos\phi = \left[1 - \frac{\sin^2\theta}{(2n^2)}\right], \text{ taking 1 to left side gives}$$

$$(1 - \cos\phi) = \frac{\sin^2\theta}{2n^2} \tag{2}$$

substituting in (1) gives the expression for displacement.

4.2.1 Displacement of Piston (X_p)

The expression for displacement of piston P,

$$X_p = r\left[(1 - \cos\theta) + n\left(\frac{\sin^2\theta}{2n^2}\right)\right]$$

$$= r\left[(1 - \cos\theta) + \frac{\sin^2\theta}{2n}\right] \tag{3}$$

4.2.2 Velocity of Piston (v_p)

Differentiating the displacement expression with respect to time gives velocity,

$$v_p = \frac{dX_p}{dt} = \left(\frac{dX_p}{d\theta}\right)\left(\frac{d\theta}{dt}\right)$$

$$= \omega \frac{dX_p}{d\theta}; \quad as \frac{d\theta}{dt} = \omega$$

$$v_p = \omega \frac{d}{d\theta}\left[r\left\{(1-\cos\theta) + \frac{\sin^2\theta}{2n}\right\}\right]$$

$$\text{Velocity of piston } P, \quad v_p = \omega r \left[\sin\theta + \frac{\sin 2\theta}{2n}\right] \quad (4)$$

4.2.3 Acceleration of Piston (a_p)

Differentiating the velocity expression with respect to time gives acceleration,

$$a_p = \frac{dv_p}{dt} = \left(\frac{dv_p}{d\theta}\right)\left(\frac{d\theta}{dt}\right)$$

$$= \omega \frac{dv_p}{d\theta}; \quad as \frac{d\theta}{dt} = \omega$$

$$a_p = \frac{d}{d\theta}\left[\omega r \left(\sin\theta + \frac{\sin 2\theta}{2n}\right)\right]$$

$$\text{Acceleration of piston } P, \quad a_p = \omega^2 r \left[\cos\theta + \frac{\cos 2\theta}{n}\right] \quad (5)$$

For $\theta = 0°$, i.e. when the crank C is at C', inner dead centre (I.D.C.), then

$$a_p = \omega^2 r \left[1 + \frac{1}{n}\right]$$

and for $\theta = 180°$, i.e. when the crank C is at C'', outer dead centre (O.D.C.), then

$$a_p = \omega^2 r \left[-1 + \frac{1}{n}\right]$$

and when θ is greater than $180°$, then

$$a_p = -\omega^2 r \left[-1 + \frac{1}{n}\right]$$

$$= \omega^2 r \left[\frac{1}{n} - 1\right]$$

4.2.4 Angular Velocity of Connecting Rod (ω_c)

Angular velocity of connecting rod,

$$\omega_c = \frac{d\phi}{dt} = \left(\frac{d\phi}{d\theta}\right)\left(\frac{d\theta}{dt}\right)$$

$$= \omega \frac{d\phi}{d\theta}; \quad \text{as } \frac{d\theta}{dt} = \omega$$

As seen above, $\sin\phi = \dfrac{\sin\theta}{n}$. Differentiating both sides with respect to time t,

$$\frac{d}{dt}(\sin\phi) = \frac{d}{dt}\left(\frac{\sin\theta}{n}\right) \text{ gives}$$

$$\cos\phi \frac{d\phi}{dt} = \omega \frac{\cos\theta}{n}$$

$$\omega_c = \frac{\omega \cos\theta}{n \cos\phi}$$

gives using (2) for $\cos\phi$.

Angular velocity of connecting rod,

$$\omega_c = \frac{\omega \cos\theta}{\left(n^2 - \sin^2\theta\right)^{1/2}} \tag{6}$$

4.2.5 Angular Acceleration (α_c)

Angular acceleration of connecting rod,

$$\alpha_c = \frac{d^2\phi}{dt^2} = \frac{d\omega_c}{dt}$$

$$= \left(\frac{d\omega_c}{d\theta}\right)\left(\frac{d\theta}{dt}\right) = \omega \frac{d\omega_c}{d\theta}$$

gives after differentiating and simplification

$$\alpha_c = -\omega^2 \frac{\sin\theta \left(n^2 - 1\right)}{\left(n^2 - \sin^2\theta\right)^{3/2}} \tag{7}$$

The negative sign shows that the sense of the acceleration of the connecting rod is such that it tends to reduce the angle ϕ.

As $\sin^2\theta$ is small compared to n^2, it can be neglected and accordingly the equations (6) and (7) reduce to

$$\omega_c = \frac{\omega \cos\theta}{n} \tag{8}$$

and

$$a_c = -\omega^2 \frac{\sin\theta(n^2 - 1)}{(n^2 - \sin^2\theta)^{3/2}}$$

again taking $(n^2 - 1)$ as n^2 and neglecting $\sin^2\theta$, gives

$$a_c = -\omega^2 \frac{\sin\theta}{n} \tag{9}$$

W E 4.1: The crank of a slider crank mechanism rotates clockwise at a constant speed of 300 r.p.m. The crank is 150 mm and the connecting rod is 600 mm long. Determine: 1. linear velocity, 2. acceleration of the piston, and 3. angular acceleration of the connecting rod by the analytical method for a crank angle of 45° from inner dead centre position. Compare the results obtained by instantaneous centre method and by relative velocity method shown in W E 3.1, 3.5 and 3.9.

Given:
$N = 300$ r.p.m.; $\omega = \dfrac{2\pi N}{60} = \dfrac{2\pi 300}{60} = 31.42$ rad/sec; $r = 150$ mm $= 0.15$ m

$L = 600$ mm $= 0.6$ m; hence $n = \dfrac{L}{r} = \dfrac{600}{150}$ mm $= 4$; $\theta = 45°$.

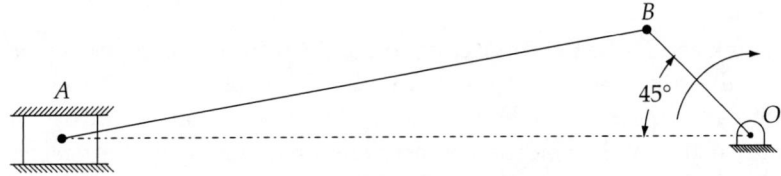

Fig. 4.4

Analytical method:

Velocity of piston: Using the expression given at (4) above,

$$v_A = \omega r \left[\sin\theta + \frac{\sin 2\theta}{2n}\right]$$
$$= 31.2 \times 0.15 \left[\sin 45° + \frac{\sin 90°}{2 \times 4}\right]$$
$$= 3.92 \text{ m/s}$$

Acceleration of piston: Using the expression given at (5) above,

$$a_a = \omega^2 r \left[\cos\theta + \frac{\cos 2\theta}{n}\right]$$
$$= (31.2)^2 \times 0.15 \left[\cos 45° + \frac{\cos 90°}{4}\right]$$
$$= 104.71 \text{ m/s}^2$$

Angular acceleration of the connecting rod AB:

$$\alpha_{AB} = -\omega^2 \frac{\sin\theta}{n} = -(31.2)^2 \frac{\sin 45°}{4} = 174.88 \text{ rad/s}^2$$

Comparison of results:

The results obtained by the Relative method:
The velocity of piston A, $v_A = 4.02$ m/s and the acceleration of piston A, $a_A = 104.0$ m/s².
Angular acceleration of the connecting rod AB, $\alpha_{AB} = 169.83$ rad/s².

The results obtained by the Instantaneous centre method:
The velocity of piston $v_A = 3.96$ m/sec. Acceleration is not possible by this method.

Note: The results are very close. Little variation may be due to scale adopted and mistakes while making measurements in the graphical methods and approximations adopted in analytical methods.

4.3 KLEIN'S CONSTRUCTION FOR RECIPROCATING MECHANISMS

This is a graphical method that can be applied to only reciprocating mechanisms which can be constructed directly on the space diagram itself to determine both velocities and accelerations of any point on any link.

Let OC be the crank and PC be the connecting rod of the reciprocating mechanism shown in the Fig. 4.5(a). Let θ be the angle made by the crank with the line of stroke PO and the crank rotates with a uniform angular velocity of θ rad/sec in the clockwise direction.

The Klein's construction of the velocity and acceleration diagrams on the space diagram of the reciprocating mechanism is explained below.

First construct the space diagram choosing a proper scale as shown in Fig. 4.5(a).

4.3.1 Klein's Velocity Diagram

1. First draw OM, perpendicular to OP at O such that it intersects the PC line produced at M. The triangle OCM represents the Klein's velocity diagram.
2. In this triangle OCM, CO is the crank.
3. The velocity diagram for the configuration by the relative velocity method is a triangle *ocp* shown at Fig. 4.5(b). Let this be rotated by 90° in the anticlockwise direction. It is shown by the dotted lines and named as oc_1p_1.
4. Compare the triangles oc_1p_1 and OCM.

    oc_1 or oc represents v_{co}, i.e. velocity of C with respect to $O = \omega.OC$. It is parallel to OC.
    op_1 or op represents v_{po}, i.e. velocity of P with respect to $O = \omega.OM$. It is similar to OM.
    c_1p_1 or cp represents v_{pc}, i.e. velocity of P with respect to $C = \omega.CM$. It is similar to CM.
5. Both the triangles are similar and the sides are proportional to ω as shown below:

$$\frac{oc_1}{OC} = \frac{c_1p_1}{CM} = \frac{p_1O}{MO} = w$$

6. Hence, obtain the velocities, by measuring the lengths OM, OC and CM on the space diagram itself as per the scale of the space diagram and multiply with the scale factor ω.

4.3.2 Klein's Acceleration Diagram

In order to get the accelerations by Klein's construction, the velocity diagram OCM constructed above is to be extended as explained below.

1. Draw a circle with C as centre and CM as radius.
2. Draw another circle with CP as diameter. Both these two circles intersect at K and L as shown. Thus KL intersects CP at Q and OP at N as shown.

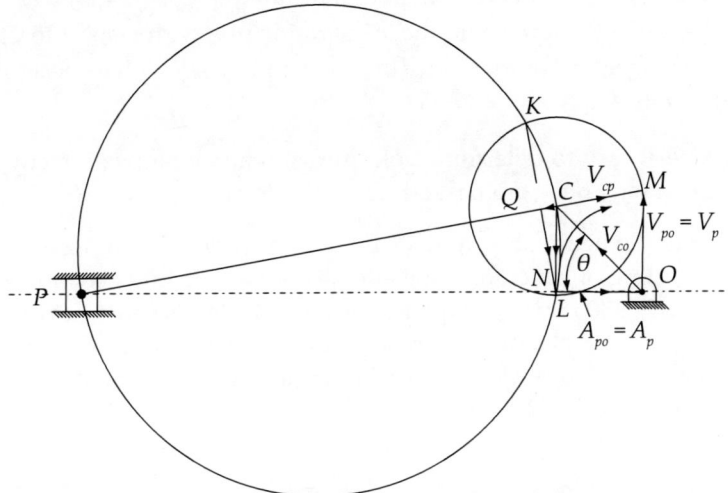

(a) Klein's construction on space diagram

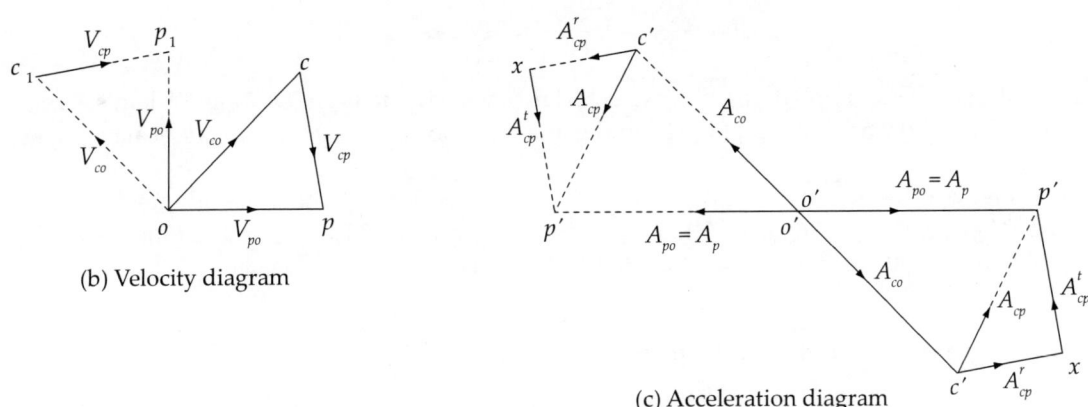

(b) Velocity diagram

(c) Acceleration diagram

Fig. 4.5 Klein's construction of velocity and acceleration diagrams.

3. Now OCQN represents the Klein's acceleration diagram similar to acceleration diagram obtained by relative method $o'c'xp'$ as shown at Fig. 4.5(c).

4. Rotate $o'c'xp'$ by 180° in anticlockwise direction and compare the sides. Both are similar and their sides are proportional to ω^2 as below.

$$\frac{o'c'}{OC} = \frac{c'x}{CQ} = \frac{xp'}{QN} = \frac{o'p'}{ON} = \omega^2$$

The acceleration of C with respect to O, i.e. $o'c'$ is equal to $\omega^2 \cdot OC$. Hence, OC represents acceleration of C with respect to O. To obtain the actual acceleration, measure OC and multiply with ω^2. Similarly, apply to all other sides.

5. The crank OC on the space diagram represents the radial component of acceleration of the crankpin C with respect to O or acceleration of the crankpin C, $a^r_{CO} = a_C = \omega^2 OC$.
6. CQ represents the radial component of the acceleration of P with respect to C, $a^r_{CP} = \omega^2 CQ$.
7. QN represents the tangential component of the acceleration of P with respect to C, $a^t_{CP} = \omega^2 QN$.
8. Acceleration of piston P, $a_{PO} = a_P = \omega^2 NO$.

Note: This method is very easy to determine velocities and accelerations directly from the space diagram. Not only time saving but graphical errors will be less.

W E 4.2: The same example again shown at W E 4.1 is considered here again for this method. The crank rod and connecting of a reciprocating engine are 150 mm and 600 mm respectively. The crank is rotating clockwise at 300 r.p.m. Find with the help of Klein's construction: (i) Velocity and acceleration of the piston P, (ii) Velocity and acceleration of the midpoint of the connecting rod D, (iii) Angular velocity and angular acceleration of the connecting rod at the instant when the crank is at 45° to I.D.C. (inner dead centre)?

Given:
OC = 150 mm = 0.15 m;   PC = 600 mm = 0.6 m;   N = 300 r.p.m. and so

$$\omega = \frac{2\pi N}{60} = \frac{2\pi 300}{60} = 31.42 \text{ rad/s.}$$

Draw the space diagram and then construct Klein's velocity diagram OCM and Klein's acceleration diagram OCQN as shown in Fig. 4.6 and explained above choosing a suitable scale for space diagram.

Then measure the lengths of OM, CM, CQ, QN and NO using the scale adopted for space diagram. They are OM = 126 mm = 0.126 m; CM = 108.4 mm = 0.108 m; CQ = 39.2 mm = 0.0392 m; QN = 107 mm = 0.107 m; and NO = 106 mm = 0.106 m. The scale of the velocity diagram is ω and that of the acceleration diagram is ω^2.

(i) **Velocity and acceleration of the piston P**

(a) *Velocity of the piston*

$$v_p = \omega \cdot OM$$
$$= 31.42 \times 0.126 = 3.96 \text{ m/s.}$$

(b) *Acceleration of the piston P*,

$$a_p = \omega^2 \times NO = 31.422 \times 0.106 = 104.7 \text{ m/s}^2.$$

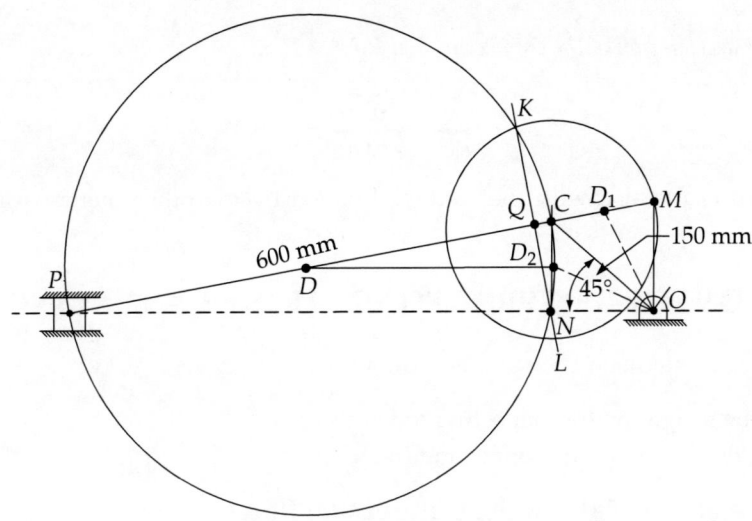

Fig. 4.6

(ii) **Velocity and acceleration of the midpoint D of the connecting rod CP**
Find the midpoint of CM and name it as D_1. Join O to D_1 and measure its length as per the scale of space diagram, $OD_1 = 128$ mm $= 0.128$ m.

(a) *Velocity of the midpoint D*,

$$v_D = \omega OD_1 = 31.42 \times 0.128 = 4.02 \text{ m/s}$$

Similarly, find the midpoint of CN and name it as D_2 by drawing a horizontal line from the midpoint D on CP. Then join O to D_2 and measure its length as per the scale,

$$OD_2 = 119 \text{ mm} = 0.119 \text{ m}$$

(b) *Acceleration of D*,

$$a_D = \omega^2 OD_2 = 31.42^2 \times 0.119 = 117.55 \text{ m/s}^2$$

(iii) **Angular velocity and angular acceleration of the connecting rod CP**
Velocity of the connecting rod PC,

$$v_{PC} = \omega CM = 31.42 \times 0.1084 = 3.41 \text{ m/s}$$

(a) *Angular velocity of the connecting rod CP*,

$$\omega_{PC} = \frac{v_{PC}}{PC} = \frac{3.41}{0.6} = 5.6 \text{ rad/s}$$

The tangential component of the acceleration of connecting rod P with respect to C,

$$a^t_{CP} = \omega^2 QN = 31.42^2 \times 0.107 = 105.63 \text{ m/s}^2$$

(b) *Angular acceleration of the connecting rod PC,*

$$\alpha_{CP} = \frac{a^t_{PC}}{PC} = \frac{105.63}{0.6} = 176.05 \text{ rad/s}^2.$$

Compare these results also with the results obtained by the other methods in the previous chapter.

4.4 FORCES ON THE RECIPROCATING PARTS OF AN ENGINE

The forces can be calculated on two aspects as follows:

1. Neglecting the weight of the connecting rod.
2. Considering the weight of the connecting rod.

4.4.1 Neglecting the Weight of the Connecting Rod

The various forces acting on the reciprocating parts of a horizontal engine are shown in Fig. 4.7. The various expressions of these forces neglecting the weight of the connecting rod are derived as follows:

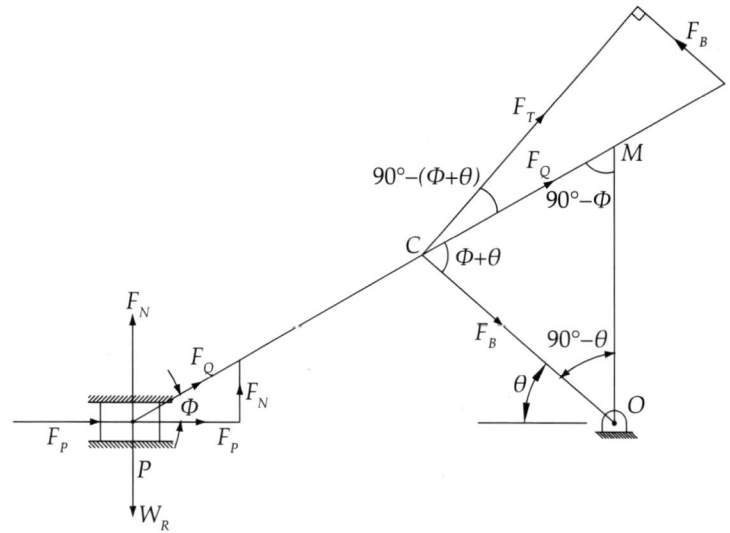

Fig. 4.7 Forces on the reciprocating parts.

(i) Piston effort (F_p):

Net force acting on the piston or cross-head pin along the line of stroke is called the piston effort, and is denoted by F_P in Fig. 4.7.

$$F_P = F_L \pm F_I - R_F \tag{1}$$

For a horizontal engine where F_L is the net load acting on the piston. For a single cylinder $p \times A$ and for a double acting $(p_1 A_1 - p_2 A_2)$ or $[p_1 A_1 - p_2(A_1 - a)]$ where p, p_1, p_2 represent pressures and A, A_1, A_2 and a represent cross-sectional areas of the cylinder on either side of the piston and a area of cross section of the piston rod. In case the piston rod area 'a' is neglected or negligible, then net load or force on the piston

$$F_L = (p_1 - p_2)A_1$$

F_I is the inertia force of the reciprocating parts and R_F is the frictional resistance.

The inertia forces F_I due to the acceleration of the reciprocating parts opposes the accelerating force F_P on the piston. Use negative sign (–ve) for θ between 0° and 180°, gives acceleration and positive sign (+ve) for θ between 180° and 360°, gives retardation.

Using the expressions derived for the accelerations above, the inertia force is given by

$$F_I = m_R \cdot a_R = m_R \omega^2 r \left\{ \cos\theta + \frac{\cos 2\theta}{n} \right\}$$

where m_R is the mass of the reciprocating parts and a_R is the acceleration of the reciprocating parts. r is the crank radius and θ is the crank angle.

$$F_P = (F_L \pm F_I \pm W_R - R_F) \tag{2}$$

for a vertical engine. Here the weight of reciprocating parts assists the piston effort while moving downwards and opposes while moving upwards. So take $+W_R$ for downward and $-W_R$ for upward.

(ii) Force along connecting rod (F_Q)

$$F_Q = \frac{F_P}{\cos\phi} = \frac{F_P}{\sqrt{\left[1 - \dfrac{\sin^2\theta}{n^2}\right]}} \tag{3}$$

From the geometry of the figure and $L\sin\phi = r\sin\theta$ or $\sin\phi = \dfrac{\sin\theta}{n}$; $\cos\phi = \sqrt{(1 - \sin^2\phi)}$. Hence,

$$\cos\phi = \sqrt{\left[1 - \frac{\sin^2\theta}{n^2}\right]}$$

(iii) Thrust on the sides of the cylinder walls or normal reaction on the guide bar (F_N):

$$F_N = F_Q \sin\phi = \frac{F_P \sin\phi}{\cos\phi} \qquad \text{since } F_Q = \frac{F_P}{\cos\phi}$$

$$F_N = F_P \tan \phi = \frac{F_P \sin \phi}{\sqrt{\left[1 - \frac{\sin^2 \theta}{n^2}\right]}} \qquad (4)$$

(iv) Crankpin effort (F_T) and thrust on crankshaft bearings (F_B)

The component of the F_Q perpendicular to the crank is called the crankpin effort indicated by F_T and the component along the crank which produces a thrust on the crankshaft bearing is F_B.

$$F_T = F_Q \sin(\theta + \phi) = \frac{F_P \sin(\theta + \phi)}{\cos \phi} \qquad (5)$$

$$F_B = F_Q \cos(\theta + \phi) = \frac{F_P \cos(\theta + \phi)}{\cos \phi} \qquad (6)$$

(v) Crank effort or turning moment or torque on the crankshaft (T)

The product of the crankpin effort (F_T) and the crank radius (r) is known as crank effort or turning moment (T)

$$T = F_T \cdot r = \frac{F_P \sin(\theta + \phi) \cdot r}{\cos \phi}$$

$$= F_P \frac{(\sin \theta \cos \phi + \cos \theta \sin \phi) \cdot r}{\cos \phi}$$

$$T = F_P (\sin \theta + \cos \theta \cdot \tan \phi) \cdot r$$

since $\tan \phi = \dfrac{\sin \theta}{\sqrt{(n^2 - \sin^2 \theta)}}$

Crank effort or torque on the crankshaft

$$T = F_p \cdot r \left[\sin \theta + \frac{\sin 2\theta}{\left(2\sqrt{(n^2 - \sin^2 \theta)}\right)}\right]$$

since $\sin^2 \theta$ is very small compared to n^2, therefore neglecting $\sin^2 \theta$.

$$T = F_p r \left[\sin \theta + \frac{\sin 2\theta}{2n}\right] = F_P \cdot OM \qquad (7)$$

where $OM = r\left[\sin \theta + \dfrac{\sin 2\theta}{2n}\right]$

Find the length OM graphically using Klein's construction and multiply with F_P to obtain the torque.

W E 4.3: *(Horizontal Engine)* The crankpin circle radius of a horizontal engine is 150 mm. The mass of the reciprocating parts is 250 kg. When the crank has travelled 45° from I.D.C., the difference between the driving and the back pressure is 0.35 N/mm². The connecting rod length between centres is 0.6 m and the cylinder bore is 0.25 m. If the engine runs at 300 r.p.m. and if the effect of piston rod diameter is neglected, calculate: 1. Pressure in the slide bars, 2. Thrust in the connecting rod, 3. Tangential force on the crankpin, and 4. Turning moment on the crankshaft.

Given:
$r = 150$ mm $= 0.15$ m; $m_R = 250$ kg; $\theta = 45°$; $p_1 - p_2 = 0.35$ N/mm²; $L = 0.6$ m;
$D = 250$ mm $= 0.25$ m; $N = 300$ r.p.m. or $\omega = \dfrac{2\pi 300}{60} = 31.42$ rad/s; $m_R = 250$ kg;
$n = \dfrac{L}{r} = \dfrac{0.6}{0.15} = 4.$

$$\text{Piston effort } (F_P) = F_L - F_I$$

where $F_L = (p_1 - p_2)A = 0.35\pi\dfrac{(250)^2}{4} = 17187.5$ N.

$$F_I = m_R\omega^2 r\left[\cos\theta + \dfrac{\cos 2\theta}{n}\right]$$

$$= 250(31.42)^2\, 0.15\left[\cos 45° + \dfrac{\cos 90°}{4}\right]$$

$$= 250 \times 104.69$$

$$= 26173.57 \text{ N}$$

where 104.69 is the acceleration of the reciprocating masses.

$$\text{Piston effort } F_P = F_I - F_L$$
$$= 26173.57 - 17187.5$$
$$= 8986 \text{ N} = 8.986 \text{ kN}$$

1. **Pressure on slide bars (F_N)**

$$F_N = F_P \tan\phi$$

where $\sin\phi = \dfrac{\sin\theta}{n} = \dfrac{\sin 45°}{4} = \dfrac{0.707}{4} = 0.1768$ or $\phi = 10.18°$.

Therefore, $F_N = 8.986 \times \tan 10.18° = 1.614$ kN.

2. **Thrust in the connecting rod (F_Q)**

$$F_Q = \dfrac{F_P}{\cos\phi} = \dfrac{8.986}{\cos 10.18°}$$

$$= \dfrac{8.986}{0.9843} = 9.129 \text{ kN}$$

3. Tangential force on the crankpin (F_T)

$$F_T = F_Q \sin(\theta + \phi)$$
$$= 9.129 \sin(45° + 10.18°)$$
$$= 9.129 \sin(55.18°)$$
$$= 9.129 \times 0.8209 = 7.4943 \text{ kN}$$

4. Turning moment on the crankshaft (T)

$$T = F_T \cdot r$$
$$= 7.4943 \times 0.15$$
$$= 1.124 \text{ kN-m}$$

Note: There is a difference between the horizontal engine and the vertical engine in calculating the piston effort and then turning moment from it. It is explained by taking the same above problem as follows.

W E 4.4: *(Vertical Engine)* A vertical engine is having the acceleration of the piston 105 m/s² when the crank has moved 45° from the inner dead centre position. The net effective pressure on the piston is 0.35 N/mm² and the frictional resistance is equivalent to a force of 600 N. The diameter of the piston is 250 mm and the mass of the reciprocating parts is 250 kg. If the length of the crank is 150 mm and the ratio of the connecting rod length to the crank length is 4, find: 1. Reaction on the guide bars, 2. Thrust on the crankshaft bearing, and 3. Turning moment on the crankshaft.

Note: The data is same as the above worked example except the engine is vertical and having a frictional resistance.

Given:
$a_P = 105 \text{ m/s}^2$; $\quad \theta = 45°$; $\quad p = 0.35 \text{ N/mm}^2$; $\quad R_F = 600 \text{ N}$; $\quad D = 250 \text{ mm}$; $\quad m_R = 250 \text{ kg}$;
$r = 150 \text{ mm} = 0.15 \text{ m}$; $\quad n = \dfrac{L}{r} = 4$

$$\text{Piston effort } (F_P) = F_L - F_I$$

$$F_L = \frac{p \pi d^2}{4}$$
$$= \frac{0.35 \pi (250)^2}{4} = 17187.5 \text{ N}$$

$$F_I = m_R \times a_P$$
$$= 250 \times 105 = 26250 \text{ N}$$

$$F_P = F_I - F_L - R_F$$
$$= 26250 - 17187.5 - 600$$
$$= 8462.5 \text{ N} = 8.4625 \text{ kN}$$

1. **Pressure on slide bars (F_N)**

$$F_N = F_P \tan \phi$$

where $\sin \phi = \dfrac{\sin \theta}{n} = \dfrac{\sin 45°}{4} = 0.1768$ or $\phi = 10.18°$.

Therefore,

$$F_N = 8.4625 \tan 10.18°$$
$$= 8.4625 \times 0.1796 = 1.5198 \text{ kN}$$

2. **Thrust on the crankshaft bearing (F_B)**

$$F_B = \dfrac{F_P \cos(\theta + \phi)}{\cos \phi}$$

$$= \dfrac{8.4625 \cos(45° + 10.18°)}{\cos 10.18°}$$

$$= \dfrac{8.4625 \times 0.571}{0.984}$$

$$= 8.4625 \times 0.58 = 4.908 \text{ kN}$$

3. **Turning moment on the crankshaft (T)**

$$T = F_T \cdot r$$

$$= \dfrac{F_P \sin(\theta + \phi) \cdot r}{\cos \phi}$$

$$= \dfrac{8.4625 \sin(45° + 10.18°) \cdot 0.15}{\cos 10.18°}$$

$$= 1.0586 \text{ kN}$$

Note: The turning moment can be observed less in vertical engine compared to horizontal engine for the same parameters.

4.4.2 Considering the Weight of the Connecting Rod

In a reciprocating engine, the three links other than the fixed one follow three different motions. The crank purely rotates, piston reciprocates to and fro linearly and the connecting rod oscillates along with their masses. The forces due to accelerations of rotating and reciprocating masses can be found. But the motion of the mass of the connecting rod is not linear since it is an oscillating mass. Hence, the mass of the connecting rod is partly shifted to crank end side and partly to piston end side. By doing so, the accelerating forces can be easily calculated. This process of shifting the masses can be done in two ways: (i) Conveniently shifting the masses to crank end and piston end of the connecting rod (ii) Partly shifting to piston end and the balance towards crank end. Both the above two methods must satisfy certain conditions. The technique of shifting is called a **dynamically equivalent system** as explained below separately for the two cases.

4.5 EQUIVALENT DYNAMICAL SYSTEM

In order to determine the motion of a rigid body under the action of external forces, it is usually convenient to replace the single mass of the rigid body especially in the case of connecting rod by two masses placed at a fixed distance apart, in such a way that

1. The sum of the two masses is equal to the total mass of the connecting rod.
2. The combined centre of mass coincides with that of the connecting rod.
3. The sum total of mass moment of inertia of the two masses about the combined centre of mass is equal to that of the connecting rod.

When these three conditions are satisfied, then it is said to be a **dynamical equivalent system**.

4.5.1 Dynamically Equivalent System

Let m be the total mass of the connecting rod at G and k be the radius of gyration as shown in Fig. 4.8. Let m_a and m_b be the two masses placed at A and B at distances a and b for an equivalent system. But the actual connecting is AC.

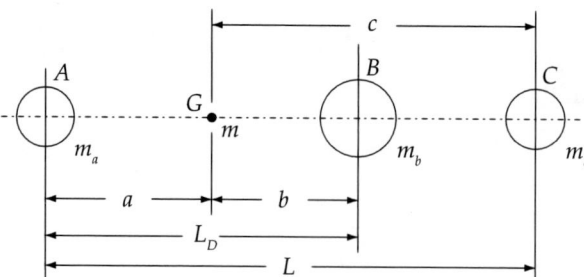

Fig. 4.8 Actual and equivalent system.

Let L_D be the total distance between the points A and B, i.e. $L_D = (a + b)$ and L be the total distance between the points A and C as shown.

For two masses to be dynamically equivalent according to the above three conditions:

$$m_a + m_b = m \qquad (1)$$

$$m_a \cdot a = m_b \cdot b \qquad (2)$$

$$m_a \cdot a^2 + m_b \cdot b^2 = I = m \cdot k^2 \qquad (3)$$

where k is the radius of gyration of the connecting rod about its centre of gravity G.

From equations (1) and (2)

$$m_a + \left(m_a \times \frac{a}{b}\right) = m$$

$$m_a \left[1 + \left(\frac{a}{b}\right)\right] = m$$

$$m_a\left[\frac{(a+b)}{b}\right] = m$$

$$m_a = \frac{m \cdot b}{(a+b)} \qquad (4)$$

and similarly

$$m_b = \frac{m \cdot a}{(a+b)} \qquad (5)$$

Substituting these values of m_a and m_b in equation (3) and simplifying gives

$$k^2 = a \cdot b \qquad (6)$$

From (6), it is clear that either the distance a or the distance b may be chosen but not both the distances. Based on the assumed value of either distance a or b, the other distance has to be obtained using the above relations. Sum of the two distances, $L_D = (a+b)$ is known as equivalent length of a dynamically equivalent system.

4.5.2 Determination of Dynamically Equivalent System of Two Masses Placed Arbitrarily (Analytically)

Let us assume that the masses are arbitrarily placed at convenient distances say a and c (assume c greater than b) from the centre of mass of the connecting rod to piston end A and crank end C as shown in the Fig. 4.8. The conditions (1) and (2) will only be satisfied but not the third condition (3). Hence, a correction couple has to be applied in order to make it a dynamically equivalent system.

Let the sum of the distances be $(a+c) = L$, the length of the connecting rod, m be the mass of the connecting rod and m_a and m_c be the masses shifted to ends A and C. Then

$$m_a + m_c = m$$

Let α_c be the angular acceleration of the connecting rod AC, ΔT be the correction couple

$$\Delta T = \alpha_c(m \cdot a \cdot c - m \cdot a \cdot b) \text{ or } m \cdot a \cdot \alpha_c(c-b)$$

Rewriting it

$$m \cdot a \cdot \alpha_c[(a+c) - (a+b)] = m \cdot a \cdot \alpha_c(L - L_D)$$

The correction couple may be produced by two equal, parallel and opposite forces F_Y, acting at the two ends perpendicular to the line of stroke, i.e. at A and C.

The force at A, i.e. at the piston end of the connecting rod is taken by the reaction of guides. But the force at crank end C of the connecting rod causes torque on crankshaft which is called the correction torque (T_C).

$$T_C = F_Y \times r \cos \theta$$

but $F_Y = \dfrac{\Delta T}{L \cos \phi}$, hence

$$T_C = \left(\dfrac{\Delta T}{L \cos \phi}\right) . r \cos \theta$$

$$= [m \cdot a \cdot \alpha_c (L - L_D)] \dfrac{\cos \theta}{n \cos \phi}$$

Substitute for $\cos \phi$ as shown above in the analytical expressions for velocity and accelerations. The final expression for T_C is

$$T_C = [m \cdot a \cdot \alpha_c (L - L_D)] \dfrac{\cos \theta}{\sqrt{(n^2 - \sin^2 \theta)}} \tag{1}$$

Also due to the weight of the mass shifted to C, a torque T_m exerted on the crankshaft which is given by

$$T_m = (m_c \times g) r \cos \theta \tag{2}$$

Let T_I be the torque exerted on the crankshaft due to inertia forces of mass acting at piston end A and the expression will be similar to one shown for turning moment on crankshaft,

$$T_I = m_A \cdot r \left[\sin \theta + \dfrac{\sin 2\theta}{2n}\right] = m_A \cdot OM \tag{3}$$

where m_A represents the total mass at A, i.e. reciprocating mass m_R plus the part of the mass of connecting rod shifted to A and $OM = r \cdot \left[\sin \theta + \dfrac{\sin 2\theta}{2n}\right]$.

The net torque T on the crankshaft will be the algebraic sum of the torques T_C, T_I and T_m.

4.5.3 Determination of Dynamically Equivalent System of Two Masses Placed Arbitrarily (Graphically)

Let the centre of gravity of the connecting rod where total mass m is acting be at G as shown in Fig. 4.9. This mass m is replaced by two masses m_a and m_b located at two points A and B on the connecting rod so that the system becomes dynamically equivalent. The distance between the mass m_a at A, i.e. at the piston side and G be a. The various steps are:

1. Mark the given distance a from A to locate G on the connecting rod.
2. Erect a perpendicular at G and mark on it a length equal to radius of gyration k and name it as GC.
3. Join A to C and draw perpendicular to AC which will cut AG line extended at B.
4. Measure the distance GB which represents the distance b. Thus, the position of the second mass m_b is to be determined for dynamically equivalent system.
5. Consider the two similar triangles ACG and BCG from the Fig. 4.9 the ratio of the sides of the similar triangles are $\dfrac{k}{a} = \dfrac{b}{k}$ or $k^2 = a \cdot b$

Thus, the square of the length of radius of gyration is equal to the product of the distances of the two masses from the centre of the total mass.

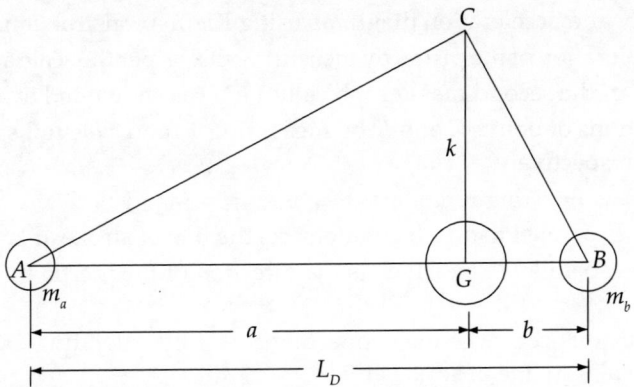

Fig. 4.9 Dynamically equivalent system by graphical method.

4.6 INERTIA FORCES IN A RECIPROCATING ENGINE

(Considering the mass of the connecting rod)

4.6.1 Graphical Method

In a reciprocating engine, let *OC* be the crank and *PC* be the connecting rod whose centre of gravity lies at *G*. The inertia forces in a reciprocating engine may be obtained graphically as follows:

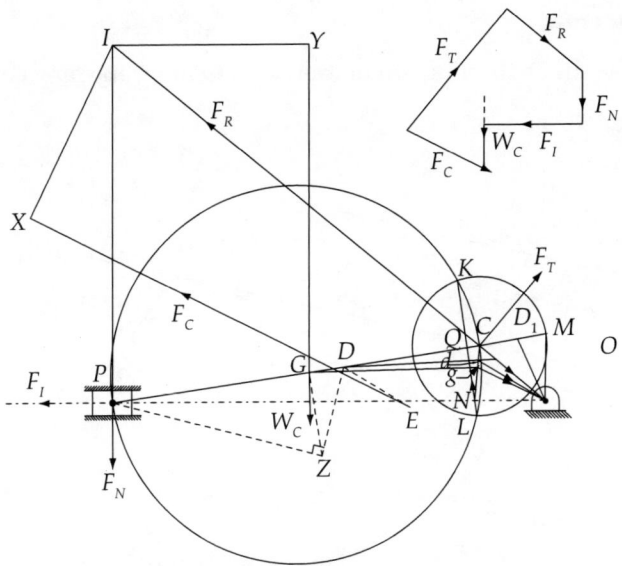

Fig. 4.10 Inertia forces by graphical method.

The various steps are as follows:

1. Draw the space diagram choosing a suitable scale as shown in Fig. 4.10.
2. Construct velocity and acceleration diagrams using Klein's construction.
3. Calculate the acceleration of the piston by measuring ON as per the scale adopted, $a_P = \omega^2 \cdot ON$.
4. Find the position of the second mass graphically, i.e. D as shown in Fig. 4.10.
5. Find the accelerations of points G and D by measuring og and od lengths, they are $a_G = \omega^2 \cdot og$ and $a_D = \omega^2 \cdot od$, respectively.
6. Inertia force of the connecting rod of mass m_C is $F_C = m_C \omega^2 \cdot go$.
7. Draw a line from D parallel to og which intersects the line of stroke at E.
8. Draw a line from E parallel to og which is the direction of the inertia force of connecting rod F_C.
9. Find the instantaneous centre of the connecting rod, I by extending OC and by drawing a normal perpendicular to line of stroke at P.
10. Drop the perpendicular from I to intersect the line of action of F_C at X.
11. Similarly, extend W_C and drop a perpendicular from I to intersect at Y.
12. Measure the distances IX, IY, IP and IC.
13. Then taking moments about I gives,

$$F_T \cdot IC = F_I \cdot IP + F_C \cdot IX + W_C \cdot IY$$

14. Knowing the distances and the forces F_I, F_C and W_C, find the tangential force due to the inertia at the crankpin, F_T. Then calculate the torque T due to the inertia forces acting on the crankshaft T, knowing the radius r of the crank using, $T = F_T \cdot r$.

4.6.2 Analytical Method

The effect of the inertia of the connecting rod on the crankshaft torque may be obtained as given in the following steps.

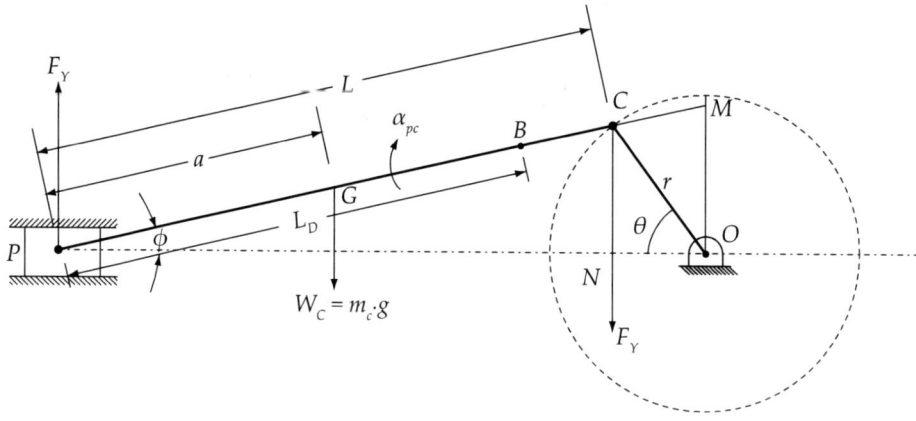

Fig. 4.11

1. The mass of the connecting rod m_C is divided into two masses. One mass is placed at the piston pin P, be m_{piston} and the other at the crankpin C, be m_{crank} as shown in Fig. 4.11, so that the centre of gravity of these two masses coincides with the centre of gravity of the connecting rod G.
2. The inertia force due to the mass at C, m_{crank} acts in radial direction outwards along the crank OC, therefore, it has no effect on the crankshaft torque.
3. The inertia force due to the total mass at P can be obtained as follows:
   The portion of the connecting rod mass m_C transferred to piston end P, $m_{piston} = \dfrac{m_C(L-a)}{L}$.
   So, the total reciprocating mass acting at P is

$$m_R + m_{piston} = m_R + \frac{m_C(L-a)}{L}$$

Therefore, the inertia force due to the total reciprocating masses be F_I which is equal to

$$F_I = \left[m_R + \frac{m_C(L-a)}{L}\right] \times a_P$$

where $a_P = \omega^2 r\left[\cos\theta + \dfrac{\cos 2\theta}{n}\right]$

Hence, $F_I = \left[m_R + \dfrac{m_C(L-a)}{L}\right] \times \omega^2 r\left[\cos\theta + \dfrac{\cos 2\theta}{n}\right]$ and the corresponding torque exerted on the crankshaft is,

$$T_I = F_I \cdot OM = F_I\left[\sin\theta + \frac{\sin 2\theta}{2\sqrt{(n^2 - \sin^2\theta)}}\right] \quad (1)$$

Usually, the value of OM is measured by drawing the perpendicular to PO at O which intersects PC produced at M.

4. As the masses are arbitrarily placed at P and C, a correcting torque T_C must be applied given by $m_C[a \cdot (L - L_D].a_C$ as in Fig. 4.9, $\alpha_P = \dfrac{\omega^2 \sin\theta}{n}$ and $\alpha_C = \dfrac{-\omega^2 \sin\theta}{n}$.
   The corresponding torque T_C can be applied on the system by two equal and opposite forces F_Y acting through P and C.
   Therefore, $F_Y \cdot PN = T_C$ or $F_Y = \dfrac{T_C}{PN}$ and the corresponding torque on the crankshaft,

$$T_{crank} = F_Y \cdot NO = \frac{T_C \cdot NO}{PN}$$

From Fig. 4.11,
$NO = OC \cdot \cos\theta = r \cdot \cos\theta$ and $PN = PC \cos\phi = L \cos\phi$
Therefore,

$$\frac{NO}{PN} = \frac{r\cos\theta}{L\cos\phi} = \frac{\cos\phi}{n\cos\phi}$$

but $\cos\phi = \sqrt{\left(1 - \dfrac{\sin^2\theta}{n^2}\right)}$ after replacing

$$\dfrac{NO}{PN} = \dfrac{\cos\theta}{\sqrt{(n^2 - \sin^2\theta)}}$$

since $\sin^2\theta$ is very small as compared to n^2, therefore neglecting $\sin^2\theta$, the ratio

$$\dfrac{NO}{PN} = \dfrac{\cos\theta}{n}$$

Hence, the torque on the crankshaft due to the correcting torque T_C is given by T_{crank}

$$T_{crank} = F_Y \cdot NO$$
$$= \dfrac{T_C NO}{PN}$$
$$= \dfrac{T_C \cos\theta}{n}$$
$$= m_C \cdot a \cdot (L - L_D) \dfrac{(-\omega^2 \sin\theta)}{n} \dfrac{\cos\theta}{n}$$

$$T_{crank} = -m_C.a.(L - L_D)\omega^2 \dfrac{\sin 2\theta}{2n^2} \qquad (2)$$

5. The equivalent mass of the connecting rod m_C acting at C, be m_{crank} equal to $m_C \cdot \dfrac{a}{L}$ and the torque exerted on the crankshaft due to this mass,

$$T_m = m_{crank} g.r \cos\theta \qquad (3)$$

6. The total torque T_{total} exerted on the crankshaft due to the inertia of the moving parts is the algebraic sum of the above three torques T_I, T_{crank} and T_m.

W E 4.5: The following data refer to a steam engine: Diameter of piston = 240 mm; stroke = 600 mm; length of connecting rod = 1.5 m; mass of the reciprocating parts = 300 kg; mass of connecting rod = 250 kg; speed = 125 r.p.m.; centre of gravity of connecting rod from crankpin = 500 mm; radius of gyration of the connecting rod about an axis through the centre of gravity = 650 mm.

Determine the magnitude and direction of the torque exerted on the crankshaft when the crank has turned through 30° from inner dead centre.

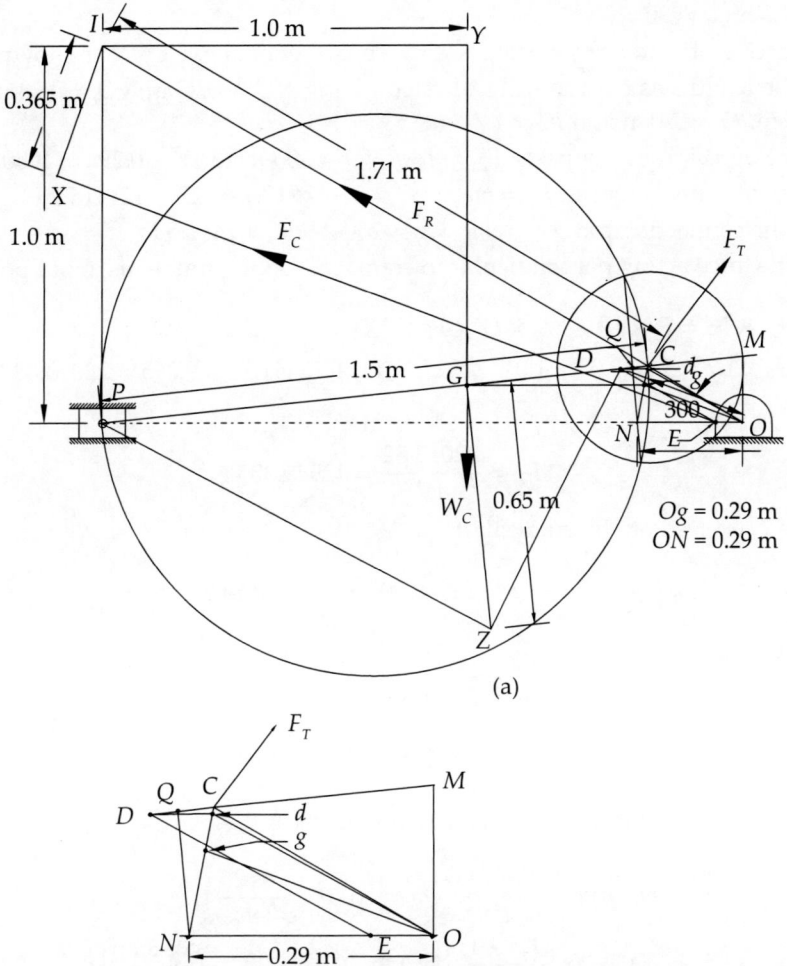

(b) Close up view of velocity and acceleration diagram

Fig. 4.12

Given:
Diameter of piston, D_P = 240 mm; stroke $(2r)$ = 600 mm or r = 300 mm = 0.3 m; L = 1.5 m;
$n = L/r = 5$; m_R = 300 kg; m_C = 250 kg; N = 125 rpm, $\omega = \dfrac{2\pi 125}{60} = 13.1$ rad/s;
Distance of G from C, = 500 mm = 0.5 m; k_G = 650 mm = 0.65 m; $\theta = 30°$.

Inertia torque can be calculated either by graphical method or by analytical method.

I. Graphical method:

The various steps of constructing the diagram shown at Fig. 4.12(a) are:

1. First draw the mechanism to some suitable scale with $OC = r$ = 300 mm = 0.3 m;

$PC = L = 1.5$ m; $\theta = 30°$.

2. Draw the acceleration diagram using Klein's construction, i.e. $OCQN$ as shown.
3. Measure the lengths as per the scale adopted giving $NO = 0.28$ m; $gO = 0.28$ m; $IP = 1.0$ m; $IX = 0.365$ m: $IY = 1.0$ m and $IC = 1.71$ m.
4. Inertia force of reciprocating parts, $F_I = m_R \omega^2 NO = 300 \times (13.1)^2 \times 0.29 = 14930.07$ N
5. Inertia force of connecting rod $F_C = m_C \omega^2 gO = 250 \times (13.1)^2 0.29 = 12441.725$ N
6. Weight of the connecting rod, $W_C = m_C \times g = 250 \times 9.81 = 2452.5$ N
7. Let F_T be the force acting perpendicular to crank OC, taking moments about point I, then

$$F_T \times IC = F_I \times IP + W_C \times IY + F_C \times IX$$
$$F_T \times 1.7 = 14930.07 \times 1.0 + 2452.5 \times 1.0 + 12441.73 \times 0.365 = 22013.82 \text{ N-m}$$

therefore,
$$F_T = \frac{22013.82}{1.7} = 12949.305 \text{ N}$$

8. The torque T, exerted on the crankshaft

$$T = F_T \times r = 12949.305 \times 0.3 = 3884.78 \text{ N-m}$$

II. Analytical method:

The distance of the CG of the connecting rod from P, i.e.

$$PG = PC - GC = 1.50 - .50 = 1.0 \text{ m}$$

Inertia force due to the total mass of the reciprocating parts at P,

$$F_I = m_R + m_C \frac{(PC - PG)}{L} \left[\omega^2 r \left(\cos\theta + \frac{\cos 2\theta}{n} \right) \right]$$
$$F_I = 300 + 250 \frac{(1.5 - 1)}{1.5} \left[13.1^2 \times 0.3 \left(\cos 30° + \frac{\cos 60°}{5} \right) \right]$$
$$= 19064 \text{ N}$$

since $n = \dfrac{L}{r} = \dfrac{1.5}{0.3} = 5$

Corresponding torque due to F_I,

$$T_I = F_I \times OM = F_I r \left[\sin\theta + \frac{\sin 2\theta}{\left[2\sqrt{(n^2 - \sin^2\theta)} \right]} \right]$$
$$T_I = 19064 \times 0.3 \left[\sin 30° + \frac{\sin 60°}{5} \right]$$
$$= 5719.2 \times 0.587 = 3357 \text{ N-m (ACW)} \tag{1}$$

L_D, equivalent length of a simple pendulum when swung about an axis through P,

$$L_D = \frac{\left[(k_G)^2 + (PG)^2\right]}{PG}$$

$$= \frac{\left[(0.65)^2 + (1.0)^2\right]}{1.0} = 1.42 \text{ m}$$

The torque at crank due to the correcting torque is,

$$T_{\text{crank}} = m_C \cdot PG \cdot (PC - PD) \omega^2 \frac{\sin 2\theta}{2n^2}$$

$$= 250 \times 1.0\,(1.5 - 1.42)\,13.1^2 \frac{\sin 60°}{2 \times 5^2}$$

$$T_{\text{crank}} = 59.5 \text{ N-m (ACW)} \tag{2}$$

$$T_m = m_{\text{crank}} \cdot g \cdot r \cos\theta$$

$$= m_C \times \frac{GC}{L} g \times r \cos\theta = 250 \times \frac{0.5}{1.5} \times 9.81 \times 0.3 \cos 30°$$

$$= 25 \times 9.81 \times 0.866 = 212.4 \text{ N-m (ACW)} \tag{3}$$

Total torque exerted on the crankshaft,

$$T = T_I + T_{\text{crank}} + T_m$$
$$= 3357 + 59.5 + 212.4 = 3628.9 \text{ N-m} \tag{4}$$

Note: The difference in the values found by the two methods is mainly due to the errors in measurements depending on the scale adopted in graphical method and approximate expression used in analytical method.

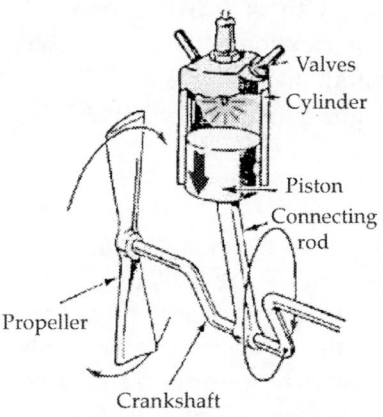

Pic. 4.2 Single cylinder.

Pic. 4.3   Multicylinder petrol engine.

4.7 EXERCISE

4.7.1 Short Answer Questions

1. Define 'inertia force' and 'inertia torque'.
2. What are the requirements of an equivalent dynamical system?
3. State D'Alembert's principle and explain equivalent dynamical system.
4. How will you determine the maximum load coming on the connecting rod of a high-speed diesel engine? What is the nature of this load? What will be the effect on stress distribution across the cross section of the connecting rod due to friction at the turning pairs?
5. Explain the procedure to construct Klein's construction to determine the velocity and acceleration of a slider crank mechanism.
6. What are free body diagrams of a mechanism? How are they helpful in finding various forces acting on various members of a mechanism?
7. Discuss the equilibrium of two and three force members.
8. Derive the expression for the acceleration of the piston of a reciprocating engine.
9. Explain inertia force, inertia torque and piston effort.
10. Explain clearly how inertia torque on the crankshaft of a reciprocating engine mechanism be calculated.

4.7.2 Problems

1. A vertical steam engine has a bore of 15 cm and stroke of 30 cm. The connecting rod is 52.5 cm long and the reciprocating parts weigh 1200 N and the engine is running at 240 r.p.m. When the crank is 60 degrees past its top dead centre, the steam pressure on the cover side of the piston is 60 N/cm^2 while that on the crank side is 10 N/cm^2. Neglecting the area of the piston rod, determine the net force on the piston rod and the turning moment on the crankshaft.

2. A gas engine working on Otto cycle is provided with two flywheels each weighing 5800 N and radius of gyration 56 cm. The diameter of the cylinder is 24 cm, stroke 27 cm and the mean speed 250 r.p.m. The mean pressure acting during the cycle reckoned above the atmosphere is given below:

   Suction stroke : atmospheric
   Compression   : 10.6 N/cm^2
   Firing   : 62 N/cm^2
   Exhaust   : 3 N/cm^2

   If the resistance is constant, find the percentage variation of speed of the engine.

3. The crank and connecting rod of a petrol engine, running at 1800 r.p.m. are 50 mm and 200 mm respectively. The diameter of the piston is 80 mm and the mass of the reciprocating parts is 1 kg. At a point during the power stroke, the pressure on the piston is 0.7 N/mm^2, when it has moved to 10 mm from the inner dead centre. Determine: (a) Net load on the gudgeon pin (b) Thrust in the connecting rod (c) Reaction between the piston and cylinder.

4. The crank and connecting rod lengths of an engine are 125 mm and 500 mm respectively. The mass of the connecting rod is 60 kg and its centre of gravity is 275 mm from the cross-head pin centre, the radius of gyration about its centre of gravity being 150 mm. If the engine is 600 r.p.m. for a crank position of 45° from IDC, determine: 1. The acceleration of the piston. 2. The magnitude, position and direction of inertia force due to the mass of connecting rod.

5. For the reciprocating engine mechanism, following data is given: Length of the crank = 10 cm; Length of the connecting rod = 30 cm; Distance of centre of gravity of link 2 from the main bearing = 5 cm; Distance of centre of gravity of link 3 from the crankpin = 15 cm; Crank angle from line of stroke = 60°; Crank speed = 1800 r.p.m. counterclockwise; Mass of link 2 = 2.5 kg; Mass of link 3 = 3.2 kg; Mass of link 4 = 4 kg; Mass moment of inertia of link 2 = 60 kg.cm^2. Mass moment of inertia of link 3 = 500 kg.cm^2. Determine the magnitude and direction of inertia of forces of links 2, 3 and 4.

6. The following data relate to a four-bar mechanism:

Link	Length	Mass	Moment of inertia about an axis through centre of
AB	60 mm	0.2 kg	80 kg.mm^2
BC	200 mm	0.4 kg	1600 kg.mm^2
CD	100 mm	0.6 kg	400 kg.mm^2
AD	140 mm		

   AD is the fixed link. The centre of mass for the links BC and CD lie at their midpoints whereas the centre of mass for link AB lies at A. Find the drive torque in the link AB at the instant when it rotates at an angular velocity of 47.5 rad/sec counterclockwise and angle ∠DAB is 135°. Neglect gravity effects.

7. The following data refer to a steam engine:

    Diameter of piston = 24 cm;
    Stroke = 48 cm;
    Length of connecting rod = 90 cm;
    Weight of the reciprocating parts = 1500 N;
    Weight of the connecting rod = 1000 N;
    Speed = 150 r.p.m.;
    Centre of gravity of the connecting rod from the crankpin = 32 cm;
    Radius of gyration of connecting rod about the centre of gravity = 40 m.

    Determine the magnitude and direction of the inertia torque on the crankshaft when the crank has turned through 45° from the inner dead centre.

8. A high-speed engine has a connecting rod length 5 times the crank which is 6 cm. It weighs 30 N has a centre of gravity 10 cm from the big end bearing. When suspended in bearings, it makes 50 complete oscillations in 52 seconds. The reciprocating parts weigh 15 N. Determine the torque exerted on the crankshaft due to the inertia of the moving parts when the crank makes an angle of 135° with the top dead centre when the speed of rotation is 1200 r.p.m.

4.7.3 Multiple Choice Questions

1. A pair of action and reaction forces acting on a body are known as

    (a) applied forces (b) constraint forces
    (c) accelerating forces (d) inertia forces [Ans. (b)]

2. In static equilibrium the vector sum of all the forces acting on the body and all the moments about _____ point is zero.

    (a) a fixed (b) a particular (c) an arbitrary (d) a permanent. [Ans. (c)]

3. If the lines of action of three or more forces intersect at a point, it is known as the _____ point

    (a) equilibrium (b) central (c) zero (d) concurrency [Ans. (d)]

4. A part isolated from the mechanism _____ be in equilibrium.

    (a) may (b) may or may not (c) must [Ans. (c)]

5. Acceleration of the piston of a reciprocating engine is _____

    (a) $r\omega^2 \left[\sin\theta + \left(\dfrac{\sin 2\theta}{n}\right)\right]$ (b) $r\omega \left[\cos\theta + \left(\dfrac{\cos 2\theta}{n}\right)\right]$

    (c) $r\omega^2 \left[\cos\theta + \left(\dfrac{\cos 2\theta}{4\pi}\right)\right]$ (d) $r\omega^2 \left[\cos\theta + \left(\dfrac{\cos 2\theta}{n}\right)\right]$ [Ans. (b)]

6. Crank effort is the net force applied at the crankpin _____ to the crank which gives the required turning moment on the crankshaft

   (a) parallel   (b) perpendicular   (c) at 45°   (d) 135°   [*Ans.* (b)]

7. In a dynamically equivalent system, a uniformly distributed mass is divided into _____ point masses.

   (a) two   (b) three   (c) four   (d) five   [*Ans.* (a)]

8. Any distributed mass can be replaced by two point masses to have the same dynamical properties if

   (a) the sum of the two masses is equal to the total mass
   (b) the combined centre of mass coincides with that of the rod
   (c) the moment of inertia of two-point masses about the perpendicular axis through their combined centre of mass is equal to that of the rod
   (d) all of the above   [*Ans.* (c)]

9. The maximum fluctuation of energy is the

   (a) ratio of maximum and minimum energies   (b) sum of maximum and minimum energies
   (c) difference of maximum and minimum energies   (d) difference of maximum and minimum energies from mean energy.   [*Ans.* (c)]

10. The maximum fluctuation of energy in a flywheel is equal to

    (a) $I\omega(\omega_1 - \omega_2)$ (b) $I\omega^2 k$ (c) $2KE$ (d) all   (e) none.   [*Ans.* (d)]

Turning Moment Diagrams and Design of Flywheel

5

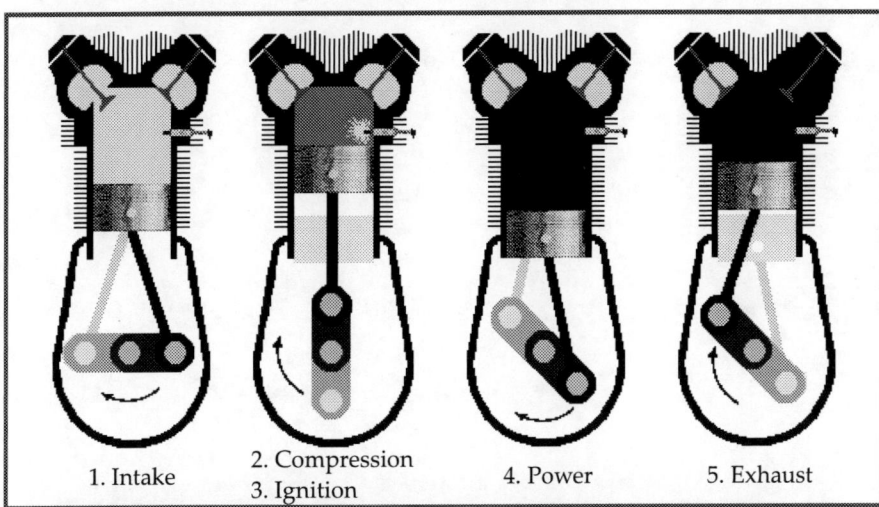

1. Intake
2. Compression
3. Ignition
4. Power
5. Exhaust

Turning Moment Diagrams and Design of Flywheel

5.1 INTRODUCTION

The diagram indicating the variation in the turning moment or torque (T) due to the pressure variation in the cylinder for one complete revolution of the crankshaft is known as the turning moment diagram. Since the turning moment (T) is crank effort (F_T) times crank radius (r), a plot of the crank effort vs. crank angle (θ) diagram is identical to turning moment diagram. Hence, it is the graphical representation with the turning moment or crank effort on the ordinate and different crank angles from 0° to 360° degrees on the abscissa. The turning moment diagrams or $T - \theta$ diagrams for different types of engines are being given below. The area under the curve represents the total work done per cycle. The diagrams for single cylinder double acting steam engine, single cylinder and multi-cylinder I.C. engines, the fluctuation of energy, design of flywheel, etc. are discussed in this chapter.

5.2 SINGLE-CYLINDER DOUBLE-ACTING STEAM ENGINE

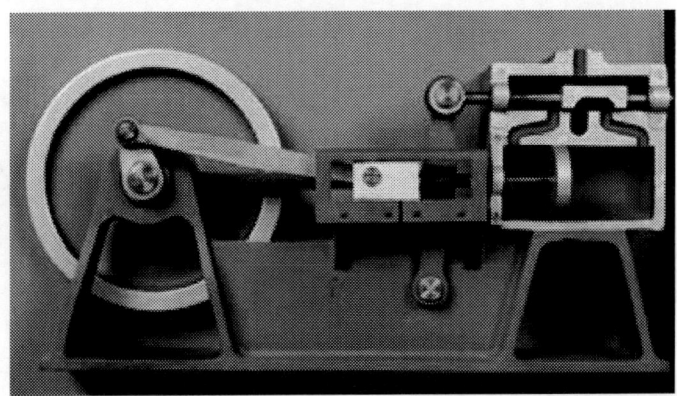

Pic. 5.1

The single-cylinder double-acting steam engine is shown in the picture Pic. 5.1 above. The $T - \theta$ diagram of it is shown in Fig. 5.1. The turning moment, T along y-axis and crank angle, θ along x-axis are represented. The expression for the turning moment $T \rightarrow$ as seen from the inertia forces chapter is $F_P \cdot r \left[\sin\theta + \dfrac{\sin 2\theta}{\sqrt{(n^2 - \sin^2\theta)}} \right]$ where F_P is the piston effort. The torque T is zero when θ is 0 and 180°. The torque T is maximum at θ, 90° and 270°. Area under the $T - \theta$ curve $abcde$ represents the work done per cycle.

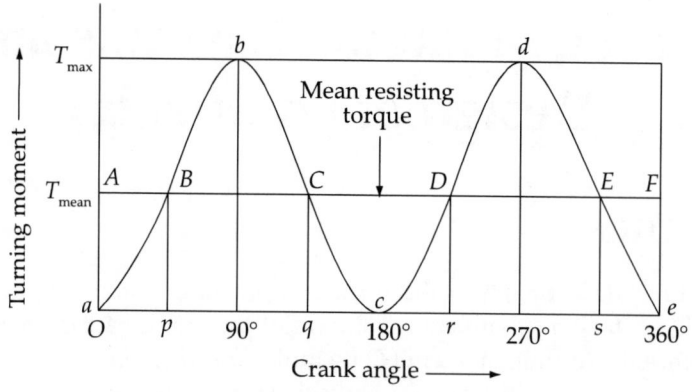

Fig. 5.1

Area under the mean torque line *ABCDEF* represents the work done against mean resisting torque (T_{mean}). Work done is positive, when the engine torque T is more than the mean resisting torque, i.e. between *BC* and *DE*, then the crankshaft accelerates. Work done is negative, when the engine torque T is less than mean resisting torque, i.e. between *AB*, *CD* and *EF*, then the crankshaft retards. Therefore, the difference between the varying torque and the mean torque ($T - T_{mean}$) gives the accelerating torque provided if it is positive and retarding torque provided if it is negative.

5.3 FOUR-STROKE CYCLE INTERNAL COMBUSTION ENGINE

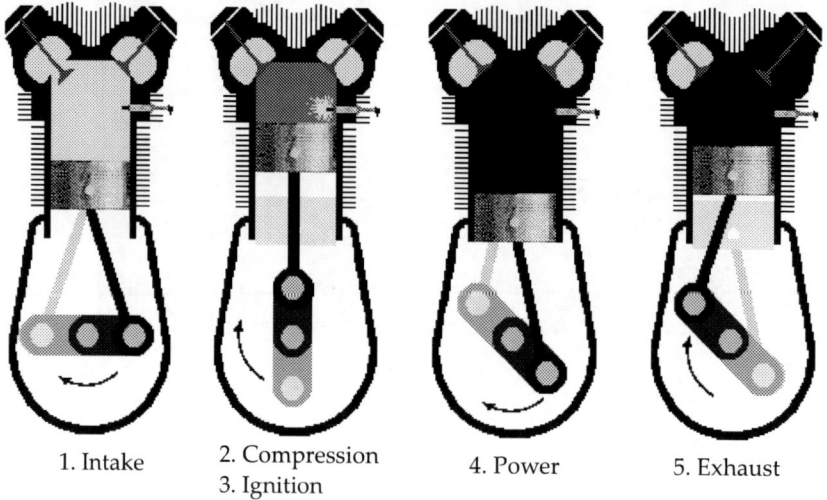

Pic. 5.2 Four-stroke internal combustion engine.

The Pic. 5.2 shows the single cylinder four-stroke internal combustion engine above. Figure 5.2 shows the $T - \theta$ diagram for the four strokes of an *I.C.* engine. During the power stroke the area under the curve can be observed more and also as positive.

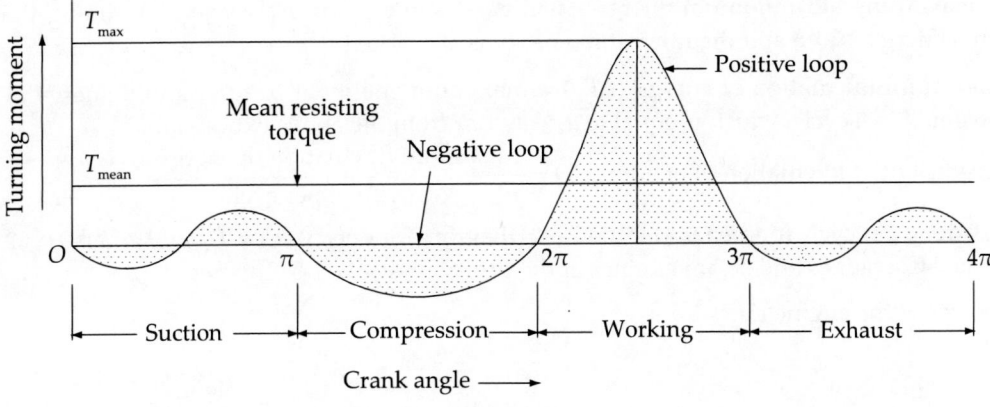

Fig. 5.2

5.3.1 Fluctuation of Energy

The difference between the maximum energy and the minimum energy in a cycle is called the maximum fluctuation of energy. It is determined using the turning moment diagram for one complete cycle of operation. The calculation of fluctuation of energy is explained by Fig. 5.3 below.

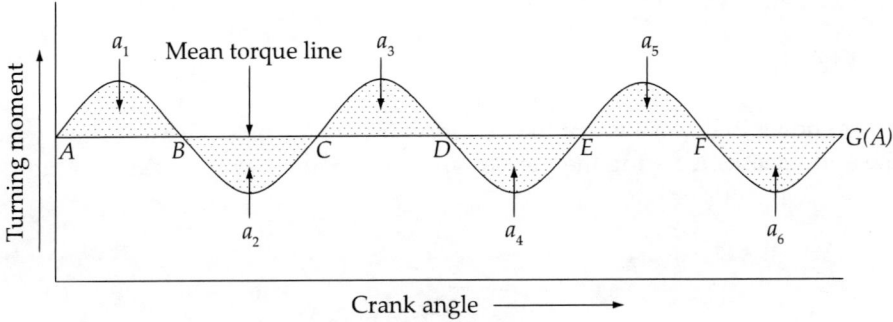

Fig. 5.3

In the figure, the $T - \theta$ curve can be seen intersecting with the mean torque line at A, B, C, D, E, F and G. Let the areas above and below the mean torque line be represented by $a_1, a_2, \ldots$ as shown. Let the total energy at each point be calculated as follows:

Energy at $A = E_n$,
Energy at B = Energy at A + area $a_1 = E_n + a_1$
Energy at C = Energy at B − area $a_2 = E_n + a_1 - a_2$,
Energy at D = Energy at C + area $a_3 = E_n + a_1 - a_2 + a_3$,
Energy at E = Energy at D − area $a_4 = E_n + a_1 - a_2 + a_3 - a_4$,
Energy at F = Energy at E + area $a_5 = E_n + a_1 - a_2 + a_3 - a_4 + a_5$ and
Energy at G = Energy at F − area $a_6 = E_n + a_1 - a_2 + a_3 - a_4 + a_5 - a_6$
   = Energy at A as the cycle repeats again with E_n.

The maximum and minimum energies will be at some point in a cycle. Assuming that the maximum energy is at B and the minimum energy is at E. Then

1. Maximum fluctuation of energy (ΔE_n) = maximum energy at B − minimum energy at E as assumed in a cycle, which is equal to $a_2 - a_3 + a_4$ from the above expression.
2. Coefficient of fluctuation of energy $(C_E) = \dfrac{\text{Maximum fluctuation of energy}}{\text{Work done per cycle}}$

Work done per cycle (or area under the $T - \theta$ diagram for one cycle) = $T_{mean}.\theta$ where θ is 2π for two-stroke I.C. engines and steam engines and 4π for four-stroke I.C. engines.

Horsepower of the engine (HP)

$$HP = \frac{2\pi N T_{mean}}{(60 \times 75)} \text{ or}$$

The power (P) in N-m

$$P = \frac{2\pi N T_{mean}}{60}$$

$$\therefore T_{mean} = \frac{60P}{2\pi N} = \frac{P}{\omega}$$

Work done per cycle = $\dfrac{P60}{n}$ where n is the number of power strokes/minute. $n = N$ for steam engine and two-stroke engine and $n = \dfrac{N}{2}$ for four-stroke I.C. engine.

5.4 FLYWHEEL

The flywheel as shown in Fig. 5.4 is just like a reservoir which stores the excess energy supplied by the engine and releases it during the period when the requirement of energy is more than the supply.

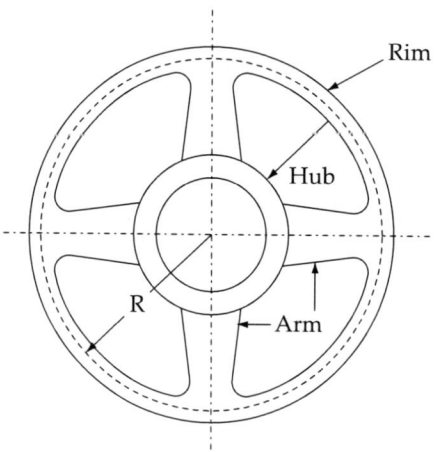

Fig. 5.4 Flywheel.

The energy is produced only during the power stroke. But the engine has to run for the whole cycle on that energy produced during the single power stroke. When the flywheel absorbs energy, the speed increases and when it releases energy the speed decreases. Flywheel reduces the fluctuations in speed during the cycle of operation by absorbing the excess energy and supplying when there is deficit in the energy. Let N_1 (or ω_1) and N_2 (or ω_2) represent the maximum and minimum speeds during the cycle, N (or ω) the mean speed.

5.4.1 Coefficient of Fluctuation of Speed

The coefficient of fluctuation of speed

$$(C_s) = \frac{(N_1 - N_2)}{N} \quad \text{or} \quad \frac{(\omega_1 - \omega_2)}{\omega} = C_s$$

where N is the mean speed equal to $\frac{(N_1 + N_2)}{2}$ or ω is $\frac{(\omega_1 + \omega_2)}{2}$. It can also be given as $\frac{(v_1 - v_2)}{v}$ in terms of linear speed.

$$\text{Coefficient of steadiness} = \left(\frac{1}{C_s}\right) = \frac{N}{(N_1 - N_2)}$$

5.4.2 Energy Stored in the Flywheel (E)

Let m be the mass of the flywheel in kg, k be the radius of gyration in metres and I be the mass moment of inertia (mk^2) in kg-m^2. Then energy stored in the flywheel

$$E = \frac{I\omega^2}{2} \quad \text{or} \quad \frac{mk^2\omega^2}{2}$$

ΔE = maximum kinetic energy − minimum kinetic energy

$$= \frac{I\omega_1^2}{2} - \frac{I\omega_2^2}{2} = \frac{I(\omega_1^2 - \omega_2^2)}{2}$$

$$= \frac{I(\omega_1 + \omega_2)(\omega_1 - \omega_2)}{2} = I\omega(\omega_1 - \omega_2) \tag{1}$$

Since, $\omega = \frac{(\omega_1 + \omega_2)}{2}$

Multiplying and dividing (1) by ω

$$\Delta E = I\omega^2 \left(\frac{\omega_1 - \omega_2}{\omega}\right)$$

$$= I\omega^2 C_s = 2EC_s \tag{2}$$

since $C_s = \frac{(\omega_1 - \omega_2)}{\omega}$ and $E = \frac{I\omega^2}{2}$

Thickness of rim will be generally small compared to radius of the flywheel R, therefore by taking the radius of gyration k as equal to R, moment of inertia $I = m.R^2$

$$\Delta E = m.R^2 \omega^2 C_s = mv^2 C_s \quad (3)$$

since $\omega.R = v$

As ω is equal to $\dfrac{2\pi N}{60}$ and $I = \dfrac{1}{2}mv^2$ or $mv^2 = 2I$ $\qquad \therefore \Delta E = 2IC_s$

From (1)
$$\Delta E = I \times \dfrac{2\pi N}{60} \left[\dfrac{2\pi N_1}{60} - \dfrac{2\pi N_2}{60} \right]$$
$$= \dfrac{4\pi^2 IN(N_1 - N_2)}{3600}$$
$$\Delta E = \dfrac{\pi^2 mk^2 N^2 C_s}{900} \quad (4)$$

since $C_s = \dfrac{(N_1 - N_2)}{N}$

5.4.3 Design of Flywheel

Determination of the various dimensions of the flywheel such as rim thickness t, rim width b and radius R, etc.

Pic. 5.3 Flywheels.

The Pic. 5.3 shows the flywheels generally used. Their shapes, arms and rim sizes, etc. can be seen clearly to get an idea of them. Consider a small elemental volume as shown in Fig. 5.5 equal to $A.Rd\theta$ where A is the area of cross section of the rim.

Then the elemental mass $(dm) = \rho A.Rd\theta$,

Elemental centrifugal force (dF) due to the elemental mass rotating at ω

$$dF = dm\omega^2 R = \rho A.R^2\omega^2 d\theta$$

Vertical component of it is $dF.\sin\theta$.

Total vertical force tending to burst the rim = $\int dF.\sin\theta$ between the limits 0 and π is $dF(-\cos\theta)$ between the limits 0 and $\pi = 2\rho A.R^2\omega^2$.

This must be equal to resisting force $2P = 2\sigma A$ where σ is the safe centrifugal stress or hoop stress or circumferential stress or tensile stress (Newton/metre^2) = $\rho.R^2\omega^2 = \rho v^2$.

Therefore,

$$v = \sqrt{\frac{\sigma}{\rho}}$$

Mass of rim (m) = $\pi DA\rho$: hence area of cross section of rim (A) = $\dfrac{m}{\pi D\rho} = b \cdot t$.

If $\dfrac{b}{t}$ ratio is known, then b and t can be determined.

Note: D is diameter equal to $2R$

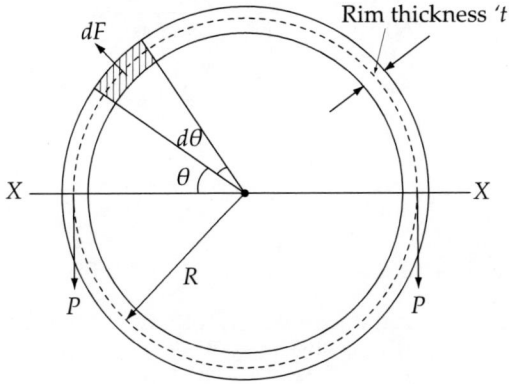

Fig. 5.5

5.5 TYPICAL WORKED EXAMPLES

There are mainly four types of problems.

1. $T - \theta$ diagram with areas of loops given.
2. $T - \theta$ diagram's description given.
3. Equation of the $T - \theta$ diagram given.
4. $T - \theta$ diagram shapes such as triangles with base and altitudes given.

W E 5.1: $T - \theta$ diagram with areas of loops given: The turning moment diagram for a multi-cylinder engine has been drawn to a scale of 1 mm = 500 N-m torque and 1 mm = 6° of crank displacement.

The intercepted areas between output torque curve and mean resistance line taken in order from one end, in sq. mm are −30, +410, −280, +320, −330, +250, −360, +280, −260 sq. mm when the engine is running at 800 r.p.m.

The engine has a stroke of 300 mm and the fluctuation of speed is not to exceed ± 2% of the mean speed. Determine a suitable diameter and cross section of the flywheel rim for a limiting value of the safe centrifugal stress of 7 MPa. The material density may be assumed as 7200 kg/m^3. The width of the rim is to be 5 times the thickness.

Given:

$N = 800$ r.p.m. or $\omega = 2\pi 800/60 = 83.8$ rad/sec; stroke = 300 mm = 0.3 m;

$\omega_1 - \omega_2 = 4\%\ \omega$ or $C_S = \dfrac{(\omega_1 - \omega_2)}{\omega} = 4\% = 0.04$; $\sigma = 7$ MPa $= 7 \times 10^6$ N/m^2; $\rho = 7200$ kg/m^3

Let the diameter of the flywheel rim be D in metres and v be the peripheral velocity of the flywheel rim in m/s.

The centrifugal stress (σ)

$$\sigma = 7 \times 10^6 = \rho v^2 = 7200 v^2$$

or

$$v^2 = 7 \times \dfrac{10^6}{7200} = 972.2$$

Therefore, $v = 31.2$ m/sec; since $v = \dfrac{\pi DN}{60}$; $D = \dfrac{v.60}{\pi.N} = \dfrac{31.2 \times 60}{(\pi.800)} = 0.745$ m or 745 mm.

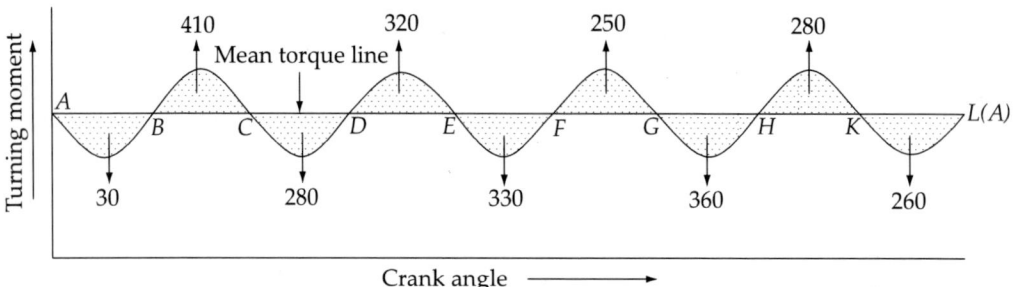

Fig. 5.6

Cross section of the flywheel rim: Let t be the thickness of the rim and b be the width of the flywheel rim in metres which is given as $5t$, therefore, the cross-sectional area of flywheel rim, $A = b.t = 5t \times t = 5t^2$.

Mass of the flywheel rim (m): From the turning moment diagram shown at Fig. 5.6, the scale in the y-axis is 1 mm = 500 N-m and crank angle scale is 1 mm = 6° = $\dfrac{\pi}{30}$ radians, therefore, one square millimetre (1 mm^2) on the turning moment diagram = $500 \times \dfrac{\pi}{30} = 52.37$ N-m.

Let the energy at $A = E_n$, then referring to Fig. 5.6

Energy at $B = E_n - 30$; — Here energy is minimum.
Energy at $C = E_n - 30 + 410 = E_n + 380$;

Energy at $D = E_n + 380 - 280 = E_n + 100$;
Energy at $E = E_n + 100 + 320 = E_n + 420$; — Here energy is maximum.
Energy at $F = E_n + 420 - 330 = E_n + 90$;
Energy at $G = E_n + 90 + 250 = E_n + 340$;
Energy at $H = E_n + 340 - 360 = E_n - 20$;
Energy at $K = E_n - 20 + 280 = E_n + 260$;
Energy at $L = E_n + 260 - 260 = E_n =$ Energy at A.

The maximum fluctuation of energy ΔE = Maximum energy − Minimum energy = $(E_n + 420) - (E_n - 30) = 450$ mm^2 = $450 \times 52.37 = 23566$ N-m.

The maximum fluctuation of energy ΔE is also equal to $mv^2 C_s = m(31.2)^2 \times 0.04 = 39m$. Therefore, $m = \dfrac{23566}{39} = 604$ kg.

Mass of the flywheel rim (m), 604 = volume × density = $\pi D A \rho = \pi \times 0.745 \times 5t^2 \times 7200 = 84{,}268 t^2$.

Therefore, $t^2 = \dfrac{604}{84{,}268} = 0.00717$ m^2 or $t = 0.085$ m = 85 mm and $b = 5t = 5 \times 85 = 425$ mm.

W E 5.2: $T - \theta$ diagram's description given: A shaft fitted with a flywheel rotates at 250 r.p.m. and drives a machine. The torque of machine varies in a cyclic manner over a period of 4 revolutions. The torque rises from 800 N-m to 3000 N-m uniformly during first half revolution and remains constant for one revolution. During the next one revolution, the torque decreases uniformly from 3000 N-m to 800 N-m. During the remaining revolutions, the torque remains constant. Thus, the cycle is completed. A flywheel of mass 1800 kg and radius of gyration of 500 mm is fitted to the shaft.

Determine the power required for driving the machine and percentage fluctuation in speed.

Given:
$N = 250$ r.p.m., or $\omega = \dfrac{2\pi 250}{60} = 26.2$ rad/s; $\quad m = 1800$ kg; $\quad k = 500$ mm = 0.5 m.

The turning moment diagram for the complete cycle is shown in Fig. 5.7.

Work done for one complete cycle = Area under the curve $OABCDEF$

$\qquad$ = Area of rectangle $OAEF$ + Area of triangle ABL+

$\qquad$ Area of rectangle $LBCM$ + Area of triangle MCD

$\qquad$ = $OF \times OA + \dfrac{AL \times BL}{2} + LM \times CM + \dfrac{MD \times CM}{2}$

$\qquad$ = $8\pi \times 800 + \dfrac{\pi(3000 - 800)}{2} + 2\pi(3000 - 800)$

$\qquad\quad + \dfrac{2\pi(3000 - 800)}{2}$ $\hfill$ (1)

$\qquad$ = $14{,}100\pi$ N-m. $\hfill$ (2)

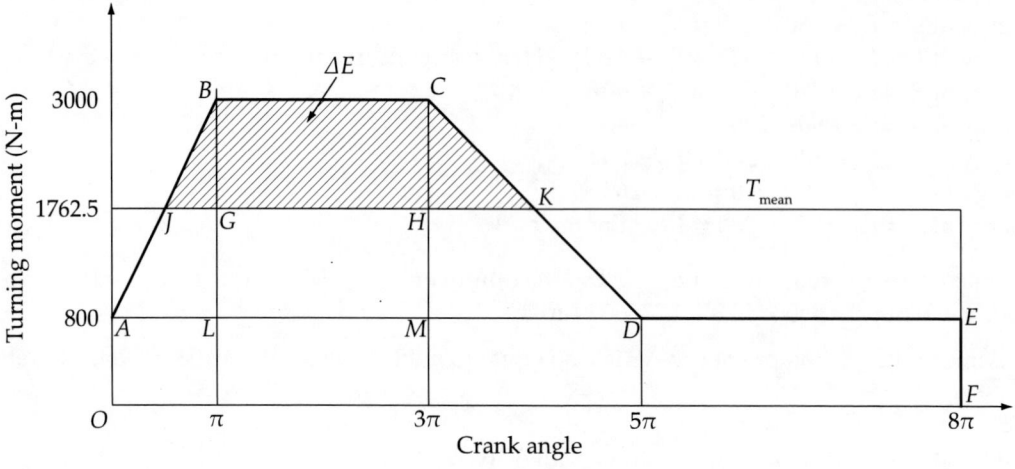

Fig. 5.7

Let T_{mean} be the mean torque in N-m, then work done for one complete cycle

$$\text{Work done} = T_{mean} 8\pi \text{ N-m} \qquad (3)$$

By equating the above two equations $T_{mean} = \dfrac{14{,}100\pi}{8\pi} = 1762.5$ N-m

Power required to drive the machine

$$P = T_{mean} \times \omega$$
$$= 1762.5 \times 26.2 = 46{,}142 \text{ W}$$
$$= 46.142 \text{ kW}$$

Coefficient of fluctuation of speed (C_S): To find the values of JG and HK, consider the similar triangles ABL and JBG,

$$\dfrac{AL}{JG} = \dfrac{BL}{BG} \quad \text{or}$$

$$\dfrac{\pi}{JG} = \dfrac{(3000 - 800)}{(3000 - 1762.5)} = \dfrac{2200}{1237.5} \quad \text{or}$$

$$JG = 1.767$$

From the similar triangles CMD and CHK, $\dfrac{MD}{HK} = \dfrac{CM}{CH}$ or

$$\dfrac{2\pi}{HK} = \dfrac{(3000 - 800)}{(3000 - 1762.5)}$$
$$= \dfrac{2200}{1237.5} \quad \text{or}$$
$$HK = 3.534$$

Since the area above the mean torque line represents the maximum fluctuation of energy,

$$\Delta E = \text{Area } JKCB = \text{Area } JGB + \text{Area } GHCB + \text{Area } CHK$$
$$= \frac{JG \times BG}{2} + GH \times BG + \frac{CH \times HK}{2}$$
$$= (3000 - 1762.5)\left[\frac{1.767}{2} + 2\pi + \frac{3.534}{2}\right]$$
$$= 11.055 \text{ N-m}$$

The maximum fluctuation of energy (ΔE)

$$\Delta E = mk^2\omega^2 C_S$$
$$= 1800(0.5)^2(26.2)^2 \cdot C_S$$
$$\therefore \quad C_S = 0.0358 \text{ or } 3.58\%$$

W E 5.3: Equation of the $T - \theta$ diagram given: The turning moment curve of a two-stroke engine (cycle repeats for 180° of crank rotation) is represented by the equation $T = (8000 + 1000 \sin 2\theta - 2000 \cos 2\theta)$ N-m, where θ is the angle moved by the crank from inner dead centre. The mass of the flywheel is 500 kg and its radius of gyration is 75 cm. The engine speed is 300 r.p.m. Assuming external resistance as constant, determine: (1) power developed by the engine; (2) the total percentage fluctuations in speed; and (3) the maximum angular retardation of the flywheel.

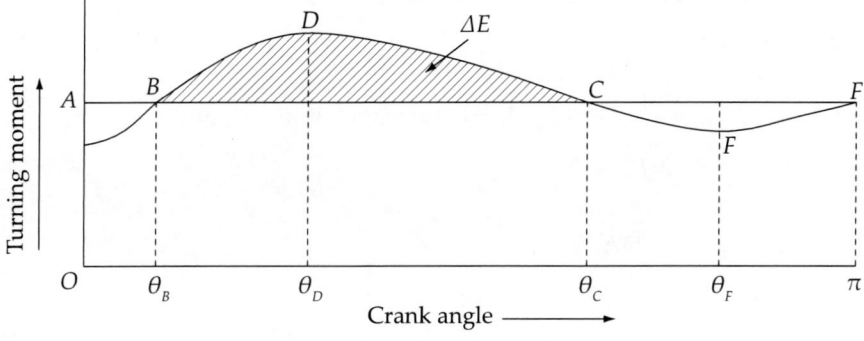

Fig. 5.8

Given:

$$T = (8000 + 1000 \sin 2\theta - 2000 \cos 2\theta) \text{ N-m}$$
$$N = 300 \text{ r.p.m.; or}$$
$$\omega = \frac{2\pi 300}{60} = 31.42 \text{ rad/sec.}$$

The total fluctuation of speed $(\omega_1 - \omega_2)$ and the coefficient of fluctuation of speed, $C_S = \frac{(\omega_1 - \omega_2)}{\omega}$ where ω_1, ω_2 and ω are the maximum, minimum and mean angular velocities.

1. Power developed by the engine

Work done per revolution $= \int T d\theta$ between the limits 0 and π

$$= \int_0^\pi (8000 + 1000 \sin 2\theta - 2000 \cos 2\theta) d\theta$$

$= [8000\theta - 500 \cos 2\theta - 1000 \sin 2\theta]$ between the limits 0 and π

$= 8000\pi$ N-m

$\therefore T_{mean} = 8000$ N-m

Power developed by the engine $= T_{mean} \times \omega = 8000 \times 31.42 = 2{,}51{,}000$ W $= 251$ kW

2. Total percentage fluctuations in speed

Mean torque

$$T_{mean} = \frac{8000\pi}{\pi}$$
$$= 8000 \text{ N-m}$$

From Fig. 5.8, $T_{mean} = T$ at points B and C.

Therefore, $8000 = 8000 + 1000 \sin 2\theta - 2000 \cos 2\theta$ or

$\tan 2\theta = 2$ or $2\theta = 63°26'$ or $180° + 63°26'$ that is $\theta_B = 31°43'$, $\theta_C = 121°43'$

Maximum fluctuation of energy,

$$\Delta E_{max} = \int (T - T_{mean}) d\theta \text{ between the limits } \theta_C \text{ and } \theta_B$$
$$= \int_{\theta_B}^{\theta_C} (8000 + 1000 \sin 2\theta - 2000 \cos 2\theta - 8000) d\theta$$
$$= 2237.3 \text{ N-m}$$

The maximum fluctuation of energy (ΔE):

$$2237.3 = I\omega^2 . C_S = I(31.42)^2 C_S$$
$$\therefore C_S = \frac{2237.3}{[500 \times (0.75)^2 \times (31.42)^2]}$$
$$= 0.0081 \text{ or } 0.81\%$$

3. Angular acceleration of the flywheel (α)

In order to determine the angular acceleration of the flywheel, the maximum torque and minimum torque are required which can be obtained by differentiating T with respect to θ and equating to zero.

i.e. $\dfrac{dT}{d\theta} = 0$;

$$(2000 \cos 2\theta + 4000 \sin 2\theta) = 0$$
$$\tan 2\theta = -0.5$$
$$2\theta = 153°26' \text{ or } 333°26'$$

By substituting in the expression for T, these values give the

$$T_{max} = 10239 \quad \text{and} \quad T_{min} = 5764 \text{ N-m}$$

The retardation occurs from T_{mean} to T_{min}, hence $T_{mean} - T_{min} = I\alpha_{max}$ where α_{max} is the maximum retardation.

$$\alpha_{max} = \dfrac{(8000 - 5764)}{[500 \times (0.75)^2]} = 7.95 \text{ rad/sec}^2$$

W E 5.4: $T - \theta$ diagram shapes such as triangles with base and altitudes given: The turning moment diagram of a four-stroke cycle gas engine may be assumed for the sake of simplicity to be represented by four triangles in each cycle. The areas of these triangles are from the line of zero pressure are as follows: suction stroke = 0.45×10^{-3} m²; compression stroke = 1.7×10^{-3} m²; expansion stroke = 6.8×10^{-3} m²; exhaust stroke = 0.65×10^{-3} m². Each m² of area represents 3M N-m of energy. Assuming the resisting torque to be uniform, find the mass of the rim of a flywheel required to keep the speed between 202 and 198 r.p.m. The mean radius of the rim is 1.2 m.

Given:

$a_1 = 0.45 \times 10^{-3}$ m²; $\quad a_2 = 1.7 \times 10^{-3}$ m²

$a_3 = 6.8 \times 10^{-3}$ m² and $a_4 = 0.65 \times 10^{-3}$ m²

$N_1 = 202$ r.p.m.; $N_2 = 198$ r.p.m.; $R = 1.2$ m.

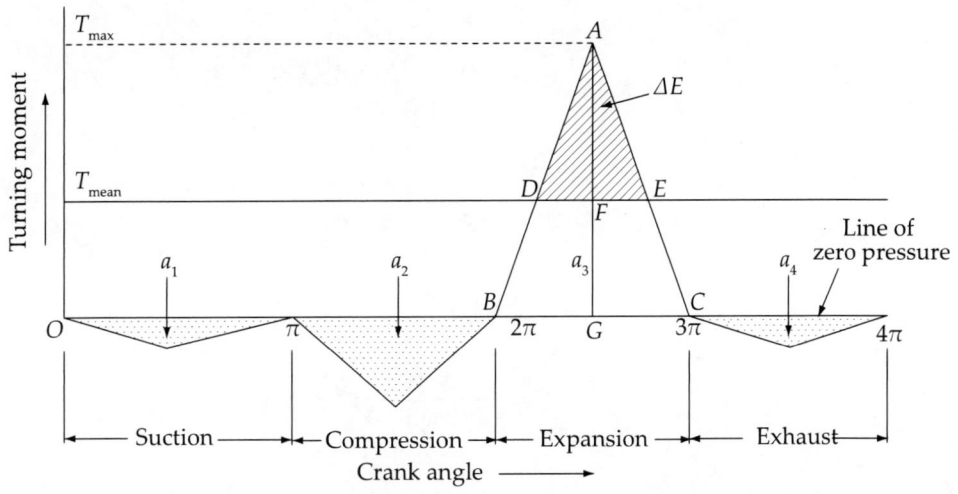

Fig. 5.9

From the turning moment and crank angle diagram of four-stroke engine shown at Fig. 5.9. The areas below the zero line of pressure are taken as negative ($-ve$) while the areas above the zero line are taken as positive ($+ve$).

Net area = $a_3 - (a_1 + a_2 + a_4) = 4 \times 10^{-3}$ m^2;

One square metre = 3×10^6 N-m of work.

$$\text{Therefore, net work done per cycle} = 4 \times 10^{-3} \times 3 \times 10^6 = 12{,}000 \text{ N-m} \tag{1}$$

$$\text{But work done per cycle} = T_{mean} \times 4\pi \text{ N-m} \tag{2}$$

Therefore,

$$T_{mean} = \frac{12{,}000}{4\pi} = 955 \text{ N-m}$$

$$\text{Work done during expansion stroke} = a_3 \times \text{scale} = 6.8 \times 10^{-3} \times 3 \times 10^6$$
$$= 20.4 \times 10^3 \text{ N-m} \tag{3}$$

But work done during expansion stroke = Area of triangle ABC

$$= \frac{AG \times BC}{2} = \frac{AG \times \pi}{2} \tag{4}$$

Therefore, from (3) and (4)

$$AG = \frac{20.4 \times 10^3}{(\pi/2)}$$
$$= 12{,}985 \text{ N-m} = \text{Maximum torque } (T_{max})$$

$$\text{Excess torque } (T_{excess}) = T_{max} - T_{mean}$$
$$= AG - FG = 12985 - 955$$
$$= 12{,}030 \text{ N-m}$$

From the similar triangles ADE and ABC,

$$\frac{DE}{BC} = \frac{AF}{AG}$$

$$DE = \frac{AF \times BC}{AG}$$
$$= \frac{12{,}030 \times \pi}{12{,}985} = 2.9 \text{ radians.}$$

Maximum fluctuation of energy

$$\Delta E = \text{Area } ADE = \frac{DE \times AF}{2}$$

$$= \frac{2.9 \times 12030}{2} = 17444 \text{ N-m}$$

Mass of the rim of the flywheel (m):

$$\text{Mean speed, } N_{mean} = \frac{(N_1 + N_2)}{2}$$

$$= \frac{(202 + 198)}{2} = 200 \text{ r.p.m.}$$

Max. fluctuation of energy (ΔE)

$$\Delta E = 17444 = I\omega^2 C_s$$

$$17444 = \pi^2 \times m \times R^2 \times N_{mean} \frac{(N_1 - N_2)}{900}$$

$$17444 = \pi^2 \times m \times (1.2)^2 \times 200 \frac{(202 - 198)}{900}$$

$$17444 = 12.63 m.$$

Hence, mass $m = \dfrac{17444}{12.36} = 1381$ kg.

5.6 FLYWHEEL IN PUNCHING PRESS

Differences between the Engine and Machine

	Torque	Load	Speed
Engine	Varies	Constant	Fluctuates
Machine	Constant	Varies	Fluctuates

Generally machines are driven by motors or by IC Engines. Example: Riveting or cutting machines are driven by motors.

The energy supplied by motor is constant. When the load is applied on the machine, the speed decreases and when the load decreases, the speed increases. Drop in speed will be large when the load is suddenly applied.

Maximum shear force F_S = Area sheared × Ultimate shear stress = $\pi\, dt\, \tau_u$ where τ_u is the shear stress, d is the diameter of the hole, t is the thickness of the plate. As the hole is punched, the shear force decreases uniformly from maximum value to zero.

Therefore, work done or energy required for punching a hole = $\dfrac{F_s t}{2} = E_1$ = Energy supplied or required per revolution for punching one hole.

Assuming one punching operation per revolution, the energy supplied to the shaft per revolution should be equal to E_1. The energy supplied by the motor to the crankshaft during actual punching operation, $E_2 = \dfrac{E_1(\theta_2 - \theta_1)}{2\pi}$.

Balance energy required for punching

$$E_1 - E_2 = E_1 - \dfrac{E_1(\theta_2 - \theta_1)}{2\pi}$$

$$= E_1\left[1 - \dfrac{(\theta_2 - \theta_1)}{2\pi}\right]$$

This energy is to be supplied by the flywheel by the decrease in its kinetic energy when its speed falls from maximum to minimum. Thus, maximum fluctuation of energy,

$$\Delta E = E_1 - E_2 = E_1\left[1 - \dfrac{(\theta_2 - \theta_1)}{2\pi}\right]$$

The energy remaining during $[2\pi - (\theta_2 - \theta_1)]$,

$$E_1 - E_2 = E_1\left[1 - \dfrac{(\theta_2 - \theta_1)}{2\pi}\right]$$

which is to be supplied by the flywheel.

Since $\left[\dfrac{(\theta_2 - \theta_1)}{2\pi}\right] = \dfrac{t}{2s} = \dfrac{t}{4r}$ where s is the stroke $= 2r$. One revolution $= 2s = 4r$.

Therefore, fluctuation of energy

$$\Delta E = E_1\left[1 - \dfrac{t}{(4r)}\right]$$

where $E_1 = \dfrac{F_s t}{2}$ and F_S is the shear force.

Knowing ΔE, mass and size of flywheel has to be determined.

W E 5.5: A machine punching 38 mm holes in 32 mm thick plate requires 7 N-m of energy per sq. mm of sheared area, and punches one hole in every 10 seconds. Calculate the power of the motor required. The mean speed of the flywheel is 25 m/s. The punch has a stroke of 100 mm. Find the mass of the flywheel required, if the total fluctuation of speed is not to exceed 3% of the mean speed. Assume that the motor supplies energy to the machine at uniform rate.

Given:
$d = 38$ mm; $\quad t = 32$ mm; $\quad E_1 = 7$ N-m/mm² of sheared area; $\quad v = 25$ m/s; stroke, $s = 100$ mm; $\quad v_1 - v_2 = 3\% = 0.03v$; $\quad C_s = \dfrac{v_1 - v_2}{v} = 0.03$.

Power of the motor required:
Sheared area $A = \pi d t = \pi \times 38 \times 32 = 3820 \text{ mm}^2$;

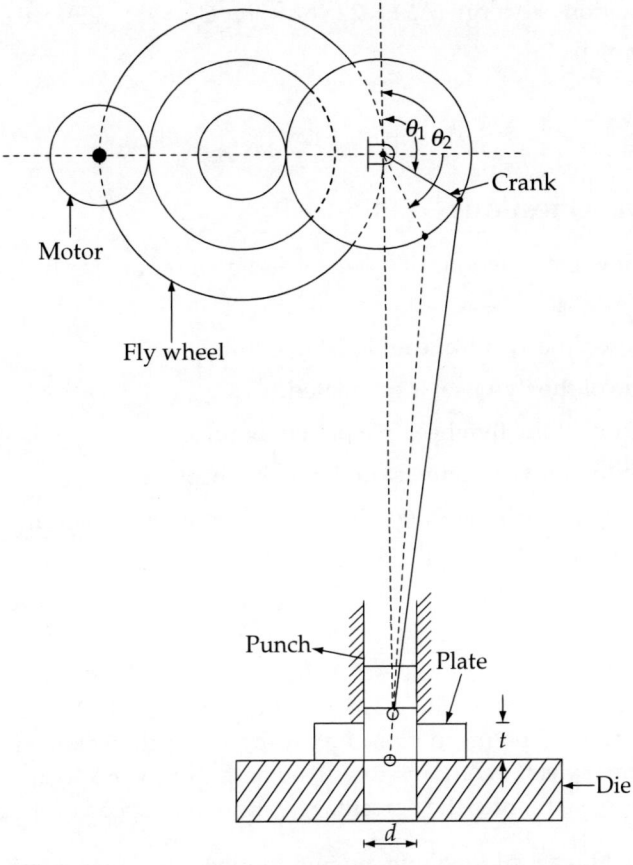

Fig. 5.10

Since the energy required for punching the hole = 7 N-m/mm², therefore, total energy required per hole, $E_1 = 7 \times 3820 = 26740$ N-m.

Time required to punch the hole is 10 seconds, therefore energy required for punching work per second = $\dfrac{26740}{10} = 2674$ N-m/s, therefore, the power of the motor required = 2674 W = 2.674 kW.

Mass of the flywheel (m): The stroke of the punch is 100 mm and it punches one hole in every 10 seconds, therefore, the time required to punch a hole in a 32 mm thick plate

$$= \frac{10 \times 32}{2 \times 100} = 1.6 \text{ sec}$$

Energy supplied by the motor in 1.6 seconds, $E_2 = 2674 \times 1.6 = 4278$ N-m.

Energy to be supplied by the flywheel during punching or the maximum fluctuation of energy

$\Delta E = E_1 - E_2 = 26{,}740 - 4278 = 22{,}462$ N-m;

Coefficient of fluctuation of speed $C_S = 0.03$.

Since maximum fluctuation of energy $(\Delta E) = 22{,}462 = mv^2 C_S = m \times (25)^2 \times 0.03 = 18.75\, m$, therefore mass, $m = \dfrac{22{,}462}{18.75} = 1198$ kg.

5.7 EXERCISE

5.7.1 Short Answer Questions

1. Explain clearly how the functions of flywheel and governor differ from each other in a steam engine.
2. Explain inertia force, inertia torque and piston effort.
3. State how the size of the flywheel is calculated.
4. What is the function of the flywheel in a punching press?
5. Derive an expression for the energy stored in a flywheel.
6. Write short notes on the following: (i) Turning moment diagram (ii) Piston effort and (iii) Co-efficient of fluctuation of speed.

5.7.2 Problems

1. A gas engine working on Otto cycle is provided with two flywheels each weighing 5800 N and radius of gyration 56 cm. The diameter of the cylinder is 24 cm, stroke 27 cm and the mean speed 250 r.p.m. The mean pressure during the cycle reckoned above atmosphere is: suction stroke-atmospheric, compression–10.6 N/cm², firing–62 N/cm², Exhaust–3 N/cm². If the resistance be constant, find the percentage variation of speed of the engine.

2. The torque exerted by a multi-cylinder engine running at a mean speed of 240 r.p.m. against a uniform resistance is as below. Find the power of the engine and the minimum weight of the flywheel if its radius of gyration is 90 cm and the maximum fluctuation of speed is to be ± 1% of the mean.

$$T = 3500 + 5600 \sin\theta + 840 \sin 2\theta + 84 \sin 3\theta \text{ N-m}$$

3. The torque exerted on the crankshaft of an engine is given by the equation $T = 17{,}150 - 8400 \sin\theta + 2100 \sin 3\theta$ N-m where θ is the crank angle. Find the radius of gyration of the flywheel whose weight is 12.5 kN, if the variation in speed is not to exceed ± 1.25% of the mean speed of 250 r.p.m. Also find the power developed.

4. An Otto cycle engine develops 50 kW at 150 r.p.m. with 75 explosions per minute. The change of speed from the commencement to the end of power stroke must not exceed 0.5% of mean on either side. Find the mean diameter of the flywheel and a suitable rim cross section having width four times the depth so that the hoop stress does not exceed 4 MPa. Assume that the

flywheel stores 16/15 times the energy stored by the rim and the work done during power stroke is 1.40 times the work done during the cycle. Density of rim material is 7200 kg/m^3.

5. A single cylinder four-stroke oil engine develops 15 kW at a speed of 400 r.p.m. and drives a machine at 750 r.p.m. The engine shaft carries a flywheel with moment of inertia of 114 kg-m^2. The machine shaft also carries a flywheel with the moment of inertia of 8 kg-m^2. If the fluctuation of energy is 80 and if the coefficient of fluctuation of speed is required to be lowered to a total value of 1%, what is the moment of inertia of the additional rotating mass to be fitted to the machine shaft?

Punching machine

1. A punching machine operates at the rate of 600 holes per hour. It does 45 N-m of work per mm^2 of sheared area in cutting 25 mm diameter hole in a 3 mm thick plate. A constant torque motor operates the machine. The speed of the machine fluctuates between 250 and 230 r.p.m. The frictional losses are 20% of work done during punching and actual punching time per hole is 2 seconds. The radius of gyration of flywheel as 500 mm. Find: (i) The power required to drive punching machine. (ii) The maximum fluctuation of energy. (iii) The mass of the flywheel required keeping the speed fluctuation in the given range.

2. A punching machine is required to punch 2 cm diameter holes in 1.5 cm thick plates having ultimate shear stress of 3200 kg/cm^2. If 30 holes are to be punched per minute and if punching operation requires 1/10th of a second, find moment of inertia of a suitable flywheel in order that the speed lies between 141 and 159 r.p.m.

5.7.3 Multiple Choice Questions

1. The maximum fluctuation of energy in a flywheel is equal to
   (a) $I\omega(\omega_1 - \omega_2)$ (b) $I\omega^2 C_S$ (c) $2EC_S$ (d) all of these. [**Ans.** (b)]

2. The ratio of maximum fluctuation of energy to the _____ is called coefficient of fluctuation of energy.
   (a) minimum fluctuation of energy   (b) work done per cycle
   (c) both   (d) none. [**Ans.** (b)]

3. The ratio of the maximum fluctuation of speed to the mean speed is called
   (a) fluctuation of speed   (b) maximum fluctuation of speed
   (c) coefficient of fluctuation of speed (d) none of these. [**Ans.** (c)]

4. In a turning moment diagram, the variation of energy above and below the mean resisting torque line is called
   (a) fluctuation of energy   (b) maximum fluctuation of energy
   (c) coefficient of fluctuation of energy (d) none of the above. [**Ans.** (a)]

5. The maximum fluctuation of energy is the

   (a) sum of maximum and minimum energies
   (b) difference between the maximum and minimum energies
   (c) ratio of the mean resisting torque to the work done per cycle.
   (d) product of the mean resisting torque to the work done per cycle. [*Ans.* (b)]

6. The function of the flywheel is

   (a) to match the input energy to the engine at constant revolutions per minute.
   (b) to smoothen cyclic variations of speed of the engine
   (c) to smoothen the cyclic variations of energy output from the engine
   (d) to maintain constant uniform speed of the engine. [*Ans.* (b)]

7. Safe peripheral velocity of a flywheel is given by

   (a) $V = \left[\dfrac{f}{\rho}\right]^{1/2}$ (b) $V = \left[\dfrac{\rho}{f}\right]^{1/2}$ (c) $V = \left[\dfrac{f}{mg}\right]^{1/2}$ (d) $V = \left[\dfrac{mg}{f}\right]^{1/2}$ [*Ans.* (a)]

8. For the same maximum fluctuation of energy of a flywheel if the mean speed of rotation is more

   (a) the size of flywheel is reduced
   (b) the size of flywheel is increased
   (c) the size of flywheel is not dependent on the speed and is unaffected.
   (d) maximum fluctuation of energy of flywheel cannot be same. [*Ans.* (a)]

9. For the same mass of flywheel

   (a) the disc type flywheel is preferable
   (b) the rim type flywheel is preferable
   (c) any type is equally preferable.
   (d) preference will depend upon the type of the prime mover. [*Ans.* (b)]

10. Which one is the correct statement?

    (a) The flywheel influences the mean speed of prime mover.
    (b) The flywheel influences the variation of load demand on prime mover.
    (c) The flywheel influences the cyclic variation of the prime mover.
    (d) The flywheel influences the mean torque developed by the prime mover. [*Ans.* (c)]

Friction 6

Friction

6.1 INTRODUCTION

The force offering resistance to relative motion between two surfaces in contact is called friction. In simple words, friction opposes motion. Friction is present between two solids in contact. There are both advantages as well as disadvantages of friction.

Note: Friction opposes motion but there is no motion without friction. For example, scooters, cars, etc., and even walking cannot take place without friction as shown in Pic. 6.1.

Pic. 6.1 More energy is needed to make the bicycle move faster.

The cause of friction: At the atomic level every surface will have depressions (d) and elevations (e) as shown in Fig. 6.1(a). When one surface is in contact with another, the elevations in one surface will set into the depressions in the other as shown in Fig. 6.1(b).

This interlocking of two surfaces opposes or hinders or resists the relative motion between the two surfaces. This resisting force which acts in the opposite direction of the movement is called force of friction or simply friction. Hence,

1. The force of friction acts in the opposite direction to that of the relative motion.
2. It will be tangential to the surfaces of the two bodies at the point of contact.

Friction always causes wear and tear and also causes loss or waste of energy which can be reduced only by polishing the surfaces or by the introduction of some lubricants between the surfaces to minimize friction so as to slide more freely and easily. There are two types of frictions:

1. **Static friction:** It is the friction experienced by a body when it is at rest or the opposing force on the object when it is stationary or when the object is not moving.

2. **Dynamic friction:** It is the friction experienced by a body when it is in motion or opposing force arising due to the motion of the object.

Again the dynamic friction is of two types:

1. **Sliding friction:** It is the friction, experienced by a body when it slides over another body.
2. **Rolling friction:** It is the friction, experienced by a body when it rolls over another body.

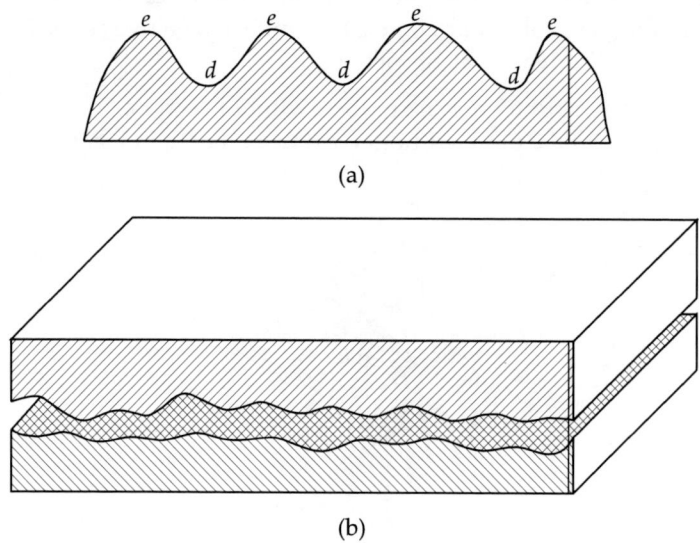

Fig. 6.1

6.2 LAWS OF FRICTION

Consider a block resting on the floor as shown in Fig. 6.2. Due to the gravitational force, contact force arises between the two surfaces. These are action and reaction forces equal and opposite. In considering the laws which govern the friction between two surfaces, it is therefore necessary to distinguish between the three possible states of the surfaces: (i) dry; (ii) greasy or partially lubricated; and (iii) film or completely lubricated.

When two solid bodies with smooth, dry surfaces are in contact, the least force required in order to make one body to slide over the other, obeys approximately the following laws:

1. The friction force is directly proportional to the normal load between the surfaces for a given pair of materials.
2. The friction force depends upon the material of which the contact surfaces are made.
3. The friction force is independent of the area of the contact surfaces for a given normal load.
4. The friction force is independent of the velocity of sliding of the one body relative to the other body.

6.2.1 Friction between Dry Surfaces

As shown in Fig. 6.2(a), the contact forces will be along the common normal to the surfaces. The downward force exerted by the body on the floor is its weight (W), i.e. mg. The upward force exerted by the floor on the body is called the normal reaction R_n. Hence, $R_n = mg$.

6.2.2 Friction between Rough Surfaces

When the surfaces are rough, the upward contact force will be inclined to the common normal as in Fig. 6.2(b).

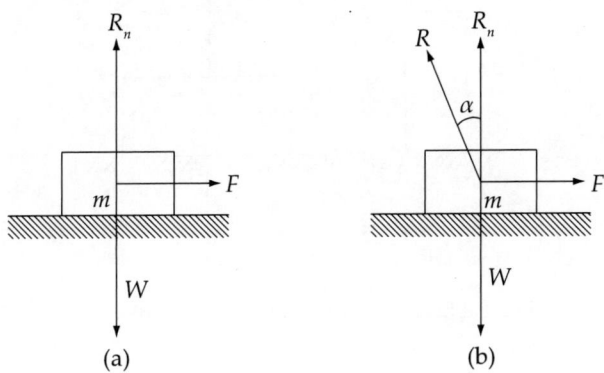

Fig. 6.2

Here the contact force R has a component along the upward normal $R \cos \alpha$. This is called the normal reaction R_n and is balanced by the vertically downward weight (w) = mg of the body. The horizontal component of R parallel to the surface $R \sin \alpha$ is the frictional force F that opposes the motion of body towards right. Hence, the contact force R can be called the resultant of R_n and F. The angle α between R and R_n is called the angle of friction.

6.2.3 Friction is Self Adjusting

Suppose a small force of dF is applied on the body and try to move it towards right. The body will not move because of the friction that opposes the motion, until dF exceeds a certain value. The body will be in static. As the force dF increases, then the friction also increases. As long as the block is in a static condition, the friction at this stage is called the static friction. Maximum value of the static friction is $F = F_{max}$ and when the applied force increases over and above the maximum value F_{max}, the body then starts moving and the friction is called the kinetic friction.

If the body is tried to move left, then the friction will be towards right side opposing the motion. Thus, the static friction adjusts its magnitude and direction.

6.2.4 Angle of Friction (ϕ)

When the static friction F attains its maximum value F_{max}, the corresponding angle α between R and R_n is called the angle of friction α is indicated by ϕ. From Fig. 6.2(b) $\dfrac{F_{max}}{R_n} \tan \phi$. This ratio $\left(\dfrac{F_{max}}{R_n}\right)$ is called the coefficient of static friction μ_s. Hence, μ_s is equal to $\tan \phi$ and it depends on (i) the materials of the surfaces in contact and (ii) the roughness of the surfaces. For a given weight of the body w (mg), F does not depend on the area of contact.

In Fig. 6.3 two bodies of equal masses having different areas of contact with the surface are shown. Even though the area of contact A_1 is less than A_2, the frictional force will be the same and the normal reaction R_n will be equal to mg in both the cases.

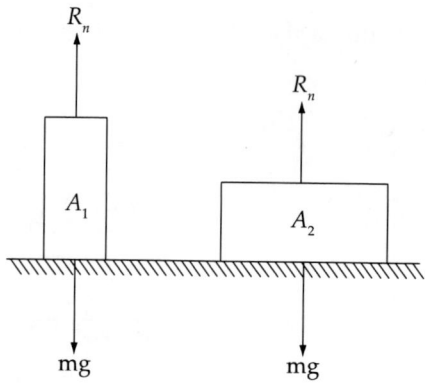

Fig. 6.3 Bodies of different areas of contact.

Note: The friction is independent of the area of contact and the speeds with which they move.

6.2.5 Rolling Friction

The resistance encountered by a rolling body on a surface is called the rolling friction. Let F_r be the component of the applied force parallel to the surface sufficient to roll the body with a constant velocity. The coefficient of rolling friction $\mu_r = \dfrac{F_r}{R_n}$.

There are both advantages and disadvantages of friction. Where friction is advantageous, its value should be higher such as for transfer of motion from one shaft to another through belts, in the friction clutches and in brake shoes of automobiles for reducing the speed of vehicles, etc.

Where friction is disadvantageous, its value should be minimum so that the power loss due to friction will be minimum (i.e., in the form of heat), especially in the bearings. The friction also causes wear and tear. In order to reduce this loss due to friction, ball bearings are used. Frictional losses also can be reduced by polishing the surfaces.

W E 6.1: A body of weight 300 N is lying on a rough horizontal plane having a coefficient of friction as 0.3. Find the magnitude of the force, which can move the body, while acting at an angle of 25° with the horizontal.

Given:
Weight of body W = 300 N;   Coefficient of friction μ = 0.3 and angle made by the force with the horizontal (θ) = 25°.

Let P be the magnitude of the force which can move the body and F be the force of friction.

Then resolving horizontally
$$F = P \cos 25° = P \times 0.9063 \tag{1}$$

Resolving vertically
$$R = W - P \sin \theta = 300 - P \sin 25°$$
$$= 300 - P \times 0.4226 \tag{2}$$

The force of friction $F = \mu R$

then from (1) and (2)
$$0.9063P = 0.3(300 - 0.4226P) \text{ giving } P = 87.1 \text{ N}$$

6.3 EQUILIBRIUM OF BODY ON A ROUGH INCLINED PLANE

Consider a body on a rough inclined plane at an angle α with the horizontal to analyse the various forces acting on the body while sliding (a) up the plane and (b) down the plane and their relationships with the force required to move the body with and without friction.

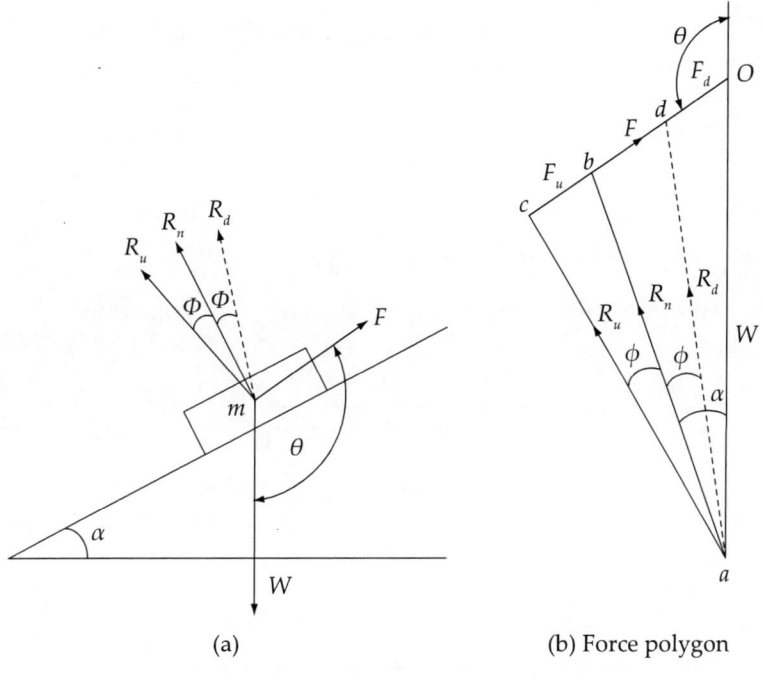

Fig. 6.4

Let W be the weight of the body; m be the mass of the body; α be the inclination of the plane to the horizontal; θ inclination of applied force F; ϕ be the limiting angle of friction; F, F_u, F_d be the forces acting on the body to slide it with uniform velocity without friction, with friction up the plane and down the plane respectively; R_n, R_u, R_d are the normal and resultant reactions when the body is at stationery, moving up and down the plane respectively as shown in the Fig. 6.4(a).

The force polygon is showed in Fig. 6.4(b) with friction for both up and down the plane.

Note: $F_u = cO; \quad F = bO; \quad F_d = dO$

6.3.1 Motion Up the Plane

Let F, W and normal reaction R_n are the three forces neglecting friction. From the triangle of forces oab as shown in Fig. 6.4(b),

$$\frac{F}{W} = \frac{bo}{oa} = \frac{\sin \angle oab}{\sin \angle oba} = \frac{\sin \alpha}{\sin(\theta - \alpha)} \tag{1}$$

When friction is considered let F_u be the force required to move up the plane. Then the reaction R_u between the plane and the body is inclined to the normal at the friction angle ϕ and the triangle of forces is oac, from Fig. 6.4(b),

$$\frac{F_u}{W} = \frac{co}{oa} = \frac{\sin \angle oac}{\sin \angle oca} = \frac{\sin(\alpha + \phi)}{\sin[\theta - (\alpha + \phi)]} \tag{2}$$

The ratio of force required without friction to the force required with friction is called the efficiency of the inclined plane as a machine and is given by:

$$\eta = \frac{F}{F_u} = \frac{\text{Equation (1)}}{\text{Equation (2)}}.$$

On simplifying gives

$$\eta = \frac{[\cot(\alpha + \phi) - \cot \theta]}{[\cot \alpha - \cot \theta]} \tag{3}$$

If $\theta = 90°$, i.e. when force is applied horizontally, then from equation (2):

$$\frac{F_u}{W} = \frac{\sin(\alpha + \phi)}{\sin[90° - (\alpha + \phi)]}$$

$$= \tan(\alpha + \phi) \tag{4}$$

and

From (3), since $\cot 90° = 0$, efficiency

$$\eta = \frac{F}{F_u} = \frac{\cot(\alpha + \phi)}{\cot \alpha} = \frac{\tan \alpha}{\tan(\alpha + \phi)} \tag{5}$$

6.3.2 Motion Down the Plane

Neglecting friction, the triangle of forces is oab, just as motion up the plane. Let F_d be the force required to move down the plane and R_d be the corresponding resultant reaction with friction as shown in Fig. 6.4(a). The triangle of forces is *oad* as shown in Fig. 6.4(b),

$$\frac{F_d}{W} = \frac{do}{oa} = \frac{\sin \angle oad}{\sin \angle oda} = \frac{\sin(\alpha - \phi)}{\sin[\theta - (\alpha - \phi)]} \quad (6)$$

The efficiency of the inclined plane as a machine is given by $\frac{F}{F_d}$, i.e. $\frac{(1)}{(6)}$.

On simplification gives

$$\eta = \frac{\cot(\alpha - \phi) - \cot\theta}{(\cot\alpha - \cot\theta)} \quad (7)$$

Note: For down the plane W acts as an effort and F or F_d acts as the resistances.

Efficiency of the inclined plane

$$\eta = \frac{F_d}{F} = \frac{(\cot\alpha - \cot\theta)}{\cot(\alpha - \phi) - \cot\theta} \quad (8)$$

If θ is 90°, i.e. when force applied is horizontal,

$$\eta = \frac{\cot(\alpha)}{\cot(\alpha - \phi)} = \frac{\tan(\alpha - \phi)}{\tan\alpha} \quad (9)$$

Angle of repose: It is the angle of inclination of the plane at which the body starts sliding down the plane due to its own weight and in other words the limiting angle of friction.

6.3.3 Maximum Efficiency

For given values of θ and ϕ there is a particular value of α which gives maximum efficiency.

For up the plane, rewriting the equation (3) as below gives

$$\text{Efficiency, } \eta = \frac{\cos(\theta - \phi - 2\alpha) - \cos(\theta - \phi)}{\cos(\theta - \phi - 2\alpha) - \cos(\theta + \phi)} \quad (10)$$

This will be maximum when $\cos(\theta - \phi - 2\alpha)$ is maximum, i.e. $(\theta - \phi - 2\alpha) = 0$. Hence, $\alpha = \frac{(\theta - \phi)}{2}$;

Maximum efficiency

$$\eta_{max} = \frac{1 - \cos(\theta - \phi)}{1 - \cos(\theta + \phi)} \quad (11)$$

When $\theta = 90°$; $\alpha = \frac{(90° - \phi)}{2} = 45° - \frac{\phi}{2}$

Maximum efficiency

$$\eta_{max} = \frac{(1 - \sin\phi)}{(1 + \sin\phi)} \quad (12)$$

W E 6.2: A body of weight 500 N is placed on a rough plane inclined at an angle of 25° with the horizontal. It is supported by an effort parallel to the plane. Determine the minimum and maximum values of the effort for which the equilibrium can exist, if the angle of friction is 20°.

Given:
$W = 500$ N and $\alpha = 25°$; and $\phi = 20°$. Here $\theta = 90° + \alpha$

1. Minimum force F_d which will keep the body in equilibrium when it is at the point of sliding downwards. Taking equation (6)

$$\frac{F_d}{W} = \frac{\sin(\alpha - \phi)}{\sin[\theta - (\alpha - \phi)]} = \frac{\sin(\alpha - \phi)}{\cos\phi}$$

Hence,
$$F_d = \frac{500 \sin 5°}{\cos 20°} = \frac{500 \times 0.872}{0.9397} = 46.4 \text{ N}$$

2. Maximum force F_u which will keep the body in equilibrium when it is at the point of sliding upwards. Taking equation (2)

$$\frac{F_u}{W} = \frac{\sin(\alpha + \phi)}{\sin[\theta - (\alpha + \phi)]} = \frac{\sin(\alpha + \phi)}{\cos\phi}$$

since $\theta = 90°$.

Hence, $F_u = \dfrac{W \sin 45°}{\cos 20°} = \dfrac{500 \times 0.7071}{0.9397} = 376.2$ N

6.4 SCREW FRICTION

The screws, bolts, studs, nuts, etc., are widely used in various applications such as machines and structures for fastening. These fastenings have screw threads, which are made by cutting a continuous helical groove on a cylindrical surface. These screw threads are mainly of two types, viz. V-threads and square threads. The friction between screw and nut depends on the type of thread.

6.4.1 Square Thread

The development of the screw thread when unwound from the body of the screw is just like an inclined plane, the inclination of the plane being equal to the helix angle of the thread as shown in Fig. 6.5. Let L be the lead of the thread or helix, i.e. the axial distance through which the nut moves in one complete turn is given on a fixed screw; let r be the mean radius of the thread and α be the lead angle or inclination of the equivalent inclined plane. Then $\tan\alpha = \dfrac{L}{2\pi r}$

Note: The rotation of the screw in the nut or the nut on the screw is equivalent to sliding on an inclined plane.

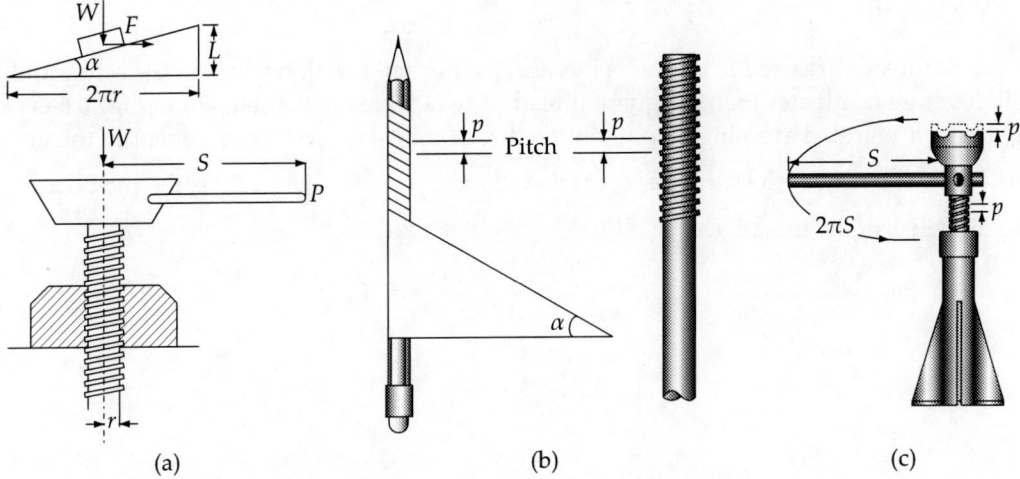

Fig. 6.5 Screw and screw jack.

6.4.2 Relation Between Effort and Weight Lifted by a Screw Jack

Figure 6.5(c) shows the screw jack with lever of length S. Let W be the axial load, F be the tangential force required at the mean radius of the screw in a plane normal to the axis of the screw; α be the inclination of the developed plane and ϕ be the angle of friction and $\tan \alpha = \dfrac{L}{2\pi r}$; where L is the lead equal to pitch of the screw p for single start and $2p$ for double start, etc.

Applying the same conditions as in the inclined plane for $\theta = 90°$, i.e. when the force is applied horizontally for up the plane by rotating the nut, the force required is F_u as seen above. The turning moment which has to be exerted on the nut is therefore $F_u r$. If an effort P_u is applied through a lever at a distance S from the axis of the screw, then

$$P_u S = F_u r = Wr \cdot \tan(\alpha + \phi)$$

If the nut is rotated in the opposite sense, the load in effect moves down the inclined plane, and the effort needed for down the incline be P_d on the lever. Then

$$P_d S = -Wr \cdot \tan(\alpha - \phi)$$

If $\alpha > \phi$, P_d will be negative, i.e. the nut will not remain at rest under the axial load W unless a torque is applied to it in order to prevent rotation. The maximum efficiency occurs when the lead angle α is $\left[45° - \left(\dfrac{\phi}{2}\right)\right]$.

6.4.3 V-Thread

In practice, V-threads are used in many screws and the normal reaction between the screw and the nut will therefore be greater than in square thread. The axial load W as shown in Fig. 6.6 is to be assumed as concentrated at a single point on the thread. Since the axial component of the normal reaction R_n must be equal to W, hence $R_n \cos \beta = W$ or $R_n = \dfrac{W}{\cos \beta}$ where 2β is the included angle between the sides of the thread.

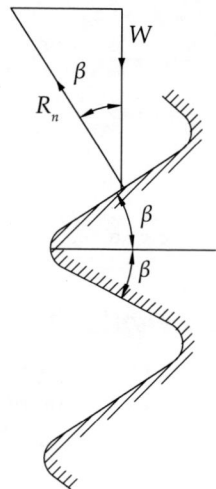

Fig. 6.6 V-thread.

But the frictional force which acts tangentially to the surface of the threads is given by $\mu R_n = \left(\dfrac{\mu}{\cos \beta}\right) W = \mu_v W$, where $\mu_v = \dfrac{\mu}{\cos \beta}$ may be regarded as a virtual coefficient of friction and the corresponding friction angle is ϕ_v in the equations already given for the square threaded screw.

6.4.4 Mechanical Advantage

Mechanical Advantage (M.A.) = $\dfrac{\text{Load lifted}}{\text{Effort applied}} = \dfrac{W}{F} = \cot(\alpha + \phi)$ for $\theta = 90°$

Velocity Ratio (V.R.) = $\dfrac{\text{Distance moved by effort}}{\text{Distance moved by load}}$ per one revolution = $\dfrac{2\pi S}{L}$ (where S is length of lever and L is the pitch of the screw).

$$\dfrac{2\pi S}{L} = \dfrac{2\pi S}{2\pi r \tan \alpha} = \dfrac{S}{r. \tan \alpha}$$

since $L = 2\pi r \tan \alpha$

$$\text{Mechanical Efficiency} = \frac{\text{M.A}}{\text{V.R}} = \frac{\tan\alpha}{\tan(\alpha+\phi)} \text{ or } \frac{\cot(\alpha+\phi)}{\cot\alpha} \text{ (for up the plane);}$$

$$= \frac{\tan(\alpha-\phi)}{\tan\alpha} \text{ or } \frac{\cot\alpha}{\cot(\alpha-\phi)} \text{ (for down the plane)}$$

W E 6.3: The mean radius of the screw of a square threaded screw jack is 25 mm. The pitch of thread is 7.5 mm. If the coefficient of friction is 0.12, what effort applied at the end of lever 60 cm length is needed to raise a weight of 2 kN. Find also the efficiency of the screw jack.

Given:
Mean radius of the screw (r) = 25 mm = 0.025 m; pitch of the thread (L) = 7.5 mm; coefficient of friction (μ) = 0.12 = $\tan\phi$; length of the lever (S) 60 cm and weight (W) to be raised = 2 kN = 2000 N.

Let P_u be the effort required at the end of the 60 cm long lever to raise the weight; and α be the helix angle. Hence, $\tan\alpha = \dfrac{L}{(2\pi r)} = \dfrac{0.75}{(2\pi 2.5)} = 0.048$ and F_u be the effort required at the mean radius r of the screw to raise the weight,

$$P_u \cdot S = F_u \cdot r = Wr \cdot \tan(\alpha + \phi)$$

$$F_u = W\tan(\alpha + \phi)$$

$$= \frac{W(\tan\alpha + \tan\phi)}{(1 - \tan\alpha \cdot \tan\phi)}$$

$$= \frac{2000(0.048 + 0.12)}{(1 - 0.048 \times 0.12)} = 2000 \times 0.169 = 338 \text{ N}$$

Now the effort applied to lift the load at the end of the lever P_u, may be found out from the relation,

$$P_u \times 60 = F_u \times 2.5 = 338 \times 2.5 = 845$$

$$P_u = \frac{845}{60} = 14.1 \text{ N}$$

Efficiency of the screw jack

$$\eta = \frac{\tan\alpha}{\tan(\alpha + \phi)}$$

$$= \frac{0.048}{0.169} = 0.2857 = 28.57\%$$

6.5 PIVOT AND COLLAR FRICTION

In many cases the rotating shafts are subjected to axial thrust and so a bearing surface must be provided in order to take this thrust for preserving the shaft in its correct axial position.

Examples: Propeller shafts of ships, the shafts of steam turbines and vertical machine shafts, etc.

The surface or surfaces on which the thrust is carried may be plane surfaces at right angles to the axis of rotation or occasionally conical surfaces, in which the axis of the cone coincides with the axis of rotation. For the relative motion to take place between the contact surfaces, a torque or a couple has to be applied. The magnitude of the torque required is determined approximately as follows:

As shown in Fig. 6.7, W is the axial load supported by a conical bearing surface with an apex angle 2α. The extreme radii of the actual area of contact be r_1 and r_2.

Consider a ring of bearing surface of radius r, radial width δr and width δl parallel to the conical surface.

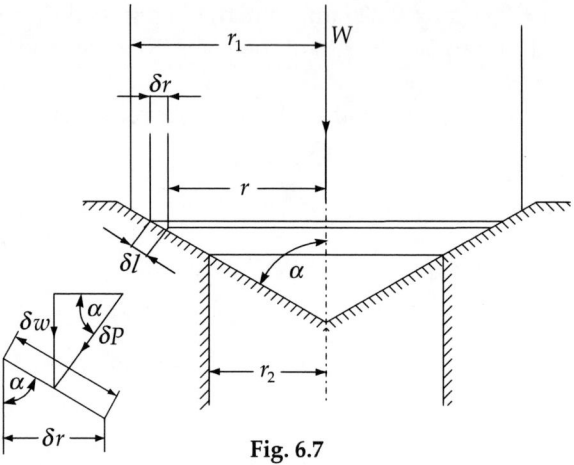

Fig. 6.7

Let p be the normal intensity of pressure between the surfaces at radius r. Let μ be the coefficient of friction between the surfaces at radius r. Then the area of the ring of bearing surface be

$$\delta A = 2\pi r \cdot \delta l \qquad (1)$$

Let δP be the normal load supported by the ring

$$p \delta A = p \cdot 2\pi r \cdot \delta l \qquad (2)$$

The axial load supported by the ring

$$\delta W = \delta P \sin \alpha = p \cdot 2\pi r \delta l \sin \alpha \qquad (3)$$

But $\delta l \sin \alpha = \delta r$. Therefore, $\delta W = p \cdot 2\pi r \cdot \delta r$ and the total axial load

$$W = \int_{r_2}^{r_1} p \cdot 2\pi r \cdot dr \qquad (4)$$

The friction force on the elemental ring of bearing surface

$$\delta F = \mu \delta P = \mu p \cdot 2\pi r \cdot \delta l$$

and the friction moment

$$\delta M = \delta F \cdot r = \mu p 2\pi r^2 \delta l = \frac{\mu p 2\pi r^2 \delta r}{\sin \alpha}$$

Total frictional moment or torque which resists the rotation of the shaft

$$M \text{ or } T = \frac{\int \mu p \cdot 2\pi r^2 \cdot dr}{\sin \alpha} \tag{5}$$

Before integrating equations (4) and (5) between the radii r_1 and r_2, the way in which μ and p vary with the radius r must be either known or assumed. Hence, by assuming that the coefficient of friction is same at all points on the bearing surface and that either

(a) the intensity of pressure is uniform, i.e. $p = a$ constant C or
(b) the rate of wear is uniform, i.e. product $p.r = a$ constant C

Since wear depends on the intensity of pressure and also the velocity of rubbing which again depends on the radius r. So the rate of wear is proportional to $p \cdot r$.

6.5.1 Uniform Intensity of Pressure

Pressure is constant, i.e. ($p = a$ constant C) from equation (4)

$$W = 2p\pi \int r \cdot dr \text{ between the limits } r_1 \text{ and } r_2 = p\pi(r_1^2 - r_2^2) \tag{6}$$

and from equation (5)

$$M = 2\pi \left(\frac{\mu}{\sin \alpha}\right) p \int r^2 .dr \text{ between the limits } r_1 \text{ and } r_2$$

$$= \left(\frac{2}{3}\right)\left(\frac{\mu}{\sin \alpha}\right) p\pi(r_1^3 - r_2^3) \tag{7}$$

On substituting from (6) for $p = \dfrac{W}{[\pi(r_1^2 - r_2^2)]}$ in (7)

$$M = \frac{2\mu}{3 \sin \alpha} \frac{W(r_1^3 - r_2^3)}{(r_1^2 - r_2^2)} \tag{8}$$

For a flat pivot or collar $\alpha = 90°$ and $\sin \alpha = 1$

$$M = \frac{2\mu}{3} \frac{W(r_1^3 - r_2^3)}{(r_1^2 - r_2^2)} \tag{9}$$

The friction moment for a conical pivot is therefore identical with that for a flat pivot which has a higher coefficient of friction $\mu_1 = \dfrac{\mu}{\sin \alpha}$.

6.5.2 Uniform Rate of Wear

Wear is the product of pressure and radius, i.e. (pr = a constant C)

Substituting C for pr in equation (4) $W = 2\pi C \int dr$ between r_1 and r_2 gives

$$W = 2\pi C(r_1 - r_2) \tag{10}$$

Similarly, substituting C for pr in equation (5) $M = \dfrac{2\pi\mu C}{\sin\alpha} \int r.dr$ between r_1 and r_2 gives

$$T \text{ or } M = \dfrac{\pi\mu C}{\sin\alpha}(r_1^2 - r_2^2) \tag{11}$$

Substituting for C from (10) $\dfrac{W}{[2\pi(r_1 - r_2)]}$ in (11) gives

$$T \text{ or } M = \dfrac{\mu W}{\sin\alpha} \dfrac{(r_1 + r_2)}{2} \tag{12}$$

For a flat pivot equation (10) remains unchanged, while equation (12) reduces to

$$T \text{ or } M = \mu W \dfrac{(r_1 + r_2)}{2} \tag{13}$$

Power lost in friction = [Moment or Torque × ω] N-m/sec or watts = $\dfrac{2\pi NT}{60}$ watts where N is revolutions per minute.

Note: 1. Which of the two assumptions is to be adopted in any given problem depends on the usage of friction. On the safer side, for example, (i) whether it is used in clutches, brakes, etc. (ii) whether it is used in power transmission such as bearings, belts, etc. One assumption gives higher moment than the other.

2. Use the one which gives the lower value for positive usage of friction as in (i) and the one which gives higher value for negative usage as in (ii) to be on the safer side.

The various types of pivot and collar bearings are shown in Fig. 6.8 such as flat, conical, truncated, single flat collar and multi-collar from (a) to (e).

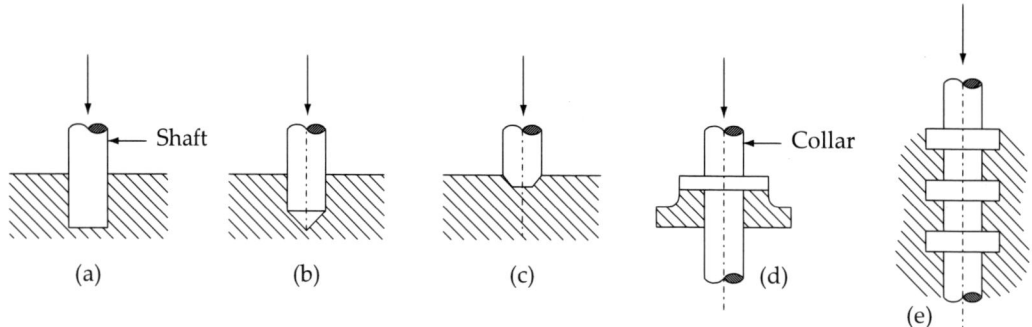

Fig. 6.8 Types of pivots and collars.

As shown in Fig. 6.8 the various types of pivots and collars, the expressions for W and M or T vary as follows:

Types of pivots	Uniform pressure ($p = C$)	Uniform wear ($pr = C$)
(a) Flat pivot: $r_2 = 0$ and $\alpha = 90°$	$W = \pi r_1^2 p$ Newtons	$W = 2\pi C r_1$ Newtons
	$T = \dfrac{2\mu W r_1}{3}$ N-m	$T = \mu \pi C r_1^2$ N-m
(b) Conical pivot: $r_2 = 0$	$W = \dfrac{\pi r_1^2 p}{\sin \alpha}$ Newtons	$W = \dfrac{2\pi C r_1}{\sin \alpha}$ Newtons
	$T = \dfrac{2\mu W r_1}{3 \sin \alpha}$ N-m	$T = \dfrac{\mu \pi C r_1^2}{\sin \alpha}$ N-m
(c) Truncated pivot	W = as in (6)	W = as in (10)
	T = as in (8)	T = as in (12)
(d) Single flat collar: $\alpha = 90°$	$W = \pi(r_1^2 - r_2^2)p$ Newtons	$W = 2\pi C(r_1 - r_2)$ Newtons
	$T = \dfrac{2\mu W(r_1 - r_2)}{3}$ N-m	$T = \mu \pi C(r_1^2 - r_2^2)$ N-m
(e) Multiple flat collars	Number of collars say n	The values will be n times given at (d)

W E 6.4: A load of 20 kN is supported by a conical pivot. The angle of cone is 120° and intensity of pressure is not to exceed 3.5 bar. The external radius is 3 times the internal radius. Find the diameter of the bearing surface. If coefficient of friction is 0.06 r.p.m. of the shaft is 120, what power in kW is absorbed by friction? 1 bar = 10^5 N/m².

Given:
W = 20 kN = 20,000 N; $2\alpha = 120°$; $r_1 = 3r_2$; $p = 3.5 \times 10^5$ N/m²
Axial load, $W = \int p \, 2\pi r \cdot dr = p \cdot \pi(r_1^2 - r_2^2)$
Hence, $20,000 = 3.5 \times 10^5 \pi (9r_2^2 - r_2^2)$.
Therefore,
$$r_2^2 = \frac{20000}{(3.5 \times 10^5 \times \pi \times 8)}$$

$r_2 = 0.0477$ m = 4.77 cm and

$r_1 = 3 \times 0.0477 = 0.1431$ m = 14.31 cm

$$\text{Total torque, } T = M = \left(\frac{2\mu}{3 \sin \alpha}\right) \frac{W(r_1^3 - r_2^3)}{(r_1^2 - r_2^2)}$$

$$= \left(\frac{2 \times 0.06}{3 \times \sin 60°}\right) \times \frac{2000(14.31^3 - 4.77^3)}{(14.31^2 - 4.77^2)}$$

$$= 143.19 \text{ N-m}$$

$$\text{Power} = \frac{2\pi NT}{60}$$
$$= \frac{2\pi \times 120 \times 143.19}{60}$$
$$= 1799 \text{ watts} = 1.799 \text{ kW}.$$

W E 6.5: Calculate power lost in overcoming friction and number of collars required for the thrust bearings whose contact surfaces are 20 cm external radius and 15 cm in internal radius. The co-efficient of friction is 0.08. The total axial load is 30 kN. Intensity of pressure is not to exceed 3.5 bar. Speed of the shaft is 420 r.p.m. Note: 1 bar = 10^5 N/m².

Given:
External radius $r_1 = 20$ cm and internal radius $r_2 = 15$ cm;   Coefficient of friction $\mu = 0.08$;
Total axial load $W = 30$ kN;   Intensity of pressure $p = 3.5$ bar;   Speed $N = 420$ r.p.m.

Assuming uniform pressure because it is to determine loss of power due to friction:

$$\text{Torque } (T) \text{ or Moment } (M) = \left(\frac{2}{3}\right) \frac{\mu W(r_1^3 - r_2^3)}{(r_1^2 - r_2^2)}$$

$$= \left(\frac{2}{3}\right) \frac{0.08 \times 30 \times 1000 \times (20^3 - 15^3)}{(20^2 - 15^2)}$$

$$= \left(\frac{2}{3}\right) \times 0.08 \times 30000 \left(\frac{4625}{175}\right)$$

$$= 42285 \text{ N-cm} = 42.285 \text{ N-m}$$

$$\text{Power lost in friction} = \frac{42.285 \times 2\pi \times 420}{60}$$

$$= 1860 \text{ watts}.$$

$$\text{Load per collar} = p\pi(r_1^2 - r_2^2)$$

$$= \frac{3.5 \times 10^5 \times \pi(400 - 225)}{10^4}$$

$$= 35\pi(175) = 19400 \text{ N}$$

$$\text{Number collars} = \frac{\text{Total load}}{\text{Load per collar}}$$

$$= \frac{30000}{19400} = 1.56 = 2 \text{ collars}.$$

6.6 CLUTCHES

This is a device used to connect or disconnect the driving shaft and the driven shaft almost instantaneously as desired by the operator or driver. These are widely used in the automobiles to stop for a while, with the engine still running without any load. These are called friction clutches as friction is used. There are two types of clutch plates (i) plate or disc clutches (ii) conical clutches just as flat collar and conical pivot collar bearings. Again plate or disc clutches are of two types: (a) single plate and (b) multi-plate clutches showed at Fig. 6.9(a) and (b).

The problems are to be solved in the same lines as bearings except the choice of assumption, i.e. uniform wear is used as the friction is made used here for positive purpose of transmitting power. So materials of high friction are used.

6.6.1 Single-plate Clutch

The clutch showed at Fig. 6.9(a) is a single plate clutch. The flywheel A is bolted to a flange on the driving shaft B. The plate C is fixed to a boss free to slide axially along the driven shaft D, but by means of splines, it is compelled to revolve with the shaft D. Two rings G of special friction material are riveted to A and E or, alternatively to the plate C.

The presser plate E is bushed internally, so as to revolve freely on the driven shaft D, and is integral with the withdrawal sleeve F. A number of spiral springs are arranged round the clutch, as shown at S, so as to provide the axial thrust between the friction surfaces.

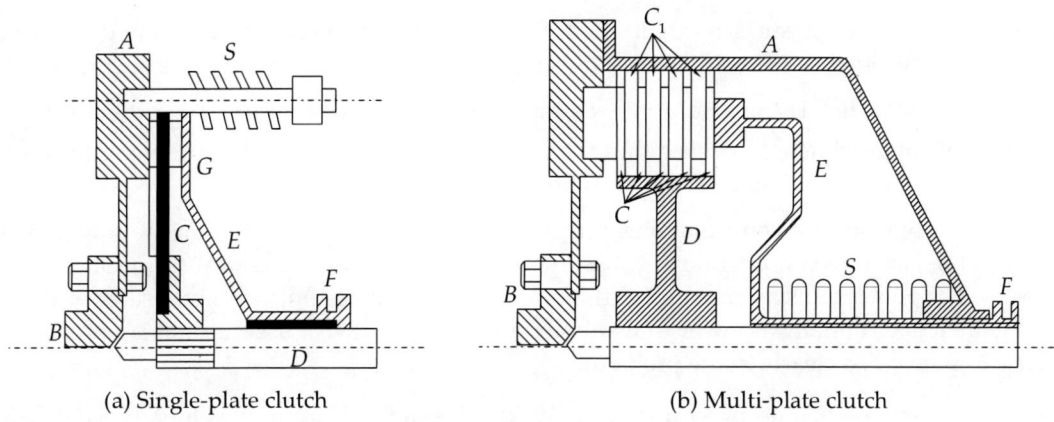

(a) Single-plate clutch (b) Multi-plate clutch

Fig. 6.9

The action of the clutch is as follows. When the withdrawal sleeve is displaced to the right, there is no axial pressure between the friction surfaces and the flywheel A revolves freely, while the plate C and the shaft D remains at rest. When the withdrawal force is removed from the sleeve F, the springs S force the pressure plate E against the rings G and the friction between the contact surfaces of the rings G and the plate C transmits a torque to the shaft D, the shaft D will revolve. This type of clutch therefore enables the driven shaft to be started or stopped at will.

6.6.2 Multi-plate Clutch

The clutch shown in Fig. 6.9(b) is similar to that at (a) except that the number of surfaces at which slip can take place is increased. The outer casing A is bolted as before to the driving shaft B, but has a number of axial grooves cut on the internal surface. Alternate discs have tongues or projections on the outer edge. The internal diameter of each disc C_1 is greater than the diameter of the driven member D. The other discs C have an external diameter less than the internal diameter of the casing A and are provided with tongues on their inner edges. Thus, the discs C_1 must revolve with the casing A and the discs C must revolve with the driven member D. The axial thrust exerted by the spring S forces the discs C_1 and C into contact and the friction between the contact surfaces enables a torque to be transmitted from the casing A to the shaft D. Displacement of the withdrawal sleeve F to the right removes the axial load from the discs and allows the outer casing to revolve freely, while the shaft D remains at rest. The friction torque between each pair of contact surfaces should be determined from $\dfrac{\mu W(r_1 + r_2)}{2}$ where W is the axial load equal to $p \cdot \pi(r_1^2 - r_2^2)$; r_1 and r_2 are the external and internal radii of the contact surfaces. The total torque transmitted from the driving to the driven shaft is equal to n times the torque given by the above equation where n is the number of pairs of surfaces between which sliding can take place. For single-plate clutch $n = 2$, since there are two contact surfaces corresponding to the two sides of the plate C and the adjacent surfaces of the ring G. For the multi-plate clutch shown with 5 such clutch plates. $n = 10$ since both faces of each of the outer discs is also effective.

6.6.3 Cone Clutch

In a cone clutch the contact surfaces are in the form of frustum of cones. The working principle is same as the above clutches. The advantage of cone clutch is that the normal force on the contact surface is increased. If F is the axial force, F_n the normal force is equal to $\dfrac{F}{\sin \theta}$ where θ is the semi-cone angle of the clutch. However, these cones are not popular in use. The problems are taken up just as conical bearings except the assumption used.

Note: In actual practice, for computations, the assumption which gives result on the safer side is preferred. Thus, for power absorption or loss due to friction, uniform pressure assumption gives safer result. For power transmission using friction, uniform wear assumption gives safer result.

Since clutch is essentially a power transmission unit, uniform wear assumption is used unless otherwise it is specified clearly in the problem.

W E 6.6: A car engine has its rated output of 10 kW. Maximum torque developed is 100 N-m. The clutch used is of single plate type, having two active surfaces. Axial pressure is not to exceed 0.85 bar. External diameter of the friction plate is 1.25 times the internal diameter. Determine the dimension of the friction plate and the axial force exerted by the springs. Assume uniform wear and coefficient of friction is 0.3.

Given:
$T = 100$ N-m;   Coefficient of friction = 0.3;
Intensity of pressure will be more at the inner radius (r_2) for uniform wear assumption.
Hence,
$$p_2 r_2 = C = 0.85 \times 10^5 \times r_2$$

Axial load, $W = 2\pi C(r_1 - r_2)$
$= 2\pi \times 0.85 \times 10^5 \times r_2(1.25 r_2 - r_2)$
$= 1.335 \times 10^5 r_2^2$ N

Torque due to both active surfaces

$$T = 2\mu W \frac{(r_1 + r_2)}{2}$$
$$= 2 \times 0.3 W \frac{(1.25 + 1) r_2}{2}$$
$$= 0.675 W r_2$$
$$\therefore T = 0.675 \times 1.335 \times 10^5 r_2^3 \text{ N-m}$$

From the data given

$$100 = 0.675 \times 1.335 \times 10^5 r_2^3$$

$$\therefore r_2 = \left[\frac{100}{0.675 \times 1.335 \times 10^5}\right]^{1/3}$$

$$= 0.1035 \text{ m} = 10.35 \text{ cm}$$

and $r_1 = 1.25 \times 10.35 = 12.94$ cm.

Axial force exerted by springs $W = 1.335 \times 10^5 \times r_2^2$
$= 1.335 \times 10^5 \times (0.1035)^2 = 1430$ N.

W E 6.7: A multiple disc clutch has five plates having four active parts of frictional surfaces. Determine the maximum axial intensity of pressure between the discs for transmitting 18 kW at 500 r.p.m. if the outer and inner radii of the friction surface are 12.5 cm and 7.5 cm respectively. Assume uniform wear and the coefficient of friction as 0.3.

Given:
Power = P = 18 kW = 18 × 1000 W; N = 500 r.p.m.; Outer radius r_1 = 12.5 cm
Inner radius r_2 = 7.5 cm; μ = 0.3

$$\text{Power } (P) = \frac{2\pi N T}{60}$$
$$\text{Torque } (T) = \frac{P \times 60}{2\pi N} = \frac{18 \times 1000 \times 60}{2\pi 500}$$
$$= 343.77 \text{ N-m}$$

This torque is transmitted by four active pairs of surfaces. Therefore, torque transmitted by each pair is given by

$$T = \frac{343.77}{4} = 85.9425 \text{ N-m} \qquad (1)$$

Torque for uniform wear

$$T = \frac{\mu W(r_1 + r_2)}{2}$$

$$T = \frac{0.3W(12.5 + 7.5)10^{-2}}{2} \text{ N-m} \qquad (2)$$

Equating (1) and (2)

$$W = \frac{85.94 \times 100 \times 2}{0.3 \times 20} = 2864.75 \text{ N}$$

Let p_2 be the maximum intensity of pressure, then

$$W = 2\pi p_2 r_2 (r_1 - r_2)$$

$$p_2 = \frac{2864.75}{2\pi \times 7.5 \times (12.5 - 7.5) \times 10^{-4}}$$

$$= 1.2158 \times 10^5 \text{ N/m}^2$$

6.7 BRAKES AND DYNAMOMETERS

6.7.1 Introduction

The application of friction for positive purpose (friction is made use) to stop or retard a moving vehicle by absorbing its kinetic energy in a short time using friction brakes. A device to measure the frictional resistance is called the dynamometer. This is used to determine the power developed by any machine running at rated speed.

The clutch is used to connect two moving members of a machine whereas the brake connects a moving member to a stationary member.

6.7.2 Types of Brakes

A brake is an appliance used to apply frictional resistance to a moving body to stop or retard it by absorbing its kinetic energy. In all types of motions, there is always some amount of resistance which retards the motion and is sufficient to bring the body to rest. But the time it takes is too large. By providing brakes, the external resistance is considerably increased and the period of retardation reduces.

There are mainly four types:

1. Block or shoe brake
2. Band brake
3. Band and block brake
4. Internal brake

1. Block or shoe brake

This consists of a block or shoe having friction layer on it, is pressed against a rotating wheel called drum. A lever is used to increase the force required to press the block or shoe as shown in Fig. 6.10(a) and (b). But a single lever produces a side thrust on the bearing of the drum shaft. This is prevented by using two blocks on two sides of the drum as shown in Fig. 6.10(c). The double levers provide double the braking effect. The blocks are made of wood or rubber for slow moving vehicles and cast iron for heavy and fast ones. In Fig. 6.10(b), the pivot O is on the other end of the lever and the force is to be applied downwards as against in Fig. 6.10(a).

Let r be the radius of the drum; μ be the coefficient of friction; F_R be the radial force applied on the drum; R be the normal reaction on the block; F be the force applied on the lever; F_f be the frictional force (μR) and T_B be the braking torque.

Assumptions:
1. Normal reaction acts at the midpoint of the block.
2. Frictional force also acts at the midpoint of the block.

The force F is applied in the direction indicated on the lever as shown in Fig. 6.10. Then the braking torque on the drum (T_B)

$$T_B = \text{Frictional force} \times \text{Radius} = \mu R \times r \tag{1}$$

In order to obtain the reaction force R, the equilibrium of the block has to be considered.

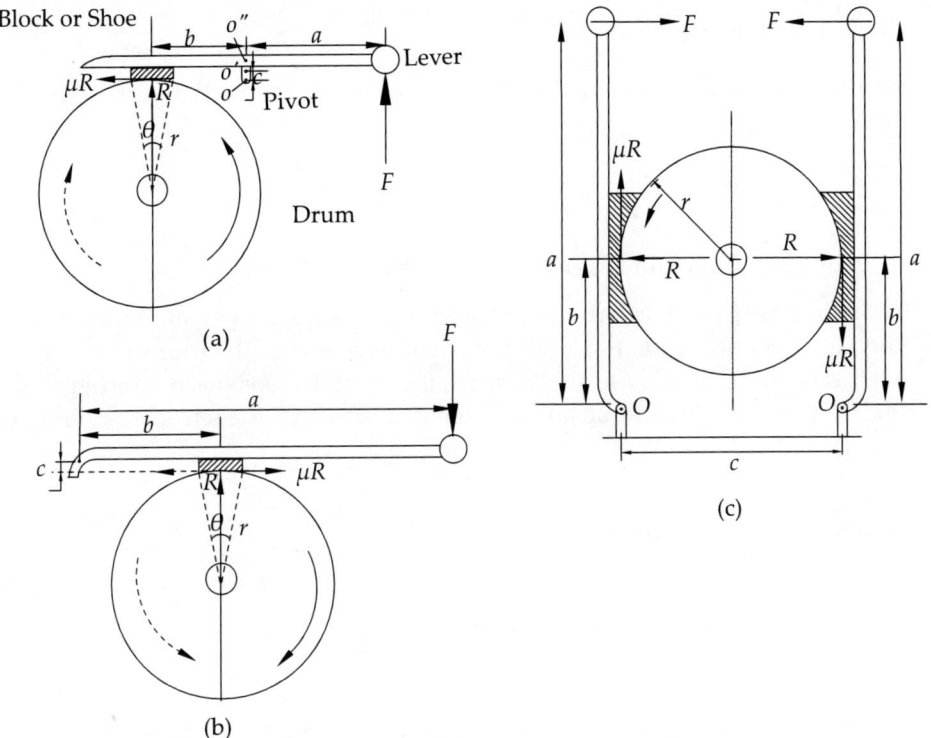

Fig. 6.10

Note: The direction of the frictional force on the drum is shown. But it will be in the opposite direction on the lever.

Taking moments about the pivot O of Fig. 6.10 of (a) or (b), gives the equation

$$R = \frac{F \times a}{(b - \mu c)} \qquad (2)$$

$$F = \frac{R \times (b - \mu c)}{a} \qquad (3)$$

(i) When $b = \mu c$, $F = 0$ which implies that the force needed to apply the brake is virtually zero. In other words, the brake is automatically applied once the contact between the block and the drum takes place. Such a brake is known as a self-locking brake. It is also called self-energised brake as the moment due to the applied force F and the moment due to the frictional force are in the same direction.

(ii) When the direction of rotation is reversed, the force required F increases.

(iii) If the pivot lies on the line of action, i.e. O', then $c = 0$ and then force $F = \frac{Ra}{b}$. It is same irrespective of the direction of rotation.

(iv) If the pivot is O'', then c becomes negative, then $\frac{R \times (b + \mu c)}{a}$ for counterclockwise and $\frac{R \times (b - \mu c)}{a}$ for clockwise.

In case the angle of contact θ is more than 45°, then coefficient of friction μ changes to μ' given by $\mu \left[\frac{4 \sin\left(\frac{\theta}{2}\right)}{\theta + \sin \theta} \right]$.

2. Differential band brake or band brake

This band brake consists of a belt or rope instead of a block or shoe along with the lever as shown in Fig. 6.11. When force is applied on the lever, the frictional effect of the rope or belt causes brake on the drum. As seen in the belts and ropes chapter, there will be tensions on the tight side T_1 and on the slack side T_2. The braking torque on the drum depends on these tensions and it is given by

$$T_B = (T_1 - T_2) \cdot r \qquad (1)$$

where r is the effective radius of the drum.

Similarly, the ratio of the tensions on the tight side and slack side is given by

$$\frac{T_1}{T_2} = e^{\mu \theta} \qquad (2)$$

The effectiveness of the force F applied depends upon three important parameters:

(i) the direction of rotation of the drum
(ii) the ratio of the distances of pivot and belt or rope ends where they are tied to the lever and
(iii) the direction of the applied force F.

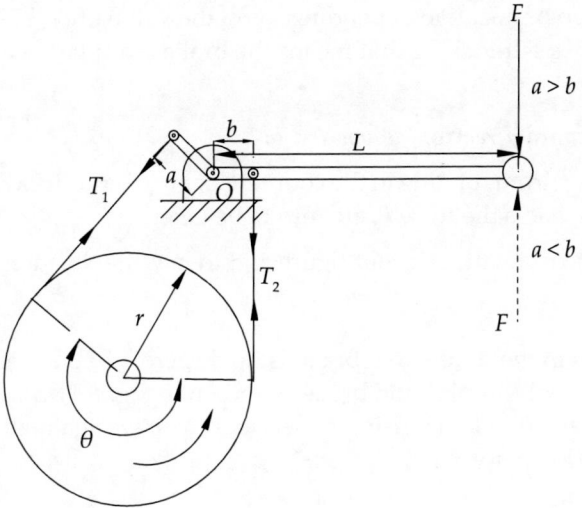

Fig. 6.11(a)

To stop the drum from rotation, a force F has to be applied on the lever such that the band becomes tight on the drum. This is possible if

(a) F is applied in the downward direction when $a > b$.
(b) F is applied in the upward direction when $a < b$.

(a) F is applied in the downward direction when $a > b$:

Here again the force F depends on the direction of rotation of the drum since the tight and slack sides of the band depend on it. Hence, both clockwise (CW) and counterclockwise (CCW) rotation of the drum has to considered separately as below.

(i) *Rotation counterclockwise*: As shown in the Fig. 6.11 (a) the tight and slack sides with the force F applied downwards, taking moments about the pivot gives

$$F \times L = (T_1 a - T_2 b)$$
$$F = \frac{(T_1 a - T_2 b)}{L} \qquad (3)$$

Here $T_1 > T_2$ and $a > b$ under all conditions for the counterclockwise rotation of drum.

(ii) *Rotation clockwise*: When the direction of rotation changes the tight and slack sides get interchanged and with the force F applied still downwards, taking moments about the pivot gives

$$F \times L = (T_1 b - T_2 a)$$
$$F = \frac{(T_1 b - T_2 a)}{L}$$

Here $T_1 > T_2$ and $a > b$ under all conditions for the clockwise rotation of drum.

The force F will be zero or negative depending upon the value of $(T_1 b - T_2 a)$. Zero or negative means, the system becomes self-locking that means the brake is applied without any force F being applied.

(b) F is applied in the upward direction, when a < b:

Similarly, for this case also, drum rotating in the counterclockwise and clockwise is to be considered. When a and b are equal, what is the force F, etc. has to be taken up.

Note: It is always better to draw the relevant figure and then write the moment equation.

3. Simple band brake

When the values of a or b in the differential brake is made zero, i.e. one end of the band is tied to pivot, then that brake is called simple band brake as shown in Fig. 6.11(b).

Simple band brakes can be self-energising or self-locked. Here again all the different cases can be considered for calculation of force F.

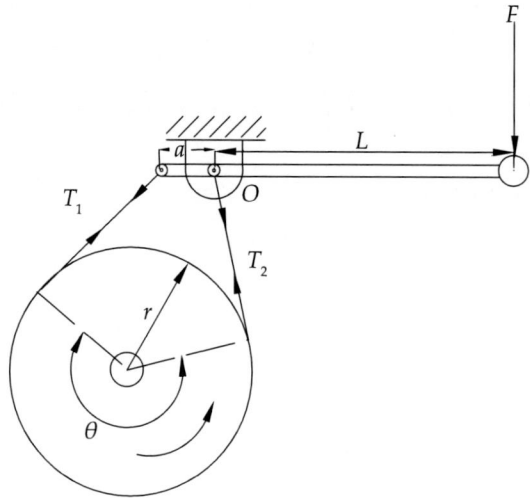

Fig. 6.11(b)

The brake is said to be more effective when maximum braking force is applied with the least effort F.

The advantage of self-locking is taken in hoists and conveyers where motion is permissible in only one direction. If the motion gets reversed somehow, the self-locking is engaged which can be released only by reversing the applied force. The differential band brake can be seen more effective only for one direction of rotation.

4. Two-way brand brake

This two-way band brake is equally effective for both the directions of rotations. As shown in Fig. 6.11(c), the two lever arms are made equal. For both directions of rotation of the drum, $F \times L - T_1 \times a - T_2 \times a = 0$ or $F = (T_1 + T_2) \times \dfrac{a}{L}$.

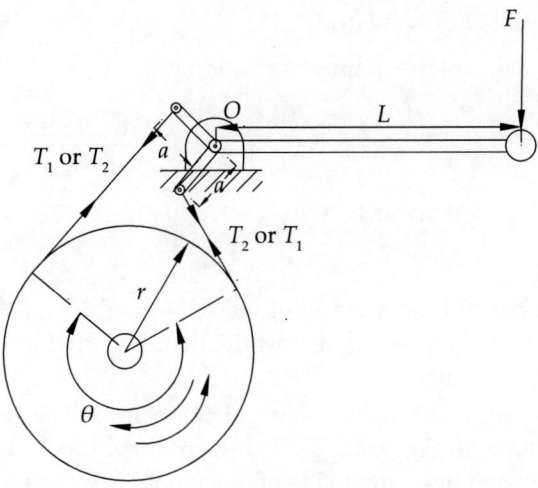

Fig. 6.11(c)

W E 6.8: A differential band brake acting on the 3/4th of the circumference of a drum of 450 mm diameter is to provide a braking torque of 500 N-m. One end of the band is attached to a pin 100 mm from the fulcrum of the lever and the other end to another pin 25 mm from the fulcrum on the other side of it where the operating force is also acting. If the operating force is applied at 500 mm from the fulcrum and the coefficient of friction is 0.45, find the two values of the operating force corresponding to the two directions of rotation of the drum.

Given:
Lever $L = 0.5$ m;   Distances $a = 0.1$ m and $b = 0.025$ m;   Radius of drum $r = 0.225$ m;
$\theta = 270°$; $m = 0.45$;   Braking torque $T_B = 500$ N-m.

$$T_B = (T_1 - T_2)r$$

$$(T_1 - T_2) = \frac{T_B}{r} = \frac{500}{0.225} = 2222.22 \text{ N}$$

$$\frac{T_1}{T_2} = e^{\mu\theta} = e^{(0.45 \times 270 \times (\pi/180))} = 8.343$$

Solving (1) and (2) gives
$$T_1 = 2524.85 \quad \text{and} \quad T_2 = 302.63$$

For anticlockwise rotation of the drum, the moment equation is

$$F \times L - T_1 a + T_2 b = 0 \quad \text{or} \quad F = \frac{(T_1 a - T_2 b)}{L}$$

Substituting gives, $F = 489.82$ N downward.

For clockwise rotation of the drum, the moment equation is

$$F \times L + T_1 b - T_2 a = 0 \quad \text{or} \quad F = \frac{(T_1 b - T_2 a)}{L}$$

Substituting gives, $F = 65.72$ N downward.

5. Band and block brake

This brake consists of number of blocks made of wood secured inside the flexible steel band as shown in Fig. 6.12. The blocks are pressed against the drum when the brake is applied. The two sides become tight and slack as usual.

Each block subtends an angle 2θ at the centre of the drum. The frictional force acts in the direction of the drum as shown in Fig. 6.12 (b). Let the number blocks be n; T_0 be the tension on the slack side; T_1 be the tension on the tight side of block one; $T_2 \ldots T_n$ be the tensions on the tight side of blocks $2, 3 \ldots n$; μ be the coefficient of friction; R be the normal reaction.

The forces acting on block are shown in Fig. 6.12(b).

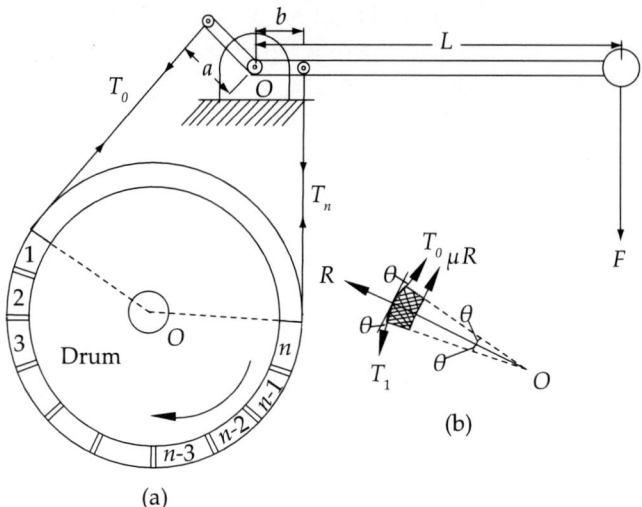

Fig. 6.12

For equilibrium, $(T_1 - T_0) \cos \theta = \mu R$ and $(T_1 + T_0) \sin \theta = R$ or

$$\frac{(T_1 - T_0)}{(T_1 + T_0)} = \mu \tan \theta \qquad (1)$$

Adding to equation (1) both sides +1, gives

$$\frac{[(T_1 - T_0) + (T_1 + T_0)]}{(T_1 + T_0)} = (1 + \mu \tan \theta) \qquad (2)$$

and subtract equation (1) from +1 both sides, gives

$$\frac{[(T_1 + T_0) - (T_1 - T_0)]}{(T_1 + T_0)} = (1 - \mu \tan \theta) \tag{3}$$

Dividing (2) by (3) gives after simplification, the tension ratio for block 1,

$$\frac{T_1}{T_0} = \frac{(1 + \mu \tan \theta)}{(1 - \mu \tan \theta)} \tag{4}$$

the tension ratio for block 2,

$$\frac{T_2}{T_1} = \frac{(1 + \mu \tan \theta)}{(1 - \mu \tan \theta)}$$

Similarly for all other blocks. Thus, the tension ratio is same for each block. The product of tension ratios for n number gives the overall tension ratio as follows:

$$\frac{T_n}{T_0} = \left[\frac{1 + \mu \tan \theta}{1 - \mu \tan \theta}\right]^n \tag{5}$$

The braking torque on the drum with the mean radius r_m is given by

$$T_B = (T_n - T_0) r_m \tag{6}$$

where r_m is given by mean radius r of drum plus half the thickness of blocks. Force F to be applied is to be calculated taking moments of the forces acting on the lever as follows.

$F \times L - T_0 \times a + T_n \times b = 0$. Knowing the values of a, b, L, T_0 and T_n where n represents the nth block, the force F can be calculated.

W E 6.9: In the band and block brake showed at Fig. 6.12, the band is lined with 10 blocks. Each block subtends an angle of 16° at the centre of the wheel. Calculate maximum force F required at lever end for the brakes to absorb 25 kW at 300 r.p.m. Take $\mu = 0.4$. Effective diameter is 80 cm. The values of a and b are 15 and 6 cm. Length of lever is 25 cm.

Given:
Number of blocks $n = 10$; $2\theta = 16°$; $\mu = 0.40$; Effective or mean diameter, $d_m = 0.8$ m; or $r_m = 0.4$ m; $L = 0.25$ m; $N = 300$ r.p.m.; Power to be absorbed = 225 kW.

Tension ratio,

$$\frac{T_{10}}{T_0} = \left[\frac{(1 + 0.4 \tan 8°)}{(1 - 0.4 \tan 8°)}\right]^{10} = 3.11 \tag{1}$$

$$\text{Power } (P) = \frac{2\pi N (T_{10} - T_0) r_m}{60 \times 1000}$$

$$T_{10} - T_0 = \frac{P \times 1000 \times 60 \times 100}{2\pi 300 \times 0.4} = 17904 \text{ N} \tag{2}$$

Therefore, solving (1) and (2),
$$T_{10} = 26389 \text{ N}, T_0 = 8485 \text{ N}$$

Taking moments about the pivot, $8485 \times 15 = 26389 \times 3 + F \times 25$, hence $F = 1982$ N.

6. Internal expanding shoe brake

In all the two wheelers, automobiles and other vehicles nowadays only internal expanding shoe brakes are used. It consists of two semicircular shoes as shown in Fig. 6.13 which are lined with a friction material.

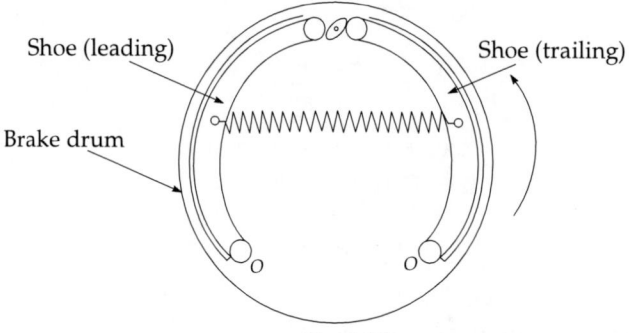

Fig. 6.13

The shoes are pressed against the inner flange of the drum when the brakes are applied. There will be a little gap between the shoe and the drum when brake is not applied. The actual force F is applied by hydraulic cylinders with pistons. Depending upon the direction of rotation, one of the two shoes will act as leading and the other one as trailing shoe. The leading shoe is self-energising whereas the trailing shoe is not. Under any circumstances they should not be self-locking.

6.7.3 Dynamometers

A dynamometer is a brake incorporated device used to measure the frictional resistance applied. This is used to determine the power developed by the machine or torque developed while maintaining its speed at the rated value. There are different types such as absorption type and transmission type. Prony brake dynamometer and rope brake dynamometers are examples of absorption type as they absorb energy and dissipate in the form of heat energy.

1. Prony brake

This is the simplest form of dynamometer made in different forms. It consists of two wooden blocks clamped together on a revolving pulley carrying a lever as shown in Fig. 6.14. The grip between the blocks and drum can be varied by tightening or loosening the bolts. This can be used upto a speed (N) of 1000 r.p.m. and 100 kW of power.

Frictional torque, $T_F = W \times L$ where W is the load and L is the length of the lever.

Power of the machine,
$$P = T_F \times (2\pi N/60)$$

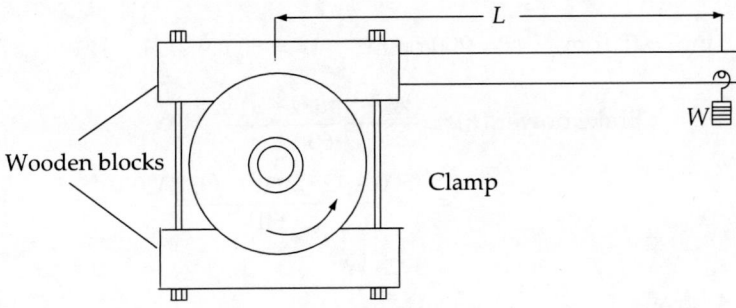

Fig. 6.14

Note: The power is independent of the drum size and coefficient of friction.

2. Rope brake dynamometer

This is most commonly used dynamometer. Rope or belt is wrapped around the drum keyed to the shaft of the engine. The size of the rope or belt will depend on the power of the machine. As shown in Fig. 6.15, the upper end is connected to a spring balance whereas the lower end carries the necessary weights.

Power of the machine,

$$P = \text{Torque } (T) \times \text{Angular velocity of brake drum } (\omega)$$

Knowing the weight W, spring balance reading S, radius of the drum and speed of rotation N, the power absorbed is given by

$$P = (W - S) \times r \times \omega$$

where ω is $\dfrac{2\pi N}{60}$.

W E 6.10: In a rope brake the diameter of the flywheel is 1.0 m and diameter of rope is 10 mm. The engine speed is 200 r.p.m. Weight on the brake is 500 N and spring balance reading 125 N. Calculate the brake power of the engine.

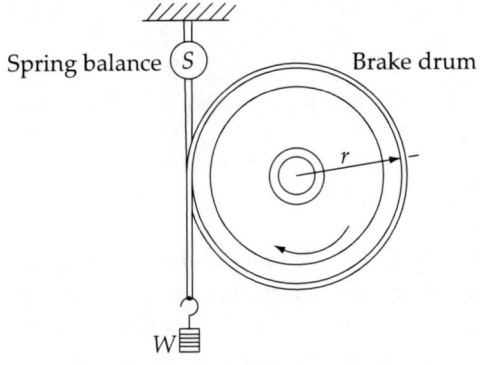

Fig. 6.15

Given:
$D = 1.0$ m; $d = 10$ mm $= 0.01$ m; $N = 200$ r.p.m.; $W = 500$ N; $S = 125$ N

$$\text{Brake power } (P) = \frac{(W - S)\pi(D + d)N}{60}$$

$$= \frac{(500 - 125)\pi(1.0 + 0.01)200}{60}$$

$$= 3928 \text{ W}$$

6.7.4 Types of Frictions

There are several types of frictions such as

1. Rolling friction
2. Anti-friction bearings
3. Greasy friction
4. Greasy friction at a journal and a friction circle

1. Rolling friction

When a ball rolls over a flat surface, the contact is theoretically at a point and similarly for a cylinder it a line contact. As they possesses weight and due to the pressure of the same, deformation of the flat surface or of the rolling object or of both takes place. The amount of deformation depends upon the elasticity of the materials in contact and the pressure. The deformation causes the surfaces to have area of contact rather than point or line of contact. Usually, the balls or cylinders are made of hard materials and so the other surface on which they roll get wear and contact area between them increases which increases the resistance to motion that is friction as shown in Fig. 6.16. This friction is known as rolling friction.

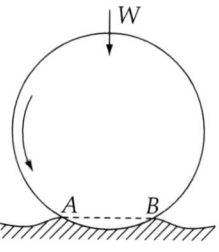

Fig. 6.16

2. Anti-friction bearings

When a shaft revolves in a bearing there will be a sliding motion also. Balls or rollers made of hardened materials such as chromium steel or chrome nickel steel are used between the shaft and the journals. The balls or rollers are mounted between two hardened races called inner and outer. The inner is fitted on to the shaft and the outer one is a tight fit into the bearing housing. Hence, there will be no relative motion between the shaft and the inner race or between the outer race and the housing. Balls or rollers are kept at a distance by means of brass cage. Though the friction increases slightly by lubrication but to prevent rust formation a small amount is used. Figure 6.17(a) shows the bearing, (b) ball bearings and (c) roller bearings. The pictures of the ball and roller bearings are shown at Pic. 6.2(a) and (b) with inner, outer and cages. The Pic. 6.3 shows the tapered roller bearings.

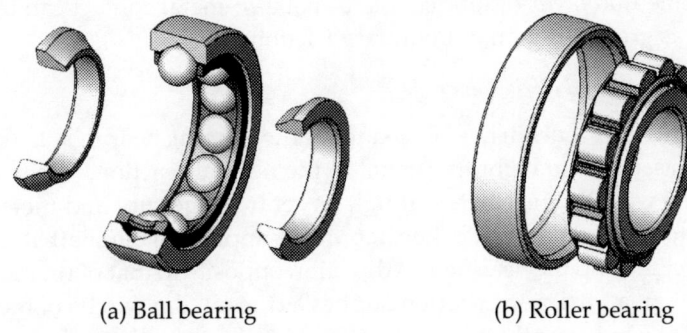

(a) Ball bearing   (b) Roller bearing

Pic. 6.2

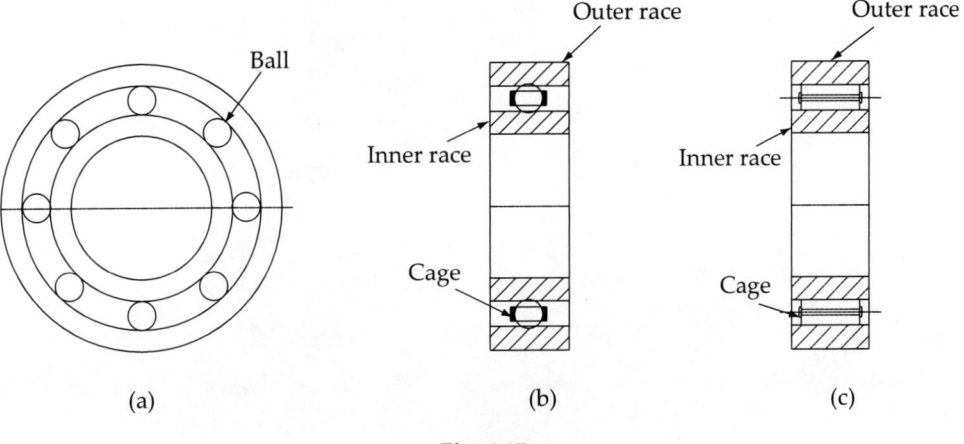

Fig. 6.17

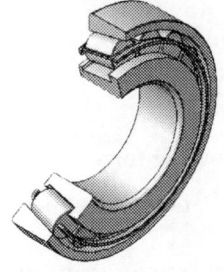

Pic. 6.3 Tappered rollers.

3. Greasy friction

When two metallic surfaces are wetted with a small amount of lubricant, a very thin film of the same is formed on each of the surfaces called adsorbed film. The coefficient of friction reduces considerably because of this film called oiliness surface. The friction of two surfaces, when they are wetted with an extreme thin layer of lubricant and metal-to-metal contact can take place between high spots is known as greasy friction or boundary friction.

4. Greasy friction at a journal and a friction circle

Greasy or boundary friction occurs in heavily loaded, slow running bearings. In this type of friction, the frictional force is assumed to be proportional to the normal reaction. When a shaft rests in its bearing because of the weight which acts through its centre of gravity and metal-to-metal contact takes place at the bottom of the bearing. When a torque is applied to the shaft, it rotates and the seat of pressure creeps or climbs up the bearing in a direction opposite to that of rotation. Metal-to-metal contact still exists and greasy friction criterion applies as the oil film will be of molecular thickness. The common normal at B between the two surfaces in contact passes through the centre of the shaft.

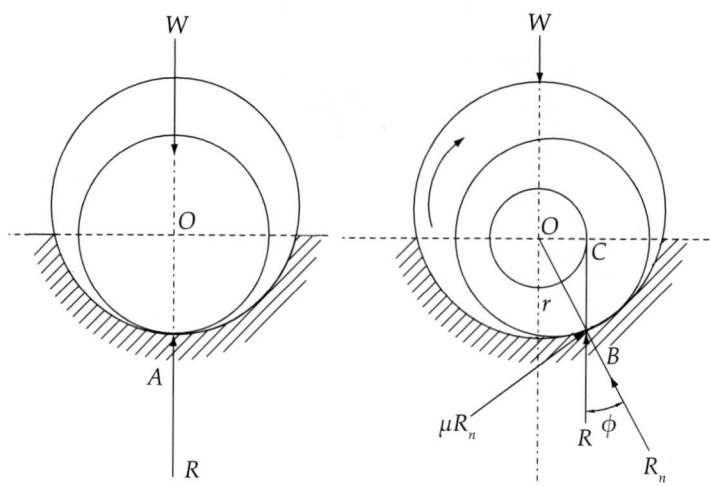

Fig. 6.18

Let R_n be the normal (radial) reaction at B.

μR_n be the frictional force tangential to the shaft.

R be the resultant of the above two inclined at ϕ to R_n.

The shaft is in equilibrium under the following forces:

(i) Weight W acting vertically downwards and (ii) Reaction R.

For equilibrium, R must act vertically upwards and must be equal to W. However, the two forces W and R will be parallel and constitute a couple.

Let OC = Perpendicular to R from O,

Friction couple (torque) = $W \times OC = Wr \sin\phi$ since $\sin\phi = \dfrac{OC}{r}$
$\approx Wr \tan\phi$ as ϕ is small
$\approx Wr\mu$ ($\mu = \tan\phi$)

This couple must be equal and opposite to the couple or torque producing motion. A circle drawn with OC (or $r \sin\phi \approx r \tan\phi \approx r\mu$) as radius is known as the friction circle of the journal as shown at Fig. 6.18. Thus, the effect of friction is equivalent to displacing the reaction through a distance equal to $r \sin\phi$ or such that it is tangential to the friction circle.

6.8 EXERCISE

6.8.1 Short Answer Questions

1. What do you mean by friction circle? Explain.
2. Derive the expression for the force required to move the body up through the inclined plane.
3. Explain that the coefficient of friction for film or viscous friction depends upon the square root of velocity of body and inversely proportional to the intensity of bearing pressure.
4. What is meant by the following: friction, friction force, coefficient of friction, limiting friction, angle of friction and angle of repose?
5. Derive an expression for the efficiency of the inclined plane, when the body is moving up the plane.
6. What is a friction circle? Derive an expression for its radius.
7. What do you understand by the term friction?
8. How do you distinguish between static and dynamic friction?
9. State the laws of friction?
10. Explain the term angle of friction.
11. Define coefficient of friction and limiting friction.
12. What is a screw jack? Explain the principle, on which it works.
13. Establish a relation between the effort and load, when a square threaded screw is used for lifting purpose, taking friction into account.
14. In a screw jack, the helix angle is α and the angle of friction is ϕ. Show that its efficiency is maximum, when $2\alpha = 90° - \phi$.
15. Derive from first principle an expression for the effort required to raise a load with a screw jack taking friction into consideration.
16. Neglecting collar friction, from first principles prove that the maximum efficiency of a square threaded screw moving in nut is $\dfrac{(1 - \sin\phi)}{(1 + \sin\phi)}$ where ϕ is the friction angle.
17. Derive the expression for the torque transmitting capacity of a single plate clutch by considering uniform pressure.

18. Describe with a neat sketch the torsion dynamometer.
19. What are various types brakes used?
20. What are the differences between the brakes and clutches?
21. What is the purpose of the dynamometer and name any two types.
22. What is the advantage of differential band brake?

6.8.2 Problems

Inclined plane

1. A load of 500 N is lying on an inclined plane, whose inclination with the horizontal is 30°. If the coefficient of friction between the load and the plane is 0.4, find the minimum and maximum horizontal force, which will keep the load in equilibrium. [*Ans.* 72.05 N, 635.4 N]

Screw jack

1. A load of 10 kN is raised by means of a screw jack, having a screw threaded screw of 12 mm pitch and diameter 50 mm. If a force of 100 N is applied at the end of a lever to raise the load, what should be the given length of the lever used? Take coefficient of friction = 0.15. What should be the mechanical advantage obtained? State whether the screw is self-locking.

2. A screw jack has a square thread of mean diameter 6 cm and pitch 0.8 cm. The coefficient of friction at the screw thread is 0.09. A load of 3 N is to be lifted through 12 cm. Determine the torque required and the work done in lifting the load through 12 cm. Find the efficiency of the jack also.

3. A screw jack has a screw thread 7.5 cm mean diameter and 1.5 cm pitch. The load on the jack revolves with the screw. The coefficient of friction at the screw thread is 0.05.
Find
(i) the tangential force to be applied to the jack at 36 cm radius so as to lift a load of 600 N.
(ii) State whether the jack is self-locking. If it is, find the torque necessary to lower the load. If not, find the torque which must be applied to keep the load from descending.

4. The mean diameter of a square threaded screw jack is 50 mm. The pitch of the thread is 10 mm. The coefficient of friction is 0.15. What force must be applied at the end of a 0.7 long lever, which is perpendicular to the longitudinal axis of the screw to raise a load of 20 kN and lower it?

5. A square threaded screw jack of mean diameter 25 mm and a pitch of 6 mm is used to lift the load of 1500 N. Find the force required at the mean circumference if the coefficient of friction between the screw and nut is 0.02. [*Ans.* 144.2 N]

6. A square threaded screw jack of mean diameter 50 mm has 3° angle of inclination of the thread and coefficient of friction 0.06. Find the effort required at the end of handle 450 mm long (i) to raise a load of 20 kN, and (ii) to lower the same load. [*Ans.* 126 N; 8.4 N]

Friction 183

7. A screw jack has a square thread of mean diameter 60 mm and pitch of 8 mm. The coefficient of friction at the screw thread is 0.09. A load of 3 kN is to be lifted through 12 cm. Determine the torque required and the work done in lifting the load through 12 cm. Find the efficiency of the jack also.

8. Find the force to be applied at the end of one metre long handle of a screw jack, so that 5 tonne load is lifted with constant velocity. The screw has a single start square thread having a pitch of 2 cm. The rod diameter is 5 cm. The coefficient of friction is 0.15. Find the mechanical efficiency of the jack when the load is being raised. Neglect collar friction.
   [*Ans.* Mech. efficiency = 43.6%]

9. A screw jack spindle moving in a fixed nut has single square threads on it. The pitch of the threads is 1 cm and outside diameter of the spindle is 5.5 cm. The load which does not rotate is carried on a swivel head whose bearing is 9 cm. If the coefficient of friction for thread and nut is 0.15 and that between swivel head and spindle is 0.12, find the load that can be lifted by applying a force of 20 kg at the end of 40 cm long lever. What is the efficiency of lifting?
   [*Ans.* W = 741 kg and Efficiency 14.79]

Screw friction

1. A 150 mm diameter valve, against which a steam pressure of 2 MN/m^2 is acting, is closed by means of a square threaded screw 50 mm in external diameter with 6 mm pitch. If the coefficient of friction is 0.12, find the torque required to turn the handle.

2. A square threaded bolt of root diameter 22.5 mm and pitch 5 mm is tightened by screwing a nut whose mean diameter of bearing surface is 50 mm. If the coefficient of friction for the nut and bolt is 0.1 and for nut and bearing surface is 0.16, find the force required at the end of a spanner 500 mm long when the load on the belt is 10 kN.

Collar bearings

1. The thrust on the propeller shaft of a marine engine is taken up by 8 collars whose external and internal diameters are 600 mm and 420 mm respectively. The thrust pressure is 0.4 MN/mm^2 and may be assumed uniform. The coefficient of friction between the shaft and collars is 0.04. Find: (i) total thrust on the collars and (ii) power absorbed by friction on the bearing.

2. A conical pivot bearing supports a vertical shaft of 200 mm diameter. It is subjected to a load of 30 kN. The angle of cone is 1200 and coefficient of friction is 0.025. Find the power lost in friction when the speed is 140 r.p.m., assuming 1. Uniform pressure and 2. Uniform wear determine the axial force to engage the clutch and width of face if the coefficient of friction is 0.25. Assume a contact pressure of 8 N/cm^2.

3. A shaft has a number of collars integral with it. External diameter of collars is 40 cm and shaft diameter is 25 cm. If the uniform intensity of pressure is 3.5 kg/sq.cm and its coefficient of friction is 0.05, estimate (a) HP absorbed in friction when the shaft runs at 105 r.p.m. and carries a load of 15 tonnes and (b) number of collars required. [*Ans.* (a) 18.2, (b) 6]

Clutches

1. A single plate clutch both sides effective is required to transmit 32 kW at 2000 r.p.m., the pressure being applied axially by means of springs and limited to 15 N/cm². If the outer diameter of the plate is to be 30 cm, find the required inner diameter of the clutch ring and the total force exerted by the springs. Assume the wear to be uniform and the coefficient of friction of 0.3.

2. A single plate clutch is required to transmit 36 HP at 1600 r.p.m. The outside diameter of the plate is 30 cm, intensity of pressure is 0.7 kg/cm². Assuming uniform wear and a coefficient of friction of 0.3, find the required inner diameter of the plates and axial force necessary to engage the clutch. [**Ans.** $r_2 = 9$ cm and $W = 237.6$ kg]

3. A friction clutch is required to transmit 34.5 kW at 2000 r.p.m. It is to be single plate disk type with both sides of the plate effective. The pressure is being applied axially by means of springs and limited to 70 kPa on the plate. If the outer diameter of the friction limit is 1.5 times the internal diameter, find the required dimensions d_1 and d_2 of the clutch ring and the total force exerted by the springs. Assume uniform wear condition. [Coefficient of friction = 0.3]

Brakes and dynamometer

1. A differential band brake acting on the 3/4th of the circumference of a drum of 450 mm diameter is to provide a braking torque of 300 Nm. One end of the band is attached to a pin 100 mm from the fulcrum of the lever and the other end to another pin 25 mm from the fulcrum on the other side of it where the operating force is also acting. If the operating force is applied at 500 mm from the fulcrum and the coefficient of friction is 0.25, find the two values of the operating force corresponding to the two directions of rotation of the drum.

2. Calculate the required force to be applied in a single block brake shown at Fig. 6.10(c). The coefficient of friction is 0.3. It sustains 200 Nm of torque at 300 r.p.m. Take the direction of rotation a clockwise and the angle of contact as (i) 35° (ii) 100°. Take $a = 750$ mm; $b = 150$ mm and $c = 30$ mm.

3. In the band block brake shown in Fig. 6.12, the band is lined with 12 blocks and each subtends an angle of 15° at the centre of the brake drum. With the lever arrangement shown, determine the least force required for the brake to absorb 150 kW at 50 r.p.m. Assume $m = 0.4$.

4. The following data refer to a laboratory experiment with rope brake:
   Diameter of the flywheel = 1 m;
   Diameter of rope = 10 mm;
   Dead weight on the brake = 50 kg;
   Speed of the engine = 200 r.p.m.;
   Spring balance reading = 150 N. Find the power of the engine.

6.8.3 Multiple Choice Questions

1. The efficiency of a screw jack may be increased by
   (a) increasing its pitch
   (b) decreasing its pitch
   (c) increasing the load to be lifted
   (d) decreasing the load to be lifted. [*Ans.* (a)]

2. The efficiency of the screw jack is maximum when the helix angle is equal to
   (a) $45° + \frac{\phi}{2}$ (b) $45° - \frac{\phi}{2}$ (c) $\frac{\phi}{2} + 30°$ (d) $\frac{\phi}{2} - 30°$ [*Ans.* (b)]

3. The efficiency of a screw jack depends on
   (a) the pitch of threads
   (b) the lead
   (c) both pitch and lead
   (d) neither of them. [*Ans.* (a)]

4. The efficiency of a screw jack increases with a/an
   (a) decrease in lead
   (b) increase in lead
   (c) decrease in pitch
   (d) increase in the pitch [*Ans.* (d)]

5. The efficiency of a screw jack is
   (a) $\eta = \frac{\tan \alpha}{\tan(\phi - \alpha)}$ (b) $\eta = \frac{\tan(\alpha + \phi)}{\tan \alpha}$ (c) $\eta = \frac{\tan \alpha}{\tan(\phi + \alpha)}$ (d) $\eta = \frac{\tan(\alpha - \phi)}{\tan \alpha}$ [*Ans.* (c)]

6. The efficiency of a screw jack is maximum when
   (a) $\alpha = 45° - \frac{\phi}{4}$ (b) $\alpha = 45° + \frac{\phi}{2}$ (c) $\alpha = 45° + \frac{\phi}{4}$ (d) $\alpha = 45° - \frac{\phi}{2}$ [*Ans.* (d)]

7. The maximum efficiency of a screw jack is given by
   (a) $\eta = \frac{(1 + \sin \phi)}{(1 - \sin \phi)}$ (b) $\eta = \frac{(1 - \sin \phi)}{(1 + \sin \phi)}$ (c) $\eta = \frac{(1 - \sin \phi)}{(1 + \cos \phi)}$ (d) $\eta = \frac{(1 + \sin \phi)}{(1 - \cos \phi)}$ [*Ans.* (b)]

8. The flat and conical pivots, the ratio of the friction torque with uniform wear to the friction torque with uniform pressure is
   (a) $\frac{2}{3}$ (b) $\frac{3}{2}$ (c) $\frac{4}{3}$ (d) $\frac{3}{4}$ [*Ans.* (d)]

9. The frictional torque for the same diameter in a conical bearing is _____ than in a flat bearing
   (a) more (b) less (c) equal (d) may be more or less [*Ans.* (a)]

10. For a safe design, a friction clutch is designed assuming
    (a) uniform pressure (b) uniform wear theory (c) any one of them [*Ans.* (b)]

11. No force is required for downward motion of a load on a screw jack if
    (a) $\alpha < \phi$ (b) $\alpha > \phi$ (c) $\alpha < 2\phi$ (d) $\alpha > 2\phi$ [*Ans.* (b)]

12. In a multiple-friction clutch, the number of active friction surfaces is

    (a) $2n$ (b) n (c) $2(n-1)$ (d) $n-1$ [Ans. (d)]

13. The force of friction between two bodies in contact

    (a) Depends upon the area of their contact
    (b) Depends upon the relative velocity between them
    (c) Is always normal to the surface of their contact
    (d) All of the above.   [Ans. (c)]

14. The magnitude of the force of friction between two bodies, one lying above the other, depends upon the roughness of the

    (a) Upper body   (b) Lower body
    (c) Both the bodies (d) The body having more roughness.   [Ans. (c)]

15. The force of friction always acts in a direction opposite to that

    (a) In which the body tends to move   (b) In which the body is moving
    (c) Both (a) and (b)   (d) None of the above two.   [Ans. (c)]

16. Which of the above statement is correct?

    (a) The force of friction does not depend upon the area of contact
    (b) The magnitude of limiting friction bears a constant ratio to the normal reaction between the two surfaces
    (c) The static friction is slightly less than the limiting friction
    (d) All (a), (b) and (c).   [Ans. (d)]

17. Dynamometer is a device for measuring

    (a) torque   (b) speed   (c) power   (d) all.   [Ans. (a, c)]

18. Which of the following is classified as absorption dynamometer

    (a) Differential band brake   (b) Band and block brake
    (c) Proney brake   (d) All the above.   [Ans. (c)]

Governors 7

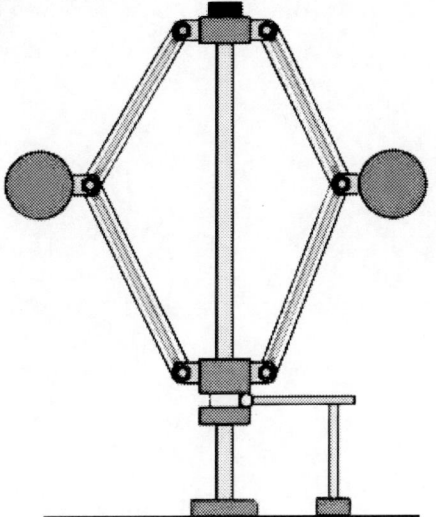

James Watt's Centrifugal Governor

Governors

7.1 INTRODUCTION

The main function of the governor is to maintain the engine speed within the prescribed limits for varying load conditions. Thus, the governor is a device which automatically controls the supply of working fluid to the engine with the varying load conditions and keeps the speed within certain limits.

Flywheel is another device which controls the speed variations caused by the fluctuations in the turning moment of the engine during each cycle of operation. The flywheel does not control the speed variation caused by a varying load. The varying demand for power is met by the governor regulating the supply of working fluid. Hence, the governor mechanism must be so designed that the variation in mean speed is as small as possible.

There are mainly two types of governors. 1. Centrifugal governors and 2. Inertia type governors. The centrifugal type works due to the variation in the centrifugal force. Examples: Pendulum or Simple Watt governor, Loaded type of governors which are again of two types (a) Dead weight type: Examples: Porter and Proell governors. (b) Spring loaded or controlled type: Examples: Hartnell and Hartung governors. Inertia governors work due to the variation in the inertia forces caused by the rate of change of speed which may be an angular acceleration or retardation of the engine shaft. Of course these are not used in practice. But simplicity of centrifugal governors makes them more popular. Hence, in this chapter only centrifugal governors are discussed.

7.2 CENTRIFUGAL GOVERNORS

A centrifugal governor in its simple form consists of two heavy balls (known as flyballs or governor balls) of equal mass attached to the arms as shown in Fig. 7.1. The arms are pivoted at their upper ends to a rotating shaft known as spindle. The flyballs are also connected to sleeve through the links. The sleeve revolves with the spindle and can also slide up and down on the spindle. The spindle is driven by the shaft of an engine either by a belt or by a gear arrangement. The change in the centrifugal forces of the rotating masses due to change in speed of the engine is used to move the sleeve upward or downward to control the throttle valve through levers. This valve in turn controls the supply of oil to the engine.

The Pic. 7.1 shows the governor with two balls. The bevel gears provided operate when the sleeve moves to lower and upper positions. An inward radial force acts on the balls, provided by a dead weight or a spring or a combination of both is called controlling force. When the governor balls are revolving at a uniform speed, the radius of rotation will be such that the outward centrifugal force is just balanced by the controlling force. When the load decreases on the engine the speed increases causing the governor balls to move outwards. Similarly, when the load increases the speed decreases causing the balls to move inwards until the centrifugal force is again balanced by the controlling force.

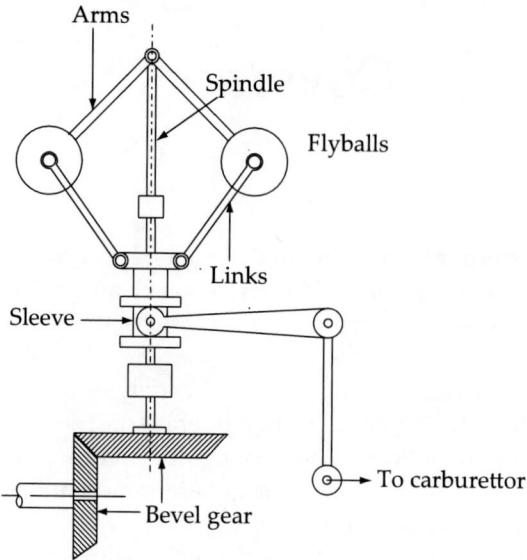

Fig. 7.1 Centrifugal governor.

This movement of the balls is transmitted by the governor mechanism to the valve which controls the amount of fuel supplied to the engine. The movement in the outward direction reduces the supply and movement in the inward direction increases the supply of fuel.

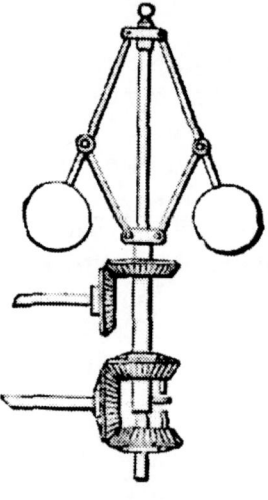

Pic. 7.1

7.3 VARIOUS PARTS AND TERMS USED IN GOVERNORS

The various parts such as arms, spindle, links, flyballs and sleeve found in the centrifugal governors are shown in Fig. 7.1.

The following are the important terms used in governors from the subject point of view:

7.3.1 Height of the Governor (h)

It is the vertical distance between two planes passing through the centre of the balls and the point where the axes of the arms (produced if necessary) intersect on the spindle axis. Figure 7.2 shows three different cases where the arms pivoted on the spindle itself as shown at (a), the arms pivoted away from the spindle axis as shown at (b) and with crossed arms intersecting the spindle axis at o as shown at (c).

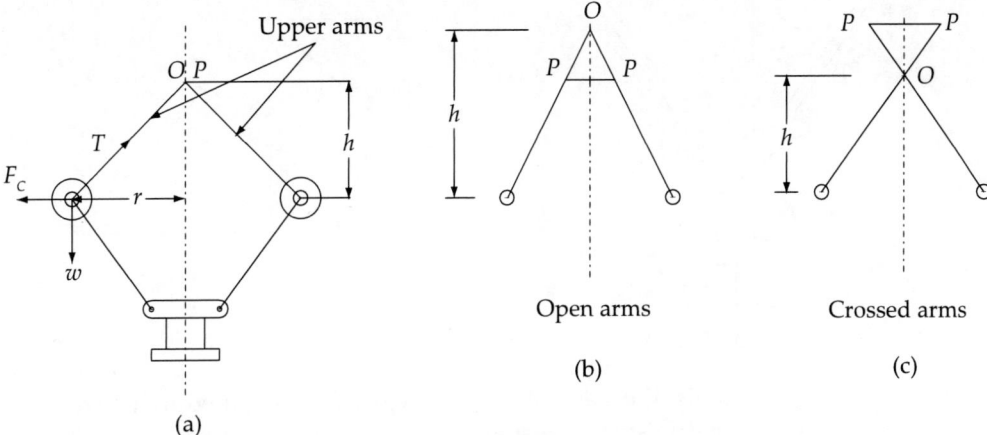

Fig. 7.2 Different types of upper arms.

7.3.2 Equilibrium Speed

It is the speed at which the governor balls, arms, links, etc. are in complete equilibrium and the sleeve does not tend to move up or downwards. Mean equilibrium speed is the speed corresponding to mean position of the balls and the sleeve. Maximum or minimum equilibrium speeds are the speeds at which the balls rotate with maximum and minimum radius of rotation.

7.3.3 Sleeve Lift

It is the vertical distance travelled by the sleeve due to the change in the equilibrium speed from minimum to maximum.

7.4 SIMPLE WATT GOVERNOR

Figure 7.3(a) shows the simplest form of a centrifugal Watt governor. It is basically a conical pendulum with links connecting flyballs and sleeve of negligible mass. Let m be the mass of each ball in kg and w (mg) be the weight of each ball in Newtons; T be the tension in the arms;

ω be the angular velocity of the flyballs about the spindle axis (rad/sec).
r be the radius of rotation of the flyballs, i.e. the horizontal distance from the centre of the balls to the spindle axis in metres.
h be the height of the governor in metres.
F or F_c be the centrifugal force acting on the balls equal to $m\omega^2 r$ in Newtons.

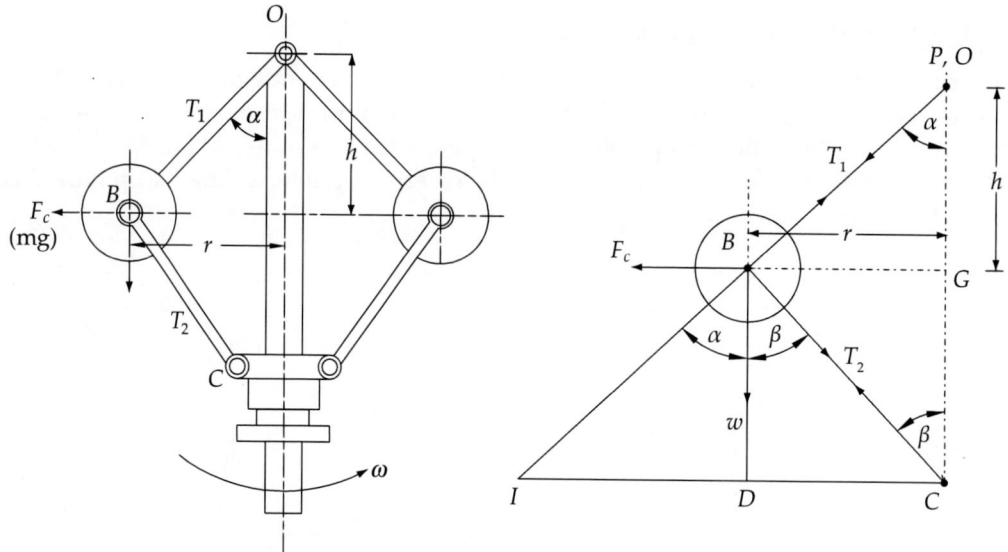

Fig. 7.3(a) Simple watt governor.　　**Fig. 7.3(b)** Line diagram of simple watt governor.

Assumptions: Weight of the arms, links and sleeve are negligible compared to the weight of the balls. As the governor is symmetric about the spindle axis, it is enough if only one half is shown in the line diagram in Fig. 7.3 (b). Assuming that the governor is revolving at uniform speed, the balls are in equilibrium under the action of the following forces acting:

1. The outward centrifugal force F_c acting on the balls,
2. The tensions T_1 or T_2 in the upper and lower arms, and
3. The gravitational force, i.e. the weight w of the balls.

Generally, the problems on governors can be done by both analytically and graphically. Both the methods are applied to simple Watt governor. But the analytical method takes lot of time compared to graphical method. Hence, it is better to practise the graphical method.

7.4.1 Analytical Method

The distances and angles are to be determined mathematically as follows:

In Fig. 7.3(b), the lengths OB, BC and radius of rotation BG, r are known.

The angles α and β have to be determined as follows. Here in this case $OB = BC$, i.e. upper arms and lower links or arms are taken equal in length so $\alpha = \beta$. So, it is enough if α is determined.

$$\sin \alpha = \frac{BG}{OB} \quad \alpha = \sin^{-1}\left(\frac{BG}{OB}\right)$$

then height of the governor $OG = h = OB \cos \alpha$.

Taking moments about O,
$$F_C \times OG = w \times BG \text{ (or)}$$
$$m\omega^2 r \times h = m \times g \times r$$

since $F_C = m\omega^2 r; w = mg$. Hence, the height of the governor 'h' (metres)

$$h = \frac{mg \cdot r}{(m\omega^2 r)} = \frac{g}{\omega^2}$$

where g is 9.81 m/s² (or) $\omega^2 = \frac{g}{h}$ or $\omega = \sqrt{\frac{g}{h}}$ radians/sec.

Speed of the governor $N = \frac{60\omega}{(2\pi)}$ in revolutions per minute.

7.4.2 Graphical Method

This method is simple compared to analytical method. Follow the procedure given below:

1. Draw the space diagram using a suitable scale.
2. Draw spindle axis OC and another parallel line BD at a distance equal to radius r of the governor.
3. Take O as centre and draw an arc with radius equal to upper arm OB to get point B representing the centre of the ball.
4. Again with B as centre, draw an arc with radius equal to lower link or arm BC to get point C on the spindle axis.
5. Join O to B and B to C. Draw a small circle at B to represent flyball.
6. Draw horizontal from B to intersect spindle axis at G. Measure height between O and G which represents height of governor 'h'.
7. Take moments about O giving $F_C \times OG = w \times BG$ or $m\omega^2 r \times h = mg \cdot r$ just as equation (1).
8. Determine instantaneous centre I of BC and take moments about I giving $F_C \times BD = w \times ID = mg \times ID$.

   Since $F_C = m\omega^2 r$ then $\omega^2 = \dfrac{mg \times ID}{(m \times r \times BD)} = \dfrac{9.81 \times ID}{(r \times BD)}$ or $\omega = \sqrt{\left[\dfrac{9.81 \times ID}{(r \times BD)}\right]}$

9. Calculate ω knowing r, ID and BD from the drawing in metres.
10. Speed $N = \dfrac{60\omega}{2\pi}$

Note:

1. Graphical method is easier than analytical method.
2. Care should be taken in choosing the scale properly as the results depend on the values measured from the diagram.
3. The height of the governor h is independent of the weight of flyballs.
4. The height is inversely proportional to square of the speed.

W E 7.1: A centrifugal Watt governor is fitted with two balls, each of mass 2.5 kg. Find the height of the governor, when it is running at 75 r.p.m. Also find the speed of the governor, when the balls (i) rise by 20 mm and (ii) fall by 20 mm. Neglect friction of the governor.

Given:
Mass of flyballs (m) = 2.5 kg and N speed of governor = 75 r.p.m.

So the angular velocity of the governor

$$\omega = \frac{2\pi N}{60} = \frac{2\pi 75}{60} = 2.5\pi \text{ rad/s}$$

Height of the governor:

$$h = \frac{g}{\omega^2} = \frac{9.81}{(2.5\pi)^2} = 0.159 \text{ m} = 159 \text{ mm. (Ans.)}$$

(i) Speed of the governor when the balls rise by 20 mm:

The height of the governor h reduces to $h_1 = 159 - 20 = 139$ mm $= 0.139$ m.

Hence, the speed N_1 corresponding to this height h_1:

$$\omega_1^2 = \frac{g}{h_1} = \frac{9.81}{0.139}$$

$$= 70.5; \omega_1 = 8.4 \text{ rad/s}$$

Speed in r.p.m., $N_1 = \dfrac{60\omega_1}{2\pi} = \dfrac{60 \times 8.4}{2\pi} = 80.2$ r.p.m. **(Ans.)**

(ii) Speed of the governor when the balls fall by 20 mm:

The height of the governor h increases to $h_2 = 159 + 20 = 179$ mm $= 0.179$ m.

Hence, the speed N_2 corresponding to this height h_2:

$$\omega_2^2 = \frac{g}{h_2} = \frac{9.81}{0.179} = 54.7$$

$$\omega_2 = 7.4 \text{ rad/s}$$

Speed in r.p.m., $N_2 = \dfrac{60\omega_2}{2\pi} = \dfrac{60 \times 7.4}{2\pi} = 70.7$ r.p.m. **(Ans.)**

7.5 PORTER GOVERNOR

Figure 7.4(a) shows the Porter governor which is similar to simple Watt governor with a dead weight W on sleeve and it moves along with the sleeve on the spindle axis up and down.

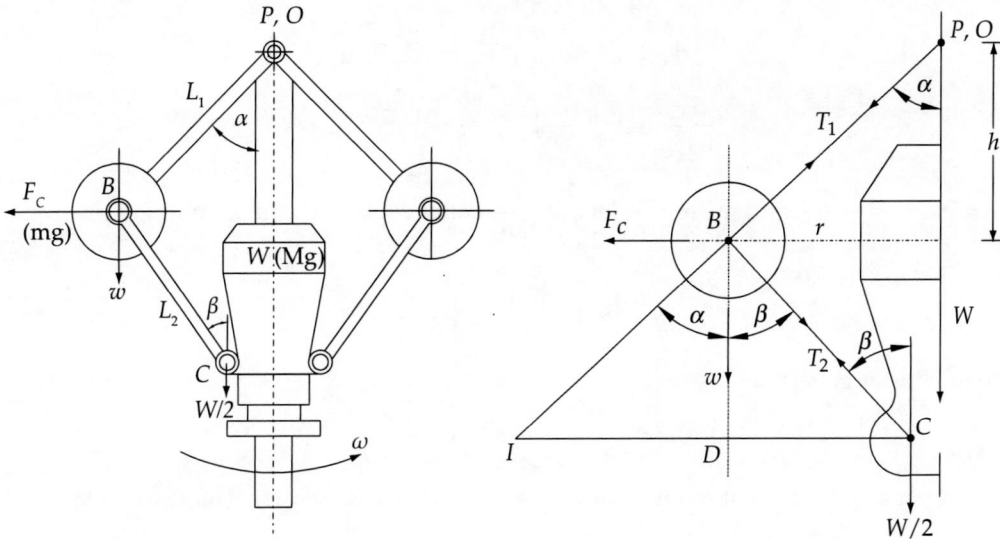

Fig. 7.4(a) Porter governor.   **Fig. 7.4(b)** Line diagram of Porter governor.

Here again neglecting the weights of the upper and lower arms and also assuming that the governor is revolving at uniform speed, the forces acting are as shown in the line diagram at Fig. 7.4(b). They are:

1. outward centrifugal force F_C
2. the gravity pull of w
3. the tension T_1 and T_2 in the upper and lower arms
4. dead weight $\dfrac{W}{2}$ acting on each side at the pin C

The equation connecting F_C, w and W is mostly conveniently derived by first finding the instantaneous centre I of the link BC and then taking moments about I as follows.

$F_C \cdot BD = w \cdot ID + \left(\dfrac{W}{2}\right) IC$. Determine the values of BD, ID and IC either by using the analytical or graphical method.

7.5.1 Analytical Method

Using the trigonometry, find the values of BD, ID and IC as follows: Rewrite the above equation as below by dividing with BD.

$$F_C = \left[\dfrac{w \cdot ID}{BD} + \dfrac{W \cdot IC}{2BD}\right]$$

$$= \left[\dfrac{w \cdot ID}{BD} + \dfrac{W}{2}\left(\dfrac{ID}{BD} + \dfrac{DC}{BD}\right)\right]$$

since $\dfrac{ID}{BD} = \tan \alpha$ and $\dfrac{DC}{BD} = \tan \beta$ from the Fig. 7.4(b).

$$F_C = w \cdot \tan \alpha + \frac{W}{2}(\tan \alpha + \tan \beta)$$

$$\frac{F_C}{\tan \alpha} = w + \frac{W}{2}\left(1 + \frac{\tan \beta}{\tan \alpha}\right)$$

since $F_C = m\omega^2 r = \dfrac{w\omega^2 r}{g}$; and $\tan \alpha = \dfrac{r}{h}$; let $q = \dfrac{\tan \beta}{\tan \alpha}$ after replacing and simplifying gives

Height of the Porter governor

$$h = \left[w + \frac{W}{2}(1+q)\right]\frac{g}{(w.\omega^2)}$$

Speed of the Porter governor

$$\omega = \sqrt{\left[w + \frac{W}{2}(1+q)\right]\frac{g}{(w.h)}} \qquad (1)$$

When upper arms are equal to lower links then $\alpha = \beta$ then $\tan \alpha = \tan \beta$ and so $q = 1$.

Height of the Porter governor,

$$h = (w + W) \cdot \frac{g}{w\omega^2}$$

Speed of the Porter governor,

$$\omega = \sqrt{\left[(w + W) \cdot \frac{g}{wh}\right]} \qquad (2)$$

Note:

(i) When W is made zero Porter governor becomes a Watt governor and the above equations (1) and (2) give the height and speed of the Watt governor.

(ii) Comparing the heights of Watt governor with the height of Porter governor, it has increased by $\left(\dfrac{w+W}{w}\right)$ times. When mass of the dead weight W is zero, the height of the governor h will become equal to that of the Watt governor, i.e. $\dfrac{g}{\omega^2}$.

Effect of friction between the sleeve and spindle

Let F_f be frictional force acting between the sleeve and the spindle, then the height of the governor,

$$h = \left[w + \left(\frac{W \pm F_f}{2}\right)(1+q)\right]\frac{g}{w\omega^2} \text{ or}$$

$$\omega^2 = \left[w + \left(\frac{W \pm F_f}{2}\right)(1+q)\right]\frac{g}{wh}$$

Note: + sign is used when sleeve moves upwards as frictional force gets added to dead weight W on the sleeve and − sign when the sleeve moves downwards as frictional force opposes the force due to dead weight W.

Range of speed = Maximum speed of the governor − Minimum speed of the governor.

7.5.2 Graphical Method

Draw the line diagram choosing proper scale as shown at Fig. 7.4(b). Extend OB and draw CI perpendicular to OC, the point of intersection gives I. Drop from B a line BD on to CI. Measure the lengths BD, ID and IC and substitute in the following equation obtained by taking moments about I, giving

$$F_C \cdot BD = w \cdot ID + (W/2)IC$$

Substitute for $F_C = m\omega^2 r$ and $w = mg$ and $W = Mg$. Then determine the speed N of the governor knowing the distances BD, ID and IC.

Hence, drawing the space diagram and measuring ID, IC and BD and using the given values of m, M and radius r, the speed can be calculated for any configuration easily.

Note: r should be in metres and masses, m and M in kilograms.

7.6 PROELL GOVERNOR

The Proell governor is similar to Porter governor but differs from it in the arrangement of the balls. The balls are provided by extending the upper arms as shown in Pic. 7.2 while in rotation. But in the case of Proell governors, the balls are carried on the extensions of the lower arms instead of the upper arms as shown in the Fig. 7.5(a).

Pic. 7.2 Governor while rotating.

The working principle and analysis is similar to that of Watt governor or Porter governor. Referring to Fig. 7.5(b), the instantaneous centre I of the lower arm BC lies at the point of intersection of OE produced with a line drawn through C at right angles to the governor axis. Then taking moments about I assuming the extension of the lower arm EB as vertical,

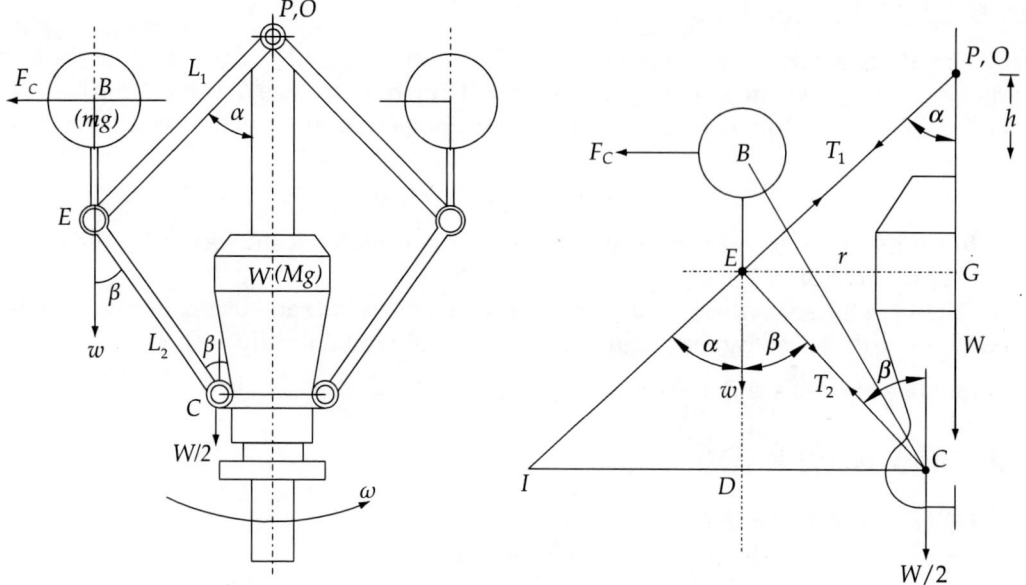

Fig. 7.5(a) Proell governor.   **Fig. 7.5(b)** Line diagram of Proell governor.

$$F_C \cdot BD = w \cdot ID + \left(\frac{W}{2}\right) \cdot IC$$

The distances BD, ID and IC can be determined by both analytically and graphically as follows.

7.6.1 Analytical Method

This method is just applied for the simple position (the extension EB is parallel to the spindle axis) of the Proell governor as shown in Fig. 7.5(b).

Divide the above equation both sides by ED giving,

$$F_C \left(\frac{BD}{ED}\right) = w \frac{ID}{ED} + \frac{W}{2}\left(\frac{IC}{ED}\right)$$

since $\dfrac{ED}{ID} = \tan \alpha$; $\dfrac{IC}{ED} = \dfrac{ID}{ED} + \dfrac{DC}{ED} = \tan \alpha + \tan \beta$ so

$$F_C \cdot \frac{BD}{ED} = m\omega^2 r \cdot \frac{BD}{ED}$$

$$= w \cdot \tan \alpha + \left(\frac{W}{2}\right) \tan \alpha + \tan \beta$$

$$= \left(w + \frac{W}{2}\right) \tan \alpha + \left(\frac{W}{2}\right) \tan \beta$$

Therefore,
$$F_C = \left(\frac{ED}{BD}\right)\left[\left(w + \frac{W}{2}\right)\tan\alpha + \frac{W}{2}\tan\beta\right]$$

This can be rewritten as
$$F_C = \left(\frac{ED}{BD}\right)\left[w + \frac{W}{2}(1+q)\right]\tan\alpha \qquad (1)$$

where q is $\dfrac{\tan\alpha}{\tan\beta}$ and if $\beta = \alpha$ then $q = 1$ and the equation (1) becomes

$$F_C = \left(\frac{ED}{BD}\right)[w + W]\tan\alpha \qquad (2)$$

$F_C = \dfrac{w\omega^2 r}{g}$ and $\tan\alpha = \dfrac{r}{h}$ substituting in (1) gives

$$\frac{w\omega^2 r}{g} = \frac{ED}{BD}\left[w + \frac{W}{2}(1+q)\right]\tan\alpha$$

Therefore,
$$\omega^2 = \left[\frac{ED}{BD}\left\{w + \frac{W}{2}(1+q)\right\}\tan\alpha\right]\frac{g}{rw}$$

If $q = 1$, then $\omega^2 = \left[\dfrac{ED}{BD}\left\{\dfrac{w+W}{w}\right\}\right]\dfrac{g}{h}$ \qquad (3)

Note: Comparing this with that of the Porter governor, it will be obvious that the effect of placing the ball at B instead of at the pin-joint E is to reduce the equilibrium speed for given values of w, W and h.

1. When $BE = 0$, then $ED = BD$; Proell governor becomes a Porter governor.
2. Balls of smaller masses are used in Proell governor than in the Porter governor in order to give the same equilibrium speeds for given values of W and h.
3. The important effect of the change in position of the balls is to reduce the increase of speed necessary in order to lift the sleeve by a given amount.

7.6.2 Graphical Method

Construct line diagram for the simple position (the extension EB parallel to spindle axis) as in Fig. 7.5(b) by drawing two parallel lines at a distance EG equal to radius r. Take O as centre and with radius equal to length of the upper arm, draw an arc and get point E. With E as centre and radius equal to length of the lower link or arm EC, draw an arc and get point C. From E, draw EB parallel to spindle axis and mark its length to get ball centre B join B and C.

Note: The length between B and C will remain fixed for all other positions as CEB is a bell crank lever.

Determine the instantaneous centre, I of the link EC as shown.

Measure the lengths ED, ID and IC. Substitute these lengths in the following equation and determine the required speed or height or any unknown.

$$F_C \cdot BD = w \cdot ID + \frac{W}{2}IC$$

Note:

1. Selection of proper scale is important as the results depend on the lengths obtained from the diagram drawn.
2. The analytical method is very tedious for other positions of the Proell governor.
3. Therefore, the graphical method is all the more convenient to do for any given radius.

Figure 7.5(c) and (d) show the line diagrams of the Proell governor for both minimum radius r_1 and maximum radius r_2 with the extension EB being inclined to spindle axis inward for minimum radius and outward for maximum radius.

Applying analytical method is more tedious and time taking for these positions. Hence, it is advisable to adopt graphical method explained below.

First determine the distance between the two points B and C of the bell crank lever CEB as indicated above by assuming the extension EB as parallel, i.e. between the ball centre B and lower arm pin joint C as shown in Fig. 7.5(b).

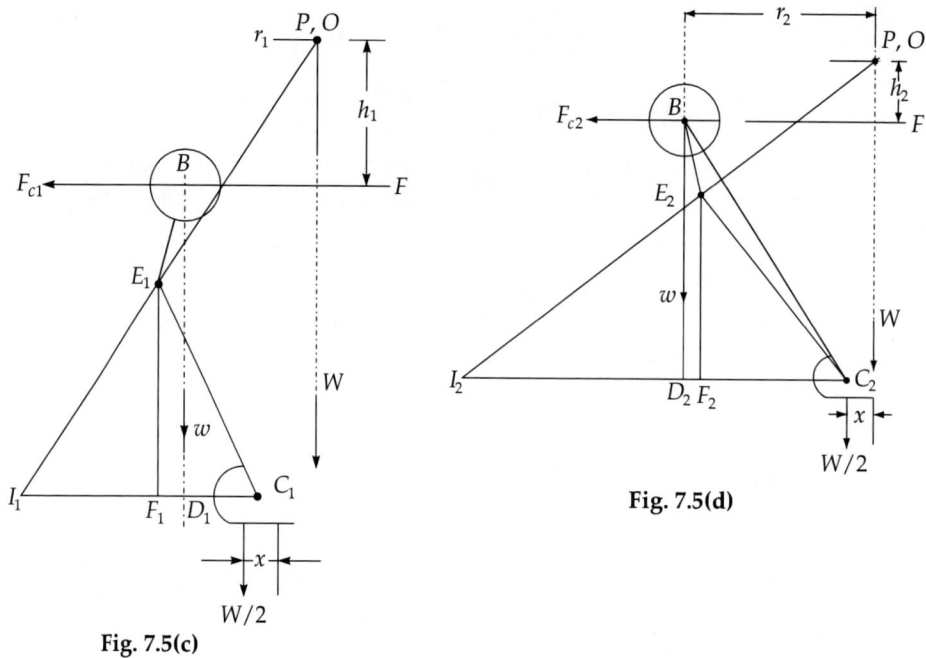

Fig. 7.5(c)

Fig. 7.5(d)

Procedure of construction

The construction of the line diagram for any other radius shown in Fig. 7.5(c and d) is as follows:

1. Draw two parallels to spindle axis with the given radius r_1 and r_2 separately.
2. Take points C_1 and C_2 as centres on the spindle axis or away from it and with CB as radius, draw arcs to get ball centres B_1 and B_2.
3. With B_1 and B_2 as centres and radius equal to extension link BE, draw two arcs.

Note: $BE = BE_1 = BE_2$; $CE = C_1E_1 = C_2E_2$; $EO = E_1O = E_2O$

4. With C_1 and C_2 as centres and radius equal to CE, draw two arcs to get E_1 and E_2 at the point of intersections.
5. With E_1 and E_2 as centres and radius equal to upper arms, EO, draw two arcs and get O_1 and O_2.
6. Join them as shown and determine the instantaneous centres I_1 and I_2.
7. From E_1, B_1, E_2 and B_2 drop verticals and name the points of intersections with C_1I_1 and C_2I_2 as D_1, F_1, D_2 and F_2 as shown.
8. Measure the distances B_1D_1, I_1D_1, I_1C_1, B_2D_2, I_2D_2 and I_2C_2 and convert according to the scale adopted.
9. Substitute these values in the equation obtained by taking moments about I_1 and I_2 giving

$$F_{C1} \times B_1D_1 = w \times I_1D_1 + W \times \frac{I_1C_1}{2} \text{ for radius } r_1 \text{ and}$$

$$F_{C2} \times B_2D_2 = w \times I_2D_2 + W \times \frac{I_2C_2}{2} \text{ for radius } r_2$$

10. Replace F_{C1} and F_{C2} by $m\omega_1^2 r_1$ and $m\omega_2^2 r_2$. Replace w and W by mg and Mg. Simplify taking out constants to get expressions for (N_1) and (N_2) as follows.
11. The expression are

$$N_1 = 29.9 \sqrt{\left[\left(\frac{I_1D_1}{B_1D_1} + \frac{M \cdot I_1C_1}{2m \cdot B_1D_1}\right)\frac{1}{r_1}\right]} \text{ and similarly for}$$

$$N_2 = 29.9 \sqrt{\left[\left(\frac{I_2D_2}{B_2D_2} + \frac{M \cdot I_2C_2}{2m \cdot B_2D_2}\right)\frac{1}{r_2}\right]}$$

Note: The radii r_1 and r_2 should be in metres and m and M in kg. All other distances are either in metres or millimetre.

W E 7.2: A Porter governor has upper arms of length 300 mm. The upper arms are pivoted on the axis of rotation. The weight of each ball is 10 kg and the load on the sleeve is 100 kg. If the extreme radii of rotation of the governor balls are 150 mm and 225 mm, find the corresponding equilibrium speeds and range of speed in each case when

(a) the lower arms are of 300 mm length and attached to the sleeve on the spindle axis
(b) the lower arms are of 350 mm length and attached to the sleeve on the spindle axis
(c) the lower arms are of 300 mm length and attached to the sleeve at distances of 40 mm from the axis
(d) crossed upper arms of 300 mm attached 40 mm on either side of spindle
(e) open upper and lower arms of 300 mm attached away by 40 mm from the spindle.

Given:
Upper arms = 300 mm;   Weight of each ball w = 10 kg;   Dead weight of the sleeve W = 100 kg;
Maximum radius r_2 = 225 mm,   Minimum radius r_1 = 150 mm.

Graphical method is easy to adopt for this Porter governor with different arrangements. The procedure is explained below:

1. For all the different cases, first choose a proper scale commonly and draw the space diagrams and determine the instantaneous centres I_1 and I_2 for both radii.
2. Name all points with suffixes 1 and 2 for easy identification of position 1 with radius r_1 and position 2 with radius r_2 as shown below for each case.
3. Measure the following distances according to the scale adopted from (i) and (ii) in each case:
$B_1D_1 = \ ; I_1D_1 = \ ; I_1C_1 = \ ; B_2D_2 = \ ; I_2D_2 = \ ;$ and $I_2C_2 = \ ;$
4. Substitute these values in the equation obtained by taking moments about I_1 and I_2. The simplified expressions after taking out constants for radius r_1 is

$$N_1 = 29.9 \sqrt{\left[\left(\frac{I_1D_1}{B_1D_1} + \frac{M \cdot I_1C_1}{2m \cdot B_1D_1}\right)\frac{1}{r_1}\right]}$$

= r.p.m.   and similarly for radius r_2 is

$$N_2 = 29.9 \sqrt{\left[\left(\frac{I_2D_2}{B_2D_2} + \frac{M \cdot I_2C_2}{2m \cdot B_2D_2}\right)\frac{1}{r_2}\right]}$$

= r.p.m.

Range of speed = $N_2 - N_1$ = r.p.m. **Ans.**

This is demonstrated in the following different cases:

Case (a): *The upper and lower arms are of 300 mm length and attached to the sleeve on the spindle axis as shown in Fig. 7.6(a).*

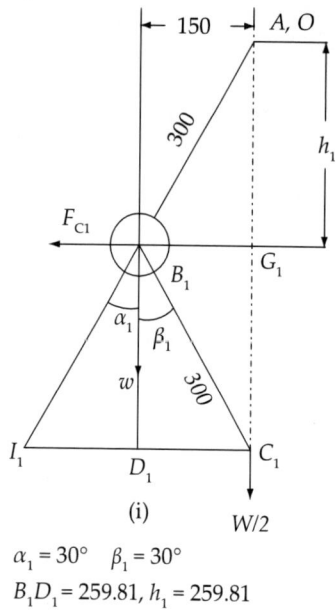

(i)   W/2

$\alpha_1 = 30°$ $\beta_1 = 30°$
$B_1D_1 = 259.81$, $h_1 = 259.81$
$I_1D_1 = 150$
$I_1C_1 = 300$

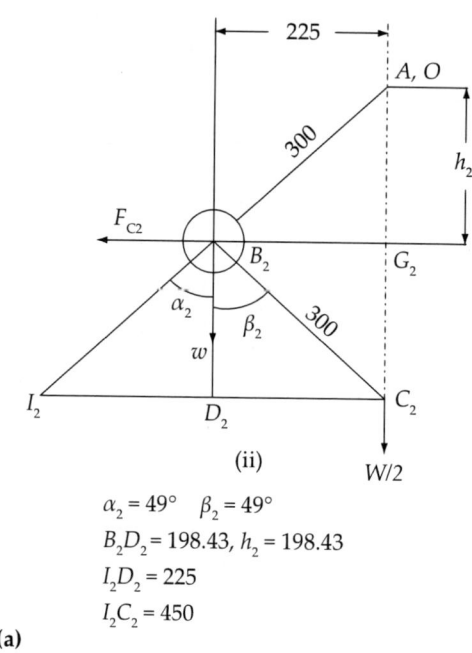

(ii)   W/2

$\alpha_2 = 49°$ $\beta_2 = 49°$
$B_2D_2 = 198.43$, $h_2 = 198.43$
$I_2D_2 = 225$
$I_2C_2 = 450$

Fig. 7.6(a)

Measure the distances according to the scale adopted from (i) and (ii):

$B_1D_1 = 259.8$ mm; $I_1D_1 = 150$ mm; $I_1C_1 = 300$ mm; $B_2D_2 = 198.43$ mm; $I_2D_2 = 225$ mm; and $I_2C_2 = 450$ mm;

Substitute these values in the equation obtained by taking moments about I_1 and I_2, and the simplified expression after taking out constants for r_1

$$N_1 = 29.9 \sqrt{\left[\left(\frac{I_1D_1}{B_1D_1} + \frac{M \cdot I_1C_1}{2m \cdot B_1D_1}\right)\frac{1}{r_1}\right]}$$

$$= 29.9 \sqrt{\left[\left(\frac{150}{259.8} + \frac{100 \times 300}{2 \times 10 \times 259.8}\right)\frac{1}{0.15}\right]}$$

$$= 194.5 \text{ r.p.m.}$$

and similarly for r_2

$$N_2 = 29.9 \sqrt{\left[\left(\frac{I_2D_2}{B_2D_2} + \frac{M \cdot I_2C_2}{2m \cdot B_2D_2}\right)\frac{1}{r_2}\right]}$$

$$= 29.9 \sqrt{\left[\left(\frac{225}{198.43} + \frac{100 \times 450}{2 \times 10 \times 198.43}\right)\frac{1}{0.225}\right]}$$

$$= 222.6 \text{ r.p.m.}$$

Range of speed = $N_2 - N_1 = 222.6 - 194.5 = 28.1$ r.p.m. **Ans.**

Case (b): *The upper arms are 300 mm and the lower arms are of 350 mm length and both attached to the sleeve on the spindle axis as shown in Fig. 7.6(b).*

Measure the distances according to the scale adopted from (i) and (ii):

$B_1D_1 = 318.92$ mm; $I_1D_1 = 181.95$ mm; $I_1C_1 = 332.57$ mm; $B_2D_2 = 268.6$ mm; $I_2D_2 = 304$ mm; $I_2C_2 = 529$ mm.

Substitute these values in the equation obtained by taking moments about I_1 and I_2 and the simplified expressions after taking out constants for r_1

$$N_1 = 29.9 \sqrt{\left[\left(\frac{I_1D_1}{B_1D_1} + \frac{M \cdot I_1C_1}{2m \cdot B_1D_1}\right)\frac{1}{r_1}\right]}$$

$$= 29.9 \sqrt{\left[\left(\frac{181.95}{318.92} + \frac{100 \times 332.57}{2 \times 10 \times 318.92}\right)\frac{1}{0.15}\right]}$$

$$= 185.7 \text{ r.p.m.}$$

and similarly for r_2

$$N_2 = 29.9 \sqrt{\left[\left(\frac{I_2D_2}{B_2D_2} + \frac{M \cdot I_2C_2}{2m \cdot B_2D_2}\right)\frac{1}{r_2}\right]}$$

$$= 29.9 \sqrt{\left[\left(\frac{304}{268.6} + \frac{100 \times 529}{2 \times 10 \times 268.6}\right)\frac{1}{0.225}\right]} = 208.9 \text{ r.p.m.}$$

Range of speed = $N_2 - N_1 = 208.9 - 185.7 = 23.2$ r.p.m. **Ans.**

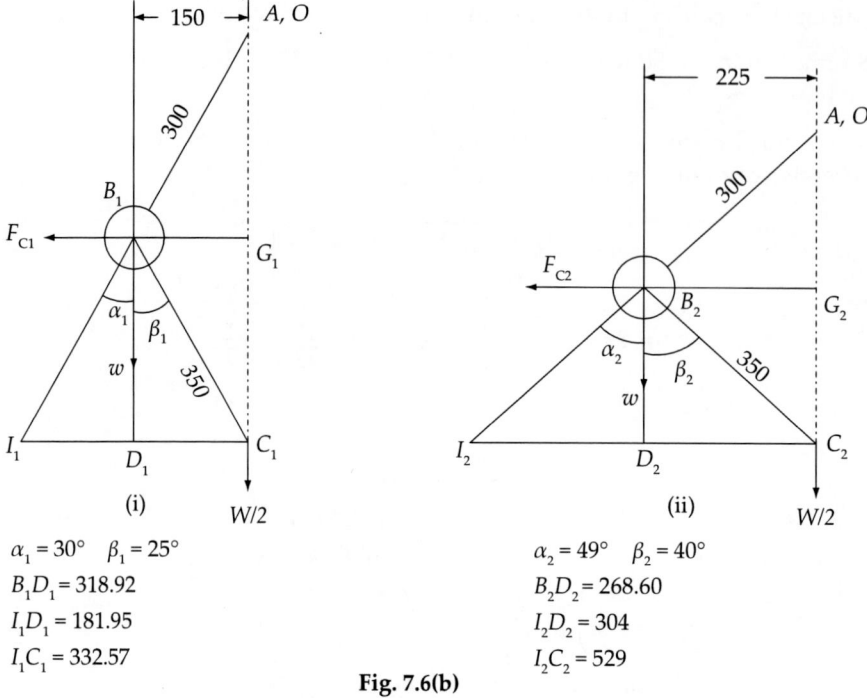

Fig. 7.6(b)

$\alpha_1 = 30°$ $\beta_1 = 25°$
$B_1D_1 = 318.92$
$I_1D_1 = 181.95$
$I_1C_1 = 332.57$

$\alpha_2 = 49°$ $\beta_2 = 40°$
$B_2D_2 = 268.60$
$I_2D_2 = 304$
$I_2C_2 = 529$

Case (c): *The upper arms on the spindle axis but the lower arms are of 300 mm length attached to the sleeve at distances of 40 mm from the axis as shown in Fig. 7.6(c).*

Measure the distances according to the scale adopted from (i) and (ii):

$B_1D_1 = 279.2$ mm; $I_1D_1 = 161.14$ mm; $I_1E_1 = 271.14$ mm; $B_2D_2 = 236.2$ mm;
$I_2D_2 = 267.2$ mm $I_2E_2 = 452.79$ mm.

Substitute these values in the equation obtained by taking moments about I_1 and I_2 and the simplified expressions after taking out constants for r_1

$$N_1 = 29.9 \sqrt{\left[\left(\frac{I_1D_1}{B_1D_1} + \frac{M \cdot I_1E_1}{2m \cdot B_1D_1}\right)\frac{1}{r_1}\right]}$$

$$= 29.9 \sqrt{\left[\left(\frac{161.14}{279.2} + \frac{100 \times 27.14}{2 \times 10 \times 279.2}\right)\frac{1}{0.15}\right]} = 179.9 \text{ r.p.m.}$$

and similarly for r_2

$$N_2 = 29.9 \sqrt{\left[\left(\frac{I_2D_2}{B_2D_2} + \frac{M \cdot I_2E_2}{2m \cdot B_2D_2}\right)\frac{1}{r_2}\right]}$$

$$= 29.9 \sqrt{\left[\left(\frac{267.16}{236.17} + \frac{100 \times 452.79}{2 \times 10 \times 236.17}\right)\frac{1}{0.225}\right]} = 206.4 \text{ r.p.m.}$$

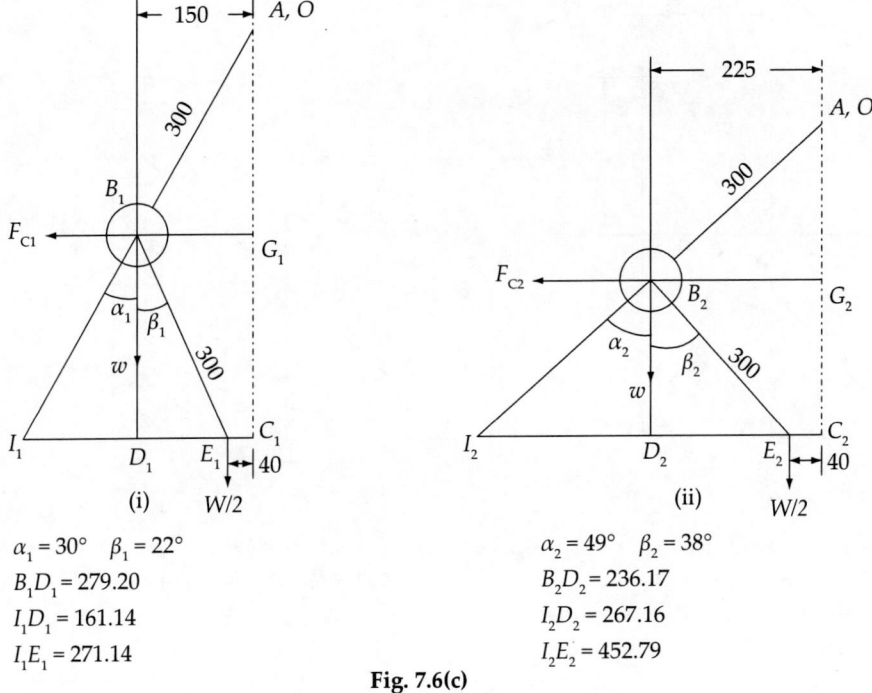

$\alpha_1 = 30°$ $\beta_1 = 22°$
$B_1D_1 = 279.20$
$I_1D_1 = 161.14$
$I_1E_1 = 271.14$

$\alpha_2 = 49°$ $\beta_2 = 38°$
$B_2D_2 = 236.17$
$I_2D_2 = 267.16$
$I_2E_2 = 452.79$

Fig. 7.6(c)

Case (d): Crossed upper arms of 300 mm attached 40 mm on either side of spindle as shown in Fig. 7.6(d).

Here in this case the upper arms intersect the spindle axis at o and extend on the other side of spindle by 40 mm upto A where it is hinged to the spindle.

Measure the distances according to the scale adopted from (i) and (ii):

$B_1D_1 = 259.8;$ $I_1D_1 = 212;$ $I_1C_1 = 362.6;$ $B_2D_2 = 198.43$ mm; $I_2D_2 = 373.3$ $I_2C_2 = 599;$

Substitute these values in the equation obtained by taking moments about I_1 and I_2 and the simplified expressions after taking out constants for r_1

$$N_1 = 29.9 \sqrt{\left[\left(\frac{I_1D_1}{B_1D_1} + \frac{M \cdot I_1C_1}{2m \cdot B_1D_1}\right)\frac{1}{r_1}\right]}$$

$$= 29.9 \sqrt{\left[\left(\frac{212}{259.8} + \frac{100 \times 362.6}{2 \times 10 \times 259.8}\right)\frac{1}{0.125}\right]} = 236.1 \text{ r.p.m.}$$

and similarly for

$$N_2 = 29.9 \sqrt{\left[\left(\frac{I_2D_2}{B_2D_2} + \frac{M \cdot I_2C_2}{2m \cdot B_2D_2}\right)\frac{1}{r_2}\right]}$$

$$= 29.9 \sqrt{\left[\left(\frac{373.3}{198.4} + \frac{100 \times 599}{2 \times 10 \times 198.4}\right)\frac{1}{0.225}\right]} = 259.9 \text{ r.p.m.}$$

Range of speed = $N_2 - N_1$ = 259.9 − 236.1 = 23.8 r.p.m. **Ans.**

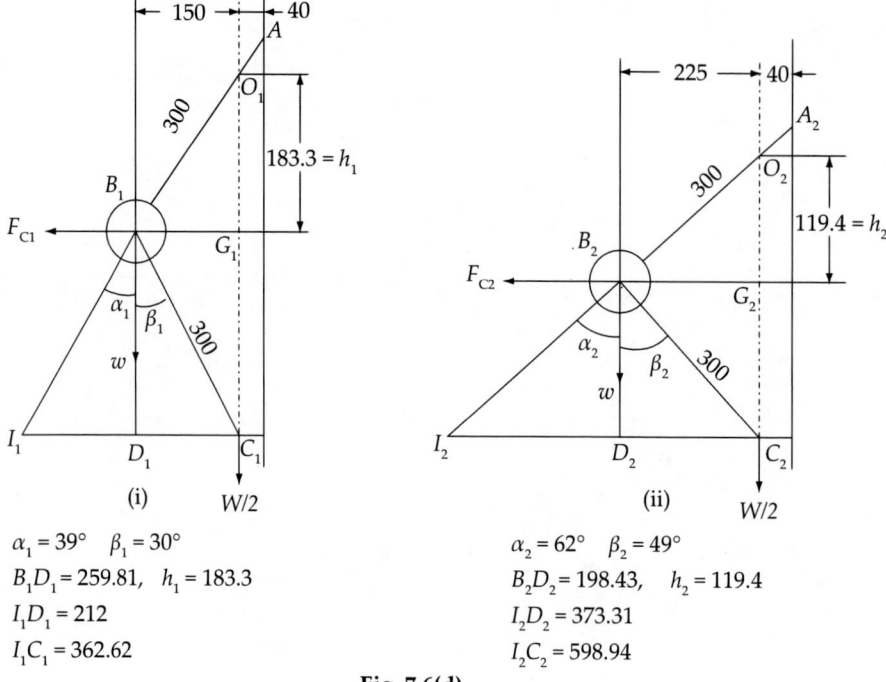

Fig. 7.6(d)

$\alpha_1 = 39°$ $\beta_1 = 30°$
$B_1D_1 = 259.81$, $h_1 = 183.3$
$I_1D_1 = 212$
$I_1C_1 = 362.62$

$\alpha_2 = 62°$ $\beta_2 = 49°$
$B_2D_2 = 198.43$, $h_2 = 119.4$
$I_2D_2 = 373.31$
$I_2C_2 = 598.94$

(e) Open upper and lower arms of 300 mm attached away by 40 mm from the spindle as shown in Fig. 7.6(e).
Here the upper arms are to be extended until they intersect the spindle axis at O.
Measure the distances according to the scale adopted from (i) and (ii):
$B_1D_1 = 279.1$; $I_1D_1 = 110$; $I_1E_1 = 220$; $B_2D_2 = 236.8$; $I_2D_2 = 185$ $I_2E_2 = 370$;
Substitute these values in the equation obtained by taking moments about I_1 and I_2 and the simplified expressions after taking out constants for r_1

$$N_1 = 29.9 \sqrt{\left[\left(\frac{I_1D_1}{B_1D_1} + \frac{M \cdot I_1F_1}{2m \cdot B_1D_1}\right)\frac{1}{r_1}\right]}$$

$$= 29.9 \sqrt{\left[\left(\frac{110}{279.1} + \frac{100 \times 220}{2 \times 10 \times 279.1}\right)\frac{1}{0.125}\right]} = 176.08 \text{ r.p.m.}$$

and similarly for r_2

$$N_2 = 29.9 \sqrt{\left[\left(\frac{I_2D_2}{B_2D_2} + \frac{M \cdot I_2E_2}{2m \cdot B_2D_2}\right)\frac{1}{r_2}\right]}$$

$$= 29.9 \sqrt{\left[\left(\frac{185}{236.8} + \frac{100 \times 370}{2 \times 10 \times 236.8}\right)\frac{1}{0.225}\right]} = 184.8 \text{ r.p.m.}$$

Range of speed = $N_2 - N_1 = 184.8 - 176.1 = 12.7$ r.p.m. **Ans.**

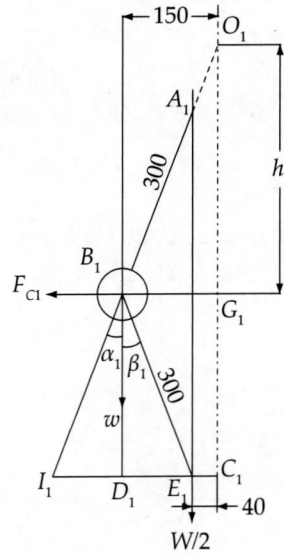
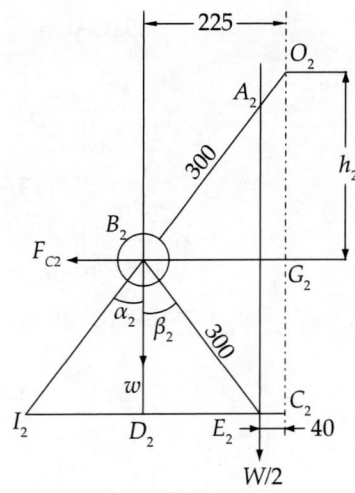

$\alpha_1 = 22°$ $\beta_1 = 22°$
$B_1D_1 = 279.11$, $h_1 = 380.60$
$I_1D_1 = 110$
$I_1E_1 = 220$ $I_1C_1 = 260$

$\alpha_2 = 38°$ $\beta_2 = 38°$
$B_2D_2 = 236.79$, $h_2 = 287.32$
$I_2D_2 = 185$
$I_2E_2 = 370$ $I_2C_2 = 410$

Fig. 7.6(e)

W E 7.3: A Proell governor shown in Fig. 7.7 has the same dimensions as the Porter governor in WE 7.2 except that the ball is carried on an extension of the lower arm at 100 mm. Determine the equilibrium speed of the Proell governor, when the radius of rotation of the ball is 150 mm and the ball centre is vertically above the pin-joint B. Find also the equilibrium speed when the radius of rotation of the balls is 225 mm. What is the range of speed?

Given:

Arms OE and $EC = 300$ mm;   Extension $EB = 100$ mm;   Radius of rotation $r_1 = 150$ mm
$r_2 = 225$ mm $w = 10$ kg $W = 100$ kg

The graphical method is easier than analytical method. Choose a proper scale and draw the line diagrams for two different radii as shown at (a) and (b).

Measure the distances according to the scale adopted from (a) and (b):

$B_1D_1 = 359.8$ mm; $I_1D_1 = 150$ mm; $I_1C_1 = 300$ mm; $B_2D_2 = 319.54$ mm; $I_2D_2 = 177.74$ mm
$I_2C_2 = 402.74$ mm.

Substitute these values in the equation obtained by taking moments about I_1 and I_2

And the simplified expressions after taking out constants for r_1

$$N_1 = 29.9 \sqrt{\left[\left(\frac{I_1D_1}{B_1D_1} + \frac{M \cdot I_1C_1}{2m \cdot B_1D_1}\right)\frac{1}{r_1}\right]}$$

208 Theory of Machines

$$= 29.9 \sqrt{\left[\left(\frac{150}{359.81} + \frac{100 \times 300}{2 \times 10 \times 359.81}\right)\frac{1}{0.15}\right]}$$

$$= 165.0 \text{ r.p.m.}$$

and similarly for r_2

$$N_2 = 29.9 \sqrt{\left[\left(\frac{I_2 D_2}{B_2 D_2} + \frac{M \cdot I_2 C_2}{2m \cdot B_2 D_2}\right)\frac{1}{r_2}\right]}$$

$$= 29.9 \sqrt{\left[\left(\frac{177.74}{319.54} + \frac{100 \times 402.74}{2 \times 10 \times 319.54}\right)\frac{1}{0.225}\right]}$$

$$= 165.0 \text{ r.p.m.}$$

Range of speed = $N_2 - N_1$ = 165 – 165 = 0 r.p.m. **Ans.**

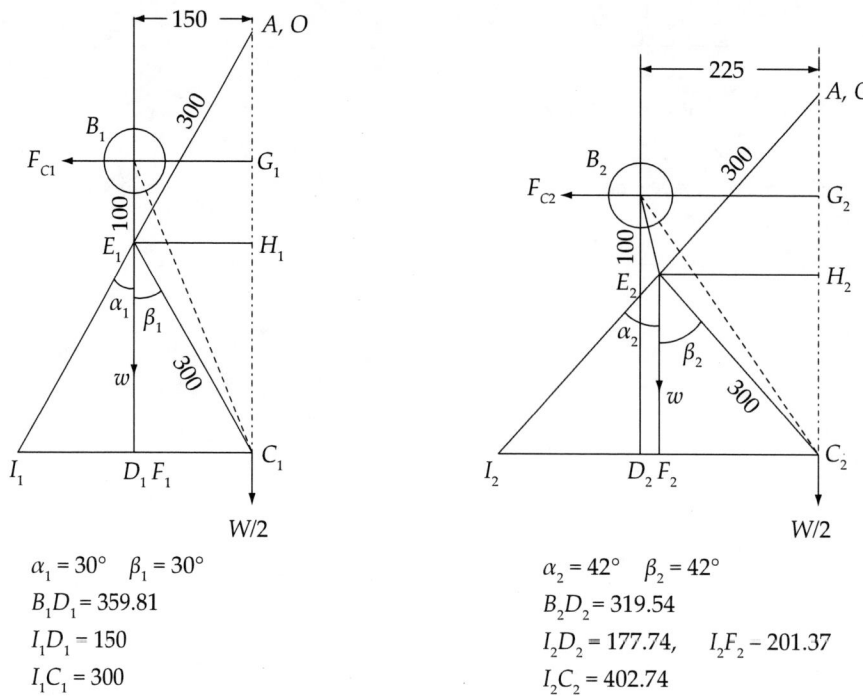

$\alpha_1 = 30°$ $\beta_1 = 30°$
$B_1 D_1 = 359.81$
$I_1 D_1 = 150$
$I_1 C_1 = 300$

$\alpha_2 = 42°$ $\beta_2 = 42°$
$B_2 D_2 = 319.54$
$I_2 D_2 = 177.74,$ $I_2 F_2 = 201.37$
$I_2 C_2 = 402.74$

Fig. 7.7 Proell governor with different radii.

Comparison of results:

Porter Governor: W E 7.2

Case (a): Arms equal and on the spindle N_{r2} = 222.6; N_{r1} = 194.5; Range of speed = 28.1 r.p.m.
Case (b): Arms unequal and on the spindle N_{r2} = 208.9; N_{r1} = 185.7; Range of speed = 23.2 r.p.m.
Case (c): Arms equal and lower arms not on the spindle N_{r2} = 206.4; N_{r1} = 179.9;
    Range of speed = 26.5 r.p.m.

Case (d): Crossed arms equal and on the spindle 12.7 $N_{r2} = 259.9$; $N_{r1} = 236.1$;
Range of speed = 23.8 r.p.m.

Case (e): Open arms equal and not on the spindle $N_{r2} = 184.6$; $N_{r1} = 176.1$;
Range of speed = 12.7 r.p.m.

Proell Governor: W E 7.3

Arms equal and on the spindle $N_{r2} = 165$ r.p.m.; $N_{r1} = 165$ r.p.m.;   Range of speed = 0 r.p.m.

7.6.3 Comparison between Flywheel and Governor

Flywheel	Governor
1. The function of the flywheel is to decrease the variations in speed due to variation in turning moment in a cycle.	1. The function of the governor is to keep the speed of a prime-mover constant due to variations in load over a number of cycles.
2. Flywheel controls rate of change of speed.	2. Governor controls change of speed.
3. Flywheel stores up energy and gives whenever required during a cycle like a capacitor in an electric circuit.	3. Governors regulate speed by regulating working agent of the prime mover. It is a device for automatic control.
4. No control on working agent either on quality or quantity. (air : fuel ratio)	4. Governor takes care of change of quality or quantity of working agent.
5. This is not an essential element for all prime movers. (Turbines)	5. This is an essential element of every prime mover.

7.7 HARTNELL GOVERNOR

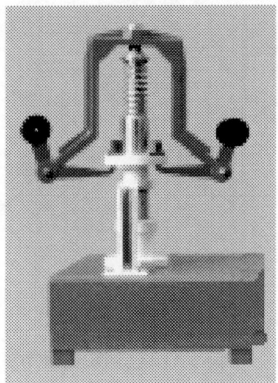

Pic. 7.3

This is an example of spring loaded governor as shown in the Pic. 7.3. The line diagram of the same is shown in Fig. 7.8. It consists of two bell crank levers pivoted to the frame which is attached to the spindle and rotates with it. Balls are fixed at one end and rollers at the other end.

Let m (or $\frac{w}{g}$) be the mass of each ball; M be the mass of sleeve;
r_1 and r_2 are the minimum and maximum radii; ω_1 and ω_2 are the minimum and maximum speeds;
S_1 and S_2 are the spring forces on the sleeve corresponding to min. and max radii;
F_{C1} and F_{C2} are the centrifugal forces;
s is the stiffness of the spring, x and y are the vertical (ball) and horizontal (sleeve) arms of the bell crank lever;
r is the radius of rotation of the balls in the middle position, i.e. with ball arm in vertical position;
h_1 and h_2 are the compressions of the spring.

Figure 7.9 (i) and (ii) show the line diagrams for two different positions

1. Mean radius r position as in Fig. 7.8
2. Minimum radius r_1 position, and
3. Maximum radius r_2 position as in Fig. 7.9(i) and (ii).

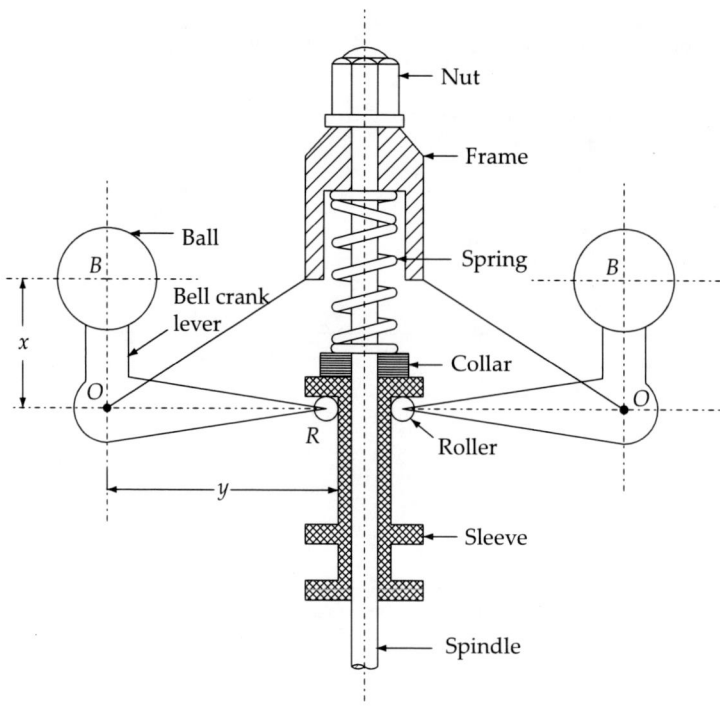

Fig. 7.8

Assuming that the effective lengths of the arms of bell crank lever in different positions be equal and so they are $x = x_1 = x_2$ and $y = y_1 = y_2$.
Considering the similar triangles from Fig. 7.9,

$$\frac{h_1}{y} = \frac{a_1}{x} = \frac{(r - r_1)}{x} \text{ since } a_1 = (r - r_1) \tag{1}$$

and
$$\frac{h_2}{y} = \frac{a_2}{x} = \frac{(r_2 - r)}{x} \text{ since } a_2 = (r_2 - r) \quad (2)$$

adding (1) and (2)
$$\frac{(h_1 + h_2)}{y} = \frac{(r_2 - r_1)}{x} \text{ or } \frac{h}{y}$$

where $h = h_1 + h_2$ hence
$$h = \frac{(r_2 - r_1)}{x} y \quad (3)$$

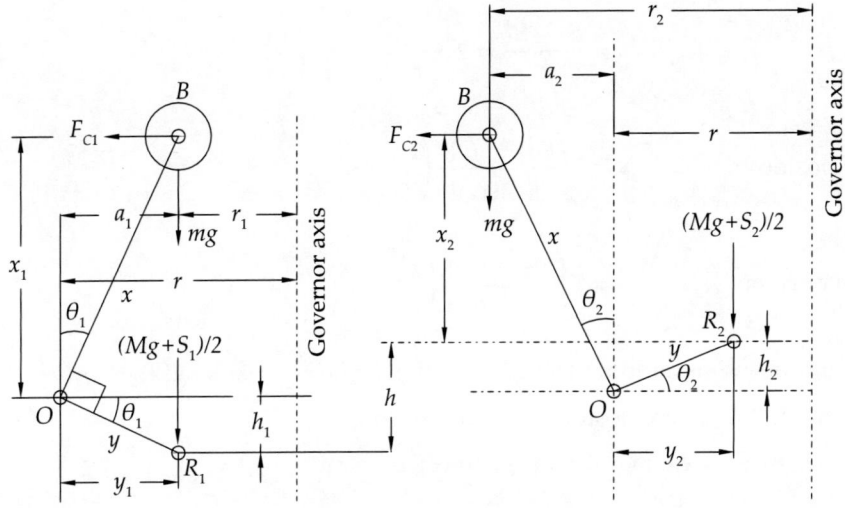

Fig. 7.9 Line diagrams.

Difference in spring forces
$$S_2 - S_1 = h.s \text{ and } h = \frac{(r_2 - r_1)}{x} y$$

Therefore,
$$s = \frac{(S_2 - S_1)}{h} = \frac{(S_2 - S_1)}{(r_2 - r_1)} \frac{x}{y} \quad (4)$$

Assuming $x = x_1 = x_2$; $y = y_1 = y_2$ and neglecting the moment due to force (mg). For minimum position, taking moments about point O,

$$\left[(Mg + S_1)\frac{y}{2}\right] = F_{C1} \cdot x$$

$$Mg + S_1 = F_{C1} \cdot \frac{2x}{y} \quad (5)$$

For maximum position, taking moments about point O,

$$\left[(Mg + S_2)\frac{y}{2}\right] = F_{C2} \cdot x$$

$$Mg + S_2 = F_{C2} \cdot \frac{2x}{y} \qquad (6)$$

Subtracting (6) from (5)

$$(S_2 - S_1) = 2(F_{C2} - F_{C1})\frac{x}{y} = s \cdot h$$

Therefore, stiffness of the spring

$$s = \frac{S_2 - S_1}{h} = 2\left[\frac{F_{C2} - F_{C1}}{r_2 - r_1}\right] \cdot \left(\frac{x}{y}\right)^2 \qquad (7)$$

For minimum position

$$s = 2\left[\frac{F_C - F_{C1}}{r - r_1}\right] \cdot \left(\frac{x}{y}\right)^2 \qquad (8)$$

For maximum position

$$s = 2\left[\frac{F_{C2} - F_C}{r_2 - r}\right] \cdot \left(\frac{x}{y}\right)^2 \qquad (9)$$

Note: For maximum and minimum positions, when friction is considered then $(Mg \pm F_f)$.
For minimum and intermediate position or mid position.
The equations (7), (8), (9) are equal since s is same. Therefore,

$$\left[\frac{F_{C2} - F_{C1}}{r_2 - r_1}\right] = \left[\frac{F_C - F_{C1}}{r - r_1}\right] = \left[\frac{F_{C2} - F_C}{r_2 - r}\right]$$

Gives

$$F_C = F_{C2} - (F_{C2} - F_{C1})\left[\frac{r_2 - r}{r_2 - r_1}\right] \qquad (10)$$

$$F_C = F_{C1} + (F_{C2} - F_{C1})\left[\frac{r - r_1}{r_2 - r_1}\right] \qquad (11)$$

For any given radius r, the F_C can be calculated knowing the minimum and maximum radius.

W E 7.4: The particulars of a Hartnell governor are the following: $m = 1.4$ kg; $x = 0.1$ m; $y = 0.05$ m; $r_1 = 0.1125$ m; $r_2 = 0.075$ m; $N_1 = 300$ r.p.m.; $N_2 = 318$ r.p.m. The axis of rotation or radius of rotation r is 0.09 m from the fulcrum. Determine the rate of the spring and equilibrium speed when radius of rotation of the ball is 0.09 m.

Centrifugal force

$$F_{C2} = m\omega^2 \cdot r_2 = 1.4\left(\frac{2\pi \times 300}{60}\right)^2 \times 7.5 = 104 \text{ N}$$

and

$$F_{C1} = F_{C2}\left(\frac{N_1}{N_2}\right)^2\left(\frac{r_1}{r_2}\right)$$

$$= 104\left(\frac{318}{300}\right)^2\left(\frac{0.1125}{0.075}\right) = 175 \text{ N}$$

use equation (7).

$$\text{Stiffness}(s) = 2\left(\frac{x}{y}\right)\left(\frac{F_{C1} - F_{C2}}{r_1 - r_2}\right)$$

$$= 2\left(\frac{0.1}{0.05}\right)\left(\frac{175 - 104}{0.1125 - 0.075}\right) = 15146 \text{ N/m}$$

Ignoring the effect of gravity

$$F_C = F_{C_2} + (F_{C_1} - F_{C_2})\left[\frac{r - r_2}{r_1 - r_2}\right]$$

where F_C is the force corresponding to radius r.

Therefore, F_C for $r = 0.09$ is $104 + (175 - 104)\left[\dfrac{0.09 - 0.075}{0.1125 - 0.075}\right]$

But F_C is also given by $m\left(\dfrac{2\pi \times N}{60}\right)^2 \times r$;

Therefore,

$$132.4 = 1.4\left(\frac{2\pi \times N}{60}\right)^2 \times 0.09$$

$$N^2 = \left(\frac{132.4 \times 91.2}{1.4 \times 0.09}\right) = 95832$$

Thus, speed $N = 309.5$ r.p.m. **Ans.**

7.8 HARTUNG GOVERNOR

The Hartung governor is shown in Fig. 7.10 with the various parts indicated on it. This governor also works with bell crank levers as in the Hartnell governor. In this, the vertical arm of the bell crank levers are fitted with spring balls which are compressed against the frame of the governor when the rollers at the horizontal arm press against the sleeve.

Figure 7.11(a, b, c) shows the three positions, i.e. Mean position, Minimum and Maximum positions of the bell crank lever.

Assuming the governor in the mid position (or mean) and neglecting the obliquity of the arms, taking moments about the fulcrum O from Fig. 7.11(a) gives:

$$F_C \cdot x = S \cdot x + Mg \cdot \frac{y}{2}$$

where $S = s$ (stiffness) × (change of radius) or (compression of the spring) and F_C is centrifugal force equal to $m\omega^2 r$. For minimum position, i.e. for minimum radius r_1 as shown in Fig. 7.11(b).

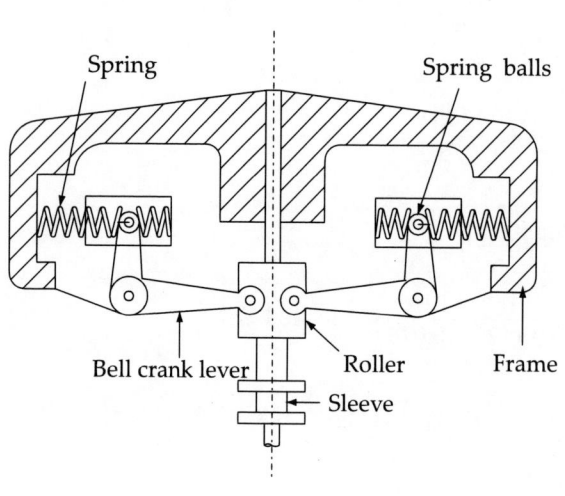

Fig. 7.10

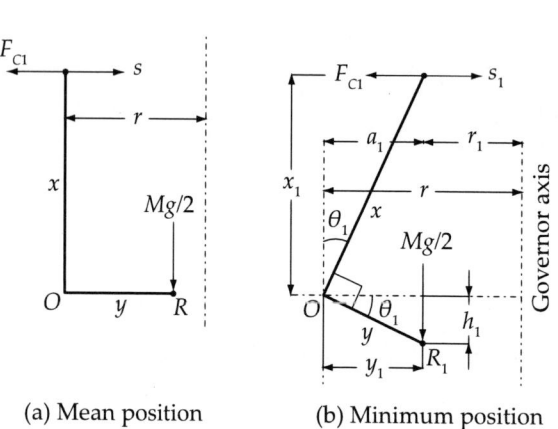

(a) Mean position   (b) Minimum position   (c) Maximum position

Fig. 7.11

$$\left(\frac{r - r_1}{h_1}\right) = \frac{x}{y}$$

$$r_1 = r - \left(\frac{h_1 x}{y}\right)$$

Therefore,
$$F_{C1}.x = S_1.x + \left(\frac{Mg.y}{2}\right)$$

where F_{C1} is $m\omega_1^2 r_1$. ω_1 can be found.

For maximum position, i.e. for maximum radius r_2 as shown in Fig. 7.11(c)

$$\left(\frac{r - r_2}{h_2}\right) = \frac{x}{y}$$

$$r_2 = r - \left(\frac{h_2 x}{y}\right)$$

Therefore,
$$F_{C2}.x = S_2.x + \left(\frac{Mg.y}{2}\right)$$

where F_{C2} is $m\omega_2^2 r_2$. ω_2 can be found.

Range of speed = $(N_2 - N_1)$ and Ratio of range of speed to mean speed = $\dfrac{(N_2 - N_1)}{N}$

Percentage of coefficient of fluctuation of speed = $\left(\dfrac{N_2 - N_1}{N}\right) 100\%$

W E 7.5: In a spring controlled governor of the Hartung type, the length of the ball and sleeve arms are 80 mm and 120 mm respectively. The load travel of the sleeve is 25 mm. In the mid position each spring is compressed by 50 mm and the radius of rotation of the mass centre is 140 mm. Each ball has a mass of 4 kg and the spring has a stiffness of 10 kN/m of compression. The equivalent mass of the governor gear at the sleeve is 16 kg. Neglecting the moment due to the revolving masses when the arms are inclined, determine the ratio of the range of speed to the mean speed of the governor. Find also the speed in the mid position.

Given:
x = 80 mm = 0.08 m; y = 120 mm = 0.12 m; h = 25 mm = 0.25 m;   radius, r = 140 mm = 0.14 m;
mass of spring ball m = 4 kg;   stiffness of spring s = 10 kN/m = 10 × 10³ N/m
Mass on the sleeve M = 16 kg;   Initial compression = 50 mm = 0.05 m

Speed of the governor in the mean position or mid position as shown at 7.11(a):

Let ω be the mean angular speed in rad/s and N be mean speed in r.p.m.

The centrifugal force
$$F_C = m\omega^2 r = 4\omega^2 0.14 = 0.56\omega^2 \text{ newtons}$$

The spring force S = stiffness × initial compression = 10 × 10³ × 0.05 = 500 N

Now taking moments about O, neglecting the moment due to the revolving masses gives

$$F_C \times x = S \times x + Mg \times \frac{y}{2}$$

$$0.56\omega^2 \times 0.08 = 500 \times 0.08 + \frac{(16 \times 9.81 \times 0.12)}{2}$$

$$= 40 + 9.42 = 49.42$$

$$\omega^2 = \frac{49.42}{(0.56 \times 0.08)} = 1103$$

$$\omega = 33.2 \text{ rad/s}$$

$$N = \frac{33.32 \times 60}{2\pi} = 317 \text{ r.p.m. } \textbf{Ans.}$$

The minimum and maximum positions are shown in Fig. 7.11(b) and (c) respectively.

To find the minimum speed N_1

From Fig. 7.11(b)

$$\frac{(r - r_1)}{h_1} = \frac{x}{y}$$

$$r_1 = r - h_1 \times \frac{x}{y}$$

$$= 0.14 - 0.025 \times \frac{0.08}{(2 \times 0.12)}$$

$$= 0.132 \text{ m}$$

since $h_1 = \frac{h}{2}$.

Centrifugal force at the minimum position,

$$F_{C1} = m\omega_1^2 r_1 = 4(\omega_1)^2 \times 0.132 = 0.528(\omega_1^2) \text{ N}$$

and spring force at the minimum position,

$$S_1 = [\text{Initial compression} - (r - r_1)] \times \text{Stiffness}$$
$$= 0.05 - (0.14 - 0.132)10 \times 10^3 = 420 \text{ N}$$

Now taking moments about fulcrum O, neglecting the obliquity of arms, i.e. taking $x_1 = x$ and $y_1 = y$,

$$F_C \times x = S_1 \times x + Mg \times \frac{y}{2}$$

$$0.528\omega_1^2 \times 0.08 = 420 \times 0.08 + \frac{(16 \times 9.81 \times 0.12)}{2}$$

$$\omega_1^2 = \frac{43.02}{(0.528 \times 0.08)} = 1019$$

$$\omega_1 = 32 \text{ rad/s}$$

$$N_1 = \frac{32 \times 60}{2\pi} = 305.5 \text{ r.p.m } \textbf{Ans.}$$

To find the maximum speed N_2

From the Fig. 7.11(c),

$$\frac{(r - r_2)}{h_2} = \frac{x}{y}$$

$$r_2 = r - h_2 \times \frac{x}{y}$$

$$= 0.14 + 0.025 \times \frac{0.08}{(2 \times 0.12)}$$

$$= 0.148 \text{ m}$$

since $h_2 = \frac{h}{2}$.

Centrifugal force at the maximum position,

$$F_{C2} = m\omega_2^2 r_2 = 4(\omega_2)^2 \times 0.148 = 0.592(\omega_1^2) \text{ N}$$

and spring force at the maximum position,

$$S_2 = [\text{Initial compression} - (r - r_2)] \times \text{Stiffness}$$

$$= 0.05 - (0.148 - 0.14)10 \times 10^3 = 580 \text{ N}$$

Now taking moments about fulcrum O, neglecting the obliquity of arms, i.e. taking $x_2 = x$ and $y_2 = y$,

$$F_{C2} \times x = S_2 \times x + Mg \times \frac{y}{2}$$

$$0.592\omega_2^2 \times 0.08 = 580 \times 0.08 + \frac{(16 \times 9.81 \times 0.12)}{2}$$

$$\omega_2^2 = \frac{55.82}{(0.592 \times 0.08)} = 1178$$

$$\omega_2 = 34.32 \text{ rad/s}$$

$$N_2 = \frac{34.32 \times 60}{2\pi} = 327.7 \text{ r.p.m. } \textbf{Ans.}$$

Range of speed = $N_2 - N_1 = 327.7 - 305.5 = 22.2$ r.p.m.

Ratio of range of speed to mean speed = $\frac{(N_2 - N_1)}{N} = \frac{22.2}{317} = 0.07$ or 7%

7.9 DEFINITIONS

7.9.1 Sensitiveness

It is defined as the ratio of the difference between the maximum and minimum speeds to the mean equilibrium speed

$$\text{Sensitiveness} = \frac{(N_2 - N_1)}{N}$$

where $N = \dfrac{(N_1 + N_2)}{2}$

$$\text{Sensitiveness} = \frac{(\omega_2 - \omega_1)}{\omega}$$

where $\omega = \dfrac{(\omega_1 + \omega_2)}{2}$

7.9.2 Stable and Unstable

A governor is said to be stable when for every speed within the working range there is a definite configuration, i.e. there is only one radius of the governor balls at which the governor is in equilibrium.

A governor is said to be unstable, if the radius of rotation decreases as the speed increases.

7.9.3 Isochronous/Isochronism

A governor is said to be isochronous when the equilibrium speed is constant (i.e., range of speed is zero) for all radii of rotation of the balls within the working range neglecting friction. So isochronous governors are not used in practice. Isochronism means $\omega_1 = \omega_2$ and thus $h_1 = h_2$.

7.9.4 Hunting

A governor is said to be hunting if the speed of the engine fluctuates continuously above and below the mean speed. Such governors are called too sensitive governors.

7.9.5 Effort

Effort of a governor is the mean force exerted at the sleeve (or mean lift of the sleeve) for a given percentage change of speed or for one per cent change of speed.

7.9.6 Power

Power of a governor is the work done at the sleeve for a given percentage change of speed. It is the product of the mean value of the effort and the distance through which sleeve moves.

$$\text{Power} = \text{Mean effort} \times \text{lift of sleeve}$$

7.9.7 Controlling Force

The inward force acting on the rotating balls (opposite of centrifugal force), i.e. centripetal force is called the controlling force ($F_C = m\omega^2 r$). The controlling force is supplied by the weight of the rotating mass in a Watt governor, the weight of the mass and that of the sleeve in a Porter governor and by the compressed spring in the case of a Hartnell governor. A graph showing the variation of the controlling force with the radius of rotation is called the controlling curve or diagram. This curve is useful in finding out the stability of a governor.

7.9.8 Coefficient of Insensitiveness

The ratio of $\dfrac{(N_1 - N_2)}{N}$, is known as the coefficient of insensitiveness where N is the corresponding speed neglecting friction.

7.10 WILSON-HARTNELL GOVERNOR

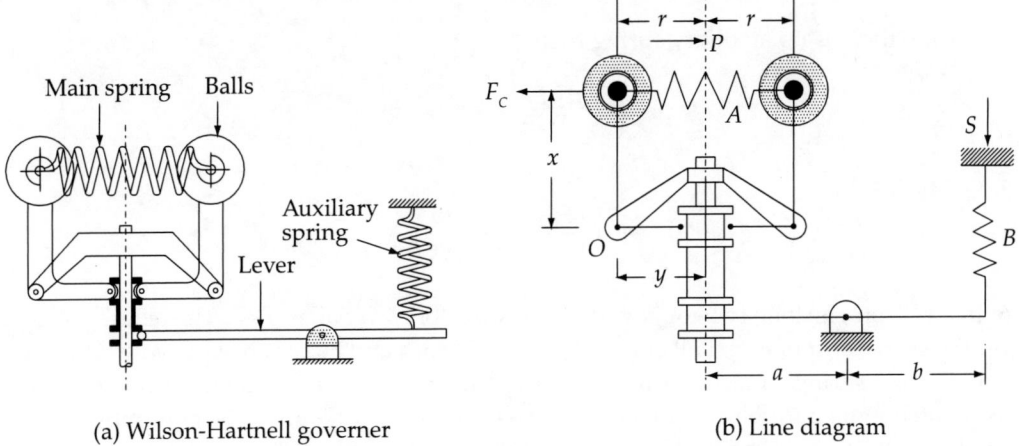

(a) Wilson-Hartnell governer   (b) Line diagram

Fig. 7.12

This is another spring loaded type of governor as shown in Fig. 7.12. P is the tension, s_a and s_b are stiffness of an auxiliary spring and ball spring, S spring force of auxiliary spring.

Total downward force on the sleeve is given by

$$(F_C - P)x = \left(Mg + \frac{Sb}{a}\right)\frac{y}{2}$$

At minimum radius

$$(F_{C1} - P_1)x = \left(Mg + \frac{S_1 b}{a}\right)\frac{y}{2} \qquad (1)$$

At maximum radius

$$(F_{C2} - P_2)x = \left(Mg + \frac{S_2 b}{a}\right)\frac{y}{2} \qquad (2)$$

Subtracting (2) from (1)

$$[(F_{C2} - F_{C1}) - (P_2 - P_1)]x = \left(\frac{S_2 - S_1}{2a}\right)\frac{yb}{2a} \qquad (3)$$

When r_1 changes to r_2, ball spring extends by $2(r_2 - r_1)$ and auxiliary spring by $(r_2 - r_1) \cdot \dfrac{yb}{xa}$

Therefore,
$$P_2 - P_1 = 4s_b(r_2 - r_1)$$

and
$$S_2 - S_1 = s_a(r_2 - r_1)\dfrac{yb}{xa}$$

Substituting for $(P_2 - P_1)$ and $(S_2 - S_1)$ in equation (3) and simplifying gives

$$\left(\dfrac{F_{C2} - F_{C1}}{r_2 - r_1}\right) = 4 \cdot s_b + \left(\dfrac{s_a}{2}\right)\left(\dfrac{yb}{xa}\right)^2 \quad (4)$$

when $s_a = 0$, i.e. there is no auxiliary spring then

$$\left(\dfrac{F_{C2} - F_{C1}}{r_2 - r_1}\right) = 4 \cdot s_b$$

Stiffness of ball spring,

$$s_b = \left(\dfrac{F_{C2} - F_{C1}}{r_2 - r_1}\right)\left(\dfrac{1}{4}\right)$$

W E 7.6: In a Wilson-Hartnell type of governor, the mass of each ball is 5 kg. The lengths of the ball arms and the sleeve arm of each bell crank lever are 100 mm and 80 mm respectively. The stiffness of each of the two springs attached directly to the balls is 0.4 N/mm. The lever for the auxiliary spring is pivoted at its midpoint. When the radius of rotation is 100 mm, the equilibrium speed is 200 r.p.m. If the sleeve is shifted by 10 mm for an increase of speed of 5%, find the required stiffness of the auxiliary spring.

Given:
$m = 5$ kg; $s = 0.4$ N/mm $= 400$ N/m; $r_1 = 100$ mm; $x = 100$ mm $y = 80$ mm;
$N_1 = 200$ r.p.m.; $\dfrac{b}{a} = 1$

We know, from equation (4)

$$\left(\dfrac{F_{C2} - F_{C1}}{r_2 - r_1}\right) = 4 \cdot s_b + \left(\dfrac{s_a}{2}\right)\left(\dfrac{yb}{xa}\right)^2$$

When $r_1 = 100$ mm, $N_1 = 200$ r.p.m. or $\omega_1 = 200 \times \dfrac{2\pi}{60} = 20.94$ rad/s;

$$F_{C1} = mr_1\omega_1^2 = 5 \times 0.1 \times (20.94)^2 = 219.2 \text{ N}$$

For 5% rise of speed,
$$\omega_2 = 20.94 \times 1.05 = 21.99 \text{ rad/s}$$

For sleeve rise of 10 mm, increase of ball radius $= 10 \times \dfrac{100}{80} = 12.5$ mm, $r_2 = 100 + 12.5 = 112.5$ mm

and so
$$F_{C2} = m.r_2\omega_2^2 = 5 \times 0.125 \times (21.99)^2 = 302.23 \text{ N}$$

Now substituting in the above equation,

$$\left(\frac{302.23 - 219.2}{0.1125 - 0.1}\right) = 4 \times 400 + \left(\frac{s_a}{2}\right)\left(\frac{0.08 \times 1}{0.1}\right)^2 = 6642.4$$

$$s_a = 15757.5 \text{ N/m or } 15.575 \text{ N/mm}$$

7.11 EXERCISE

7.11.1 Short Answer Questions

1. Define sensitiveness of governors.
2. Classify the types of governors with their suitability in a particular application.
3. What is the function of a governor?
4. Define stability.
5. Prove that the sensitiveness of a Proell governor is greater than that of a Porter governor.
6. Explain what do you mean by controlling force and controlling force curve.
7. Give sketches of open, crossed arms type of governors.
8. What is the height of a governor and how is it measured?

7.11.2 Problems

Porter

1. *Arms on the spindle:* A loaded governor of the Porter type has equal links 25 cm long pivoted at the axis: weight of each ball is 30 N and the weight of the central load is 140 N. The ball radius is 15 cm when the governor begins to lift and 20 at the maximum speed. Determine the maximum speed and range of speed. If the friction at the sleeve is equivalent to 15 N, find the maximum and minimum speed and the range of speed.

2. *Arms away from spindle:* A Porter governor carries a central load of 30 kgf and each ball weighs 4.5 kgf. The upper links are 20 cm long and lower links are 30 cm long. The points of suspension of upper and lower links are 5 cm from axis of spindle.
   Calculate: (a) The speed of the governor in r.p.m. if the radius of revolution of the governor ball is 12.5 cm and (b) the effort of the governor for the increase of speed of 1%.

3. *Arms on the spindle:* In a Porter governor, the arms and links are each 25 cm long and intersect on the main axis. Each ball weighs 3 kg and the central load 27.25 kgf. The sleeve is in the lowest position when the arms are inclined at 27° with the axis. The lift of the sleeve is 5 cm. What is the force of the friction at the sleeve if the speed at the beginning of ascent at the lowest position is equal to the speed at the beginning of descent from the highest position?

4. *Arms away from spindle with friction:* All the four arms of a Porter governor are 30 cm long and are hinged at a distance of 3 cm from the axis of rotation. Each ball weighs 80 N. The weight of the sleeve is 750 N. Find the equilibrium speed corresponding to the radius of rotation of 23 cm. Also determine the higher and lower speeds for this configuration, if a frictional force of 50 N acts against the moment of the sleeve.

Proell

1. *Lower arm away from spindle:* A Proell governor has all four arms of length 305 mm. The upper arms are pivoted on the axis of rotation and the lower arms are attached to sleeve at a distance of 38 mm from the axis. The mass of each ball is 4.8 kg and are attached to the extension of the lower arms, which are 102 mm long. The mass on the sleeve is 54 kg. The minimum and maximum radii of the governor are 165 mm and 216 mm. Assume the extensions of the lower arms are parallel to the governor axis at the minimum radius, find the corresponding equilibrium speeds.

2. *Lower arms away from spindle:* The arms of a Proell governor are 30 cm long. The upper arms are pivoted on the axis of rotation, while the lower arms are pivoted at a radius of 4 cm. Each ball weighs 5 kgf and is attached to an extension 10 cm long of the lower arm, the central weight is 60 kgf. At the minimum radius of 16 cm the extension to which balls are attached are parallel to the governor axis. Find the equilibrium speed corresponding to a radius of 16 cm.

3. *Arms away from the spindle:* A governor of the Proell type has each arm 250 mm long. The pivots of the upper and lower arms are 25 mm from the axis. The central load acting on the sleeve has a mass of 25 kg and each rotating ball has a mass of 3.2 kg when the governor sleeve is in mid position, the extension link of the lower arm is vertical and the radius of path of rotation of the masses is 175 mm. The vertical height of the governor is 200 mm. If the speed of governor is 160 r.p.m., when in mid position, find: (a) Length of the extension link and (b) Tension in the upper arm.

4. *Open arms, away from the spindle and with friction:* The weight of each ball of a Proell governor is 60 N, the central load is 1500 N and the arms are 25 cm long. The arms are open and are each pivoted at a distance of 5 cm from the axis of rotation. The extension of the lower arm to which each ball is attached is 12.5 cm long and the radius of rotation of the balls is 22.5 cm. When the arms are inclined at 45 degrees to the axis of rotation find (a) The equilibrium speed for the above configuration (b). The coefficient of insensitiveness of the friction of the governor mechanism is equivalent to at force of 20 N at the sleeve.

Spring controlled governors

1. A spring controlled governor of Hartnell type has equal arms. The weights rotate in a circle of 13 cm diameter when the sleeve is in the mid position and the weight arms are vertical. Neglecting friction, the equilibrium speed is 450 r.p.m. in this position. The maximum sleeve movement is to be 2.5 cm and the maximum variation of speed, taking friction into account, is to be 2.5 cm and the maximum variation of speed, taking friction into account, is to be $\pm 5\%$

of the mid position speed. The weight of sleeve is 4 kg and friction may be considered to be 3 kg at the sleeve. The power of the governor must be sufficient to overcome the friction by 1% change of speed either way at mid position. Neglecting the obliquity of arms, determine (i) weight of each rotating mass (ii) the spring stiffness (iii) the initial compression of spring.

2. The controlling force in a spring-controlling governor is 1500 N when radius of rotation is 200 m and 887.5 N when radius of rotation is 130 mm. The mass of each ball is 8 kg. If the controlling force curve is a straight line, then find: (i) Controlling force when radius of rotation is 150 mm. (ii) The speed of the governor when radius of rotation is 150 mm. (iii) Increase in initial tension so that governor is isochronous, and (iv) Isochronous speed.

7.11.3 Multiple Choice Questions

1. A centrifugal governor is rotating with an angular velocity of 60 r.p.m. Find the change in its vertical height when its speed increases to 61 r.p.m. [Ans. 9 mm]

2. A _____ governor is a spring loaded governor.
   (a) Watt (b) Hartnell (c) Porter (d) Proell [Ans. (b)]

3. The height of a Watt governor is
   (a) $\dfrac{g}{\omega^3}$ (b) $\dfrac{\omega^2}{g}$ (c) $g\omega^2$ (d) $\dfrac{g}{\omega^2}$ [Ans. (d)]

4. The ratio of the height of a Porter governor to that of a Watt governor when the length of the links and the arms are the same is
   (a) $\dfrac{(M+m)}{M}$ (b) $\dfrac{(M+m)}{m}$ (c) $\dfrac{M}{(M+m)}$ (d) $\dfrac{m}{(M+m)}$ [Ans. (b)]

5. A Hartnell governor is a/an _____ governor.
   (a) dead weight (b) pendulum type (c) inertia (d) spring loaded [Ans. (d)]

6. The frictional resistance at the sleeve _____ the sensitivity of the governor.
   (a) does not affect (b) increases (c) decreases (d) may increase or decrease [Ans. (c)]

7. The governor is said to be _____ when the speed of the engine fluctuates continuously above and below the mean speed.
   (a) isochronous (b) hunting (c) insensitive (d) stable [Ans. (b)]

8. If the controlling force of a spring controlled governor is expressed as $a.r + b$, where r is the radius of rotation and a and b are constants, it is _____
   (a) isochronous (b) centrifugal (c) dead weight (d) inertia [Ans. (a)]

9. In a governor if the equilibrium speed is constant for all radii of rotation of balls, the governor is said to be
   (a) stable (b) unstable (c) inertia (d) isochronous [Ans. (d)]

10. A _____ governor is a spring loaded governor.
    (a) Watt (b) Hartnell (c) Porter (d) Proell [*Ans.* (b)]

11. A Hartnell governor is a/an _____ governor.
    (a) dead weight (b) pendulum type (c) inertia (d) spring loaded [*Ans.* (d)]

12. The frictional resistance at the sleeve _____ the sensitivity of the governor.
    (a) does not effect (b) increases (c) decreases (d) may increase or decrease [*Ans.* (c)]

13. The governor is said to be _____ when the speed of the engine fluctuates continuously above and below the mean speed.
    (a) isochronous (b) hunting (c) insensitive (d) stable [*Ans.* (a)]

14. If the controlling force of a spring-controlled governor is expressed as $a.r + b$, where r is the radius of rotation and a and b are constants, it is a/an _____ governor
    (a) isochronous (b) centrifugal (c) dead weight (d) inertia [*Ans.* (a)]

15. In a governor if the equilibrium speed is constant for all radii of rotation of balls, the governor is said to be
    (a) stable (b) unstable (c) inertia (d) isochronous [*Ans.* (d)]

16. The force resisting the outward movement of balls is known as _____ of the governor
    (a) effort (b) centrifugal force (c) controlling force (d) inertia force [*Ans.* (c)]

17. In a Wilson-Hartnell governor, the balls are connected by
    (a) one spring   (b) two springs in series
    (c) two parallel springs (d) four springs [*Ans.* (c)]

18. The effort of a governor is the force exerted by the governor on the
    (a) balls (b) sleeve (c) upper links (d) lower links [*Ans.* (b)]

19. The condition of isochronism can be realised in a _____ governor
    (a) Watt (b) Porter (c) Proell (d) Hartnell [*Ans.* (d)]

Belt, Rope and Chain Drives

Belt, Rope and Chain Drives

8.1 INTRODUCTION

Belts or ropes or chains are used to transmit power or motion from one shaft to another shaft at a considerable distance by means of pulleys. The power transmitted by belts depends upon:

1. The velocity of the belt.
2. The tension under which the belt is placed on the pulley.
3. The arc of contact between the belt and the smaller pulley.
4. The conditions under which the belt is used such as:
    (a) whether the shafts are parallel
    (b) pulleys are very close or else the arc of contact on smaller pulley will be less
    (c) pulleys are too far as the belt weight falls heavily on the shaft.

Pic. 8.1 shows the pulleys on a line shaft. Line shaft is a lengthy shaft on which several pulleys are mounted. Each pulley is connected by a separate belt to different units. All the pulleys on the line shaft act as drivers. This line shaft is driven by I.C. engines or a motor.

Pic. 8.1 Pulleys on a line shaft.

8.2 TYPES OF BELTS

There are different types of belts on the basis of the pulleys as follows:

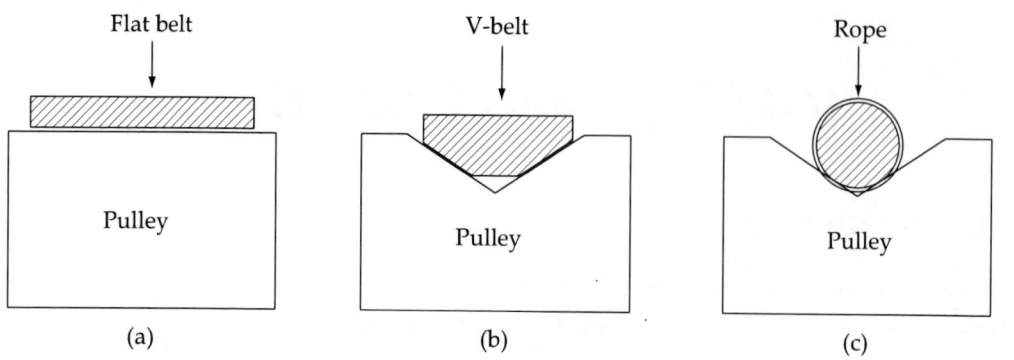

Fig. 8.1 Types of belts.

8.2.1 Flat Belt

The flat belt is mostly used in the factories and workshops, where a moderate amount of power is to be transmitted, from one pulley to another, when the two pulleys are not more than 10 m apart. Pic. 8.2 shows the pulley of flat belt.

Pic. 8.2 Pulley of flat belt.

8.2.2 V-belt

The V-belt is mostly used in the factories and workshops, where a great amount of power is to be transmitted, from one pulley to another, when the two pulleys are very near to each other. Pic. 8.3 shows the pulley of the V-belt. Pic. 8.4 shows the section of a V-belt.

Pic. 8.3 Pulley of V-belt.

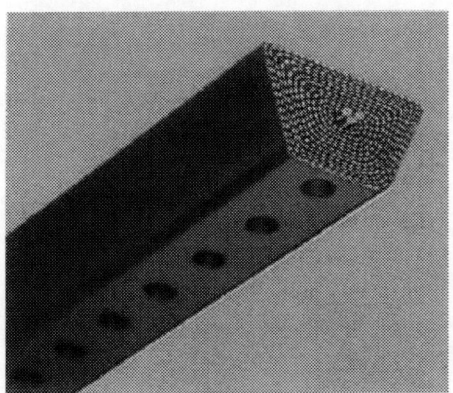

Pic. 8.4 Section of a V-belt.

8.2.3 Circular Belt or Rope

The circular belt or rope is mostly used in the factories and workshops, where a great amount of power is to be transmitted, from one pulley to another, when the two pulleys are more than 5 m apart.

If a huge amount of power is to be transmitted, then a single belt may not be sufficient. In such a case, wide pulleys (for V-belts or circular belts) with a number of grooves are used. Then one belt in each groove is provided to transmit the required power from one pulley to another. Picture 8.5 shows the pulley of a rope drive and Pic. 8.6 shows the rope details.

Pic. 8.5 Pulley of a rope.

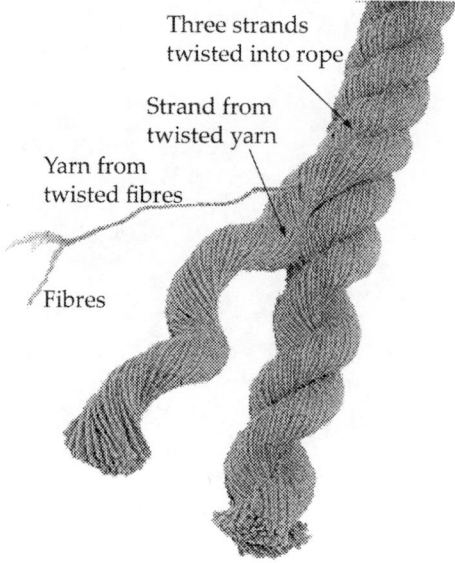

Pic. 8.6 Rope details.

Different materials such as leather, fabric, rubber, etc. are used for belts and ropes.

8.3 TYPES OF BELT DRIVES

The power is transmitted from one pulley to another by any one of the following two types of belt drives:

1. Open belt drive
2. Cross belt drive or twist belt drive as shown in Pic. 8.7 and Fig. 8.2(a, b).

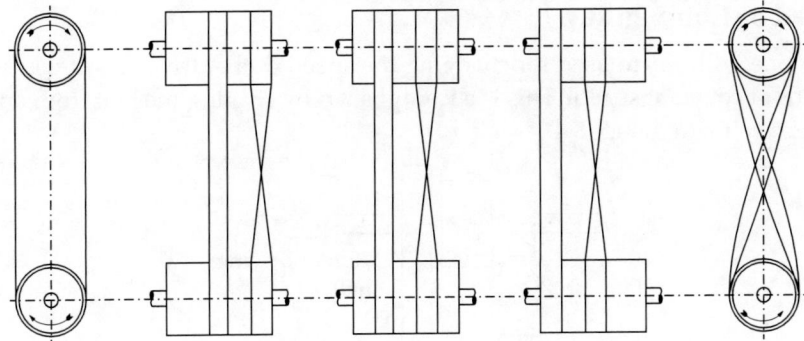

Pic. 8.7 Types of belt drives.

The open belt drive is used with shafts arranged in parallel and rotating in the same direction. The driver pulls the belt from the lower side and delivers it to the upper side. So the tension on lower side will be more than on the upper side. The lower side is called tension side or tight side and the upper side is known as slack side. In the case of crossed belt, the shafts are parallel but rotate in the opposite direction.

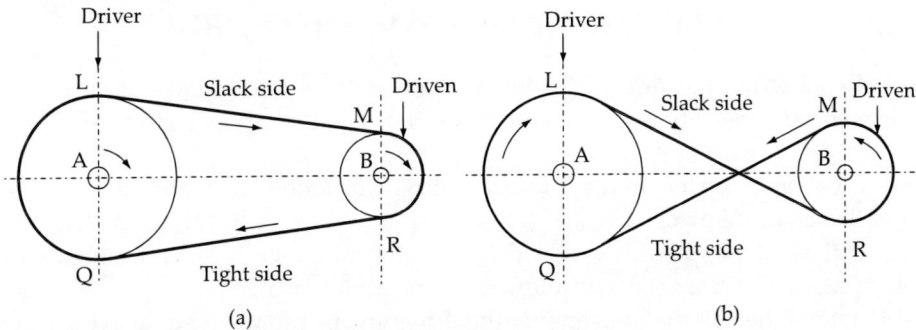

Fig. 8.2 (a) Open and (b) Crossed belts.

8.3.1 Compound Belt Drives

The compound belt drive is used when power is transmitted from one shaft to another through a number of pulleys as shown in Fig. 8.3.

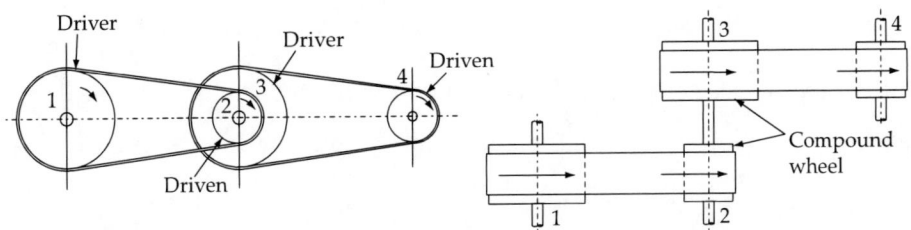

Fig. 8.3 Compound belt drive.

8.3.2 Stepped or Cone Pulley

The stepped or cone pulleys are used for changing the speed of the driven shaft while the main or driving shaft runs at constant speed. This is accomplished by shifting the belt from one part of the steps to the other as shown in Fig. 8.4.

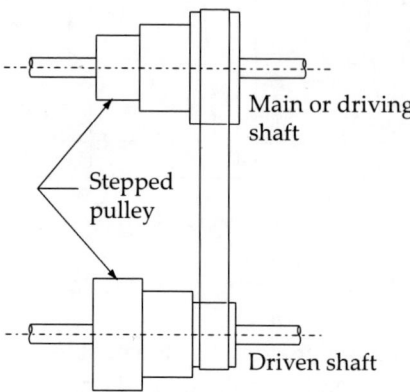

Fig. 8.4 Stepped or cone pulley.

8.4 SPEED RATIO OR VELOCITY RATIO OF A BELT DRIVE

Velocity ratio of a belt drive may be broadly defined as the ratio of the velocities of the driver and the follower or driven. The velocity ratios of simple belt drive and compound drive are to be studied separately.

Let d_1 and d_2 be the diameters of the driver and driven or follower, N_1 and N_2 are the speeds of the driver and driven or follower.

The length of the belt that passes over the driver in one minute is $\pi d_1 N_1$ and similarly the length of the belt that passes over the driven or follower in one minute is $\pi d_2 N_2$.

Since the length of the belt that passes over the driver in one minute is equal to the length of the belt that passes over the driven or follower in one minute, therefore

$$\pi d_1 N_1 = \pi d_2 N_2 \text{ or } \frac{N_2}{N_1} = \frac{d_1}{d_2}$$

The term $\frac{N_2}{N_1} = \frac{d_1}{d_2}$ is popularly known as velocity ratio.

Note: When thickness of the belt (t) is taken into account, then velocity ratio is $\frac{N_2}{N_1} = \frac{d_1 + t}{d_2 + t}$.

The velocity ratio can also be derived by equating the peripheral velocity on the driving pulley and on the driven or follower pulley. Then accordingly equating the velocities $\left(v_1 \frac{\pi d_1 N_1}{60}\right)$ and $\left(v_2 \frac{\pi d_2 N_2}{60}\right)$ gives

$$d_1 N_1 = d_2 N_2 \text{ or } \frac{N_2}{N_1} = \frac{d_1}{d_2}$$

W E 8.1: It is required to drive a shaft at 620 revolutions per minute, by means of a belt from a parallel shaft, with a pulley of 300 mm diameter on it and running at 240 revolutions per minute. What size pulley is required on the driven shaft?

Given:
Speed of the driven pulley (N_2) = 620 r.p.m.;   Diameter of driver pulley (d_1) = 300 mm
Speed of the driver pulley (N_1) = 240 r.p.m.

Let d_2 be the diameter of the driven pulley required, then

$$d_2 = d_1 \times \frac{N_1}{N_2} = 300 \times \frac{240}{620} = 116.1 \text{ mm}$$

8.4.1 Velocity Ratio of a Compound Belt Drive

When power is transmitted from one shaft to another using a compound pulley as shown in Fig. 8.3. In the figure pulley 1 drives the pulley 2. Since the pulleys 2 and 3 are keyed to the same shaft, therefore, the pulley 1 drives the pulley 3 also which in turn drives the pulley 4.

Let d_1, d_2, d_3 and d_4 be the diameters of the pulleys 1, 2, 3 and 4. Let N_1, N_2, N_3 and N_4 be the speeds of the pulleys 1, 2, 3 and 4 respectively.

We know that the velocity ratio of the pulleys 1 and 2 is $\frac{N_2}{N_1} = \frac{d_1}{d_2}$, and similarly the velocity ratio of the pulley 3 and 4 is $\frac{N_4}{N_3} = \frac{d_3}{d_4}$.

Multiplying both gives

$$\frac{N_2 \times N_4}{N_1 \times N_3} = \frac{d_1 \times d_3}{d_2 \times d_4}$$

$$\frac{N_4}{N_1} = \frac{d_1 \times d_3}{d_2 \times d_4}$$

since $N_2 = N_3$.

Thus, the speed ratio in the case of compound belt drive is

$$\frac{N_4}{N_1} = \frac{d_1 \times d_3}{d_2 \times d_4}$$

A little consideration will show that if there are six pulleys, then the velocity ratio is

$$\frac{N_6}{N_1} = \frac{d_1 \times d_3 \times d_5}{d_2 \times d_4 \times d_6} = \frac{\text{Product of diameters of drivers}}{\text{Product of diameters of followers}}$$

W E 8.2: In a workshop, an engine drives a shaft by a belt. The diameters of the engine pulley and the shaft pulley are 500 mm and 250 mm respectively. Another pulley of 700 mm diameter on the same shaft drives a pulley 280 mm in diameter of the follower. If the engine runs at 180 r.p.m., find the speed of the follower.

Given:
Diameter of the engine pulley (d_1) = 500 mm;   Diameter of the shaft pulley (d_2) = 250 mm;

Diameter of another pulley $(d_3) = 700$ mm. Diameter of the follower pulley $(d_4) = 280$ mm
Speed of the engine $(N_1) = 180$ r.p.m.

Therefore, the speed of the follower shaft,

$$N_4 = N_1 \times \frac{d_1 \times d_3}{d_2 \times d_4}$$

$$= 180 \times 500 \times \frac{700}{250 \times 280} = 900 \text{ r.p.m.}$$

8.4.2 Slip of the Belt

The motion is transmitted due to a firm frictional grip between belt and pulley. But sometimes the frictional grip becomes somewhat loose. This may cause the driver rotate without carrying the belt with it. This may also cause some forward motion of the belt without carrying the driven pulley with it. This is called slip of the belt and is expressed as a percentage.

The effect of the belt slipping is to reduce the velocity ratio of the system. As the slipping of the belt is a common phenomenon, thus, the belt should never be used where a definite velocity ratio is of importance.

Let $s_1\%$ be the slip between the driver and the belt and $s_2\%$ be the slip between the belt and the driven or follower.

Net velocity of the belt passing over the driver per minute,

$$v = \pi d_1 N_1 - \frac{\pi d_1 N_1 \times s_1}{100}$$

$$v = \pi d_1 N_1 \left[1 - \left(\frac{s_1}{100}\right)\right] \tag{1}$$

Similarly, the net velocity of the driven or follower per minute,

$$\pi d_2 N_2 = \left[v - v\left(\frac{s_2}{100}\right)\right]$$

$$= v\left[1 - \left(\frac{s_2}{100}\right)\right] \tag{2}$$

Substituting the value of v from (1) in (2),

$$\pi d_2 N_2 = \pi d_1 N_1 \left[1 - \left(\frac{s_1}{100}\right)\right] \times \left[1 - \left(\frac{s_2}{100}\right)\right]$$

$$\frac{N_2}{N_1} = \left(\frac{d_1}{d_2}\right)\left[1 - \left(\frac{s_1}{100}\right) - \left(\frac{s_2}{100}\right)\right]$$

After neglecting the term $\frac{(s_1 \times s_2)}{(100 \times 100)}$. Then the speed ratio is given by

$$\frac{N_2}{N_1} = \left(\frac{d_1}{d_2}\right)\left[1 - \left(\frac{s_1 + s_2}{100}\right)\right]$$

$$= \left(\frac{d_1}{d_2}\right)\left[1 - \left(\frac{s}{100}\right)\right] \tag{3}$$

where $s = s_1 + s_2$, i.e. total percentage of slip.

Note: When thickness of the belt (t) is taken into account, then the velocity ratio is

$$\frac{N_2}{N_1} = \frac{d_1 + t}{d_2 + t}\left(1 - \frac{s}{100}\right) \tag{4}$$

W E 8.3: An engine shaft running at 120 r.p.m. is required to drive a machine shaft by means of a belt. The pulley on the engine shaft is of 2 m diameter and that of the machine shaft is of 1 m diameter. If the belt thickness is 5 mm, find the speed of the machine shaft when (i) there is no slip, and (ii) there is a slip of 3%.

Given:
Speed of the engine shaft (N_1) = 120 r.p.m.; Diameter of the pulley on it (d_1) = 2 m;
Diameter of the machine shaft (d_2) = 1 m; Thickness of belt (t) = 5 mm = 0.005 m and slip (s) = 3%

(i) Speed of the machine shaft when there is no slip,

$$N_2 = N_1 \frac{d_1 + t}{d_2 + t}$$

$$= 120\left(\frac{2 + 0.005}{1 + 0.005}\right) = 239.4 \text{ r.p.m.}$$

(ii) Speed of the machine shaft when there is a slip of 3%, N_2

$$N_2 = N_1 \frac{d_1 + t}{d_2 + t}\left(1 - \frac{s}{100}\right)$$

$$= 120\left(\frac{2 + 0.005}{1 + 0.005}\right)\left(1 - \frac{3}{100}\right) = 232 \text{ r.p.m.}$$

8.4.3 Effect of Creep on Velocity Ratio

The extension and contraction of the belt when it passes over the pulleys from slack side to the tight side, the relative motion between the belt and the pulley surfaces is called creep. The effect is to reduce the speed of the driven or follower.

The velocity ratio, $\frac{N_2}{N_1} = \frac{d_1}{d_2}\left(\frac{E + \sqrt{\sigma_2}}{E + \sqrt{\sigma_1}}\right)$ where σ_1 and σ_2 are stresses in the belt on the tight and slack sides. E is the Young's modulus of the belt material.

8.5 LENGTH OF AN OPEN BELT

In an open belt drive, both the pulleys rotate in the same direction as shown in Fig. 8.5. Let r_1 and r_2 be the radii of the larger pulley and smaller pulley, x be the distance between the centres O_1 and O_2 of the two pulleys and L be the total length of the belt. Let the belt leave the larger pulley at E

and join at G and similarly the smaller pulley at H and F. From O_2, draw O_2M parallel to FE and perpendicular to O_1E. Let α be the angle of MO_2O_1. Length of the open belt

$$L = \text{Arc } GJE + EF + \text{Arc } FKH + HG$$
$$= 2(\text{Arc } JE + EF + \text{Arc } FK) \qquad (1)$$

From Fig. 8.5, $\sin \alpha$ is

$$\frac{O_1M}{O_1O_2} = \frac{O_1E - EM}{O_1O_2} = \left(\frac{r_1 - r_2}{x}\right)$$

Since α is very small, taking

$$\sin \alpha = \alpha(\text{radians}) = \frac{(r_1 - r_2)}{x} \qquad (2)$$

$$\text{Arc } JE = r_1\left[\left(\frac{\pi}{2}\right) + \alpha\right] \qquad (3)$$

$$\text{and arc } FK = r_2\left[\left(\frac{\pi}{2}\right) - \alpha\right] \qquad (4)$$

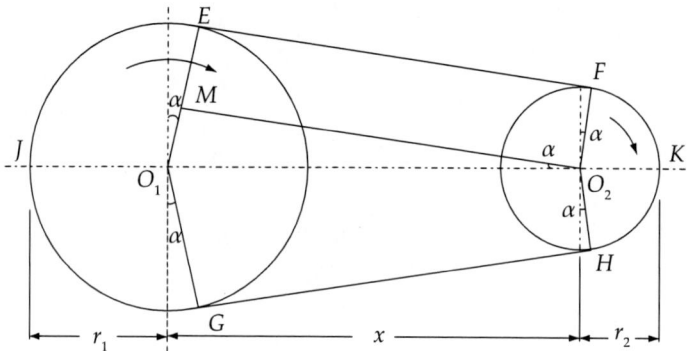

Fig. 8.5 Open belt.

$$EF = MO_2 = \sqrt{(O_1O_2)^2 - (O_1M)^2}$$
$$= \sqrt{x^2 - (r_1 - r_2)^2}$$
$$= x\sqrt{1 - \left(\frac{r_1 - r_2}{x}\right)^2}$$

Expanding this equation using binomial theorem,

$$EF = x\left[1 - \left(\frac{1}{2}\right)\left[\frac{r_1 - r_2}{x}\right]^2 + \ldots\right]$$

Consider the first two terms,

$$= x - \frac{(r_1 - r_2)^2}{2x} \qquad (5)$$

Substituting the values of arc *JE* from equation (3), arc *FK* from equation (4) and *EF* from equation (5) in equation (1). The length of the belt

$$L = 2\left[r_1\left(\frac{\pi}{2} + \alpha\right) + x - \frac{(r_1 - r_2)^2}{2x} + r_2\left(\frac{\pi}{2} - \alpha\right)\right]$$

after expanding and simplifying gives

$$L = \pi(r_1 + r_2) + 2\alpha(r_1 - r_2) + 2x - \frac{(r_1 - r_2)^2}{x}$$

Substituting the value of α from equation (2) and simplifying gives the final expression for

$$L = \left(\frac{\pi}{2}\right)(d_1 + d_2) + 2x + \frac{(d_1 - d_2)^2}{4x}$$

8.6 LENGTH OF A CROSSED BELT

In a crossed belt drive, the pulleys rotate in the opposite direction as shown in Fig. 8.6. Let r_1 and r_2 be the radii of the larger pulley and smaller pulley. x be the distance between the centres O_1 and O_2 of the two pulleys. L be the total length of the belt. Let the belt leave the larger pulley at E and join at G and similarly the smaller pulley at H and F. From O_2, draw O_2M parallel to FE and perpendicular to O_1E extension. Let α be the angle of MO_2O_1.

Length of the belt

$$L = \text{Arc } GJE + EF + \text{Arc } FKH + HG$$
$$= 2(\text{Arc } JE + EF + \text{Arc } FK) \quad (1)$$

From Fig. 8.6,

$$\sin \alpha = \frac{O_1M}{O_1O_2} = \frac{O_1E + EM}{O_1O_2} = \left(\frac{r_1 + r_2}{x}\right)$$

Since α is very small, taking

$$\sin \alpha = \alpha(\text{radians}) = \frac{(r_1 + r_2)}{x} \quad (2)$$

$$\text{Arc } JE = r_1\left(\frac{\pi}{2} + \alpha\right) \quad (3)$$

$$\text{and arc } FK = r_2\left(\frac{\pi}{2} + \alpha\right) \quad (4)$$

$$EF = MO_2 = \sqrt{(O_1O_2)^2 - (O_1M)^2}$$
$$= \sqrt{x^2 - (r_1 + r_2)^2}$$
$$= x\sqrt{\left[1 - \left(\frac{r_1 + r_2}{x}\right)^2\right]}$$

Expanding this equation by binomial theorem,

$$EF = x\left[1 - \left(\frac{1}{2}\right)\left[\frac{r_1+r_2}{x}\right]^2 + \ldots\right]$$

Consider the first two terms,

$$= x - \frac{(r_1+r_2)^2}{2x} \tag{5}$$

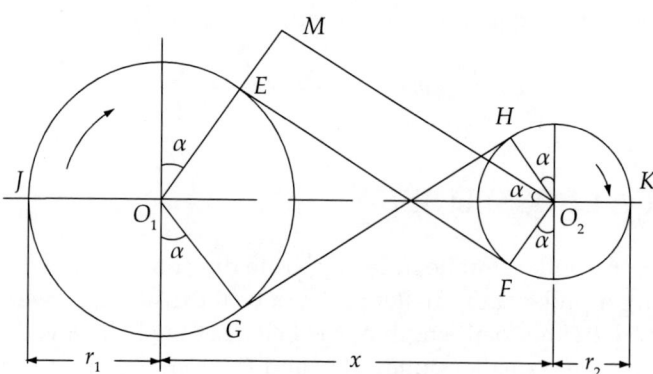

Fig. 8.6 Crossed belt.

Substituting the values of arc JE from equation (3), arc FK from equation (4) and EF from equation (5) in equation (1), gives the length of the belt

$$L = 2\left[r_1\left(\frac{\pi}{2}+\alpha\right) + x - \frac{(r_1+r_2)^2}{2x} + r_2\left(\frac{\pi}{2}+\alpha\right)\right]$$

after expanding and simplifying gives

$$L = \pi(r_1+r_2) + 2\alpha(r_1+r_2) + 2x - \frac{(r_1+r_2)^2}{x}$$

Substituting the value of α from equation (2) and again simplifying gives the final expression for the length of the crossed belt

$$L = \left(\frac{\pi}{2}\right)(d_1+d_2) + 2x + \frac{(d_1+d_2)^2}{4x}$$

W E 8.4: Find the length of the (a) open belt and (b) crossed belt necessary to drive a pulley of 500 mm diameter running parallel at a distance of 12 m from the driving pulley of diameter 1600 mm.

Given:
Diameter of the driving pulley (d_1) = 1600 mm = 1.6 m or radius (r_1) = 0.8 m
Diameter of the driven pulley (d_2) = 500 mm = 0.5 m or radius (r_2) = 0.25 m

Distance between the centres of the two pulleys $(x) = 12$ m

(a) Length of the open belt (L_{open})

$$L_{open} = \left(\frac{\pi}{2}\right)(d_1 + d_2) + 2x + \frac{(d_1 - d_2)^2}{4x}$$

$$= \left(\frac{\pi}{2}\right)(1.6 + 0.5) + 2 \times 12 + \left(\frac{(1.6 - 0.5)^2}{4 \times 12}\right)$$

$$= 27.32 \text{ m}$$

(b) Length of the open belt (L_{cross})

$$L_{cross} = \left(\frac{\pi}{2}\right)(d_1 + d_2) + 2x + \frac{(d_1 + d_2)^2}{4x}$$

$$= \left(\frac{\pi}{2}\right)(1.6 + 0.5) + 2 \times 12 + \left(\frac{(1.6 + 0.5)^2}{4 \times 12}\right)$$

$$= 27.39 \text{ m}$$

8.7 RATIO OF TENSIONS

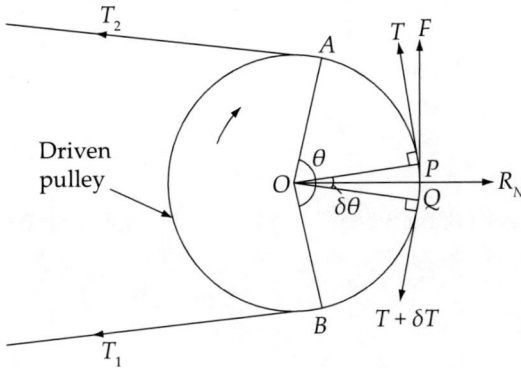

Fig. 8.7 Ratio of driving tensions for flat belt.

Consider a follower (i.e., driven) pulley rotating in the clockwise direction as shown in Fig. 8.7. Let $\delta\theta$ be the angle subtended by a small portion of the belt PQ at the centre of the pulley. Let μ be the coefficient of friction between the belt and pulley. R_n be the normal reaction or radial reaction. Let T_1 and T_2 be the tensions on the tight and slack sides. The belt PQ is in equilibrium under the following forces:

1. Tension T in the belt at P
2. Tension $(T + \delta T)$ in the belt at Q
3. Normal reaction R_n
4. Frictional force $F = \mu R_n$ where μ is the coefficient of friction between the belt and pulley

Resolving all the forces horizontally and equating the same to R_n gives

$$R_n = (T + \delta T)\left(\sin \frac{\delta\theta}{2}\right) + T \sin \frac{\delta\theta}{2} \qquad (1)$$

Since $\delta\theta$ is very small, therefore, substituting $\sin \frac{\delta\theta}{2} = \frac{\delta\theta}{2}$ in (1)

$$R_n = (T + \delta T)\left(\frac{\delta\theta}{2}\right) + T\frac{\delta\theta}{2}$$

$$= T\delta\theta \quad \text{neglecting } \delta T \cdot \frac{\delta\theta}{2} \qquad (2)$$

Now resolving the forces vertically gives,

$$F = \mu R_n = \left[(T + \delta T)\cos \frac{\delta\theta}{2}\right] - T\cos \frac{\delta\theta}{2} \qquad (3)$$

since $\delta\theta$ is very small, therefore, putting $\cos \frac{\delta\theta}{2} = 1$ in equation (3) gives

$$F = \mu R_n = [(T + \delta T)] - T = \delta T$$

$$R_n = \frac{\delta T}{\mu} \qquad (4)$$

Equating the values of R_n from (2) and (4)

$$T\delta\theta = \frac{\delta T}{\mu} \quad \text{or} \quad \frac{\delta T}{T} = \mu\delta\theta$$

Integrating both sides from A to B, between T_1 and T_2 on left side and 0 to θ on the right side,

$$\int \frac{\delta T}{T} = \int \mu\delta\theta$$

$$\log_e\left(\frac{T_1}{T_2}\right) = \mu\theta \quad \text{(or)} \quad \frac{T_1}{T_2} = e^{\mu\theta} \qquad (5)$$

The equation (5) may also be expressed in terms of corresponding logarithm to the base 10, i.e. $2.3 \log\left(\frac{T_1}{T_2}\right) = \mu\theta$.

The above expression gives the relation between the tight side and slack side tensions, in terms of coefficient of friction and the angle of contact.

Note:

1. θ in the above expression is the angle of contact at the smaller pulley.
2. $\theta = (180° - 2\alpha)$ for open belt drive and $\theta = (180° + 2\alpha)$ for cross belt drive.

8.7.1 Power Transmitted by a Belt

The driving pulley pulls the belt from one side and delivers the same to the other. The tension T_1 in the former side (tight side) will be more than the tension T_2 in the later side (i.e., slack side) as shown in Fig. 8.2. Let v be the velocity of the belt.

Effective turning force (i.e., driving force) at the circumference of the follower is the difference between the tensions (i.e., $T_1 - T_2$). Therefore,

$$\text{Work done per second} = \text{Force} \times \text{Distance}$$
$$= (T_1 - T_2) \times v \text{ N-m/s}$$

Therefore, Power = $(T_1 - T_2)v$ Watts.

Note:

1. The torque exerted on the driving pulley, $(T_1 - T_2)r_1$
2. The torque exerted on the driven pulley, $(T_1 - T_2)r_2$ where r_1 and r_2 are the radii of the driver and driven in m.

W E 8.5: The tensions in the two sides of the belt are 1000 and 800 N respectively. If the speed of the belt is 75 m/s, find the power transmitted by the belt.

Given:

$T_1 = 1000$ N and $T_2 = 800$ N and speed of the belt $v = 75$ m/s

Power transmitted by the belt,

$$P = (T_1 - T_2)v$$
$$= (1000 - 800) \times 75$$
$$= 15000 \text{ N-m/s} = 15000 \text{ W} = 1.5 \text{ kW}$$

8.7.2 Effect of Centrifugal Tension T_C on Power Transmitted

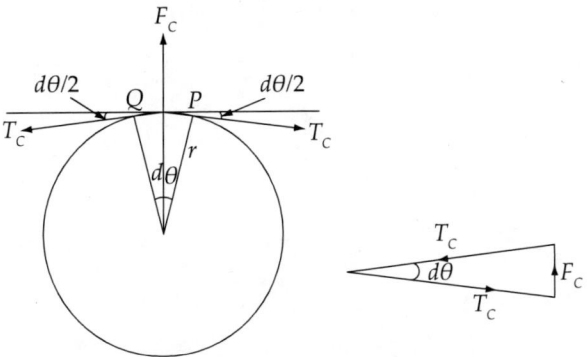

Fig. 8.8 Centrifugal tension.

The belt continuously runs over both the pulleys. It causes some centrifugal force due to mass of the belt at both the pulleys. The effect of this centrifugal force is to increase the tension on both, tight as well as the slack sides. The tension caused due to centrifugal force is called centrifugal tension depending upon the speed of rotation. But at higher speeds its effect is to be considered. The tension due to centrifugal force (T_C) is equal to mass of the belt × square of the speed (i.e., mv^2) as shown in Fig. 8.8.

Let T_C be the centrifugal tension of the belt then the tension on the tight side is

$$T_{C1} = T_1 + T_C$$

and on the slack side is $T_{C2} = T_2 + T_C$.

But the power transmitted (P) is

$$P = (T_{C1} - T_{C2}) \cdot v = [(T_1 + T_C) - (T_2 + T_C)] \cdot v = (T_1 - T_2) \cdot v$$

Hence, there is no effect of T_C on the power transmitted.

Note: The maximum tension in the belt is $(T_1 + T_C)$.

8.7.3 Condition for Maximum Power

The power transmitted by a belt
$$P = (T_1 - T_2) \cdot v \tag{1}$$

and the ratio of tensions,
$$\frac{T_1}{T_2} = e^{\mu\theta} \quad \text{or} \quad T_2 = \frac{T_1}{e^{\mu\theta}} \tag{2}$$

Substituting the value of T_2 in (1),

$$P = \left[T_1 - T_1/e^{\mu\theta}\right] \cdot v$$
$$= T_1\left[1 - \frac{1}{e^{\mu\theta}}\right] \cdot v$$
$$= T_1 \times v \times C \tag{3}$$

where $C = \left[1 - \left(\frac{1}{e^{\mu\theta}}\right)\right]$; Maximum tension, $T_{max} = (T_1 + T_C)$ or $T_1 = T_{max} - T_C$ substituting the value of T_1 in equation (3)

$$P = (T_{max} - T_C) \times v \times C = (T_{max} - mv^2) \times v \times C = (T_{max} \cdot v - mv^3)C$$

For maximum power, differentiate the above equation with respect to v and equate to zero.

$$T_{max} - 3mv^2 = 0$$
$$T_{max} - 3T_C = 0$$

$$T_{max} = 3T_C$$
$$T_C = \frac{T_{max}}{3}$$

It shows that when the maximum power is transmitted then $\frac{1}{3}$rd of the maximum tension is absorbed as centrifugal tension.

8.7.4 Effect of Initial Tension (T_0)

T_0 which is $< T_1$ and $> T_2$. Net decrease of tension on tight side = $T_1 - T_0$ and net increase in tension on the slack side = $T_0 - T_2$ or $-(T_2 - T_0)$. Let α be the coefficient of increase of length per unit force, then the increase in length on the tight side = $\alpha(T_1 - T_0)$ and decrease on the slack side = $\alpha(T_0 - T_2)$. The length of the belt remains constant at rest or in motion. Therefore, both must be equal. Hence,

$T_1 - T_0 = T_0 - T_2$ or $T_0 = \dfrac{(T_1 + T_2)}{2}$ or $\dfrac{(T_1 + T_2 + 2T_C)}{2}$ when T_C is considered.

8.8 ROPE DRIVE

Rope is used instead of belt when the distance between shafts is long and considerable power is to be transmitted. Here the rim of the pulley is grooved in which the rope runs as shown at Fig. 8.1. The effect of the groove is to increase the frictional grip of the rope on the pulley and to reduce the slipping.

Advantages over belt drive

1. Used when the distance between shafts is large.
2. Frictional grip is more.
3. The net driving tension is more.

8.8.1 Ratio of Tensions

The same principle adopted in a belt drive is used for rope drive also to determine the ratio of tensions. Figure 8.9 (a and b) shows the cross section of the rope in a groove and the V-belt. Let R_1 be the normal reaction between rope and sides of the groove, R be the total reaction in the plane of the groove, 2β be the angle of groove, and μ be the coefficient of friction between rope and sides of groove.

Resolving the reactions vertically:

$$R = R_1 \sin\beta + R_1 \sin\beta = 2R_1 \sin\beta$$
$$R_1 = \frac{R}{(2\sin\beta)}$$
$$\text{Friction force} = 2\mu R_1 = 2\mu \frac{R}{(2\sin\beta)}$$
$$= \mu \frac{R}{(\sin\beta)} = \mu R \operatorname{cosec}\beta$$

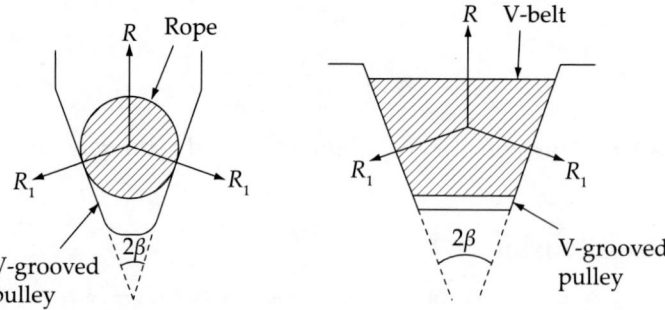

Fig. 8.9

Consider a small portion of the rope as in Fig. 8.7 subtending an angle $\delta\theta$ at the centre. The tension on one side will be T and on the other side $T + \delta T$. The frictional resistance will be $\mu R \operatorname{cosec} \beta$ instead of simply μR. The ratio of tensions just as in the case of belt $\dfrac{T_1}{T_2} = e^{\mu\theta \operatorname{cosec}\beta}$.

Therefore,
$$\frac{T_1}{T_2} = e^{\mu_1 \theta}$$

where $\mu_1 = \dfrac{\mu}{\sin\beta} = \mu \operatorname{cosec}\beta$.

The equation may also be expressed in terms of corresponding logarithm to the base 10, i.e.
$2.3 \log\left(\dfrac{T_1}{T_2}\right) = \mu\theta \operatorname{cosec}\beta$.

W E 8.6: Find the power transmitted by a rope drive from the following data: Angle of contact = 180°; Pulley groove angle = 60°; Coefficient of friction = 0.2; Mass of rope = 0.4 kg/metre length; Permissible tension = 1.5 kN; Velocity of rope = 15 m/s.

Given:
$\theta = 180° = 3.142$ radians; $2\beta = 60°$ or $\beta = 30°$; Mass of rope $m = 0.4$ kg/m; $\mu = 0.2$;
Permissible tension $T = 1.5$ kN   velocity of rope $v = 15$ m/s.

Centrifugal tension = $T_C = mv^2 = 0.4(15)^2 = 90$ N; $T_1 = T - T_C = 1500 - 90 = 1410$ N

$$2.3 \log\left(\frac{T_1}{T_2}\right) = \mu\theta \operatorname{cosec}\beta$$
$$= 0.2 \times 3.142 \times \operatorname{cosec} 30°$$
$$= 0.2 \times 3.142 \times 2.0 = 1.257.$$

Therefore,
$$\log\left(\frac{T_1}{T_2}\right) = \frac{1.257}{2.3} = 0.5465$$
$$\frac{1410}{T_2} = 3.2$$
$$T_2 = \frac{1410}{3.52} = 400 \text{ N}$$

$$\text{Power transmitted} = (T_1 - T_2)v$$
$$= (1410 - 400) \times 15$$
$$= 15150 \text{ W} = 15.15 \text{ kW}$$

8.9 CHAIN DRIVES

A chain may be regarded as a belt built up of rigid links, which are hinged together in order to provide the necessary flexibility for the wrapping action around the driving and driven wheels. These wheels have projecting teeth, which fit into suitable recesses in the links of the chain and thus enable to obtain a positive drive. They are called chain sprockets like spur gears. The Pic. 8.8 and Pic. 8.9 show the chain and the sprocket.

Pic. 8.8 Chain and sprocket.

Pic. 8.9 Chain drive.

The pitch of the chain is the distance between a hinge centre of one link and the corresponding hinge centre of the adjacent link. The pitch circle diameter of the chain sprocket is the diameter of the circle on which the hinge centres lie, when the chain is wrapped around the sprocket.

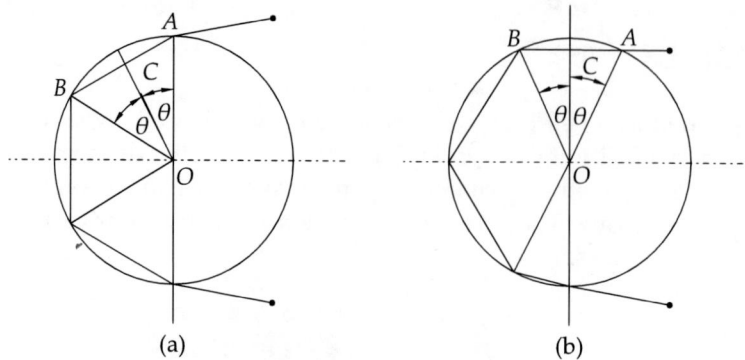

Fig. 8.10 Extreme conditions.

Referring to Fig. 8.10(a), the pitch becomes a chord AB, not an arc of the pitch circle as the chain links are rigid. The relation between the pitch circle diameter d, the pitch p and the number of teeth T on the sprocket, may be found as follows:

Let the angle subtended at the sprocket centre by one pitch $AB = \dfrac{360°}{T}$. But $AC = OA \sin \theta$, so that

$$\frac{p}{2} = \left(\frac{d}{2}\right) \sin \frac{180°}{T}$$

$$d = p \operatorname{cosec} \frac{180°}{T}$$

The chain speed and the angular velocity will vary with the angular position of the sprocket. The extreme conditions are shown in Fig. 8.10(a) and (b). The speed (v) of the chain for position (a) is $\omega \cdot OA$ and for position (b) is $\omega \cdot OC = \omega \cdot OA \cos \theta$.

8.9.1 Types of Chains

There are two types of chains in common use for transmitting power.

(a) The roller chain

The construction of this type of chain is shown in Fig. 8.11. The inner plates are held together by steel bushes, through which pass the pins riveted to the outer links. A roller surrounds each bush and the teeth of the sprockets bear on the roller. The rollers turn freely on the bushes and the bushes turn freely on the pins. All the contact surfaces are hardened so as to resist wear and are lubricated so as to reduce friction. The sprockets are so shaped that the rollers rest on the bottom of the recesses between the teeth. The centre of curvature of each recess lies on the pitch circle of the sprocket and the radius of curvature is a few thousandths of an inch larger than the radius of the roller.

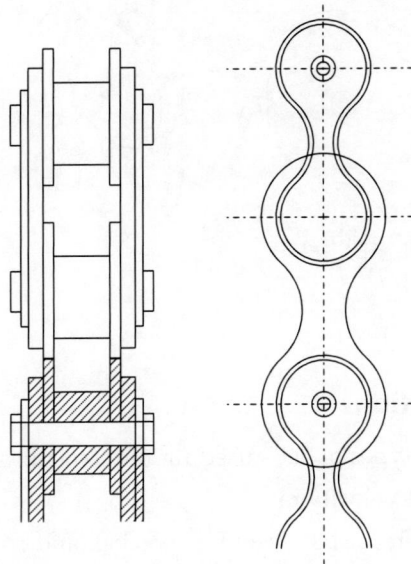

Fig. 8.11

(b) The inverted tooth or silent chain

The construction of this type of chain is shown in Fig. 8.12. It is built up from a series of flat plates, each of which has two projections or teeth. The outer faces of the teeth are ground to give an included angle of 60° or 75° in some cases. The inner faces of the link teeth take no part in the drive and are so shaped as to clear the sprocket teeth. The required width of chain is built up from a number of these plates, arranged alternately and connected together by hardened steel pins which pass through hardened steel bushes inserted in the ends of the links. The pins are riveted over the outside plates. The chain may be prevented from sliding axially across the face of the sprocket teeth by outside guide plates without teeth, or by a centre guide plate without teeth which fits into a recess turned in the sprocket.

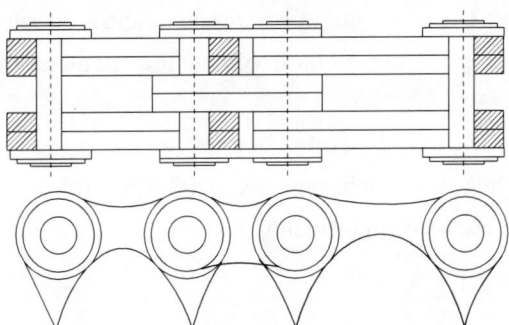

Fig. 8.12 Inverted tooth or silent chain.

The Figure 8.13 shows the different types of chains along with sprockets used in industries.

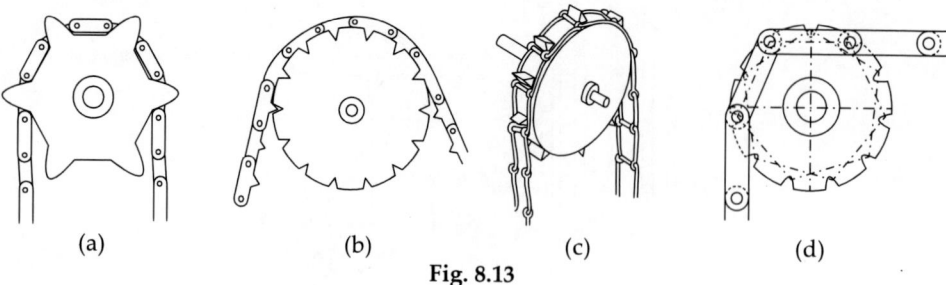

Fig. 8.13

8.10 EXERCISE

8.10.1 Short Answer Questions

1. Discuss briefly the various types of belts used for the transmission of power.
2. How does the slip affect the velocity ratio?
3. Distinguish clearly the difference between the open belt and cross belt drive.
4. Obtain an equation for the length of a belt in: (i) an open belt (ii) a cross belt drive.
5. Deduce the relation between the tension on the tight side and slack side of a belt connecting two pulleys and transmitting power. Neglect the centrifugal effect of the belt.
6. What are the advantages of a rope drive?
7. Define the term initial tension in a belt. How would you find out the initial tension in a belt?
8. Derive the expression for the length of the belt for a cross belt drive.
9. Make a note on a chain drive and friction in slider crank mechanism.
10. Draw a neat sketch of centrifugal clutch and indicate the design procedure in brief.
11. What are the desirable properties of belt materials?
12. What are the types of chain drives?
13. What do you mean by crowing of pulleys in flat belt drives? What is its use?
14. Deduce an expression for the exact and approximate lengths of belt in open belt drive.
15. Find the relation between pitch of the chain, pitch circle diameter of sprocket and number of teeth on the sprocket.
16. Derive an expression for the length of open belt drive.
17. Derive the relation for the ratio of belt tensions in a flat belt drive.
18. Derive the condition for transmitting the maximum power in a flat belt drive.

8.10.2 Problems

1. An open belt connects two flat pulleys. The smaller pulley is 30 cm diameter and runs at 200 r.p.m. The angle of lap on this pulley is 160° and the coefficient of friction between belt and pulley face is 0.25. The belt is on the point of slipping when 3.5 kW is being transmitted. Which of the following alternatives would be more effective in increasing the power which could be transmitted?
   (a) Increasing the initial tension in the belt by 10 per cent.
   (b) Increasing the coefficient of friction by 10 per cent by the application of a suitable dressing to the belt.

2. The power transmitted by a belt drive is 2.5 kW. The linear velocity of the belt is 2.5 m/s. The angle of lap on the smaller pulley is 165°. The coefficient of friction is 0.3. Determine the effect on power transmission in the following cases: (i) Initial tension in the belt is increased by 8% by suitable dressing to the friction surface of the belt.

3. Power is transmitted using a V-belt drive. The included angle of V-groove is 30°. The belt is 20 mm deep and its mean width is 20 mm. If the mass of the belt is 0.35 kg/m length and maximum allowable stress is 1.4 MPa, determine the maximum power transmitted when the angle of lap is 14°. Take $\mu = 0.15$.

I. Belts, Slip, Length

1. A diesel engine shaft having a speed of 180 r.p.m. is required to drive a machine shaft with the help of a belt. Find the speed of the machine shaft, if the diameters of the engine shaft and machine shafts are 300 mm and 200 mm respectively. [*Ans.* 270 r.p.m.]

2. In a workshop, a machine shaft is driven by an electric motor with the help of belts across a main shaft and a counter-shaft. The diameters of the driving pulleys are 500 mm, 400 mm and 300 mm respectively whereas the diameters of the driven pulleys are 250 mm, 200 mm and 150 mm respectively. Find the speed of the machine shaft when the electric motor runs at 150 r.p.m. [*Ans.* 1200 r.p.m.]

3. An engine running at 150 r.p.m. drives a line shaft by means of a belt. The engine pulley is 750 mm diameter and the pulley on the line shaft being 450 mm diameter. The 900 mm pulley on the line shaft drives a 150 mm diameter pulley keyed to a dynamoshaft. Find the speed of the dynamoshaft, when (i) there is no slip; and (ii) there is a slip of 2% at each drive. [*Ans.* 1500 r.p.m; 1380 r.p.m.]

4. Two parallel shafts 6 m apart are provided with 900 mm and 300 mm diameter pulleys and are connected by means of a cross belt. The direction of rotation of the follower pulley is to be reversed by changing over to an open belt drive. How much length of the belt has to be reduced? [*Ans.* 40 mm]

II. Power, Ratio of Tensions

1. Two pulleys of diameters 500 mm and 300 mm are connected by a belt having tensions 3 kN and 2.5 kN on the two sides of the belt connecting them. If the power transmitted is 25 kW, find the speed of the belt. [*Ans.* 50 m/s]

2. A pulley is driven by a flat belt running at a speed of 600 m/min. The coefficient of friction between the pulley and the belt is 0.3 and the angle of lap is 160°. If the maximum tension in the belt is 700 N, find the power transmitted by the belt. [*Ans.* 3.97 kW]

3. A belt connects two pulleys A and B 4 m apart. The pulley A is 1 m diameter, whereas the pulley B is 50 cm diameter. If the coefficient of friction between the belt and the pulley is 0.32, find the ratio of the tensions, when the drive is (i) an open belt drive and (ii) cross belt drive. [*Ans.* 2.623; 3·084]

4. A flat belt 7.5 mm thick and 100 mm wide transmits power between two pulleys, running at 1600 r.p.m. The mass of the belt is 0.9 kg/m length. The angle of lap on the smaller pulley is 1650° and the coefficient of friction between the belt and the maximum power transmitted; and (ii) initial tensions. [*Ans.* 15.44 kW; 710.5 N]

5. An open belt connects two flat pulleys. The smaller pulley is 30 cm diameter and runs at 200 r.p.m. The angle of lap on this pulley is 160° and the coefficient of friction between belt and pulley face is 0.25. The belt is on the point of slipping when 3.5 kW is being transmitted. Which of the following alternatives would be more effective in increasing the power which could be transmitted?
   (a) Increasing the initial tension in the belt by 10 per cent.
   (b) Increasing the coefficient of friction by 10 per cent by the application of a suitable dressing to the belt.

6. Find the maximum power that can be transmitted by a belt 15 cm ×1 cm if the ratio of tensions is 2 and the maximum tension allowed is 140 N/cm². Density of leather may be taken as 0.01 N/m³. Derive the formulae used, if any.

7. The power transmitted between two shafts 3.5 m apart by a cross belt drive around the two pulleys 600 mm and 300 mm in diameters, is 6 kW. The speed of the larger pulley (driver) is 220 r.p.m. The permissible load on the belt is 25 N/mm width of the belt, which is 5 mm thick. The coefficient of friction between the smaller pulley surface and the belt is 0.35.
Determine: 1. Necessary length of the belt; 2. Width of the belt, and 3. Necessary initial tension in the belt.

8. An open belt drive running at 2.5 m/s transmits 2.5 kW. The angle of lap on the smaller pulley is 165° and coefficient of friction between belt and pulley being 0.30. Determine the effect on power transmission if initial tension is increased by 10%.

9. An open belt drive connects two pulleys 1.2 m and 0.5 m diameter, on parallel shafts 4 m apart. The mass of the belt is 0.9 kg/m length and the maximum tension is not to exceed 2000 N. The

coefficient of friction is 0.3. The 1.2 m pulley, which is the driver, runs at 200 r.p.m. Due to belt slip on one of the pulleys, the velocity of the driven shaft is only 450 r.p.m. Calculate the torque on each of the shafts, the power transmitted and power lost in friction. What is the efficiency of the drive?

10. 2.5 kW of power is transmitted by an open belt drive. The linear velocity of the belt is 2.5 m/s. The angle of lap on smaller pulley is 165°. The coefficient of friction is 0.3. Determine the effect on power transmission in the following cases: (i) Initial tension in the belt is decreased by 8% (ii) Coefficient of friction is increased by 8% (iii) Angle of lap is increased by 8%.

11. A belt drive is required to transmit 10 kW from a motor running at 600 r.p.m. The belt is 12 mm thick and has a mass density of 0.001 gm/mm^3. Safe stress in the belt is not to exceed 2.5 N/mm^2. Diameter of the driving pulley is 250 mm, whereas the speed of the driven pulley is 220 r.p.m. The two shafts are 1.25 m apart. The coefficient of friction is 0.25. Determine the width of the belt.

12. A leather belt 120 mm wide and 6 mm thick transmits power from a pulley 800 mm diameter which rotates at 450 r.p.m. The angle of lap is 160° and coefficient of friction is 0.3. The mass of the belt is 1000 kg/m^3 and the stress is not to exceed 2.5 MPa. Find the maximum power that can be transmitted.

13. A shaft runs at 80 r.p.m. and drives another shaft at 150 r.p.m. through belt drive. The diameter of the driving pulley is 600 mm. Determine the diameter of the driven pulley in the following cases: (i) Neglecting belt thickness, (ii) Taking belt thickness as 5 mm. (iii) Assuming for case (ii) a total slip of 4% and (iv). Assume for case (ii) a slip of 2% on each pulley.

14. A belt is to transmit 40 kW from a pulley 1.5 m diameter running at 300 rpm. The angle of contact is spread over $\frac{11}{24}$th of the circumference of the pulley, and the coefficient of friction is 0.3, determine the width of the belt required, if thickness of belt is 10 m, safe working stress for the belt material is 2.5 MPa, and density of belt material is 1100 kg/m^3.

15. An open belt connects two pulleys 1.5 m and 0.5 m diameter on parallel shafts 3.5 m apart. The belt has a mass of 1 kg/m length and the maximum tension in the belt is not to exceed 2 kN. The 1.5 m pulley which is the driver runs at 250 r.p.m. Due to belt slip, the velocity of the driven shaft is only 730 r.p.m. If the coefficient of friction between the belt and the pulley is 0.25, find (a) torque on each shaft (b) power transmitted (c) power lost in friction and (d) efficiency of the drive.

8.10.3 Multiple Choice Questions

1. The length of the belt used in a crossed belt drive is _____ than that used in open belt drive, provided distance between the two pulleys and their diameters remains same.

    (a) less   (b) equal to   (c) more   [Ans. (c)]

2. The power transmitted by a belt depends upon
   (a) sum of the tensions in the tight side and slack side
   (b) difference of tension in the tight side and slack side
   (c) none of them. [Ans. (b)]

3. The relation between the tension in tight side (T_1) and slack side (T_2) of a belt drive is
   (a) $\dfrac{T_2}{T_1} = e^{\mu\theta}$ (b) $2.3 \log\left(\dfrac{T_2}{T_1}\right) = \mu\theta$ (c) both of them   (d) none of them. [Ans. (d)]

4. In a rope drive, the relation between the tension in the tight side and slack side is $2.3 \log\left(\dfrac{T_1}{T_2}\right) = \mu\theta \sec \alpha$ where 2α = angle of groove.
   (a) True   (b) False [Ans. (b)]

5. Which of the following is not a flexible type of connector?
   (a) Belt   (b) Rope   (c) Chain   (d) Gear [Ans. (d)]

6. In an open or crossed belt drive, the velocity ratio of the two pulleys is
   (a) directly proportional to their diameters
   (b) directly proportional to the square of their diameters
   (c) inversely proportional to their diameters
   (d) inversely proportional to the square of their diameters. [Ans. (c)]

7. Due to slip the velocity ratio of a belt drive
   (a) increases   (b) decreases   (c) remains same [Ans. (b)]

8. The include angle of a pulley for a V belt is
   (a) $50° - 60°$ (b) $30° - 40°$ (c) $20° - 30°$ (d) $40° - 50°$ [Ans. (b)]

9. The crowning of pulleys is done to
   (a) increase the tightness of the belt on the pulley
   (b) prevent belt running off the pulley
   (c) increase the torque transmitted
   (d) improve the shape and strength of the pulley [Ans. (b)]

10. For maximum power transmission by a belt drive the maximum tension must be
    (a) $2T_C$ (b) $3T_C$ (c) $4T_C$ (d) $5T_C$ [Ans. (b)]

11. For maximum power transmission, the velocity ratio of the belt is
    (a) $\dfrac{T}{\sqrt{m}}$ (b) $\dfrac{T}{\sqrt{2m}}$ (c) $\dfrac{T}{\sqrt{3m}}$ (d) $\dfrac{T}{\sqrt{4m}}$ [Ans. (c)]

12. The belt drive is designed on the basis of the angle of contact on the _____
    (a) large pulley (b) smaller pulley (c) any pulley [Ans. (b)]

13. The law of belting states that the centre line of the belt when it _____ a pulley must lie in the mid-plane of that pulley.
    (a) leaves (b) approaches (c) approaches as well as leaves [Ans. (b)]

14. The ratio of tight to slack side tensions in a V-belt or rope is
    (a) $e^{\mu\theta \sin\alpha}$ (b) $e^{\mu\theta/\cos\alpha}$ (c) $e^{\mu\theta \cos\alpha}$ (d) $e^{\mu\theta/\sin\alpha}$ [Ans. (d)]

15. An increase in the initial tension in the belt _____ the power transmitted.
    (a) increases (b) decreases (c) does not affect [Ans. (a)]

Gyroscope 9

Gyroscope

9.1 INTRODUCTION

Whenever a person drives around a curve or a circular track on a bicycle or a motorcycle or a scooter, one has to lean inward in order to maintain a perfect equilibrium. The angle at which a person leans with the vertical depends on the speed, radius of track or curvature and height of the centre of gravity of the vehicle and the person. The person who rides the vehicle naturally leans on his own inwards automatically whether he is educated or uneducated to neutralise the overturning effect and to avoid accident.

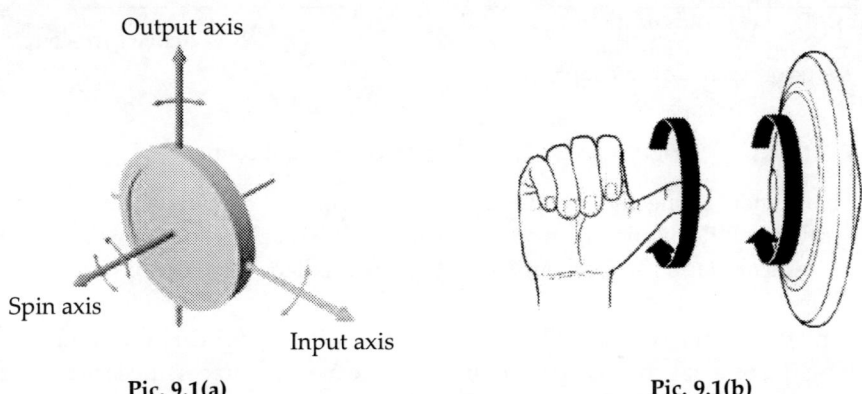

Pic. 9.1(a)   Pic. 9.1(b)

Pic. 9.1(a) shows a rotor rotating clockwise (CW) direction when viewed from behind as indicated. Its angular motion has been shown vectorially along the axis of rotation of the rotor. The sense of direction of the vector is from rotor to outwards according to the screw rule or right hand thumb rule as in Pic 9.1(b). However, if the direction of rotation of the rotor is reversed, i.e. anticlockwise (ACW) or counterclockwise (CCW), the vector would be towards the rotor. The axis about which the rotor rotates is called spin axis. The other two axes shown are called input and output axes.

Figure 9.1(a) shows three mutually perpendicular axes called,

    axis of spin along X-axis,
    axis of precession along Z-axis, and
    axis of gyro-couple along Y-axis.

There are three mutually perpendicular planes called,

    plane of spinning in Y-Z plane perpendicular to X-axis,
    plane of precession in X-Y plane perpendicular to Z-axis, and
    plane of gyro-couple in X-Z plane perpendicular to Y-axis.

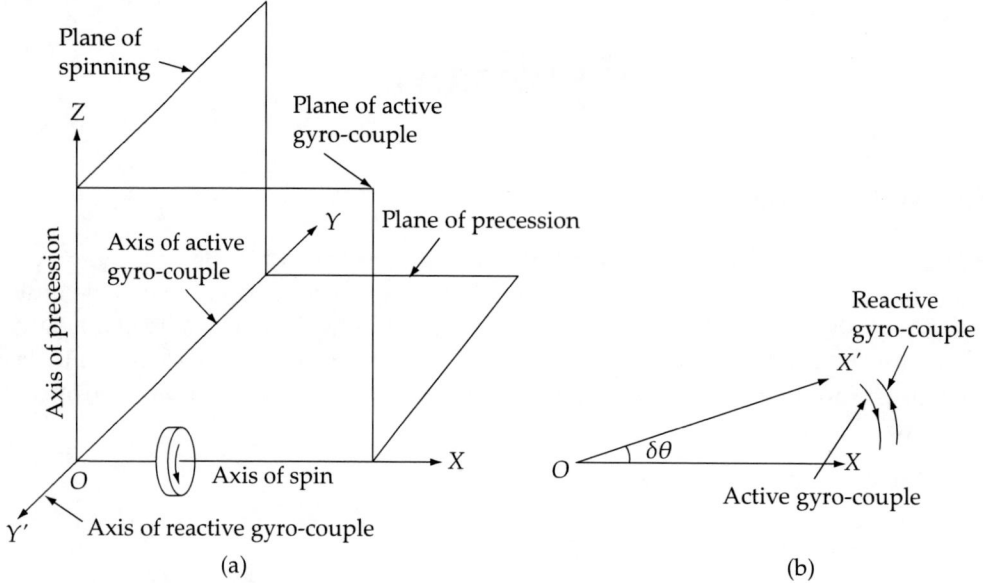

Fig. 9.1 Axes of spinning, precession and couple.

Let a disc of moment of inertia I, rotate at an angular velocity ω about the X-axis as shown. The rotation or revolving of the disc is called spinning along X-axis called spinning axis and the plane of rotation is in a vertical plane Y-Z perpendicular to X-axis called plane of spinning with an angular velocity ω radians/sec.

The rotation of spinning axis from OX to OX' as shown in Fig. 9.1 (b) by a small angle $\delta\theta$, called precession along Z-axis is called axis of precession and the plane of precession is in a horizontal plane X-Y perpendicular to Z-axis called plane of precession with an angular velocity of ω_p radians/sec.

Applying an active couple on the rotor about Y-axis, called axis of active gyro-couple and the plane of active couple is in a vertical plane X-Z perpendicular to Y axis called plane of gyro-couple. The active gyro-couple is represented by the vector along OY and the reactive gyro couple by the vector along OY', opposite to OY.

9.2 GYROSCOPIC COUPLE AND ITS EFFECT

Let the vector OX shown in Fig. 9.1(b) represent the angular momentum of the disc $I\omega$. Let it change to OX' by a small angle $\delta\theta$ in a time interval of δt in the XOY plane as shown. Assume that the angular velocity of the disc ω is constant.

The change in angular momentum XX' is equal to $OX\delta\theta = I\omega \cdot \delta\theta$.

The rate of change of angular momentum $= I \cdot \omega \times \left(\dfrac{\delta\theta}{\delta t}\right)$.

This rate of change of momentum will give rise to a couple to the axis of the rotation of the disc. Hence, if an active gyro-couple is applied to the axis of the disc, it causes precession of the rotor shaft or the axis changes from OX to OX'.

Therefore, the applied couple C or torque $T = (I\omega)\dfrac{\delta\theta}{\delta t}$. As δt tends to zero, the couple C or torque $T \approx OX.(\delta\theta/\delta t) \approx \dfrac{XX'}{dt}$ becomes $(I\omega)\omega_p$ where ω_p is called the angular velocity of precession.

This expression is similar to the expression for centrifugal force,

$$F_C = m\omega^2 r = m(\omega r)\omega = (m \cdot v) \cdot \omega$$

where mv is the linear momentum. The angular velocity of the linear momentum vector is ω. The centripetal force is required to change the linear velocity direction without changing its magnitude.

Thus, the applied gyro torque (T) or gyro-couple $(C) = (I\omega)\omega_p$ where $I\omega$ represents the angular momentum and ω_p the angular velocity of the momentum vector.

The effect produced by the reactive gyroscopic couple is known as the gyroscopic effect. Thus, aeroplanes, ships, automobiles, etc., having rotating parts in the form of wheels or rotors, gears and flywheels, etc. of engines experience this effect while taking a turn, i.e. the axis of spin being subjected to some angular motion. A gyroscope is a spinning body which is free to move in other directions under the action of external forces.

Note:

1. When the angular velocity changes in direction but remains constant in magnitude, it is known as a gyroscopic couple.
2. When the angular velocity remains constant in direction but changes in magnitude, it is known as an angular acceleration.

Direction of Spin Vector, Precession Vector and Couple Vector with forced precession using the right-hand screw rule or right-hand rule, it is easy to know the gyroscopic effect. See Pic. 9.15.

Let the four fingers of the right hand represent the rotation then the thumb shows the direction of the spin vector. Thus, the precession vector direction is also found in the same manner as follows.

Rotate the spin vector in the direction of precession by 90°, in order to get the direction of the gyro-couple or torque. The spin vector represents the direction of the applied torque or couple vector. Obtain the reaction torque or couple exerted by rotating the applied torque by 180°.

Applications

The gyroscopic principle is used in an instrument or toy known as gyroscope. The gyroscopes are installed in ships in order to minimise the rolling and pitching effects of waves. They are also used in aeroplanes, monorail cars, gyrocompasses, etc. The effect of the gyroscopic couple is discussed in the below sections.

9.3 EFFECT OF GYROSCOPIC COUPLE ON AN AEROPLANE

Pic. 9.2 shows the principle of the gyroscope with the flywheel, its axis, shaft bearings. The inner and outer gimbal rings with pivots are shown. The aeroplane with three rotations called pith, roll and yaw are also shown clearly. The effect of the gyroscope on the aeroplane is explained below.

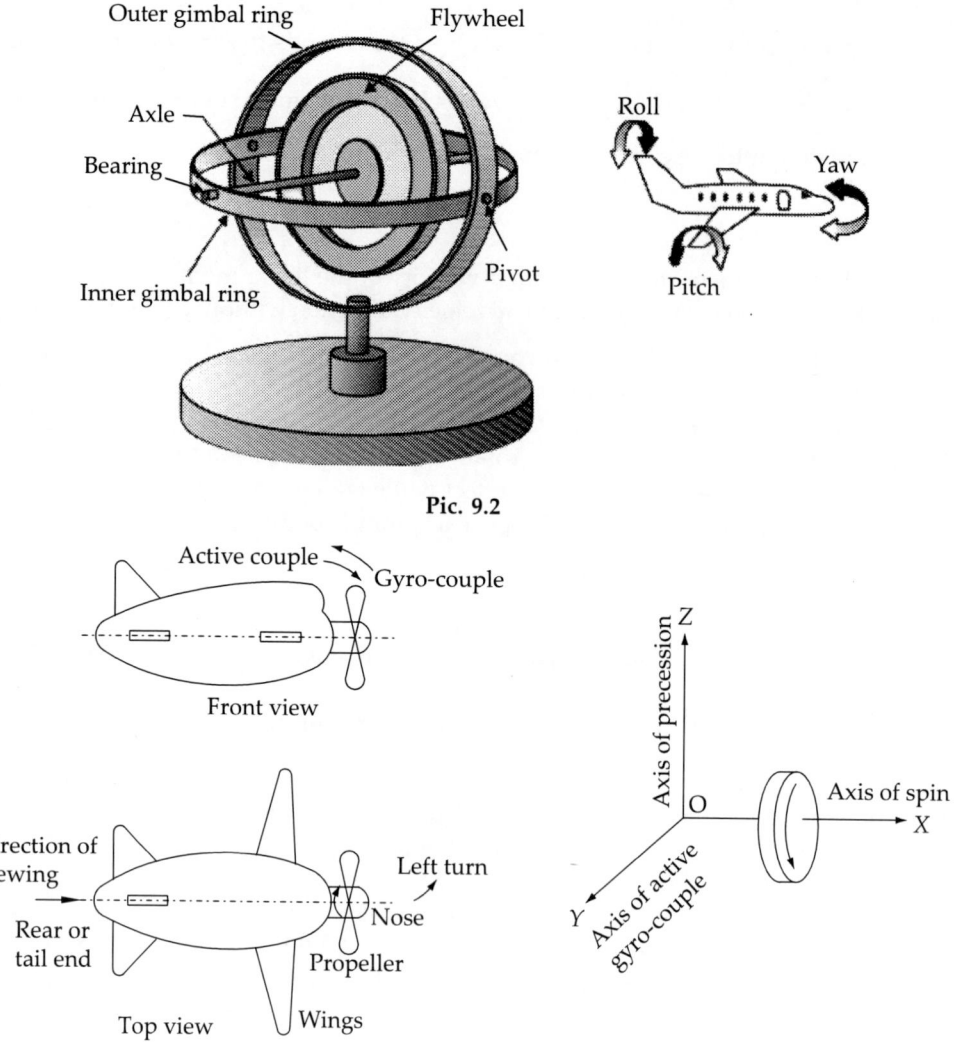

Fig. 9.2 Aeroplane taking a turn.

The front view and the top view of an aeroplane are shown in the Fig. 9.2. Assume that the engine/propeller is rotating in the clockwise direction when seen from rear and assume that the aeroplane is taking a left turn as shown.

Let ω be the angular velocity of the engine/propeller in radians/sec,
  m be the mass of the engine/propeller in kilograms (kg),
  k be the radius of gyration in metres (m)
  I be the mass moment of inertia of the engine/propeller in kg-m² = mk^2
  v be the speed or linear velocity of the aeroplane in metres/sec (m/s)
  R be the radius of the turn (curvature) in metres (m)
  ω_p be the angular velocity of precession = $\dfrac{v}{R}$ (rad/s)

Therefore, the gyroscopic couple acting on the aeroplane $C = I \cdot \omega \cdot \omega_p$.

The effect of the gyro-couple on the aeroplane depends on two aspects:

1. The rotation of the propeller or engine clockwise or anticlockwise and
2. The plane is taking a left turn or right turn as follows:

Effects:

1. When the aeroplane engine or propeller rotates clockwise when viewed from the rear.
    (a) When the plane turns towards left, the effect of the gyroscopic couple is to raise the nose upwards and dip the tail downwards.
    (b) When the plane turns towards right, the effect of the gyroscopic couple is to dip the nose downwards and raise the tail upwards.
2. When the aeroplane engine or propeller rotates anticlockwise when viewed from the rear.
    (a) When the plane turns towards left, the effect of the gyroscopic couple is to dip the nose downwards and raise the tail upwards.
    (b) When the plane turns towards right, the effect of the gyroscopic couple is to raise the nose upwards and dip the tail downwards.

Pic. 9.3 Plane taking a right turn.

The different Pics. 9.3 to 9.5 show the influence of gyroscope on the aeroplane while taking a turn.

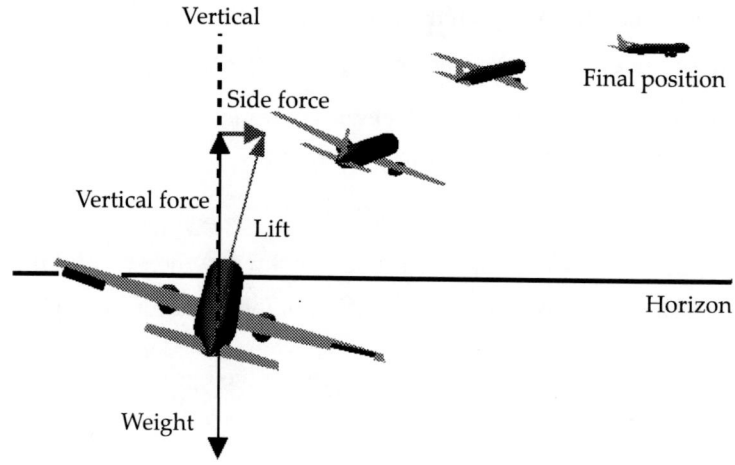

Pic. 9.4

The effects of the gyroscope are explained through the following example.

W E 9.1: An aeroplane makes a complete half circle of 50 m radius towards left, when flying at 360 km per hour. The rotary engine and the propeller of the plane have a mass of 500 kg and radius of gyration of 0.5 m. The engine rotates at 3000 r.p.m. clockwise when viewed from the rear. Find the gyroscopic couple on the aeroplane and state its effect. In what way is the effect changed when (i) the aeroplane turns towards right (ii) the engine rotates clockwise when viewed from the front (nose end) and the aeroplane turns (a) left (b) right?

Given:

$R = 50$ m; $\quad v = 360$ km/hr $= 360 \times \dfrac{1000}{3600} = 100$ m/s; $\quad m = 500$ kg; $\quad k = 0.5$ m;

$N = 3000$ r.p.m. $\quad \omega = \dfrac{2\pi 3000}{60} = 100\pi$ rad/s.

Mass moment of inertia of the engine/propeller $I = mk^2 = 500 \times (0.5)^2 = 125$ kg-m²

Angular velocity of precession $\omega_p = \dfrac{v}{R} = \dfrac{100}{50} = 2$ rad/sec.

Gyroscopic couple acting on the aeroplane

$$C = I \cdot \omega \cdot \omega_p$$
$$= 125 \cdot 100\pi \cdot \dfrac{2}{1000}$$
$$= 78.53 \text{ kN·m}$$

From Fig. 9.2, the gyroscopic effect when the aeroplane engine or propeller rotates clockwise when viewed from the rear and when the plane turns towards left is to raise the nose upwards and dip the tail downwards.

Pic. 9.5 Air show performing turns.

(i) When the aeroplane turns towards right: When the plane turns towards right, the effect of the gyroscopic couple is to dip the nose downwards and raise the tail upwards.
(ii) When the engine of the aeroplane rotates clockwise when viewed from the front or when the engine of the aeroplane rotates anticlockwise, when viewed from the rear are the same.
    (a) When the plane turns towards left, the effect of the gyroscopic couple is to dip the nose downwards and raise the tail upwards.
    (b) When the plane turns towards right, the effect of the gyroscopic couple is to lift the nose upwards and dip the tail downwards.

The pictures showed in Pic. 9.3–9.5 give an idea on how the aeroplane takes turn and how it bends to overcome the effect of gyroscopic effect.

9.4 SPECIAL TERMS USED IN SHIPS

The top view and front view of the ship are shown below in Fig. 9.3.

1. The front side of the ship is called fore-end or bow.
2. The back side of the ship is called aft or stern.

3. The left and right portions of the ship when viewed from back are called port and star-board respectively.

The ships have six degrees of freedom. Three linear motions and three angular motions about X, Y and Z axes. The three angular motions are as below.

1. **Steering:** Ship taking a turn towards left or right while moving forward as shown in Fig. 9.4.
2. **Pitching:** Ship oscillating in a vertical plane along the longitudinal axis and about the transverse axis clockwise or anticlockwise as shown at Fig. 9.5.
3. **Rolling:** Ship oscillating in a vertical plane along the transverse axis and about the longitudinal axis clockwise and anticlockwise alternately.

The effect of the gyroscopic couple is explained for all the three motions separately.

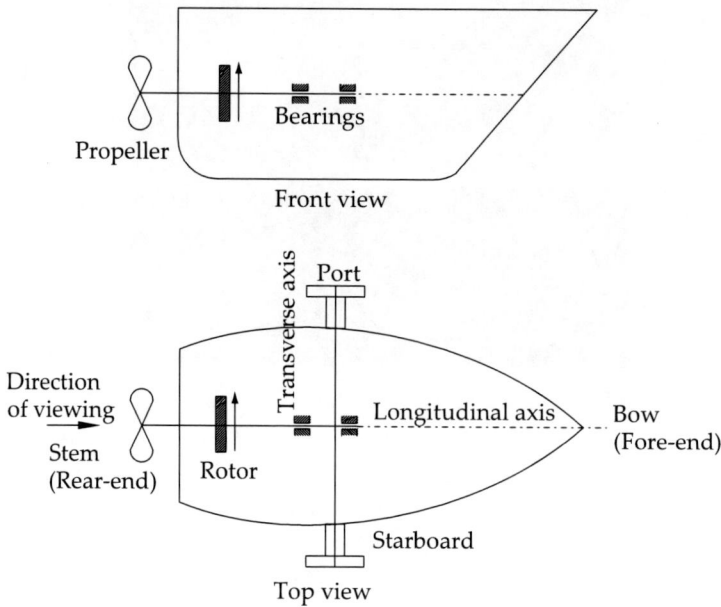

Fig. 9.3 Ship.

9.4.1 Effect of Gyroscopic Couple on the Ship During Steering

(i) When the ship takes left turn and when the rotor rotates in the clockwise when viewed from the stern as shown in Fig. 9.4.
The effect of the reactive gyroscopic couple on the ship is to raise the bow (front) and dip the stern (back of the ship).

(ii) When the ship takes right turn and when the rotor rotates in the clockwise when viewed from the stern as shown in Fig. 9.4.
The effect of the reactive gyroscopic couple on the ship is to dip the bow (front) and raise the stern (back of the ship) as shown in the Fig. 9.4.

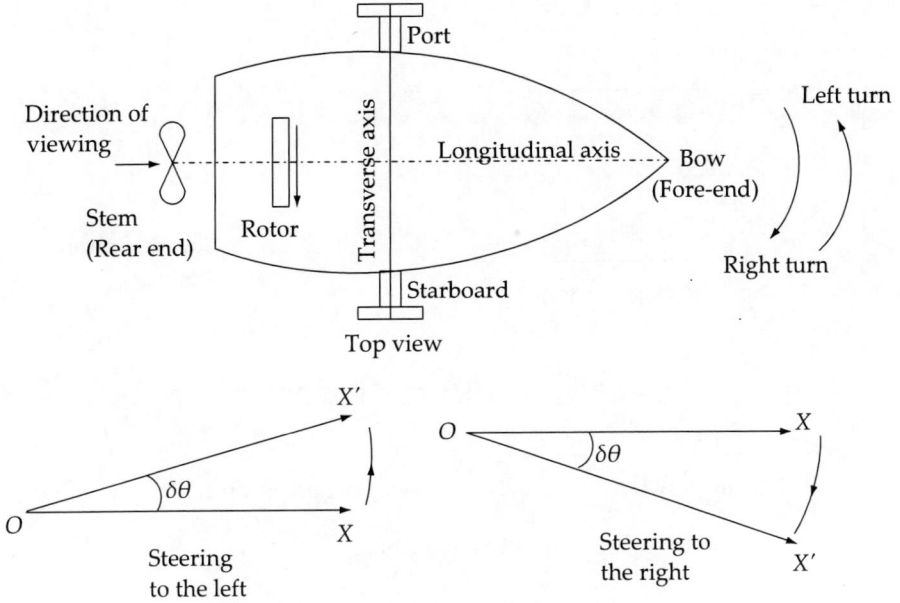

Fig. 9.4 Ship turning.

Pic. 9.6 Ship turning.

9.4.2 Effect of Gyroscopic Couple on the Ship During Pitching

The pitching of the ship is assumed to take place with simple harmonic motion (SHM), i.e. the motion of the axis of spin about the transverse axis.

Pitching types and their effect:

1. *When the pitching is upwards:* The effect of the reactive gyroscopic couple as shown in Fig. 9.5 will turn the ship towards starboard.
2. *When the pitching is downwards:* The effect of the reactive gyroscopic couple as shown in Fig. 9.5 will turn the ship towards port side.

266 Theory of Machines

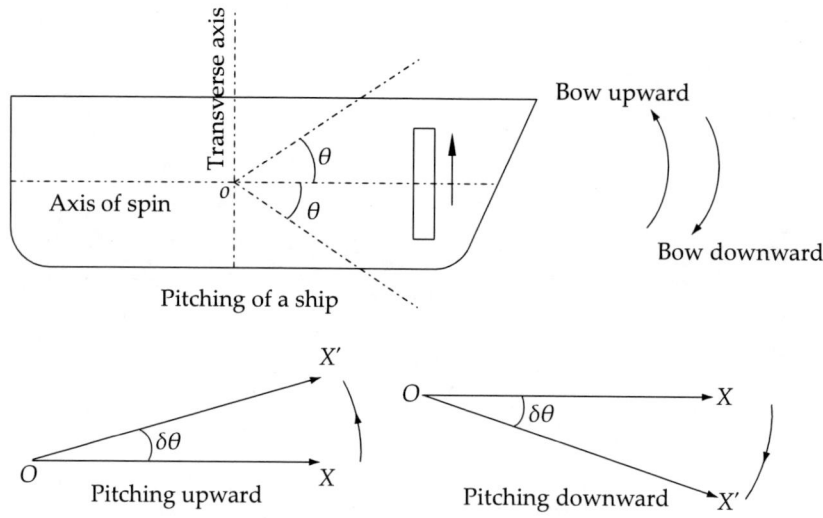

Fig. 9.5 Ship pitching.

9.4.3 Effect of Gyroscopic Couple on the Ship During Rolling

When the axis of precession is perpendicular to the axis of spin, then only there will be a gyroscopic effect on the ship.

Effect of rolling: During rolling the axes of precession and the spinning are along the longitudinal axis, i.e. both are parallel, hence there will be no effect of the gyroscopic couple on the body of the ship due to rolling.

W E 9.2: The turbine rotor of a ship has a mass of 3000 kg. It has a radius of gyration of 0.5 m and speed of 3000 r.p.m. The rotor rotates clockwise when looking from stern. Determine the gyroscopic couple and its effect upon the ship:

Pic. 9.7 Ship heel due to gyro-effect.

1. When the ship is steering to the right on a curve of 100 m radius at a speed of 36 km/h.
2. When the ship is pitching in a simple harmonic motion, the bow falling with its maximum velocity.

The period of pitching is 40 seconds and the total angular displacement between the two extreme positions of pitching is 10°.

Given:

$m = 3000$ kg; $k = 0.5$ m; $N = 3000$ r.p.m.; $\omega = \dfrac{2\pi 3000}{60} = 314.2$ rad/s.

1. When the ship is steering to the right

Given:
$R = 100$ m; $v = 36$ km/h $= 10$ m/s

Mass moment of inertia of the rotor, $I = m.k^2 = 3000(0.5)^2 = 750$ kg-m²

Angular velocity of precession, $\omega_p = \dfrac{v}{R} = \dfrac{10}{100} = 0.1$ rad/s

Gyroscopic couple, $C = I \cdot \omega \cdot \omega_p = 750 \times 314.2 \times 0.1 = 23565$ N-m $= 23.565$ kN-m

When the rotor rotates clockwise when looking from the stern and when the ship takes a right turn, the effect of the reactive gyroscopic couple is to lower the bow and raise the stern.

2. When the ship is pitching with the bow falling

Given:
$t_p = 40$ sec.

The total angular displacement between the two extreme positions of pitching is 10° (i.e., $2\theta = 10°$), therefore the amplitude of swing,

$$\theta = \dfrac{10}{2} = 5° = \dfrac{5 \times \pi}{180} = 0.0873 \text{ radians}$$

Angular velocity of the simple harmonic motion,

$$\omega_{SHM} = \dfrac{2\pi}{t_p} = \dfrac{2\pi}{40} = 0.157 \text{ rad/s}$$

The maximum angular velocity of precession,

$$\omega_p = \theta \cdot \omega_{SHM} = 0.0873 \times 0.157 = 0.0011965 \text{ rad/s}$$

Gyroscopic couple,

$$C = I\omega\omega_p = 750 \times 314.2 \times 0.0011965$$
$$= 281.96 \text{ N-m} = 0.282 \text{ kN-m}.$$

When the pitching is downward (bow is moving downward), the effect of the reactive gyroscopic couple is to move the ship towards portside or towards left, like taking left turn.

When the pitching is upward (bow is moving upward), the effect of the reactive gyroscopic couple is to move the ship towards starboard side or towards right, like taking right turn as shown in Pic. 9.7.

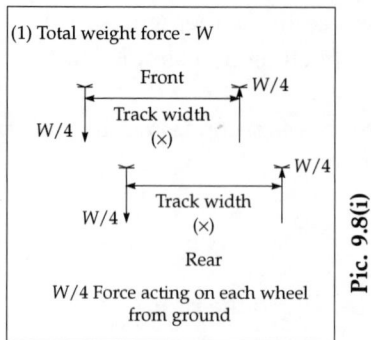
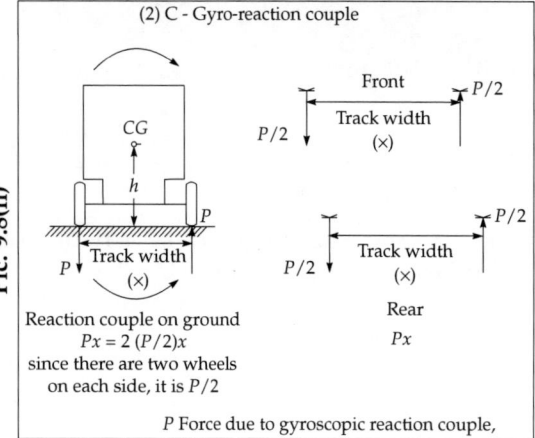
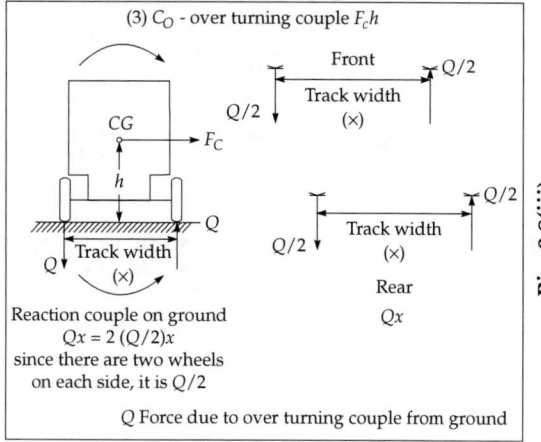
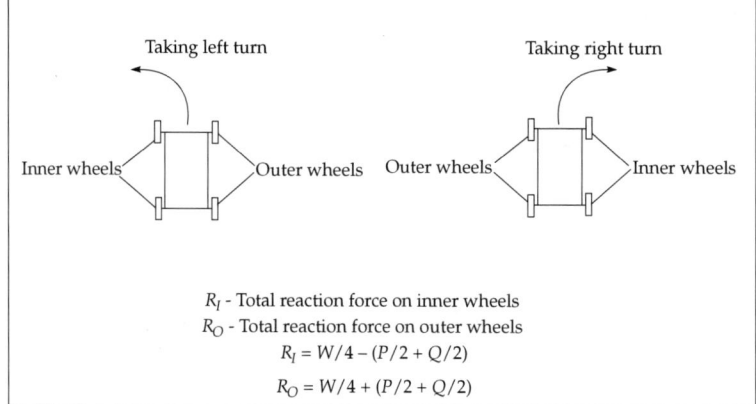

Pic. 9.8(iv) Bus taking a turn.

9.5 STABILITY OF FOUR-WHEELER

In case of a four-wheeler or an automobile as in Pic. 9.8, it is essential that no wheel is lifted off the road or ground while the vehicle takes a turn. The condition is fulfilled as long as the vertical reaction of the ground on any of the wheels is positive (or upwards).

Let A, B, C and D represent the four wheels of an automobile taking a turn towards left as shown in Fig. 9.6. The wheels A and C are inner wheels, whereas B and D are outer wheels. The centre of gravity G of the vehicle lies vertically above the road surface.

Let m be the mass of the vehicle in kg $\left(\text{i.e.,} \dfrac{W}{g}\right)$;

r_w be the radius of the wheels in metres;
R be the radius of curvature in metres $(R > r_w)$;
h be the distance of centre gravity G above the road surface in metres;
x be the width of the track (distance between tyres) in metres.
I_w be the mass moment of inertia of the each wheel in kg-m^2;
ω_w be the angular velocity of the wheels or velocity of spin in radians/sec;
I_E be the mass moment of inertia of the rotating parts of the engine in kg-m^2;
ω_E be the angular velocity of the rotating parts of the engine in rad/sec;
G be the gear ratio $= \dfrac{\omega_E}{\omega_w}$ or $\omega_E = G\omega_w$; and
v be the linear velocity or speed of the vehicle in metres/sec.

Assuming that the weight of the vehicle W is equally distributed among the four wheels, i.e. $\dfrac{W}{4}$ acting downwards on each wheel. The reaction between each wheel and the road surface is equal and opposite of $\dfrac{W}{4}$ acting upwards on the ground as indicated in Fig. 9.6.

The effect of the gyroscopic couple and the centrifugal couple on the four wheels is to be regarded as given below:

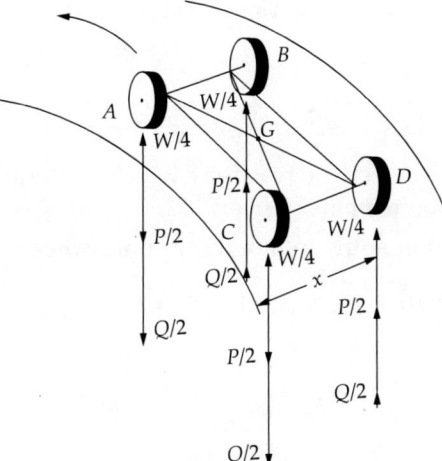

Fig. 9.6 Four-wheel vehicle.

9.5.1 Effect of the Gyroscopic Couple

When the vehicle takes a left or right turn then due to precession of the wheels and the other rotating parts of engine will give raise a gyroscopic couple.

Angular velocity of precession $\omega_p = \dfrac{v}{R}$;

Gyroscopic couple due to all the four wheels, $C_W = 4I_w\omega_w\omega_p$ and due to the rotating parts of the engine $C_E = I_E\omega_E\omega_p = I_E \cdot G\omega_w.\omega_p$, where $G = \dfrac{\omega_E}{\omega_w}$.

Net gyroscopic couple

$$C = C_W \pm C_E$$
$$= 4I_w \omega_w \omega_p \pm I_E \cdot G \omega_w \cdot \omega_p$$
$$= \omega_w \cdot \omega_p \cdot (4I_w \pm GI_E).$$

The positive sign is used when the wheels and rotating parts of the engine rotate in the same direction and negative sign when the wheels and rotating parts of the engine rotate in the opposite direction.

Due to the gyroscopic couple, vertical reaction produced on the road surface acts vertically upwards on the outer wheels and vertically downwards on the inner wheels. Let the magnitude of this reaction at the two outer or inner wheels be P Newton or on each of the inner or outer wheel is $\dfrac{P}{2}$ indicated in Fig. 9.6.

Then the couple C is the product of P and the track width x, $C = P \times x$. Therefore, vertical reaction at each of the outer or inner wheels $\dfrac{P}{2} = \dfrac{C}{2x}$

Note: When $C_E > C_W$, then C will be $-ve$. The reaction acts vertically downwards on the outer wheels and vertically upwards on the inner wheels.

9.5.2 Effects of the Centrifugal Couple

When the vehicle moves along a curved path, the centrifugal force F_C will act outwardly at the centre of gravity of the vehicle. The effect of this force is also to overturn the vehicle. The centrifugal force $F_C = \dfrac{mv^2}{R}$.

The couple that overturns the vehicle or overturning couple $C_O = F_C \cdot h = \dfrac{mv^2 h}{R}$.

This overturning couple is balanced by vertical reactions, which are vertically upwards on the outer wheels and vertically downwards on the inner wheels.

Let the magnitude of this reaction at the two outer or inner wheels be Q. Then $Q \cdot x = C_O$ or $Q = \dfrac{C_O}{x} = \dfrac{mv^2 h}{(R \cdot x)}$. Therefore, the vertical reaction at each of the outer or inner wheel, $\dfrac{Q}{2} = \dfrac{mv^2 h}{(2R \cdot x)}$ as indicated in Fig. 9.6.

Pic. 9.9

Therefore, the total vertical reaction at each of the outer wheel,
$$R_O = \frac{W}{4} + \frac{P}{2} + \frac{Q}{2}$$

and the total vertical reaction at each of the inner wheel,
$$R_I = \frac{W}{4} - \frac{P}{2} - \frac{Q}{2}$$

Note: When the vehicle is running at high speeds, R_I may be zero or even negative. This will cause the inner wheels to leave the ground thus tending to overturn the automobile as shown in Pic. 9.9. In order to have the contact between the inner wheels and the ground, the sum of $\left(\frac{P}{2} + \frac{Q}{2}\right)$ must be less than $\left(\frac{W}{4}\right)$.

W E 9.3: A rear engine automobile is travelling along a track of 100 m mean radius. Each of the four road wheels has a moment of inertia of 2.0 kg-m² and an effective diameter of 0.6 m. The rotating parts of the engine have a moment of inertia of 1.5 kg-m². The engine axis is parallel to the rear axle and the crankshaft rotates in the same sense as the road wheels. The ratio of engine speed to back axle speed is 3 : 1. The automobile has a mass of 1500 kg and has its centre of gravity 0.3 m above road level. The width of the track of the vehicle is 1.5 m.

Determine the limiting speed of the vehicle around the curve for all four wheels to maintain contact with the road surface. Assume that the road surface is not cambered and centre of gravity of the automobile lies centrally with respect to the four wheels.

Given:
$R = 100$ m; $I_w = 2.0$ kg-m²; $d_w = 0.6$ m; $r_w = 0.3$ m; $I_E = 1.5$ kg-m²; $G = \frac{\omega_E}{\omega_w} = 3$;
mass, $m = 1500$ kg; $h = 0.3$ m; $x = 1.5$ m;

Limiting speed of the vehicle be v m/s.

Road reaction over each wheel,
$$\frac{W}{4} = \frac{mg}{4} = \frac{1500 \times 9.81}{4} = 3678.75 \text{ N} \tag{1}$$

Angular velocity of the wheels,
$$\omega_w = \frac{v}{r_w} = \frac{v}{0.3} = 3.33v \text{ rad/sec}$$

Angular velocity of precession,
$$\omega_p = \frac{v}{R} = \frac{v}{100} = 0.01v \text{ rad/sec}$$

Gyroscopic couple due to 4 wheels,
$$C_W = 4 I_W \omega_w \omega_P$$
$$= 4 \times 2.0 \times \frac{v}{0.3} \times \frac{v}{100}$$
$$= 0.26v^2 \text{ N-m}$$

Gyroscopic couple due to rotating parts of the engine

$$C_E = I_E \cdot \omega_E \cdot \omega_P$$
$$= I_E \cdot G \cdot \omega_w \cdot \omega_P$$
$$= 1.5 \times 3 \times 3.33v \times 0.01v = 0.1499v^2 \text{ N-m}$$

Total gyroscopic couple

$$C = C_W + C_E = 0.26v^2 + 0.1499v^2 = 0.4099v^2 \text{ N-m}$$

Due to this gyroscopic couple, the vertical reaction on the wheels will be vertically upwards on the outer wheels and vertically downwards on the inner wheels.

The magnitude of this reaction at each wheel be

$$\frac{P}{2} = \frac{C}{2x} = \frac{0.4099v^2}{1.5 \times 2} = 0.1366v^2 \text{ N} \qquad (2)$$

The centrifugal force

$$F_C = \frac{m \cdot v^2}{R} = \frac{1500 \cdot v^2}{100} = 15v^2 \text{ N}$$

Overturning couple acting in the outward direction

$$C_O = F_C \cdot h = 15v^2 \cdot 0.5 = 7.5v^2 \text{ N}$$

This is balanced by the vertical reactions which are vertically upwards on the outer wheels and downwards on the inner wheels; the magnitude of this reaction at each wheel is

$$\frac{Q}{2} = \frac{C_O}{2x} = \frac{7.5v^2}{2 \times 1.5} = 2.5v^2 \text{ N} \qquad (3)$$

Therefore, the total vertical reaction at each of the outer wheels,

$$R_O = \frac{W}{4} + \frac{P}{2} + \frac{Q}{2} \qquad (4)$$

and the total reaction at each of the inner wheel

$$R_I = \frac{W}{4} - \frac{P}{2} - \frac{Q}{2} \qquad (5)$$

From equation (4), it can be understood that there will be contact between outer wheels and the road surface because all the three quantities are positive and vertically upwards.

From equation (5), it can be understood that there will be contact between inner wheels and the road surface provided the sum total should be either zero or positive. If the sum is negative, then the inner wheels loose contact with the ground. In order to prevent it, the sum of $\left(\dfrac{P}{2} + \dfrac{Q}{2}\right) \leq \dfrac{W}{4}$

i.e. $(0.1366v^2 + 2.5v^2) \le 3678.75$ or $2.6366v^2 \le 3678.75$

Therefore, $v_2 \le 1395.27$ or the limiting speed of the vehicle (v) should be ≤ 37.3533 m/s = 134.47 km/hr.

9.6 STABILITY OF A TWO-WHEELER

The case of a two-wheel vehicle can be taken in the same way as the four-wheel vehicle. However, it is easier to tilt inwards to nullify the overturning effect in this case and the vehicle can stay in equilibrium while taking a turn. Consider a two-wheel vehicle say a scooter or motorcycle taking a turn.

Let m be the mass of the vehicle and its rider in kg $\left(\text{i.e., } \dfrac{W}{g}\right)$;

r_w be the radius of the wheels in metres;
R be the radius of curvature in metres ($R > r_w$);
h be the distance of CG above the road surface in metres;
I_w be the mass moment of inertia of each wheel in kg-m²;
ω_w be the angular velocity of the wheels or velocity of spin in radians/sec;
I_E be the mass moment of inertia of the rotating parts of the engine in kg-m²;
ω_E be the angular velocity of the rotating parts of the engine in rad/sec;
G be the gear ratio = $\dfrac{\omega_E}{\omega_w}$
v be the linear velocity of the vehicle = $\omega_w . r_w$
θ be the angle of heel.

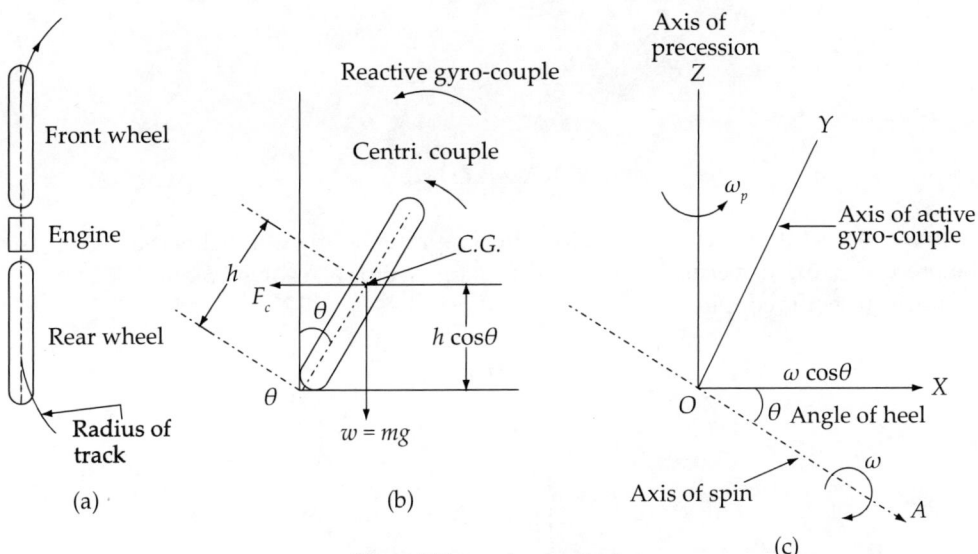

Fig. 9.7 Two-wheel vehicle.

9.6.1 Effect of the Gyroscopic Couple

Since $v = \omega_w \cdot r_w$ or $\omega_w = \dfrac{v}{r_w}$ and $\omega_E = G \cdot \omega_w = G \cdot \dfrac{v}{r_w}$.

Therefore, the total $(I \cdot \omega)$

$$I \cdot \omega = 2 \cdot I_w \cdot \omega_w \pm I_E \cdot \omega_E$$
$$= 2 I_w \cdot \dfrac{v}{r_w} \pm I_E \cdot G \cdot \dfrac{v}{r_w}$$
$$= \dfrac{v}{r_w}(2 I_w \pm G \cdot I_E)$$

and angular velocity of precession, $\omega_p = \dfrac{v}{R}$.

When the vehicle moves over the curved path, the vehicle is always inclined at an angle θ with the vertical plane (in other words the axis of spin is inclined to the horizontal at an angle (θ)) as shown in Fig. 9.7. This angle is known as angle of heel θ as shown in Pic. 9.10 below.

Pic. 9.10 Angle of heel θ.

The angular momentum vector $I\omega$ due to spin is represented by OA inclined to OX at an angle θ. But the precession axis is vertical. Therefore, the spin vector is resolved along OX.

Therefore, gyroscopic couple,

$$C_{GC} = I \cdot \omega \cdot \cos\theta . \omega_p$$
$$= \left(\dfrac{v}{r_w}\right)(2 \cdot I_w \pm G \cdot I_E)\cos\theta \dfrac{v}{R}$$
$$= \left(\dfrac{v^2}{R r_w}\right)(2 \cdot I_w \pm G \cdot I_E)\cos\theta$$

Note:

1. When the engine rotates in the same direction as that of wheels, then the positive sign,
2. When the engine rotates in the opposite direction as that of wheels, then the negative sign,
3. The gyroscopic couple acts over the vehicle outwards, i.e. in the anticlockwise direction when seen from the front of the vehicle. The tendency of this couple is to overturn the vehicle in outward direction.

9.6.2 Effects of the Centrifugal Couple

The centrifugal force $F_C = \dfrac{mv^2}{R}$ acts horizontally through the centre of gravity (CG) along the outward direction. Therefore, centrifugal couple

$$C_{CC} = F_C \cdot h \cos\theta = \dfrac{mv^2}{R} h \cos\theta$$

The effect is to overturn the vehicle, therefore, the total overturning couple,

$$C_O = C_{GC} + C_{CC}$$

$$= \left[\left(\dfrac{v^2}{Rr_w}\right)(2.I_W \pm G.I_E).\cos\theta\right] + \dfrac{mv^2}{R}h\cos\theta$$

$$= \left[\left(\dfrac{v^2}{R}\right)\left[\left(\dfrac{2.I_W + G.I_W}{r_w}\right) + m.h\right]\cos\theta\right] \quad (1)$$

$$C_B = \text{The balancing couple} = mgh\sin\theta \quad (2)$$

Pic. 9.11 Two wheelers taking heel.

For equilibrium condition (1) = (2).

From this, the angle of heel (θ) can be determined, so that the vehicle does not skid.

Pic. 9.11 shows the racers of two-wheel vehicles gradually increasing their angle of heel as they approach the curve. This happens when the vehicle is used on a ghat road as shown in Pic. 9.12 and in Pic. 9.13 heeling for each turn. Pic. 9.14 shows the mono rail also bending during turn.

Pic. 9.12 Two-wheeler on a ghat way.

Pic. 9.13 Two-wheeler heeling.

Pic. 9.14 Mono rail having a heel.

W E 9.4: Find the angle of inclination with respect to the vertical of a two-wheeler negotiating a turn: Combined mass of the vehicle with its rider 250 kg; moment of inertia of the engine flywheel 0.5 kg-m²; moment of inertia of each road wheel 1 kg-m²; speed of engine flywheel 5 times that of road wheels and in the same direction; height of centre of gravity of rider with vehicle 0.5 m; two-wheeler speed 100 km/hr; wheel radius 300 mm; radius of turn is 35 m.

Given:
$m = 250$ kg; $I_E = 0.5$ kg-m²; $I_w = 1$ kg-m²; $\omega_E = 5\omega_w$; $G = \dfrac{\omega_E}{\omega_w} = 5$; $h = 0.5$ m;
$v = 100$ km/hr $= 27.78$ m/s; $r_w = 300$ mm $= 0.3$ m; $R = 35$ m.

Let θ be the angle of inclination with respect to the vertical of a two-wheeler (angle of heel).

$$\text{Gyroscopic couple } C_{GC} = \left[\left(\frac{v^2}{Rr_w}\right)(2 \cdot I_w \pm G \cdot I_E)\cos\theta\right]$$

$$= \left[\left(\frac{27.78^2}{35 \times 0.3}\right)(2.1 + 5 \times 0.5)\cos\theta\right] = 330.74\cos\theta \text{ N-m}$$

$$\text{Centrifugal couple } C_{CC} = \frac{mv^2 h \cos\theta}{R}$$

$$= \frac{250 \times 27.78^2 \times 0.5 \cos\theta}{35} = 2756.2 \cos\theta$$

Total overturning couple

$$C_O = C_{GC} + C_{CC}$$
$$= 330.74\cos\theta + 2756.2\cos\theta = 3086.94\cos\theta \text{ N-m}$$

The balancing couple
$$C_B = mgh \sin\theta = 250 \times 9.81 \times 0.5 \sin\theta$$
$$= 1226.25 \sin\theta \text{ N-m}$$

For equilibrium condition, overturning couple = balancing couple.

Therefore, $3086.94 \cos\theta = 1226.25 \sin\theta$.

Therefore,
$$\tan\theta = \frac{\sin\theta}{\cos\theta} = \frac{3086.94}{1226.25} = 2.517 \text{ or } \theta = 68.33°$$

9.7 EXERCISE

9.7.1 Short Answer Questions

1. In what way can the angular velocity be represented by a vector?
2. What do you mean by gyroscopic couple? Derive a relation for its magnitude.
3. What do you mean by spin, precession and gyroscopic planes?
4. Explain what is meant by applied torque and reaction torque.
5. Explain in what way the gyroscopic couple affects the motion of an aircraft while taking a turn.
6. Explain the gyroscopic effect on four-wheeled vehicles.
7. What is the effect of the gyroscopic couple on the stability of a four-wheeler while negotiating a curve?
8. How do the effects of gyroscopic couple and of centrifugal force make the velocity of a two-wheeler tilt on one side? Derive a relation for the limiting speed of the vehicle.
9. What is centripetal force?
10. Explain why a cyclist has to lean inwards while negotiating a curve.
11. Discuss the effect of gyroscopic couple on an aeroplane and on a naval ship with neat sketches.
12. Give the effect of gyroscopic couple on an aircraft when taking a left turn.
13. What is the effect of gyroscopic couple when a ship is rolling?
14. Explain the principle of gyroscope.
15. Define axis of spin and axis of precession.

9.7.2 Problems

Disc on a shaft

1. A disc of 5 kg mass with radius of gyration 70 mm is mounted at mid-span on a horizontal shaft of 120 mm length between the two bearings. The shaft spins at 720 r.p.m. in clockwise direction when viewed from the right-hand bearing. If the shaft processes about the vertical axis at 30 r.p.m. in clockwise direction when viewed from the above, determine the reactions at each bearing due to mass of the disc on gyroscopic effect.

2. A disc of 5 kg mass with radius of gyration 70 mm is mounted at mid-span on a horizontal shaft of 120 mm length between the two bearings. The shaft spins at 720 r.p.m. in clockwise direction when viewed from the right-hand bearing. If the shaft processes about the vertical axis at 30 r.p.m. in clockwise direction when viewed from the above, determine the reactions at each bearing due to mass of the disc on gyroscopic effect.

Two-wheeler

1. Each wheel of a motorcycle is of 600 mm diameter and has a moment of inertia of 1.1 kg-m^2. The motorcycle and the rider together weigh 220 kg and the combined centre of mass is 620 mm above the ground level when the motorcycle is upright. The moment of inertia of the rotating parts of the engine is 0.18 kg/m^2. The engine rotates at 4.5 times the speed of road wheels in the same sense. Find the angle of heel necessary when the motorcycle is taking a turn of 35 m radius at a speed of 72 kg/h. [38.6°]

2. A motorcycle with its rider weighs 250 kg, the CG of motorcycle and rider combined being 60 cm above ground when bike is standing upright. Each road wheel has a moment of inertia of 8 kg/m^2 and a rolling diameter of 60 cm. The engine rotates 6 times the wheel speed in the same sense. Moment of inertia of rotating parts of engine is 1.5 kg/m^2. Determine the angle of wheel necessary if the bike with rider is running at 15 m/sec in a curve of 30 m.

Ship

1. A ship is propelled by a turbine rotor a mass of 6 tonnes and a speed of 2400 r.p.m. The direction of rotation of the rotor is clockwise when viewed from the stern. The radius of gyration of the rotor is 450 mm. Determine the gyroscopic effect when the (i) ship steers to the left in a curve of 60 m radius at a speed of 18 knots (1 knot = 1860 m/h). (ii) ship pitches 7.5° below the normal position and the bow is descending with its maximum velocity; the pitching motion is simple harmonic with a periodic time of 18 seconds. (iii) ship rolls and at the instant, its angular velocity is 0.035 rad/s counterclockwise when viewed from the sterns. Also find the maximum angular acceleration during pitching.

   [*Ans.* (i) 47.33 kN-m, bow is raised; (ii) 13.96 kN-m ship turns towards port side; (iii) no gyroscopic effect; 0.016 rad/s^2]

Aeroplane

1. The moment of inertia of an aeroplane air screw is 20 kg-m² and rotates at 3000 rpm clockwise when viewed from aft. If the ship pitches with angular simple harmonic motion having a periodic time of 16 seconds and an amplitude of 0.1 radian, find the (i) maximum angular velocity of the rotor axis. (ii) maximum value of the gyroscopic couple. (iii) gyroscopic effect as the bow dips. [*Ans*. (i) 0.0393 rad/s; (ii) 9261 N-m; (iii) bow swings to port as it dips.]

Four-wheeler

1. A rear engine automobile is travelling along a curved track of 120 m radius. Each of the four wheels has a moment of inertia of 2.2 kg/m² and an effective diameter of 600 mm. The gear ratio of the engine to the back wheel is 3.2. The engine axis is parallel to the rear axle and the crankshaft rotates in the same sense as the road wheels. The mass of the vehicle is 2050 kg and the centre of mass is 520 mm above the road level. The width of the track is 1.6 m. What will be the limiting speed of the vehicle if all the four wheels maintain contact with the road surface? [*Ans*. 150.2 km/h]

2. A four-wheeled trolley of total weight 20 kN running on rails of 1 m gauge rounds a curve of 30 m at 40 km/hr on a track of embankment slope of 10. The wheels have external diameter of 60 cm and each pair of axle weighs 2000 N and has a radius of gyration of 25 cm, the height of the centre of gravity of the car above the wheel base is 1 m. Determine allowing centrifugal force, gyroscopic couple acting and the pressure on each rail.

3. A four-wheeled trolly car of mass 2500 kg runs on rails, which are 1.5 m apart and travel around a curve of 30 m radius of 24 km/hr. The rails are at the same level. Each wheel of the trolly is 0.75 m in diameter and each of the two axles is driven by a motor running in a direction opposite to that of the wheels. The moment of inertia of each axle with gear and wheels is 18 kg-m². Each motor with shaft and gear pinion has a moment of inertia of 12 kg-m². The centre of gravity of the car is 0.9 m above the rail level. Determine the vertical force exerted by each wheel on the rail taking into consideration the centrifugal and gyroscopic effects.

9.7.3 Multiple Choice Questions

1. When a body is moving along a circular path the centrifugal force tends to overturn the body. The chances of overturning can be decreased by decreasing the

    (a) weight of the vehicle
    (b) speed of the vehicle
    (c) height of the centre of gravity from the road level
    (d) all of the above   [*Ans*. (c)]

2. The magnitude of the gyroscopic couple applied to a disc of moment of inertia I, spinning with an angular velocity ω and having an angular velocity of precession ω_p is

    (a) $I^2 \omega \omega_p$ (b) $I\omega^2 \omega_p$ (c) $I\omega \omega_p^2$ (d) $I\omega \omega_p$ [*Ans*. (d)]

3. The gyroscopic acceleration is given by

   (a) $\dfrac{\delta\omega}{\delta t}$ (b) $\dfrac{\omega\delta\theta}{\delta t}$ (c) $\dfrac{r\delta\theta}{\delta t}$ (d) $\dfrac{r\delta\omega}{\delta t}$ [Ans. (b)]

4. If the air screw of an aeroplane rotates clockwise when viewed from the rear and the aeroplane takes a right turn, the gyroscopic effect will

   (a) tend to raise the tail and depress the nose
   (b) tend to raise the nose and depress the tail
   (c) tilt the aeroplane about spin axis
   (d) none of the above   [Ans. (a)]

5. The axis of spin, the axis of precession and the axis of gyroscopic torque are in

   (a) two parallel planes   (b) two perpendicular planes
   (c) three perpendicular planes   (d) three parallel planes   [Ans. (c)]

6. The effect of gyroscopic torque on the naval ship when it is rolling and the rotor is spinning about the longitudinal axis is

   (a) to raise the bow and lower the stern
   (b) to lower the bow and raise the stern
   (c) to turn the ship to one side
   (d) none of the above   [Ans. (d)]

Cams 10

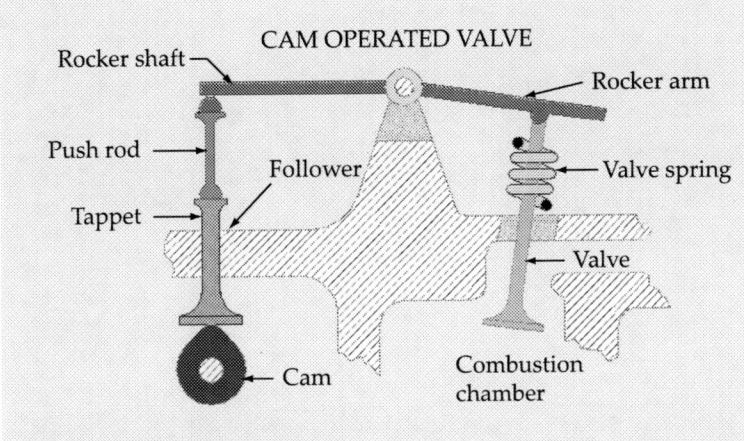

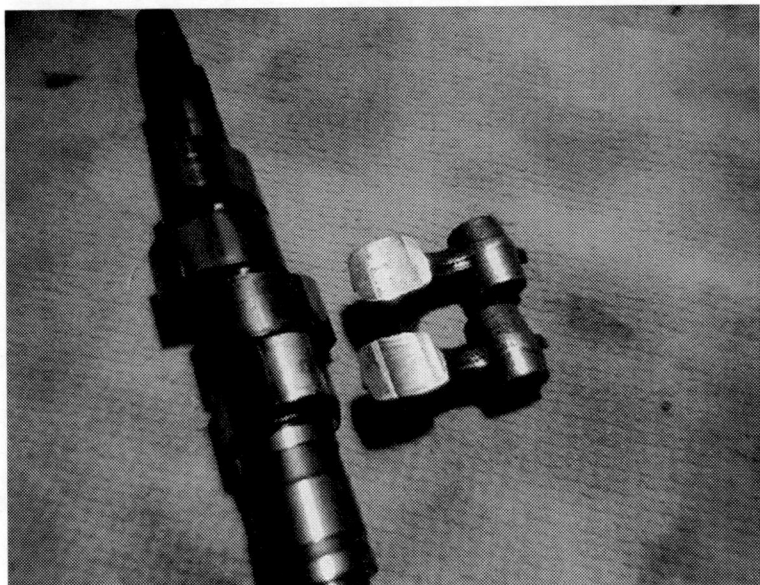

Camshaft

Cams

10.1 INTRODUCTION

A **cam** is a mechanical element used to drive another element called the **follower** through a specified motion by direct contact. Cam and follower mechanisms are simple and inexpensive, have few moving parts, and occupy a very small space. The contact between the two elements is a line contact. The prescribed exact motion required in the present-day machine components is rarely fulfilled by connected members. For these reasons cam mechanisms are used extensively in modern machinery. Cams are easy to design and the motion produced can be predicted accurately.

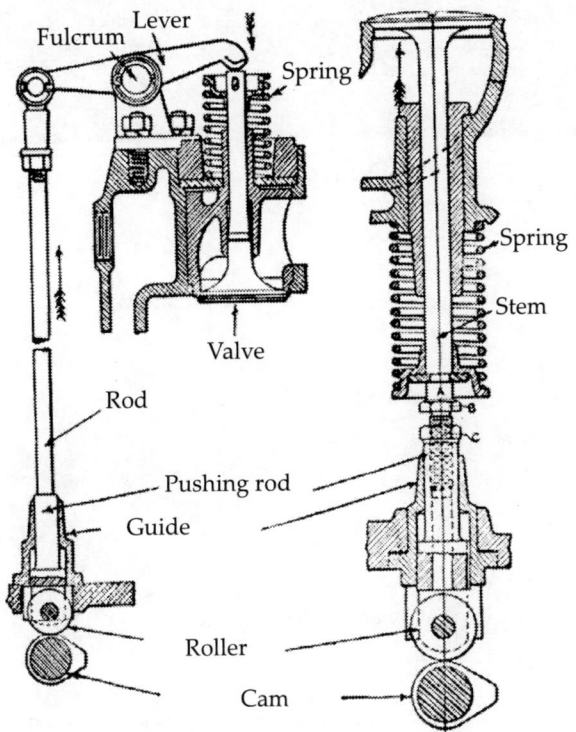

Pic. 10.1 Cam application.

Picture 10.1 shows how cam operates the inlet and outlet valves in I.C. engines. Cams are widely used in textile machine tools, in automatic machineries, in printing machines and in spinning and weaving textile machines, etc. In all the cam systems, the designer must ensure that the follower maintains contact with the cam at all times. This can be done by gravity, by inclusion of a suitable spring or a mechanical constraint as shown.

10.2 CLASSIFICATION OF FOLLOWERS

Cam systems are classified according to the follower in the following three ways:

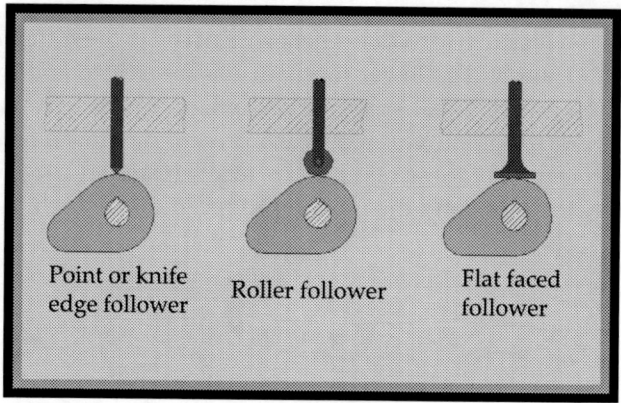

Pic. 10.2 Types of followers.

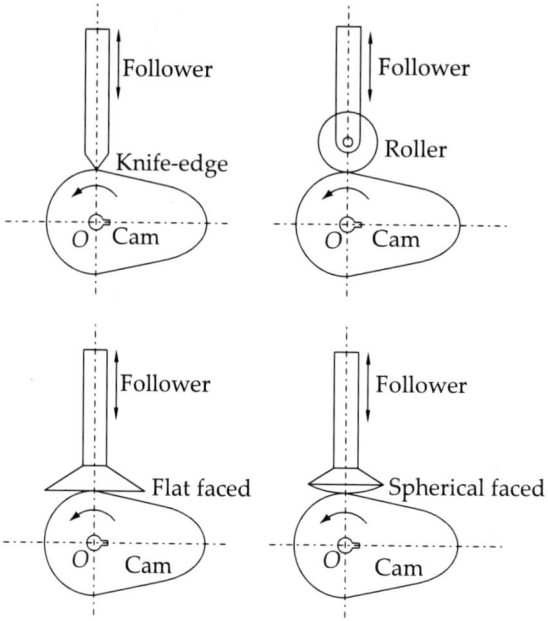

Fig. 10.1 Types of followers.

10.2.1 Based on the Surface in Contact

(a) A knife edge follower with a sharp edge having contact with the cam is shown in Fig. 10.1 which causes extreme wear of the cam surface. It is the simplest in construction but of very little in practical use.
(b) A roller follower with a cylindrical roller held by a pin to the follower assembly as shown in Fig. 10.1. This type of follower is extensively used as it provides a perfect contact and to limit wear at high speeds.
(c) A mushroom follower may be either of flat-face or spherical-face. They are as shown in Fig. 10.1. Flat face causes high stresses and to minimise these stresses spherical shape having a surface of large radius is used. Automobile engines use this type of followers because of limited space and pin weakness compared to roller followers.

All the three types of followers are also shown in Pic. 10.2.

10.2.2 Based on the Type of Movement of the Follower

Another method of classifying cams is according to the characteristic output motion allowed between the follower and the frame as follows:

(a) Reciprocating or to and fro (translating) followers as shown in Fig. 10.1.
(b) Oscillating follower is as shown in Fig. 10.2 and in Pic. 10.3.

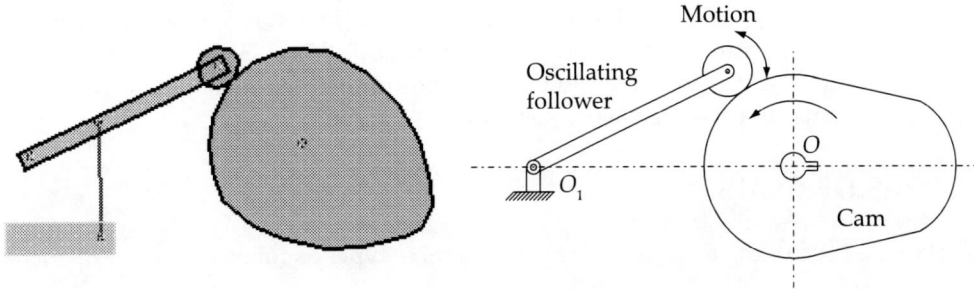

Pic. 10.3

Fig. 10.2 Oscillating roller follower.

10.2.3 Based on the Line of Motion of Follower

Further classification of reciprocating followers distinguishes whether the centre line of the follower stem relative to the centre of the cam as follows:

(a) Radial follower is one in which the axis of the follower passes through the cam centre of rotation as shown in Fig. 10.1.
(b) Offset follower is one in which the axis of the follower is displaced or kept away from the cam centre of rotation as shown in Fig. 10.3.

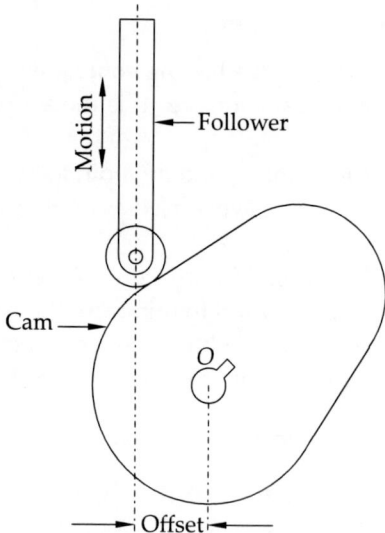

Fig. 10.3 Offset roller follower.

10.2.4 Based on the Desired Mathematical Motions

There are many possible follower motions which can be used for the rise and return, and some are preferred to others depending on the situation. There are four types of desired mathematical motions. They are (a) Uniform Velocity. (b) Simple Harmonic Motion (SHM). (c) Uniform Acceleration and Deceleration. (d) Cycloidal Motion. The details of their expressions and construction of displacement diagrams are presented separately below.

10.3 TYPES OF CAMS

Cams are classified according to their basic shapes in three types as follows:

10.3.1 Based on Follower Motion

D stands for Dwell which means the period during which the follower has no displacement even though the cam rotates. R stands for rise and return.

   (a) D-R-D-R cam means dwell rise dwell return cam. There is dwell after rise and after return.
   (b) D-R-R-D cam means dwell rise return dwell cam.
   (c) R-R-R cam means rise return rise cam. There is no dwell in this type of cams.

10.3.2 Based on the Shape of the Cam

There are two types of cams:

   (a) **Disc or plate or radial cam:** These cams are most commonly used as shown in Fig. 10.4(a) and in Pic. 10.4(b).

(b) **Cylindrical or barrel cam:** Cam made from a cylinder is shown in Fig. 10.4(b) and in Pic. 10.4(a).

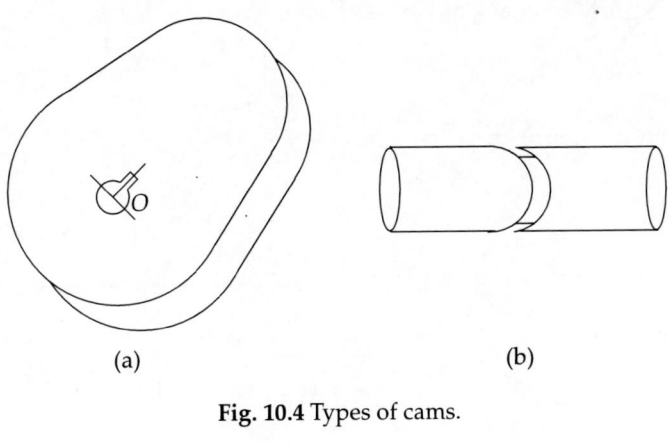

(a)   (b)

Fig. 10.4 Types of cams.

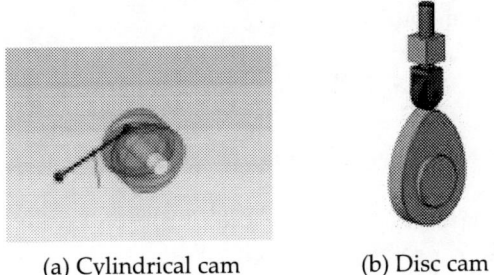

(a) Cylindrical cam   (b) Disc cam

Pic. 10.4

10.4 TERMINOLOGY

The various terms used in cams are explained and showed below in Fig. 10.5.

10.4.1 Cam Profile

It is the actual working contour or profile or curve of the cam. It is the surface having contact with the follower.

10.4.2 Base Circle

It is the smallest circle drawn on the cam profile from the centre of rotation of a radial cam. Cam size depends upon the size of the base circle.

10.4.3 Trace Point

It is a reference point located on the follower for the purpose of tracing the cam profile. It is at the knife edge in the knife edge follower, the centre of the roller in the roller follower or centre of the spherical face of the spherical, shaped follower, etc.

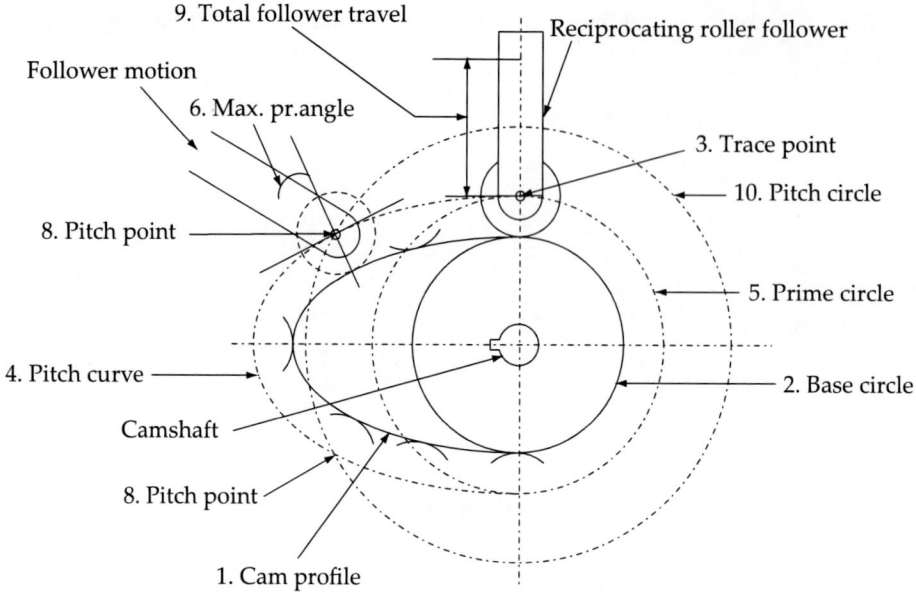

Fig. 10.5 Terminology.

10.4.4 Pitch Curve

It is the path traced out by the trace point. For the purpose of drawing the cam profile, cam is assumed as fixed and the follower rotates around it.

10.4.5 Prime Circle

It is the smallest circle drawn to the pitch curve from the centre of rotation of the cam.

10.4.6 Pressure Angle

It is the angle at any point on the pitch curve between the normal to that point on the curve and the line of motion of the follower at that instant. It indicates the steepness of the cam profile.

10.4.7 Cam Angle

It is the angle of rotation of the cam for a definite displacement of the follower.

10.4.8 Pitch Point

It is the point on the cam pitch curve having the maximum pressure angle.

10.4.9 Lift or Stroke(s)

It is the maximum displacement of the follower from the base circle of the cam. It is also sometimes called throw of the cam.

10.4.10 Pitch Circle

It is the circle with centre as the centre of the cam axis and radius such that it passes through the pitch point.

10.5 ANALYSIS OF MOTION OF THE FOLLOWER

The displacement, velocity and acceleration of the follower are expressed mathematically depending upon motion required as a function of cam angle and time.

Displacement

In general, the displacement of the follower is represented by S in metres (m) as a function of the cam rotation varying from 0° to 360° degrees or 0 to 2π radians. The displacement is expressed as $y = f(\theta)$. The maximum displacement y is equal to stroke of the follower S, i.e. $y_{max} = S$.

Velocity

The velocity of the follower (v) is represented by

$$\frac{dy}{dt} = \left(\frac{dy}{d\theta}\right)\left(\frac{d\theta}{dt}\right) = \frac{dy}{d\theta}.\omega$$

since ω is $\frac{d\theta}{dt}$.

Acceleration

The acceleration of the follower (a) is represented by

$$\frac{dv}{dt} = \left(\frac{dv}{d\theta}\right)\left(\frac{d\theta}{dt}\right) = \frac{dv}{d\theta}.\omega$$

$$= \frac{d}{d\theta}\left[\left(\frac{dy}{d\theta}\right).\omega\right].\omega = \frac{d}{d\theta}\left(\frac{dy}{d\theta}\right).\omega^2$$

Note:

1. Outward stroke means, the follower moving from lowermost position to uppermost position as cam rotates for a particular period of time called outward stroke and is indicated by the corresponding cam rotation indicated as θ_O or sometimes with the time taken as t_O.
2. Return means, the follower moving from uppermost to lowermost position as cam rotates for a particular period of time called return stroke and is indicated by the corresponding cam rotation indicated as θ_R or sometimes with the time taken as t_R.
3. Dwell outward means the follower having no displacement at the end of outward stroke even though the cam rotates and it is indicated by the cam angle as θ_{DO}. Similarly, dwell after the return stroke is indicated by the cam angle as θ_{DR}.

Analysis of four different types of motions for displacement, velocity and acceleration is given below for each type of motion.

10.5.1 Uniform Velocity

The equation for displacement $y = xt + c$ where x and c are constants;

(i) When time $t = 0$, displacement $y = 0$. Therefore, constant $c = 0$. When time $t = t_O$; $y = S_O$. Therefore, the constant $x = \dfrac{S_O}{t_O}$.

Hence, the equation for displacement y during outward stroke is

$$y_O = \left(\dfrac{S_O}{t_O}\right) \cdot t$$

and similarly for the return stroke

$$y_R = \left(\dfrac{S_R}{t_R}\right) \cdot t$$

as shown at Fig. 10.6(a).

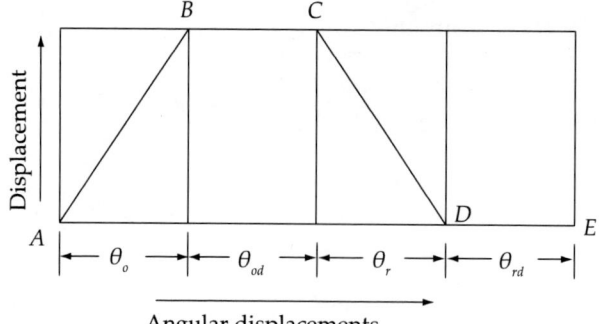

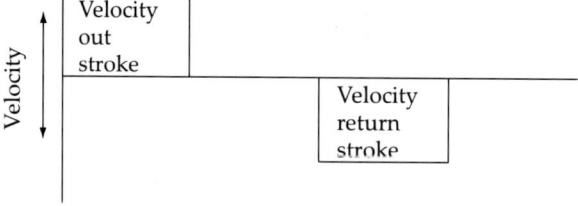

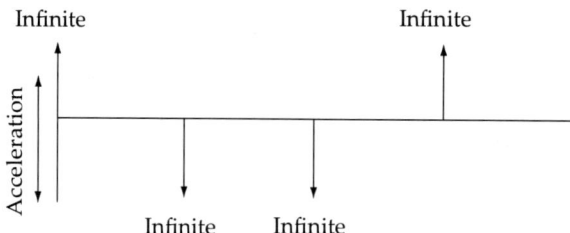

Fig. 10.6(a) Follower with uniform velocity.

Note: Stroke $S = S_O = S_R$ for a cam rotation of θ_O or θ_R or in terms of time $t = t_O = t_R$.

(ii) The velocity is the rate of change of displacement, i.e. $V = \dfrac{dy}{dt} = \dfrac{d}{dt}\left(\dfrac{S_O}{t_O}\cdot t\right)$ during outward. Thus, the velocity during outward stroke $V_O = \dfrac{S_O}{t_O}$ is a constant and similarly during return stroke $V_R = \left(\dfrac{S_R}{t_R}\right)$ is also a constant as shown at Fig. 10.6(a).

(iii) The acceleration is the rate of change of velocity and it is zero as velocity is constant. Thus, the acceleration during the outward and return strokes A_O and A_R are zeros. But the acceleration at the beginning and end of each stroke is infinite as the velocity changes suddenly from 0 to V_O or V_R to 0 in a negligible interval of time as shown at Fig. 10.6(a).

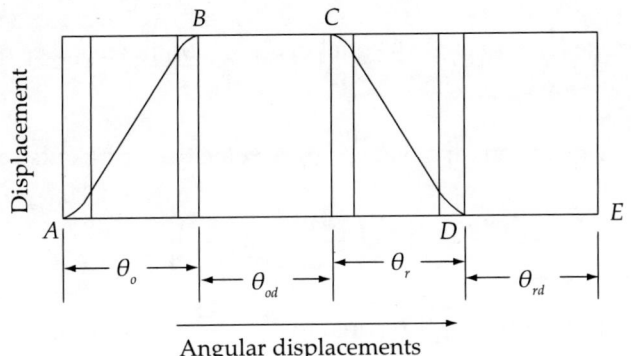

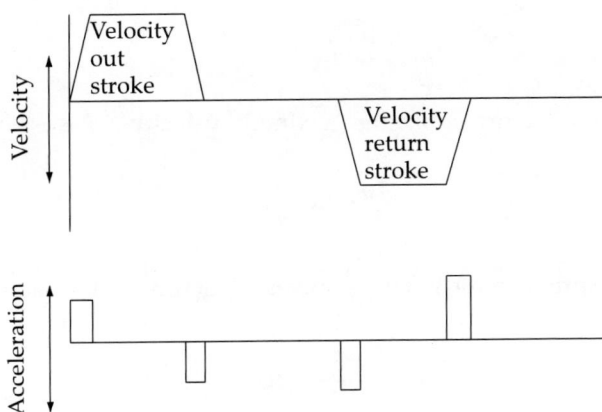

Fig. 10.6(b) Modified uniform velocity motion.

These conditions are impracticable since the acceleration and retardation of the follower at the beginning and at the end of each stroke are infinitely high. It is therefore necessary to modify the conditions which govern the follower motion, so that the acceleration and retardation are reduced to finite proportions. This can be done by rounding off the sharp corners at A, B, C and D on the displacement diagram so that the velocity of the follower increases gradually to its maximum value at the beginning of each stroke and decreases gradually to zero at the end of each stroke. The modified displacement, velocity and acceleration diagrams are as shown in Fig. 10.6(b).

In drawing these diagrams, assume that the follower is accelerated or retarded uniformly. The follower motion takes place with uniform velocity only for a short period at the beginning and at the end of each stroke. The rounded corners of the displacement diagram are parabolic arcs.

10.5.2 Simple Harmonic Motion (SHM)

(i) The displacement equation for SHM is given as a function of outward stroke as

$$y = \left(\frac{S}{2}\right)\left[1 - \cos\left(\frac{\pi\theta}{\theta_O}\right)\right]$$

During the out stroke for cam angle θ equal to $0°$, $\frac{\theta_O}{2}$, θ_O the displacement y equal to 0, $\frac{S}{2}$, S respectively $y_{O\max} = S = y_{R\max}$.

(ii) Differentiating the expression for y, i.e. $\frac{dy}{dt}$ gives velocity during outstroke

$$V = \left(\frac{\pi S \omega}{2\theta_O}\right)\left[\sin\left(\frac{\pi\theta}{\theta_O}\right)\right]$$

During the outstroke for cam angle θ equal to $0°$, $\frac{\theta_O}{2}$, θ_O the velocity V equal to 0, $\frac{\pi S \omega}{(2\theta_O)}$, 0 respectively.

The maximum velocity during outstroke

$$V_{O\max} = \frac{\pi S \omega}{2\theta_O}$$

and similarly during the return stroke the maximum velocity

$$V_{R\max} = \frac{\pi S \omega}{2\theta_R}$$

(iii) Differentiating the expression for V, i.e. $\frac{dV}{dt}$ gives acceleration during outstroke

$$A = \left(\frac{\pi^2 S \omega^2}{2\theta_O^2}\right)\left[\cos\left(\frac{\pi\theta}{\theta_O}\right)\right]$$

During the outstroke for cam angle θ equal to $0°$, $\frac{\theta_O}{2}$, θ_O the acceleration A is equal to $\left(\frac{\pi^2 S \omega^2}{2\theta_O^2}\right), 0, \left(\frac{\theta^2 S \omega^2}{2\theta_O^2}\right)$ respectively.

The maximum acceleration during outstroke

$$A_{O\,max} = \frac{\pi^2 S \omega^2}{2\theta_O^2}$$

and similarly during return stroke the maximum acceleration

$$A_{R\,max} = \frac{\pi^2 S \omega^2}{2\theta_R^2}$$

The displacement, velocity and acceleration diagrams are shown in Fig. 10.6(c).

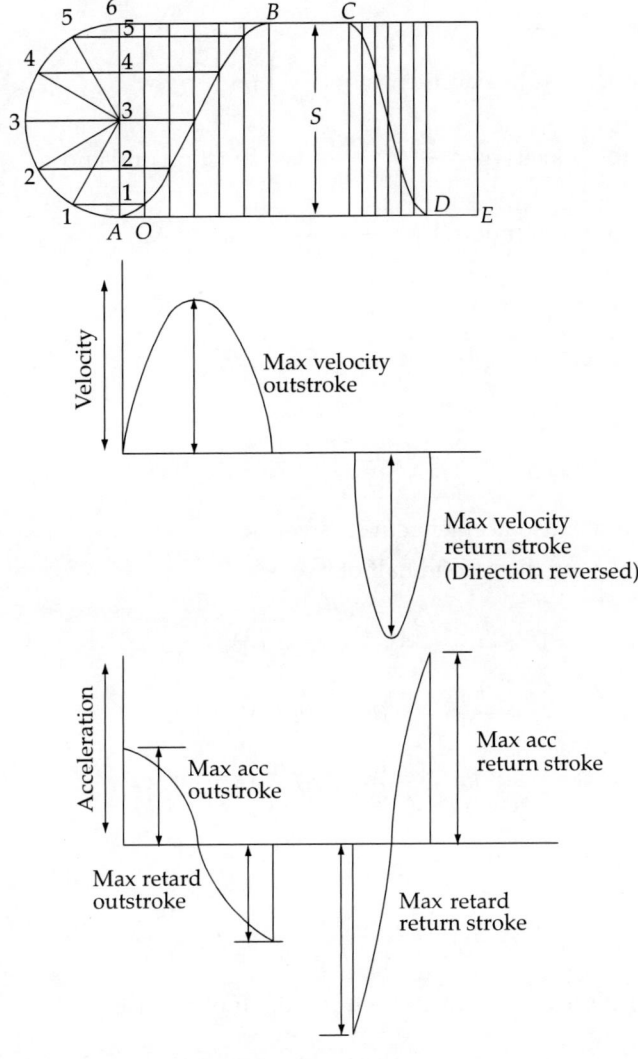

Fig. 10.6(c) SHM.

10.5.3 Uniform Acceleration and Retardation

(i) The expression for displacement y is a second order one, i.e. $y = \dfrac{ct^2}{2}$ where c is a constant and t time equal to $\dfrac{\theta}{\omega}$, so $y = \dfrac{c}{2}\left(\dfrac{\theta}{\omega}\right)^2 = K\theta^2$ where K is a constant $= \dfrac{c}{2\omega^2}$.

During the outstroke for cam angle θ equal to $0°$, $\dfrac{\theta_O}{2}$, the displacement y is equal to 0, $\dfrac{S}{2}$. On substitution in the equation for y, gives

$$\dfrac{S}{2} = K \cdot \dfrac{\theta_O^2}{4} \text{ or } K = \dfrac{2S}{\theta_O^2}$$

Hence, the equation for displacement $y = \left(\dfrac{2S}{\theta_O^2}\right) \cdot \theta^2$

(ii) The expression for velocity $V = \dfrac{dy}{dt} = \dfrac{4\omega S\theta}{\theta_O^2}$ is a function of θ and $V_{max} = \dfrac{2S\omega}{\theta_O}$ at $\theta = \dfrac{\theta_O}{2}$.

(iii) The expression for acceleration $A = \dfrac{dV}{dt} = \dfrac{4\omega^2 S}{(\theta_O^2)} = $ a constant.

For second half of the rise, the displacement $y = S - K(\theta_O - \theta)^2$ at $\theta = \dfrac{\theta_O}{2}$, $y = \dfrac{S}{2}$.

Therefore,

$$\dfrac{S}{2} = S - \dfrac{2S}{\theta_O^2}\left(\theta_O - \dfrac{\theta_O}{2}\right)^2 = \dfrac{S}{2}$$

Thus, the continuity of the displacement curve is verified.
The follower velocity during the second half of rise is

$$V = \dfrac{dy}{dt} = \dfrac{d}{dt}\left[S - K(\theta_O - \theta)^2\right]$$
$$= \dfrac{d}{dt}\left[S - K(\theta_O^2 - 2\theta_O\theta + \theta^2)\right]$$
$$= -K\left(-2\theta_O\dfrac{d\theta}{dt} + 2\theta\dfrac{d\theta}{dt}\right) = 2K(\theta_O - \theta)\omega$$

Again, at $\theta = \dfrac{\theta_O}{2}$,

$$V = 2K\left(\theta_O - \dfrac{\theta_O}{2}\right)\omega = K\theta_O\omega = \dfrac{2S\omega}{\theta_O}$$

The expression for the acceleration in the second half of the rise

$$A = -2K\omega^2 = -4\omega^2 \dfrac{S}{\theta_O^2}$$

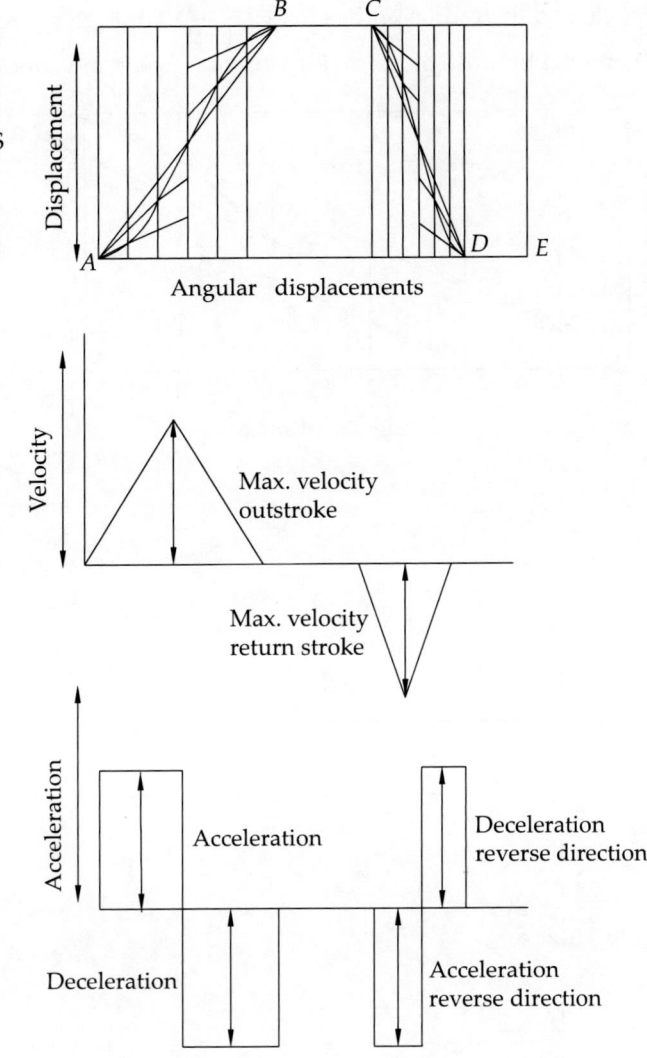

Fig. 10.6(d) Uniform acceleration and retardation.

The negative sign indicates that it is retardation which is opposite to acceleration in the first half of the rise. There is a sudden jump of acceleration to a finite value of $\dfrac{4\omega^2 S}{\theta_O^2}$ at $\theta = 0$, $\theta = \dfrac{\theta_O}{2}$ and $\theta = \theta_O$. The displacement, velocity and acceleration diagrams are shown at Fig. 10.6(d).

10.5.4 Cycloidal Motion

This curve is basically generated from a cycloid. A cycloid is the locus of a point on a circle which is rolled on a straight line. This line is the stroke of the follower which is made equal to the circumference of the circle and the radius of the circle is $\dfrac{S}{(2\pi)}$. For high speeds the cycloidal curve is the best over other contours because of its lowest vibration, wear and shock.

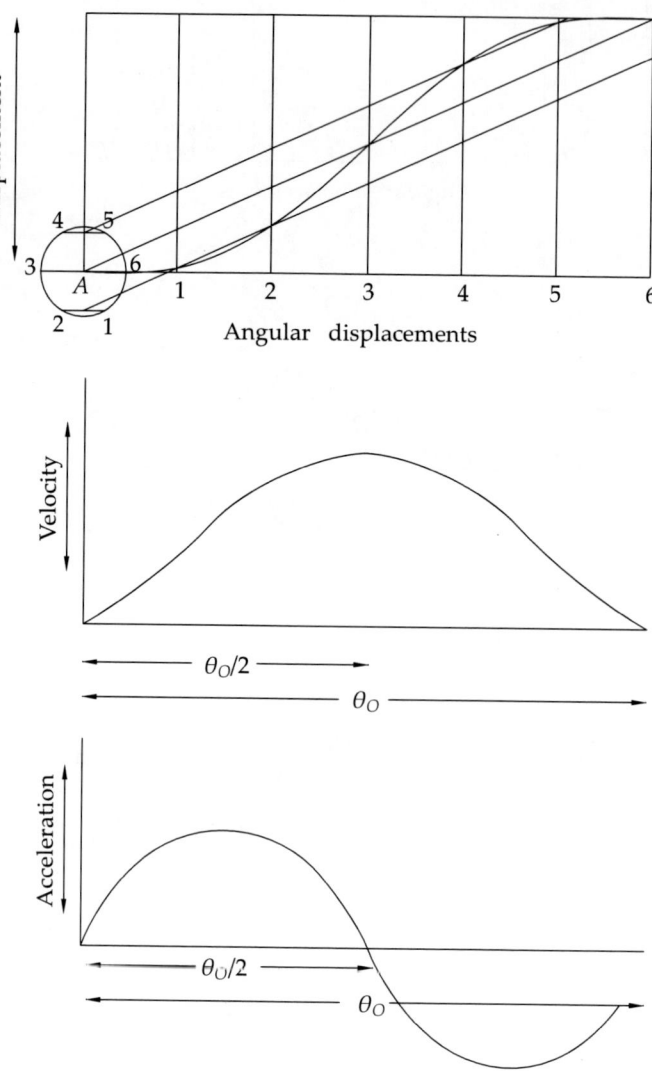

Fig. 10.6(e) Cycloidal motion.

(i) The expression for displacement during out stroke

$$y_O = \dfrac{S}{\pi}\left[\dfrac{\pi\theta}{\theta_O} - \dfrac{1}{2}\sin\left(\dfrac{2\pi\theta}{\theta_O}\right)\right]$$

The displacements for θ equal to $0, \dfrac{\theta_O}{2}, \theta_O$ are $0, \dfrac{S}{2}$ and S respectively.

(ii) The expression for velocity during the outstroke is $\dfrac{dy_O}{dt}$,

$$V_O = \dfrac{S\omega}{\theta_O}\left[1 - \cos\left(\dfrac{2\pi\theta}{\theta_O}\right)\right]$$

Velocity is maximum at $\theta = \dfrac{\theta_O}{2}$ giving $V_{O\max} = \dfrac{2\omega S}{\theta_O}$.

The expression for acceleration during outstroke, A_O is

$$A_O = \dfrac{dV_O}{dt} = \dfrac{2S\pi\omega^2}{\theta_O}\sin\left(\dfrac{2\pi\theta}{\theta_O}\right)$$

Acceleration is maximum at $\theta = \dfrac{\theta_O}{4}$ and is equal to $\dfrac{2S\pi\omega^2}{\theta_O^2}$. The displacement, velocity and acceleration curves are shown at Fig. 10.6(e).

10.6 CONSTRUCTION OF DISPLACEMENT DIAGRAMS

In order to design the cam profile for required motion of the follower, the corresponding displacement diagram has to be drawn first. During the rotation of the cam through one cycle of input motion, the follower executes a series of events as shown in graphical form in the displacement diagram. In such a diagram the abscissa represents one cycle of the input motion θ (one revolution of the cam) and is drawn to any convenient scale. The ordinate represents the follower travel y and for a reciprocating follower is usually drawn in full scale to help in layout of the cam. On a displacement diagram, it is possible to identify a portion of the graph called the rise or outward, where the motion of the follower is away from the cam centre. The maximum rise is called the lift. Periods during which the follower is at rest are referred as dwells, and the return is the period in which the motion of the follower is toward the cam centre.

Data required: Type of motion, the maximum stroke or displacement or lift, cam rotation for out stroke, dwell after out stroke, return stroke and dwell after return given in terms of either angles or in terms of time in seconds and sometimes as angles $\theta_O, \theta_{OD}, \theta_R, \theta_{RD}$ or by way of time t_O, t_{OD}, t_R and t_{RD} in seconds as time $t = \dfrac{\theta}{\omega}$. Hence, time for out stroke $t_O = \dfrac{\theta_O}{\omega}$; time for dwell at the end of outstroke $t_{OD} = \dfrac{\theta_{OD}}{\omega}$; time for the return stroke $t_R = \dfrac{\theta_R}{\omega}$; and time for dwell after return $t_{RD} = \dfrac{\theta_{RD}}{\omega}$, respectively.

Once the motion of the follower is chosen, that is, once the exact relationship between the input θ and the output y is known, the displacement diagram can be constructed precisely and it is a graphical representation of the functional relationship, $y = f(\theta)$. The construction procedure varies depending upon the type of motion as illustrated in the following worked example and showed in Fig. 10.7(a to d). In Fig. 10.7(e) all the displacement diagrams are superimposed just to have an idea on how the four different displacement curves vary.

W E 10.1: Draw the displacement diagrams for the following data assuming all the four types of motions as shown in Fig. 10.7(a to d).

Stroke $S = 50$ mm
Cam rotation for out stroke, $\theta_O = 120°$
Dwell after outstroke $\theta_{OD} = 60°$
Return stroke, $\theta_R = 120°$
Dwell after return $\theta_{RD} = 60°$.

10.6.1 Displacement Diagram for Uniform Velocity

Construction: The displacement diagram for a uniform motion or velocity is a straight line with a constant slope as shown. The displacement is directly proportional to θ for the cam rotating at uniform velocity as shown above.

Draw a horizontal line to some suitable scale to represent 2π radians or 360° for one rotation of the cam. On it mark the given angles $\theta_O, \theta_{OD}, \theta_R, \theta_{RD}$ as shown. Draw verticals. At A mark a length equal to stroke $S = 50$ mm given, to actual scale or full scale and complete the rectangle as shown.

For a cam rotation of θ_O degrees during outward stroke, draw a diagonal AG to represent the displacement diagram as shown. Similarly, if the velocity is uniform during return stroke also, i.e. during θ_R degrees of cam rotation, draw GA diagonal to represent the displacement diagram as shown. For the dwell periods θ_{OD} and θ_{RD}, the displacement remains unchanged as represented by GG and AA. Thus, the displacement curve ABCDEFG for outward and GFEDCBA for return stroke is shown in Fig. 10.7(a) as a straight line for the uniform velocity of follower.

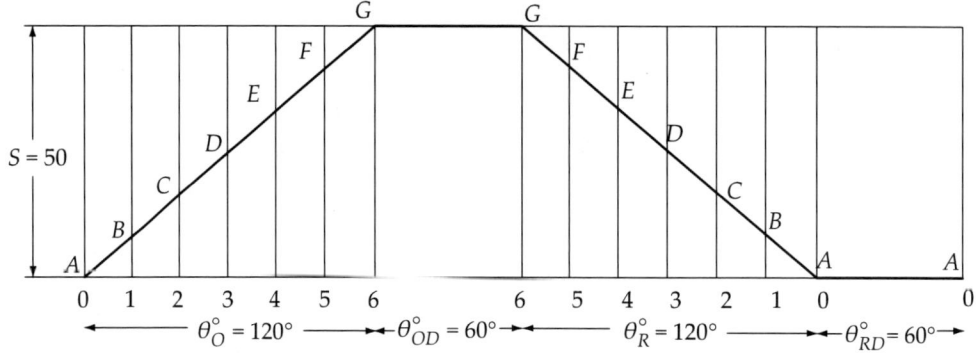

Fig. 10.7(a) Displacement diagram.

10.6.2 Displacement Diagram for Simple Harmonic Motion (SHM)

Construction: Construct the rectangle as explained above and draw a semicircle on the vertical at A and diameter equal to stroke $S = 50$ mm. Divide semicircle and the angle θ_O into same number of parts say 6 parts as shown. Draw horizontals from all the points on the semicircle on to the diameter and extend. Divide θ_O also into 6 parts and name them as 1 to 6. Draw verticals at all the six points. The horizontals drawn from a to g marked on the semicircle cut the verticals at A, B, C, D, E, F and

G respectively. Draw a smooth curve through these points to give the displacement curve for the outward stroke. Repeat the same procedure for the return stroke also and for dwell periods as shown at Fig. 10.7(b). The curve ABCDEFG represents the displacement diagram for simple harmonic motion (SHM).

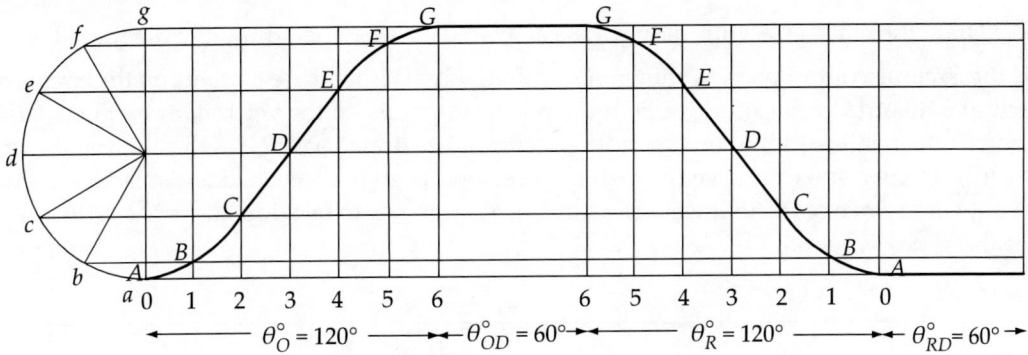

Fig. 10.7(b) Displacement diagram.

10.6.3 Displacement Diagram for Uniform Acceleration and Retardation (UAR)

Construction: Draw the rectangle of length representing 360° and height representing stroke $S = 50$ mm. Divide the outstroke into 6 parts. The uniform acceleration is for half the outward stroke $\left(\dfrac{S}{2}\right)$ and retardation for the remaining half $\left(\dfrac{S}{2}\right)$ of the outward stroke. At the midpoint, i.e. at 3, draw a vertical and divide it into 6 parts as shown and give numbers from 0 to 6. Join A to 0, 1, 2, 3 and then G to 3, 4, 5, 6 which cut the verticals drawn, at 0, 1, 2, 3, 4, 5, 6. Join all the points by a smooth curve ABCDEFG to give a uniform acceleration from A to D and retardation from D to G during the out stroke. Repeat the same procedure for the return stroke and dwell after outward to get the displacement diagram as shown at Fig. 10.7(c).

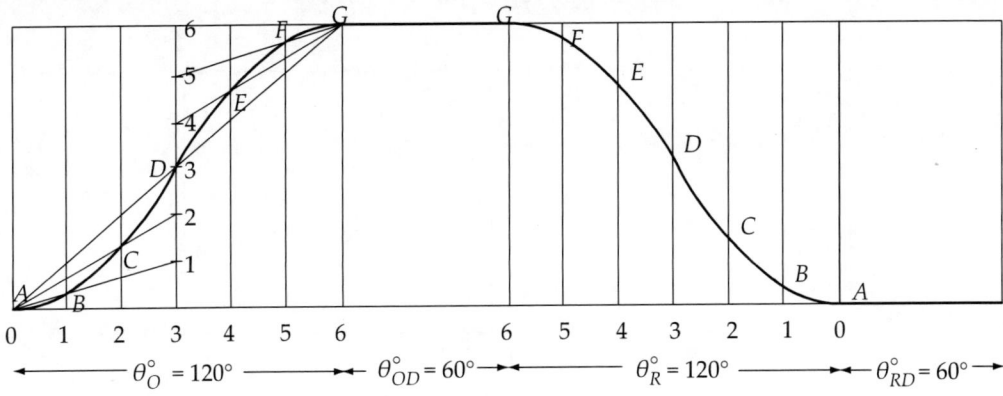

Fig. 10.7(c) Displacement diagram.

10.6.4 Displacement Diagram for Cycloidal Motion

Construction: Determine the radius of the circle of circumference equal to stroke S, i.e. 50 mm. The radius r of the circle is $\dfrac{S}{(2\pi)}$, i.e. $\dfrac{50}{2\pi}$.

Draw the rectangle as above and divide the outward stroke θ_O into 6 parts in this case always, i.e. $\dfrac{\theta_O}{6}$. Draw the circle at A with radius r as calculated above. Then draw the diagonal line AG. Divide the circle also into 6 parts as shown. Draw horizontals from these 6 points on the circle on to the vertical diametric line at A. Through these projected points on the vertical, draw lines parallel to diagonal which intersect the corresponding vertical lines drawn at 1, 2, 3, 4, 5, and 6 as shown at Fig. 10.7(d). Draw a smooth curve through all these points giving a cycloidal displacement curve for the outward stroke of the follower. The same procedure has to be adopted for the return stroke also for this type of motion.

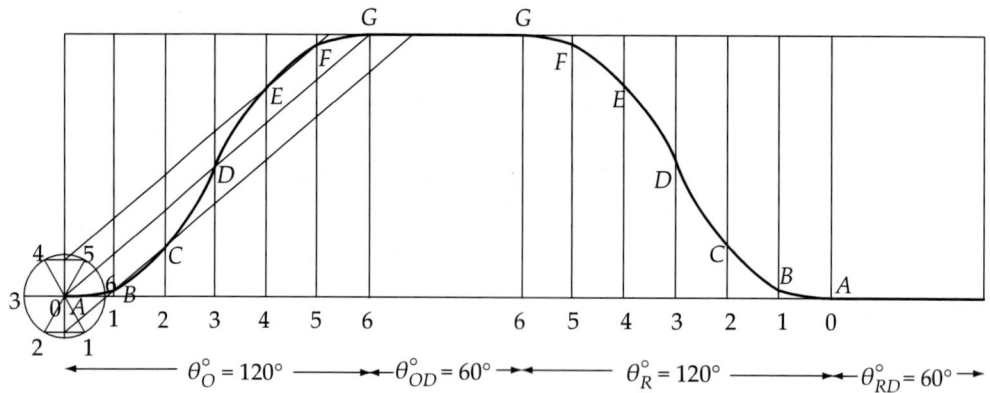

Fig. 10.7(d) Displacement diagram.

Figure 10.7(e) shows all the displacement diagrams shown superimposed so that one can easily observe the variation in the displacement diagrams of different motions.

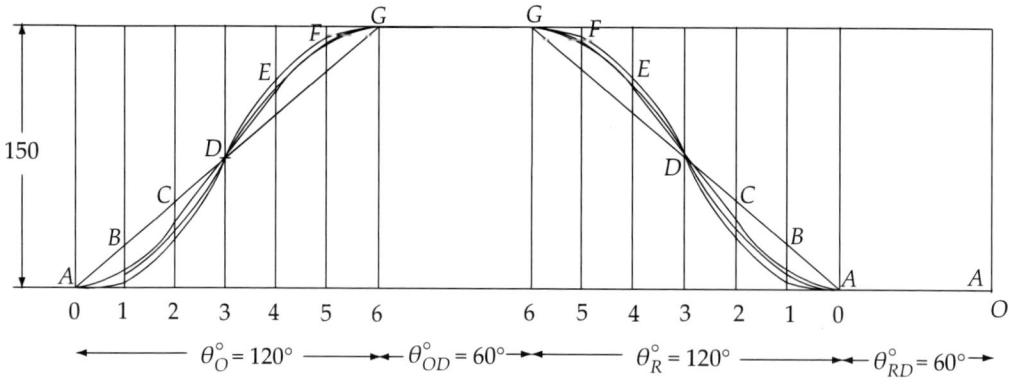

Fig. 10.7(e) Displacement diagrams superimposed.

10.7 CONSTRUCTION OF CAM PROFILES

Steps in drawing the cam profile

1. First draw the displacement diagram based on the type of motion of the follower as explained at 10.6. The X-axis indicates cam rotation (θ), from 0° to 360° and the Y-axis indicates stroke (S) of the follower in millimetres. Use suitable scale for X-axis to represent cam rotation and full scale for Y-axis.
2. Follow the construction of cam profile given below separately for different motions, for different types of followers including radial and offset. The various steps involved are explained below through a common example by varying one parameter and keeping all other parameters same to observe its influence on the cam profile.

W E 10.2: Cam with a minimum radius of 30 mm is rotating anticlockwise at a uniform speed of 1200 r.p.m. Stroke (S) of the follower is 50 mm. The cam rotations for the outward, dwell after outward, return and dwell after return are 120°, 60°, 120° and 60° respectively.

1. Draw one cam profile (a) with outward cycloidal motion and return with uniform velocity motion with a radial knife edge follower and another cam profile (b) with outward simple harmonic motion (SHM) and return uniform acceleration and retardation with a knife edge radial follower as shown at Fig. 10.8(a) and (b).
2. Draw one cam profile with outward simple harmonic motion (SHM) and return uniform acceleration and retardation with a knife edge radial follower having an offset of 10 mm to the right of cam axis as shown at Fig. 10.8(c).
3. Draw two cam profiles with the radial roller follower of radius 10 mm having outward cycloidal motion and return with uniform velocity motion (a) with the axis of the roller follower passing through the cam axis and (b) with the axis of the roller follower having an offset of 10 mm to the right of cam axis as shown at Fig. 10.8(d and e).
4. Draw one cam profile with the radial flat-faced follower with its axis passing through the cam axis having outward cycloidal motion and return with uniform velocity motion as shown at Fig. 10.8(f).

Given:
Cam angle along X-axis: $\theta_O = 120°$; $\theta_{OD} = 60°$; $\theta_R = 120°$; $\theta_{RD} = 60°$ along Y-axis stroke, S = 50 mm.

10.7.1 Cam Profile with Radial Knife Edge Follower Having Outward Cycloidal Motion and Return Uniform Velocity Motion

The various steps for the construction of cam profile showed in Fig. 10.8(a) for 1(a):

1. Draw a base circle with radius equal to the minimum radius of the cam, i.e. 30 mm taking O as centre.
2. Since it is a radial follower, the axis of the follower passes through the axis of the camshaft. Draw a vertical at O and mark a trace point A at a distance of 30 mm as shown.

3. From OA, mark angle 120° to represent oustroke, 60° to represent dwell after outward, 120° to represent return stroke and 60° to represent dwell after return in the clockwise direction about O as the cam is given as rotating in the anticlockwise direction.

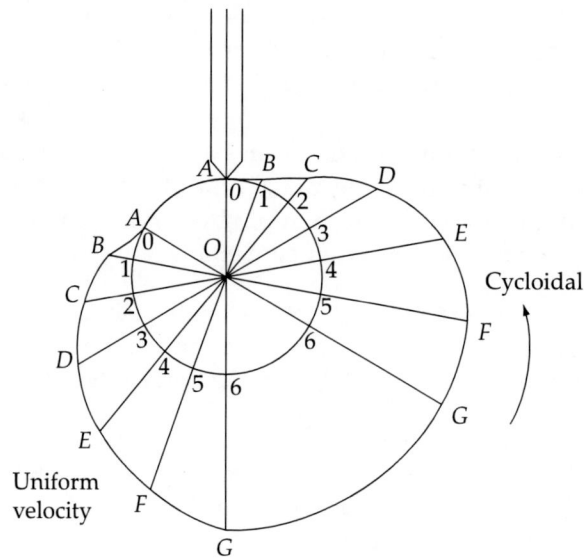

Fig. 10.8(a) Cam profile with outward cycloidal motion and return uniform velocity.

4. Divide the angular displacement during outstroke and return stroke into six equal parts as in the displacement and give numbers as shown.
5. Join points 0, 1, 2, 3, 4, 5, 6 and 6, 5, 4, 3, 2, 1, 0 to centre O and produce beyond the base circle as shown.
6. Now set off on the radial lines drawn, 0A, 1B, 2C, 3D, 4E, 5F and 6G equal to 0A, 1B, 2C, 3D, 4E 5F and 6G taken from the displacement diagram for cycloidal motion showed in Fig. 10.7(d).
7. Repeat the same procedure for the return stroke by taking the distances from the displacement diagram for uniform velocity showed in Fig. 10.7(a).
8. Join the points A, B, C, D, E, F, G and G, F, E, D, C, B and A with a smooth curve as shown at Fig. 10.8(a) to get the cam profile.

Note:

1. Scale of cam profile diagram should be same as the scale adopted for the stroke in the displacement diagram.
2. Mark angles θ_O, θ_{OD}, θ_R and θ_{RD} in the clockwise direction when the cam rotates in the anticlockwise direction and vice versa.

10.7.2 Cam Profile with a Radial Knife Edge Follower Having Outward SHM and Return Uniform Acceleration and Retardation (UAR)

The various steps for the construction of cam profile showed in Fig. 10.8(b) for 1(b).

This is also to be constructed by following the same steps explained above by taking the displacements from the displacement diagram Fig. 10.7(b) for simple harmonic motion (SHM) and from Fig. 10.7(c) for uniform acceleration and retardation.

Note: Observe the change in the cam profile due to change of motions between Fig. 10.8(a) and Fig. 10.8(b).

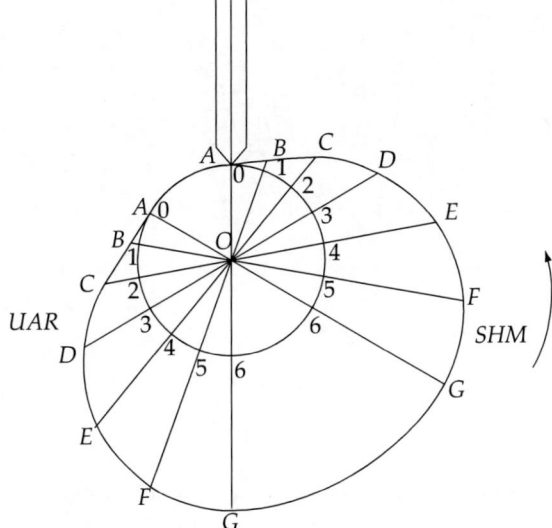

Fig. 10.8(b) Cam profile with SHM and uniform acceleration and retardation.

10.7.3 Cam Profile with an Offset Knife Edge Follower Having Outward SHM and Return UAR

The various steps for the construction of cam profile showed in Fig. 10.8(c) for 2.

1. Draw a base circle with the radius equal to the minimum radius of the cam, i.e. 30 mm with O as centre.
2. Taking given offset of 10 mm, draw the axis of the follower at 10 mm from the axis of the camshaft or centre of base circle O. This axis intersects the base circle at A as shown.
3. Join OA and draw an offset circle of radius 10 mm with O as centre.
4. From OA, mark angles 120° to represent oustroke, 60° to represent dwell after outward, 120° to represent return stroke and 60° to represent dwell after return in the clockwise direction from OA as the cam is rotating in the anticlockwise direction.
5. Divide the angular displacement during outstroke and return stroke into say six equal parts as in the displacement diagram and give 0 to 6 for outward and 6 to 0 for return on the base circle as shown.

6. Now from the points 0, 1, 2, 3, 4, 5 and 6 and again from 6 to 0 on the base circle, draw tangents to the offset circle as shown in the Fig. 10.8(c) and produce these tangents beyond the base circle as shown.
7. Now set off, $0A, 1B, 2C, 3D \ldots 6G$ taken from the displacement diagram for SHM Fig. 10.7(b) for outward stroke and again $6G \ldots 1B, 0A$ from the respective displacement diagram for UAR Fig. 10.7(c) for return stroke.
8. Join the points $A, B \ldots F, G$ and $G, \ldots B, A$ by a smooth curve as shown in Fig. 10.8(c).

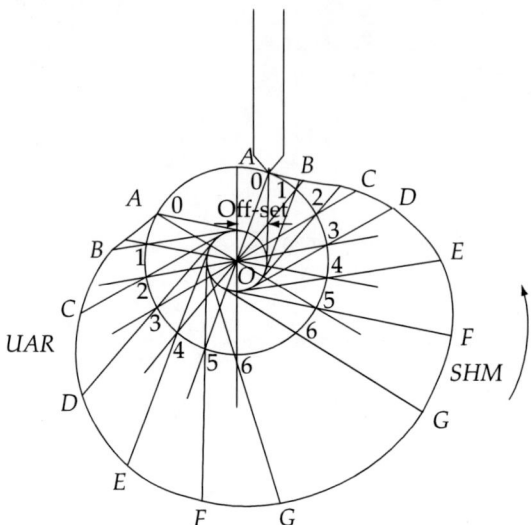

Fig. 10.8(c) Cam profile with an offset follower.

Note:

1. Observe the difference in the construction of offset follower.
2. The offset circle is to be drawn with a radius equal to offset.
3. Tangents are to be drawn from the points on the base circle to offset circle.
4. Observe the two cam profiles at Fig. 10.8(b) without offset and at Fig. 10.8(c) with offset follower with all other parameters remaining same.

10.7.4 Cam Profile with the Radial Roller Follower with Outward Cycloidal Motion and Return Uniform Velocity

The various steps for the construction of cam profile showed in Fig. 10.8(d) for 3(a):

1. Draw a base circle with O as centre and radius equal to the minimum radius of cam, i.e. 30 mm.
2. Since the axis of the follower passes through the axis of the camshaft, mark the centre of the roller circle which is the trace point, i.e. A such that $OA = 40$ mm (base circle radius + radius of roller, i.e. 30 mm + 10 mm). Draw prime circle with OA as radius shown.

3. From *OA*, mark angle 120° to represent outstroke, 60° to represent dwell after outward, 120° to represent return stroke and 60° to represent dwell after return in the clockwise direction as shown.
4. Divide the angular displacement during outstroke and return stroke into say six equal parts as in the displacement diagrams.
5. Join the points 0, 1, 2, 3, 4, 5, 6 and 6, ... 1, 0 on the prime circle to centre *O* and produce beyond the prime circle as shown.
6. Now set off 0*A*, 1*B*, 2*C*, 3*D* ... 6*G* and again 6*G* ... 1*B*, 0*A* displacements taken from the respective displacement diagrams for cycloidal from Fig. 10.7(d) and for uniform velocity from Fig. 10.7(a).
7. Now with *A*, *B*, *C* ... *G* and *G* ... *B*, *A* as centres with radius equal to roller circle radius, draw circles as shown to represent roller.
8. Draw a smooth curve touching all the roller circles as shown by a smooth curve.

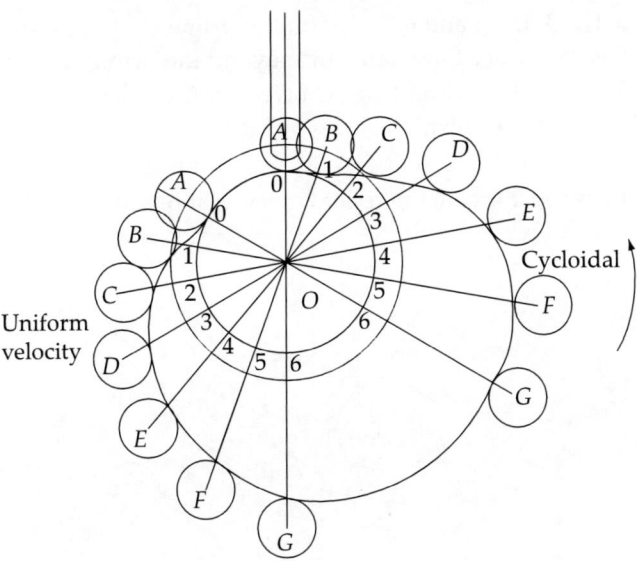

Fig. 10.8(d) Cam profile with radial roller follower.

Note:

1. Draw the prime circle of radius equal to base circle radius plus roller radius.
2. Mark distances from the prime circle to get the centres of the roller.
3. Draw rollers at each point and then draw a smooth curve tangential to all the roller circles.

10.7.5 Cam Profile with an Offset Roller Follower with Outward Cycloidal Motion and Return with Uniform Velocity

The various steps for the construction of cam profile showed in Fig. 10.8(e) for 3(b).

1. Draw a base circle with radius equal to the minimum radius of the cam, i.e. 30 mm with O as centre.
2. Since the axis of the follower is offset from the axis of the follower by 10 mm of the camshaft, draw an offset circle with O as centre and radius equal to 10 mm. Draw a tangent to the offset circle by drawing the horizontal from the centre O as shown. Draw an arc with O as centre and radius of 40 mm (base circle radius + radius of roller, i.e. 30 mm + 10 mm) to get point A as shown on the vertical tangent drawn. Join OA. Draw a prime circle with OA as radius.
3. From OA, mark angles 120° to represent outstroke, 60° to represent dwell after outward, 120° to represent return stroke and 60° to represent dwell after return in the clockwise as shown.
4. Divide the angular displacement during outstroke and return stroke into same number parts as in the displacement.
5. Mark the points $0, 1, 2, 3, 4, 5, 6$ and $6, \ldots 1, 0$ on the prime circle as shown and draw tangents to the offset circle and produce these tangents beyond the prime circle as shown.
6. Now set off $0A, 1B, 2C, 3D \ldots 6G$ and again $6G \ldots 1B, 0A$ from the respective displacement diagram to get the centres of roller.
7. Now with $A, B, C \ldots G$ and $G \ldots B, A$ as centres and radius equal to roller radius draw circles.
8. Draw a smooth curve touching all the roller circles as shown.

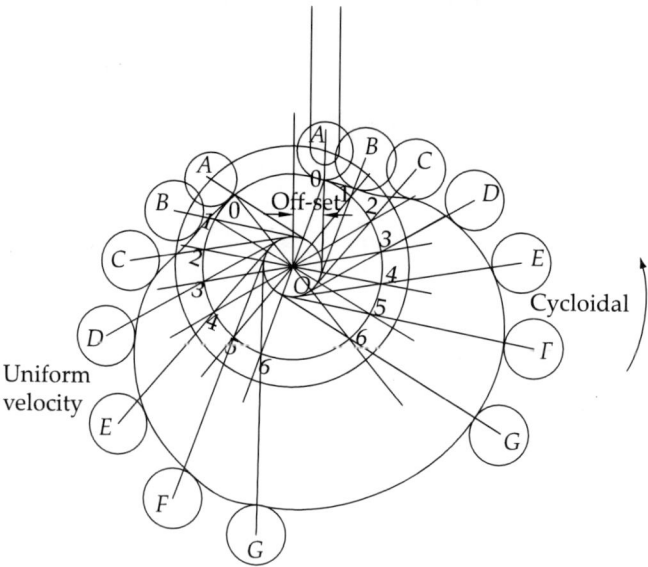

Fig. 10.8(e) Cam profile with an offset roller follower.

Note:

1. Observe the cam profiles without offset and with offset roller follower and see the difference.
2. The determination of the tracing point A initially is to be observed in the case of offset follower.

10.7.6 Cam Profile for Radial Flat Faced Radial Follower with Outward Cycloidal Motion and Return Uniform Velocity

The various steps for the construction of cam profile showed in Fig. 10.8(f) for 4.

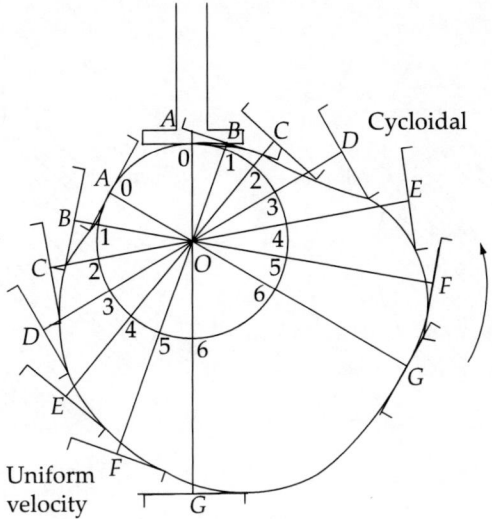

Fig. 10.8(f) Cam profile with flat faced follower.

1. Draw a base circle with a radius equal to the minimum radius of the cam, i.e. 30 mm taking O as centre.
2. Since it is a radial follower, the axis of the follower passes through the axis of the camshaft. Mark trace point A at a distance of 30 mm as shown by drawing the vertical at O. Draw a circle with OA as radius.
3. From OA, mark angle 120° to represent outstroke, 60° to represent dwell after outward, 120° to represent return stroke and 60° to represent dwell after return in the clockwise direction about O as the cam is given as rotating in the anticlockwise direction.
4. Divide the angular displacement during outstroke and return stroke into same number of parts as in the displacement diagrams.
5. Join the points 0, 1, 2, 3, 4, 5, 6 and 6, 5, 4, 3, 2, 1, 0 to centre O and produce beyond the base circle as shown.
6. Now set off on the radial lines drawn, 0A, 1B, 2C, 3D, 4E, 5F and 6G equal to 0A, 1B, 2C, 3D, 4E, 5F and 6G from the displacement diagram for cycloidal motion as shown in Fig. 10.7(d).
7. Similarly, repeat the same procedure for the return stroke by taking the distances from the displacement diagram for uniform velocity as shown in Fig. 10.7(a).
8. Draw lines representing flat faces at the points A, B, C, D, E, F, G and G, F, E, D, C, B, A perpendicular to radial lines.
9. Draw a smooth curve tangential to all the flat faces as shown at Fig. 10.8(e).

Note: Construction of the displacement diagram for cycloidal motion has to be done carefully.

Pic. 10.5 shows the cam profiles obtained above are given side by side for comparison purpose.

1. The two profiles (a) and (b) show the difference in the profiles due to offset.

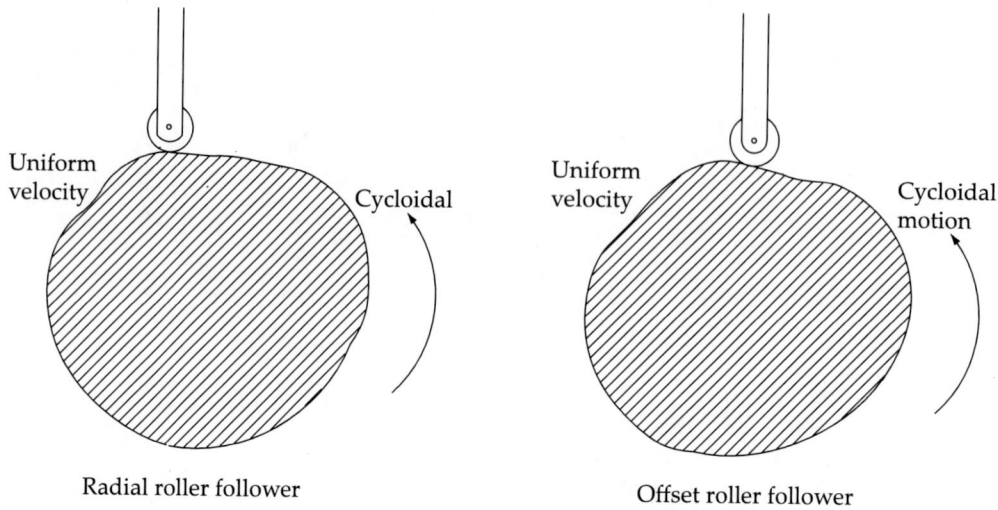

Pic. 10.5(a and b) Cam profiles without and with offset follower.

2. The two profiles (c) and (d) show the difference in the profiles due to variation in the motions.

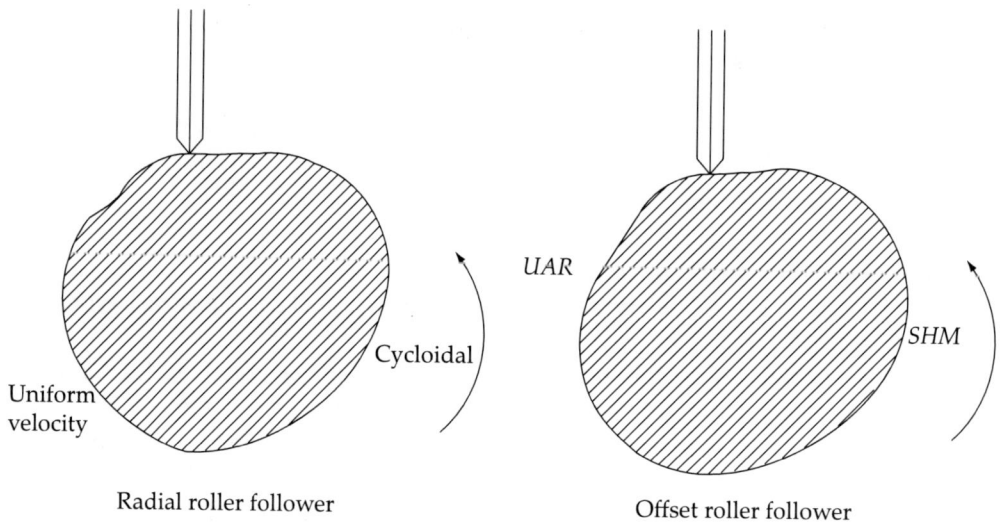

Pic. 10.5(c and d) Cam profiles with different motions of the follower.

3. The two profiles (e) and (f) show the difference in the profiles due to variation in the followers.

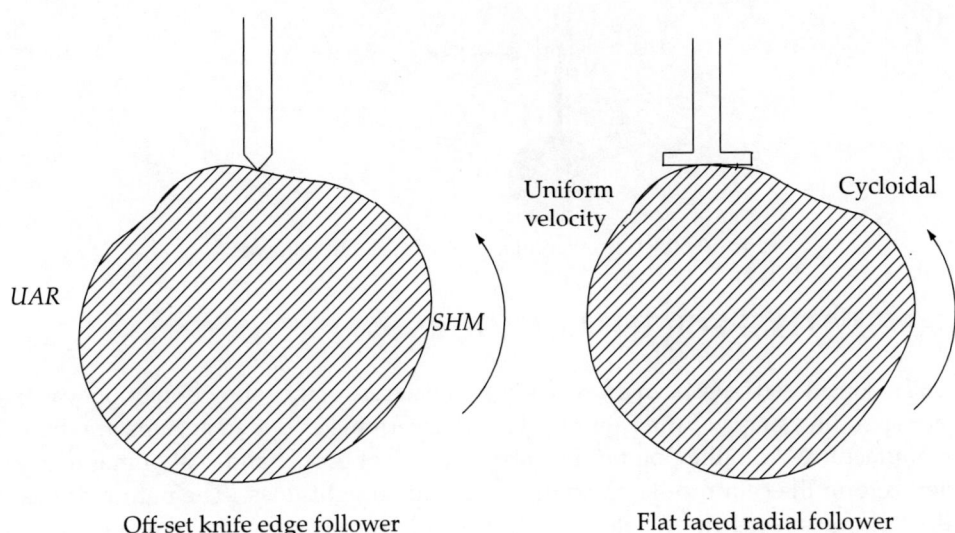

Off-set knife edge follower Flat faced radial follower

Pic. 10.5(e and f) Cam profiles with different types of followers.

Three types of cams with different followers are shown at Pic. 10.6(a, b, c).

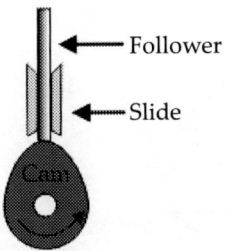

Pic. 10.6(a) Symmetric cam with radial follower.

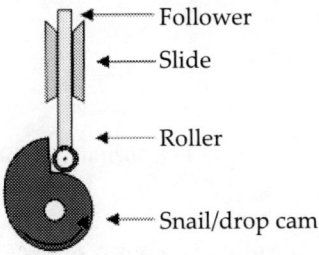

Pic. 10.6(b) Snail or drop cam with roller follower.

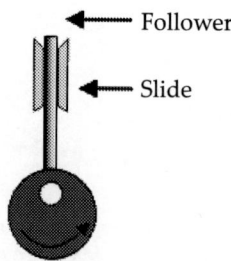

Pic. 10.6(c) Offset circular cam with radial follower.

10.8 CAMS WITH SPECIFIED CONTOURS

The cams so far considered have been those in which the nature of the follower motion was specified and the corresponding shape of the cam profile was determined. But such cams are difficult and costly to manufacture. From the point of view of accuracy of profile and cost of manufacture, it is much better to form the cam profile of circular arcs and straight lines. The nature of the motion given to the follower may then be determined. The valves of I.C. Engines are operated by cams whose profiles consist entirely of circular arcs with the followers either of flat faced or of rollers.

10.8.1 Circular Arc Cam with Flat-faced Reciprocating Follower

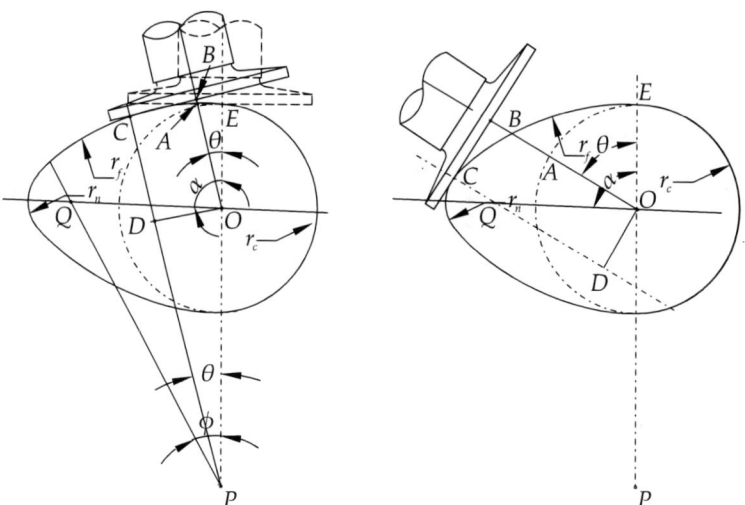

Fig. 10.9(a) and (b)

Figure 10.9 shows a circular arc cam. OE represents the minimum radius of the cam (r_c) and the flank consists of a circular arc of radius (r_f) with centre P. Assume that the cam remains fixed and the line of stroke of the follower turn in the opposite sense to that of the actual rotation of the cam, i.e. counterclockwise in the figure. When the line of stroke turns through an angle θ relative to the cam, the flat faced follower will be in contact with the cam profile at the point C, where PC is perpendicular to the face of the follower. Then the displacement of the follower is given by AB, where OB is perpendicular to BC and therefore parallel to PC. Draw OD perpendicular to PC.

1. Follower on the flank (θ is 0° to ϕ) as in Fig. 10.9(a).

Then the displacement of the follower is

$$x = AB = BO - AO = CD - EO$$

But

$$CD = PC - PD = PE - PO \cos \theta$$
$$= (OP + OE) - PO \cos \theta = OE + OP(1 - \cos \theta)$$

Therefore, displacement

$$x = OP(1 - \cos \theta)$$

since $PE = r_f$; $OE = r_c$; $\therefore OP = (r_f - r_c)$

$$x = (r_f - r_c)(1 - \cos \theta) \qquad (1)$$

The velocity of the follower,

$$v = \frac{dx}{dt} = \frac{dx}{d\theta} \cdot \frac{d\theta}{dt}$$
$$= \omega \cdot \frac{dx}{d\theta} = \omega \cdot (r_f - r_c) \cdot \sin \theta \qquad (2)$$

This is maximum when $\theta = \phi$; $v_{max} = \omega \cdot (r_f - r_c) \cdot \sin \phi$

The acceleration of the follower,

$$a = \frac{dv}{dt} = \omega \cdot \frac{dv}{d\theta}$$
$$= \omega^2 \cdot (r_f - r_c) \cdot \cos \theta \qquad (3)$$

This is maximum when $\theta = 0°$; $a_{max} = \omega^2 \cdot (r_f - r_c)$.

The above three equations (1), (2) and (3) apply only while the follower is in contact with that part of the cam profile which has the centre of curvature P, i.e. for values of θ from 0 to ϕ where ϕ is equal to angle OPQ.

2. Follower in the nose (θ is greater than ϕ) as in Fig. 10.9(b)

When $\theta > \phi$, the follower will be in contact with the nose of the cam, the centre of curvature is Q as shown. From Fig. 10.9(b), the cam and the follower are in contact at C on the nose of the cam. Let OQ be the distance between the centre of the nose and the centre of the cam be h and radius of the nose be r_n.

The displacement of the follower,

$$x = AB = OB - OA = CD - OA$$

But
$$CD = CQ + QD = CQ + OQ\cos(\alpha - \theta)$$

Therefore,
$$x = CQ - OA + OQ\cos(\alpha - \theta)$$
$$= [r_n - r_c + h\cos(\alpha - \theta)] \qquad (4)$$

In the above equation CQ, OA, OQ and α are constants and differentiating with respect to time gives the velocity of the follower,
$$v = \omega \cdot OQ \sin(\alpha - \theta)$$
$$= \omega \cdot h \cdot \sin(\alpha - \theta) \qquad (5)$$

and again differentiating gives the acceleration of the follower,
$$a = -\omega^2 \cdot OQ \cos(\alpha - \theta)$$
$$= -\omega^2 \cdot h \cdot \cos(\alpha - \theta) \qquad (6)$$

$-ve$ sign indicates retardation of the follower. The retardation of the follower is maximum when $(\alpha - \theta) = 0$. i.e. at the end of the lift which is given by $\omega^2 \cdot OQ$ or $\omega^2 h$.

The above three equations (4), (5) and (6) apply only while the follower is in contact with that part of the cam profile which has the centre of curvature Q, i.e. for values of θ from θ to α where α is equal to $\angle EOC$.

Cams of the above type are symmetrical. Occasionally the follower is given a short period of dwell at the end of the lift.

W E 10.3: The valve timing for a four-stroke petrol engine is as follows: Inlet opens 4° early and closes 50° later. Exhaust opens 50° earlier and closes 10° later. Each valve has a lift of 10.5 mm. Base circle radius of cam is 20 mm and nose radius of 2.5 mm. The cams are of circular arc type with flat faced followers. Set out the cam profile and calculate the maximum velocity, acceleration and retardation of follower. If the camshaft speed is 2000 r p m, what is the minimum force required which must be executed by spring of each valve in order to overcome the inertia of the moving parts of mass 0.25 kg for each valve.

Given:
Cam rotation is half the crank rotation and crank rotates for $(50 + 180 + 10) = 240°$ for exhaust valve.

Therefore, $\alpha = \dfrac{240°}{2} = 120°$. All particulars are given to draw the cam to scale as shown at Fig. 10.9(c) below and then measure the angle ϕ, i.e. $\angle OPQ = \phi = 18°$.

$$\text{Velocity } v = \omega \cdot OP \cdot \sin\phi$$
$$= \dfrac{2\pi \times 2000}{60} \times 60 \times \sin 18°$$
$$= 3870 \text{ mm/sec} = 3.387 \text{ m/s}$$

Maximum acceleration,

$$a_{max} = \omega^2 OP = \left(2\pi \times \frac{2000}{60}\right)^2 \times 60$$
$$= 4.4 \times 10^4 \times 60 = 2.64 \times 10^6 \text{ mm/sec}^2$$

Maximum retardation,

$$a_{max} = -\omega^2 OQ = \left(2\pi \times \frac{2000}{60}\right)^2 \times 28$$
$$= -4.4 \times 10^4 \times 28 = -1.21 \times 10^6 \text{ mm/sec}^2$$

Incidentally, the maximum retardation of the inlet valve is also same since OQ is same. Mass of the reciprocating mass = 0.25 kg. So to maintain contact at this highest retardation, force required by spring = mass × acceleration = $m \times a$ = 0.25 × 1.21 × 10³ = 302.5 N. The gravity pull of the weight = 0.25 × 9.81 = 2.425 N is neglected. Similarly, do for the inlet valve in similar lines.

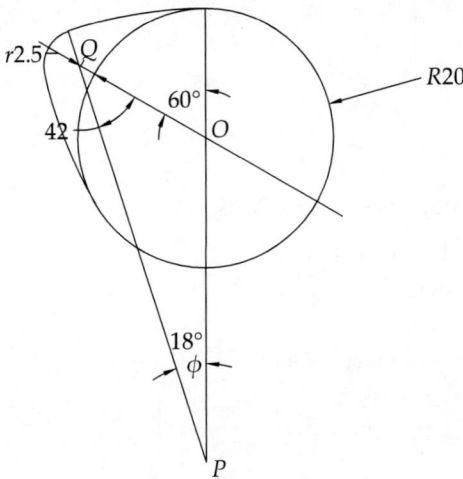

Fig. 10.9(c) Cam for exhaust valve.

10.8.2 Tangent Cam with Reciprocating Roller Follower

When the reciprocating follower is fitted with a roller, the flank of the cam has to be straight and tangential to the base circle. Then that type of cam is called tangent cam as shown in Fig. 10.10. The flanks AB and EF are straight lines tangential to the circular arc AF of base circle of cam and the arcs BU and VE respectively. UV is a circular arc drawn with centre O. The dotted line shows the path of the roller. Since the whole of the cam profile is formed of circular arcs and straight lines, it is very easy to manufacture accurately.

The outward displacement of the follower takes place partly while the roller is in contact with the straight flank AB and partly while it is in contact with the rounded corner BU. If θ is the angle through which the cam has turned from the beginning of the lift of the follower, then contact takes place between the follower and the straight flank AB for values of θ from 0 to ϕ, where $\tan \phi = \dfrac{GK}{GO}$, and between the follower and the rounded nose BU for values of θ from ϕ to α, where $\tan \alpha = \dfrac{CR}{RO}$.

The displacement, velocity and acceleration of the follower for a given value of θ may be found as follows:

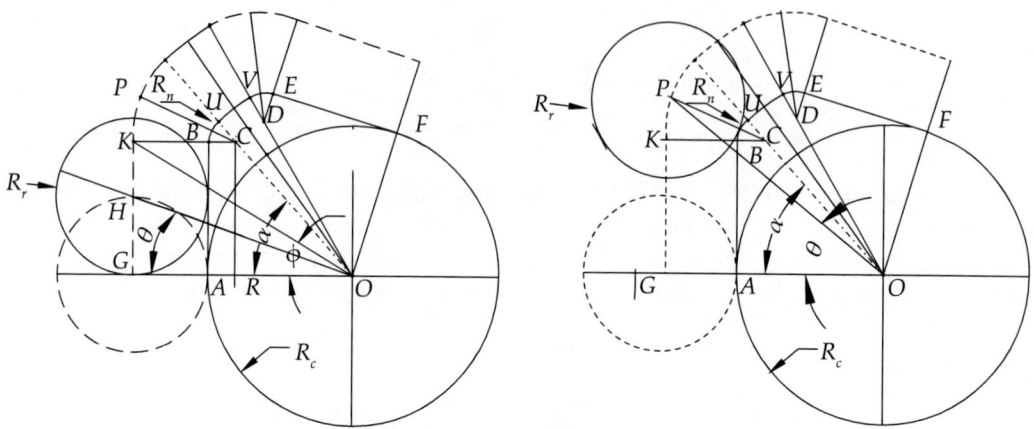

Fig. 10.10(a) and (b)

1. Contact between the roller and the straight flank AB

As shown in Fig. 10.10(a), let θ be the angle turned by the cam from the beginning of the follower displacement, x be the displacement of the follower and ω be the angular velocity of the cam. Then

$$x = OH - OG = OG\left(\dfrac{1}{\cos\theta} - 1\right)$$
$$= (R_c + R_r)\left[\dfrac{1}{\cos\theta} - 1\right] \qquad (1)$$

where R_c and R_r are the radii of the base circle of cam and roller respectively.

Differentiating equation (1) with respect to time, the velocity of the follower,

$$v = \dfrac{dx}{dt} = \dfrac{\omega dx}{d\theta}$$
$$= \omega \cdot (R_c + R_r) \cdot \dfrac{\sin\theta}{\cos^2\theta} \qquad (2)$$

velocity increases as θ increases and is maximum when $\theta = \phi$.

Differentiating equation (2) with respect to time, the acceleration of the follower,

$$a = \omega \cdot \dfrac{dv}{d\theta} = \omega^2 \cdot (R_c + R_r) \dfrac{(2 - \cos^2\theta)}{\cos^3\theta} \qquad (3)$$

Acceleration is minimum, when $\omega^2 \cdot (R_c + R_r)\dfrac{(2 - \cos^2 \theta)}{\cos^3 \theta}$ is minimum or $(2 - \cos^2 \theta)$ is minimum and $\cos^3 \theta$ is maximum at the beginning of lift, i.e. when $\theta = 0$.

These three equations apply only for the part of the follower motion during which the roller is in contact with the straight flank AB, i.e. for values of θ from 0 to ϕ.

2. Contact between the roller and the nose

As shown in Fig. 10.10(b), for values of θ from ϕ to α, the roller is in contact with the arc BU of the cam profile and the distance of the roller centre P from the centre C of the arc BU remains constant. The arrangement is kinematically equivalent to a slider-crank chain with a crank of length OC, turning at uniform speed with a connecting rod of length CP.

The displacement, velocity and acceleration of the follower may be found from the appropriate analytical expressions given in Chapter 4 for a reciprocating mechanism, where for $n = \dfrac{CP}{OC}$ and $\theta = (\alpha - \theta)$ have to be adopted.

Let R_n–nose radius, R_r–roller radius, R_c–cam radius,
  h be distance between the nose centre and the cam centre,
  L be sum of nose radius and roller radius $(R_n + R_r)$ represented by PC, and
  n be the sum of cam radius and roller radius $(R_c + R_r)$ represented by OG.

Then displacement, $x = h\cos(\alpha - \theta) + \sqrt{L^2 - h^2 \sin^2(\alpha - \theta)} - n$

Obtain by differentiating the above expression once, the velocity and differentiating twice, the acceleration.

W E 10.4: A tangent cam with a base circle diameter of 50 mm operates a roller follower 20 mm in diameter. The line of stroke of the roller passes through the axis of the cam. The angle between the tangential faces of the cam is 60°, speed of the camshaft 200 r.p.m. and the lift of the follower 15 mm. Calculate (i) the main dimensions of the cam (ii) the accelerations of the follower at (a) the beginning of lift (b) the apex of the circular nose.

Given:
Radius of base circle $R_c = 25$ mm, radius of roller $R_r = 10$ mm;
lift of the follower, $x = 15$ mm; speed $N = 200$ r.p.m.; $\omega = \dfrac{2\pi \times 200}{60} = 20.94$ rad/s
As it forms a right-angled triangle, $\alpha = (180° - 30° - 90°) = 60°$.

(i) *Referring to Fig. 10.10,*

$$h + R_n + R_r = R_c + R_r + x$$
$$h + R_n = R_c + x = 25 + 15 = 40 \tag{1}$$

$$OA = OR + R_n = R_c$$
$$h \cos 60° + R_n = 25 \text{ or } 0.5h + R_n = 25 \tag{2}$$

Subtracting (2) from (1) gives $h = 30$ mm and so $R_n = 10$ mm.
Therefore, $n = (R_c + R_r) = 35$ mm and $L = (R_n + R_c) = 35$

$$\tan \phi = \frac{h \sin \alpha}{(R_c + R_r)}$$

$$= \frac{30 \sin 60}{(25 + 10)} = 0.742 \text{ or } \phi = 36.6°$$

(ii) *Acceleration of the follower:*

(a) At the beginning of lift, i.e. roller centre at B, $\theta = 0°$

$$\text{Acceleration, } a = \omega^2 \left[\frac{n(2 - \cos^2 \theta)}{\cos^3 \theta} \right]$$

$$= 20.94^2 \left[\frac{0.035(2 - \cos^2 0)}{\cos^3 0} \right] = 14.41 \text{ m/s}^2$$

(b) When the roller is at the apex of the circular nose, i.e. at U, $\theta = \alpha$,

$$\text{Acceleration, } a = \omega^2 h \left[-1 - \frac{h}{L} \right]$$

$$= 20.94^2 \times 0.03 \left[-1 - \left(\frac{0.03}{0.02} \right) \right] = -24.43 \text{ m/s}^2$$

10.9 EXERCISE

10.9.1 Short Answer Questions

1. With the help of neat sketches, explain the types of cams and followers. Give the specific application of each type of cam.
2. Sketch displacement, velocity and acceleration vs time curves for SHM of follower.
3. Give a neat sketch of a cam with offset roller follower.
4. Give the classification of cam followers.
5. Classify cams and followers.
6. What is a cam? What type of motion can be transmitted with a cam and follower combination? What are its elements?
7. Define pitch circle, trace point, pitch point, pitch curve and pressure angle.
8. Deduce an expression for the velocity and acceleration of the follower when it moves with SHM.

9. What is the pressure angle with reference to cam and follower? What is its importance?
10. Derive an expression for the displacement, velocity and acceleration of a tangent cam with roller follower, when roller is in contact with flank.
11. What is displacement diagram? How is it helpful to form a cam profile?
12. Define prime circle, pitch circle, base circle of a cam.

10.9.2 Problems

1. *(Circular arc cam)* A symmetric circular arc cam operating a flat faced follower has the following particulars:

   Minimum radius = 30 mm; lift = 20 mm; angle of ascent = 75°; Nose radius = 5 mm, speed = 600 r.p.m.; Find the principal dimensions of the cam, the acceleration of the follower at the beginning of the lift, at the end of contact with circular flank at the beginning of contact with nose and at the apex of the nose.

2. *(SHM–Offset cam with roller follower)* Draw a cam profile to impart the following motion to a reciprocating follower: During each stroke (the upward and downward stroke) the follower is to have simple harmonic motion. The upward stroke is to be performed while the cam makes half of the revolution and the downward stroke while the cam makes one-third of a revolution. There are to be equal periods of rest at each end of the stroke. Stroke of the follower is 8 cm. Line of stroke is 1.5 cm to the left of the camshaft axis. Diameter of the roller is 2.5 cm. Minimum distance between axis of cam and the axis of roller is 5 cm measured vertically. If the cam rotates uniformly at 60 r.p.m., calculate the maximum velocity and acceleration of the follower during each stroke.

3. *(Knife edge–Offset cam with UAR)* A cam with a minimum radius of 50 mm rotating clockwise at a uniform speed, is required to give a knife edge follower the motion as described below: (i) To move outwards through 40 mm during 100° rotation of the cam (ii) To dwell for next 80°. (iii) To return to its starting position during next 90°, and (iv) To dwell for the rest period of a revolution, i.e. 90°.
   Draw the profile of the cam, when the line of stroke of the follower is offset by 15 mm. The displacement of the follower is to take place with uniform acceleration and uniform retardation.

4. *(Roller–Offset with SHM)* A cam turning with uniform angular velocity operates a reciprocating follower through a roller 50 mm diameter. The line of stroke of the follower is 25 mm from the axis of the cam, the stroke of the follower is 50 mm and the minimum radius of the cam is 50 mm. The follower is required to move outwards and inwards with simple harmonic motion, each stroke occupying 75° of cam rotation. During the remainder of the cam rotation the follower is to rest at the bottom of its stroke. Draw the outline of the working surface of the cam.

5. Construct the profile of a cam to suit the following specifications: Camshaft diameter = 40 mm; Least radius of cam = 25 mm; Diameter of roller = 25 mm; Angle of lift = 120°; Angle

of fall = 150°; Lift of the follower = 40 mm; Number of passes are two of equal interval between motions. During the lift and fall, the motion is SHM. The line of stroke passes through the centre of the cam.

6. *Tangent cam* A tangent cam with straight working faces tangential to a base circle of 120 mm diameter has a roller follower of 48 mm diameter. The line of stroke of the roller follower passes through the axis of the cam. The nose circle radius of cam is 12 mm and the angle between tangential faces of cam 90°. If the speed of the cam is 180 r.p.m., determine the acceleration of the follower when the roller just leaves the straight flank during the lift.

7. *(SHM and UAR-Radial cam with roller follower)* Draw the profile of the cam operating a roller reciprocating follower and with the following data: Minimum radius of cam = 25 mm, lift = 30 mm, roller diameter = 15 mm. The cam lifts the follower for 120° with simple harmonic motion followed by a dwell period of 30°. Then the follower lowers down during 150° of the cam rotation with uniform speed of 150 r.p.m. Calculate the maximum velocity and acceleration of the follower during the descent period.

8. *(SHM and UAR, Off–set cam with roller follower)* Draw the profile for the disc cam offset 20 mm to the right of the centre of the camshaft. The base circle diameter is 75 mm and the diameter of the roller is 10 mm. The follower is to move outward a distance of 40 mm with SHM in 140° of cam rotation to dwell for 40° of cam rotation to move inward with 150° of cam rotation with uniform acceleration and retardation. Calculate the maximum velocity and acceleration of the follower during each stroke if the camshaft rotates at 90 r.p.m.

10.9.3 Multiple Choice Questions

1. The cam follower used in automobile engine is
   (a) roller   (b) flat-faced   (c) spherical-faced   (d) knife-edged   [*Ans.* (c)]

2. In a radial cam the follower moves in a direction
   (a) parallel to the cam axis   (b) perpendicular to the cam axis
   (c) along the cam axis.   [*Ans.* (b)]

3. The cam follower used in aircraft engines is a _____ follower.
   (a) roller   (b) flat-faced   (c) spherical-faced   (d) knife-edged   [*Ans.* (a)]

4. The reference point on the follower to lay the cam profile is known as the
   (a) cam centre   (b) pitch point   (c) trace point   (d) prime point   [*Ans.* (c)]

5. The circle drawn to the cam profile with the minimum radius is called
   (a) prime circle   (b) cam circle   (c) pitch circle   (d) base circle   [*Ans.* (d)]

6. The size of the cam depends on
   (a) pitch circle   (b) prime circle   (c) base circle   (d) pitch circle   [*Ans.* (c)]

7. The angle between the axis of the follower and the normal to the pitch curve is known as
   (a) base angle   (b) pressure angle   (c) pitch angle   (d) prime angle   [*Ans.* (b)]

8. The pressure angle of the cam _____ with increase in the base circle diameter.
   (a) decreases   (b) increases
   (c) does not change   (d) may decrease or increase   [*Ans.* (a)]

9. The point on the cam with the maximum pressure angle is known as
   (a) cam centre   (b) pitch point   (c) trace point   (d) prime point   [*Ans.* (a)]

10. The path described by the trace point is known as
    (a) pitch curve   (b) pitch circle   (c) prime circle   (d) prime curve   [*Ans.* (a)]

11. The most suitable follower motion for high-speed engine is
    (a) uniform acceleration and deceleration   (b) uniform velocity
    (c) simple harmonic motion   (d) cycloidal.   [*Ans.* (d)]

Toothed Gearing

11

Toothed Gearing

11.1 INTRODUCTION

In the transmission of rotary motion or power between two shafts using belt or rope causes slip which reduces velocity ratio. In precession machines, a definite velocity ratio is important which is possible only by means of gears or toothed wheels. The toothed gearing is adopted, when the distance between the shafts is small and to provide an invariable velocity ratio between the shafts called positive drive and also to transmit a large torque from the driving shaft to the driven shaft.

Hence, the advantages of gear drive are:

1. Gives exact velocity ratio (positive drive),
2. Transmits large power,
3. Gives high efficiency,
4. Provides reliable service, and
5. System will be most compact.

The disadvantages of gear drive are:

It requires special tools to manufacture and any errors in making them causes vibrations and noise.

11.2 CLASSIFICATION OF TOOTHED GEARING

The classification of toothed gearing used for transmission of motion or power is done in many ways as follows:

11.2.1 According to Axes

The axes of the driving and driven shafts may be:

1. **Parallel:** Spur gearing in Fig. 11.1(a) or helical gearing in Fig. 11.1(b), the axes of the shafts are parallel as shown.

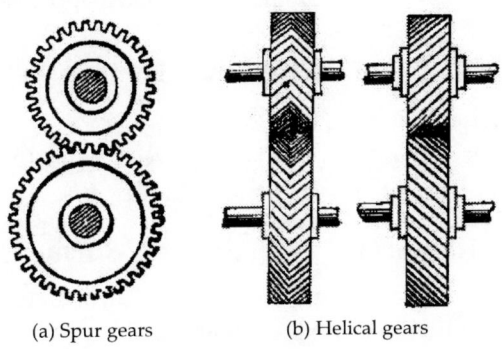

(a) Spur gears (b) Helical gears

Fig. 11.1 Axes–parallel.

2. **Intersecting:** Bevel gearing in which the axes of the shafts intersect as shown in the Fig. 11.2(a).

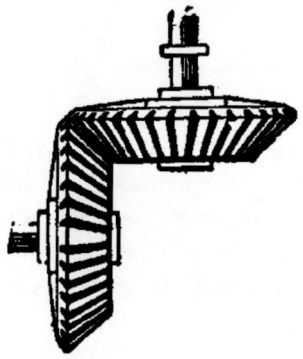

Fig. 11.2(a) Axes—intersecting.

3. **Non-intersecting and non-parallel:** Skew or spiral gearing in which the axes of the shafts are non-parallel and non-intersecting as shown in Fig. 11.2(b).

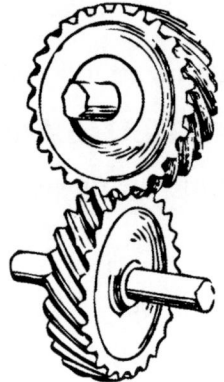

Fig. 11.2(b) Axes—non-intersecting and non-parallel.

11.2.2 According to the Range of Peripheral Velocity

1. **Low velocity:** The peripheral velocity is less than 3 m/sec;
2. **Medium velocity:** The peripheral velocity is in between 3 and 15 m/sec;
3. **High velocity:** The peripheral velocity is greater than 15 m/sec.

11.2.3 According to Position of Teeth on the Gear Surface

1. **Teeth parallel to axis:** In spur gears shown in Fig. 11.1(a) with external teeth and in Pic. 11.1 with internal teeth are parallel to the axis of rotation.

Pic. 11.1

2. **Teeth inclined to axis:** The helical gears shown in Fig. 11.1(b), the teeth are inclined to the axis of rotation. Double helical gearing is also called Herringbone gears.
3. **Curved:** In spiral gears the teeth are curved over the rim surface as shown in Pic. 11.2.

Pic. 11.2

11.2.4 According to Type of Gearing

1. **External gearing:** In external gearing, the gears of the two shafts mesh externally with each other as shown in Fig. 11.1(a). The larger of the two wheels is called wheel and the smaller one pinion. In external gearing, the direction of rotation of the two wheels is always in the opposite directions.

2. **Internal gearing:** In internal gearing the gears of the two shafts mesh internally with each other as shown in Fig. 11.3. The larger of the two wheels is called annular wheel. In internal gearing the motion of the two wheels is always in the same direction.

3. **Rack and pinion:** The gears mesh externally with one gear in a straight line as shown in Pic. 11.3. The straight line gear is called rack. This is used in automobiles for steering purpose.

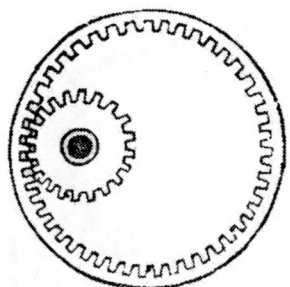

Fig. 11.3

Pic. 11.3

11.2.5 According to Materials Used for Gears

1. **Metals:** Cast iron, steel, bronze.
2. **Non-metals:** Wood, nylon (less noise).

11.3 TERMINOLOGY USED IN GEARS

The gear teeth nomenclature or terminology of gear teeth are shown in Fig. 11.4(a, b, c) and their definitions are given below:

11.3.1 Pitch Circle

The pitch circle is a theoretical circle on which all calculations are usually based. The pitch circles of a pair of mating gears are tangent to each other. It is the diameter of the imaginary circle which by pure rolling action transmits the motion as that of the actual gear. The size of the gear is usually specified by the pitch circle diameter and is called the pitch diameter (D).

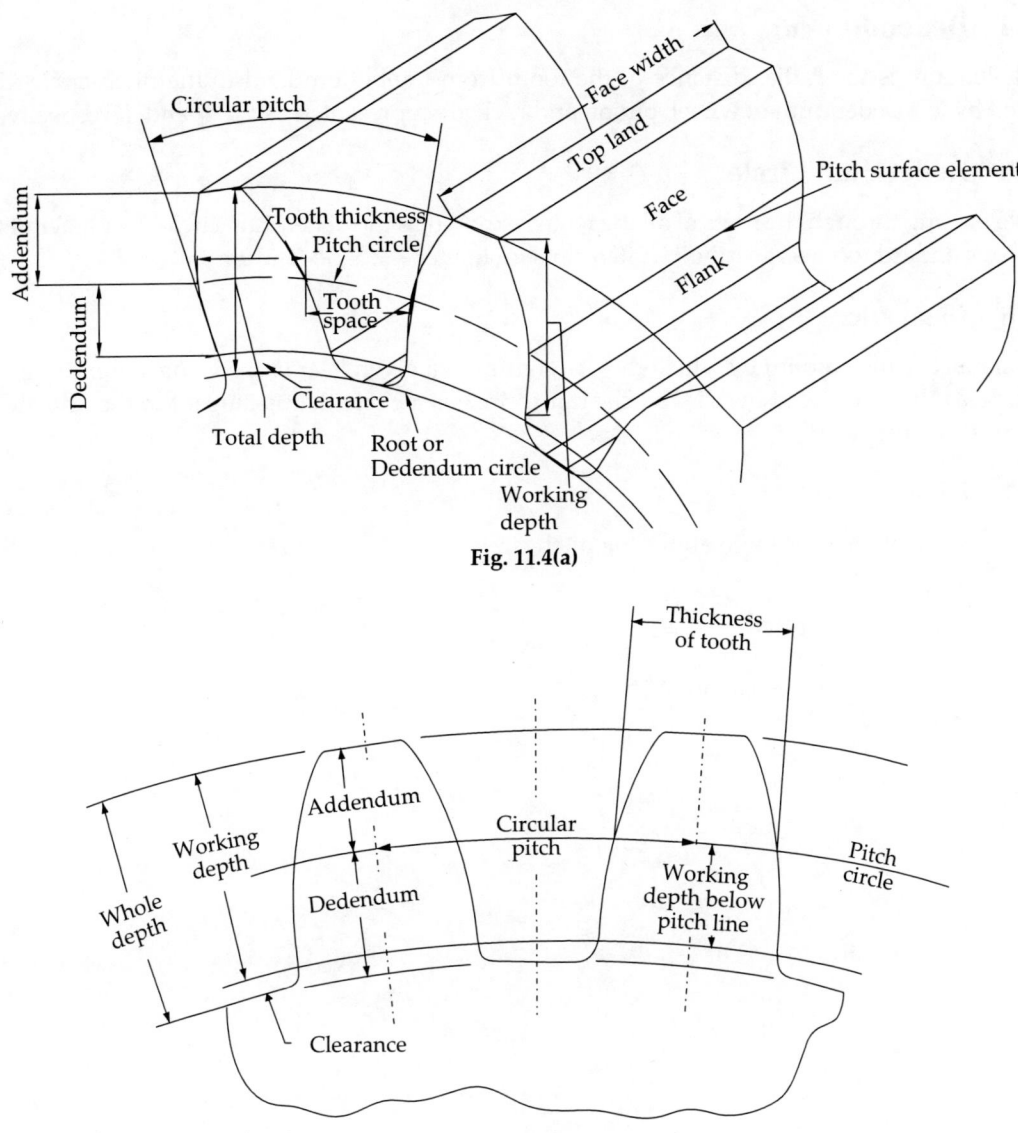

Fig. 11.4(a)

Fig. 11.4(b)

11.3.2 Addendum (*a*)

The addendum is the radial distance of the tooth between the top land and the pitch circle. It is denoted by '*a*'. Addendum of wheel, pinion and rack are represented as a_w, a_p and a_r respectively.

11.3.3 Addendum Circle

A circle passing through the tips of all the teeth is known as the addendum circle. Diameter of this circle for standard toothed gearing = (Pitch circle diameter + 2 × addendum).

11.3.4 Dedendum (*d*)

The dedendum is the radial distance of the tooth from the bottom land to the pitch circle. It is denoted by '*d*'. Dedendum of wheel, pinion and rack are represented as d_w, d_p and d_r respectively.

11.3.5 Dedendum Circle

A circle passing through the roots of all the teeth is known as the dedendum circle. Diameter of this circle for standard toothed gearing = (Pitch circle diameter – 2 × dedendum).

11.3.6 Clearance

The clearance is the amount by which the dedendum in a given gear exceeds the addendum of its mating gear. It is indicated by '*c*'. The clearance circle is a circle that is tangent to the addendum circle of the mating gear.

11.3.7 Face

It is the surface of the gear tooth above the pitch circle.

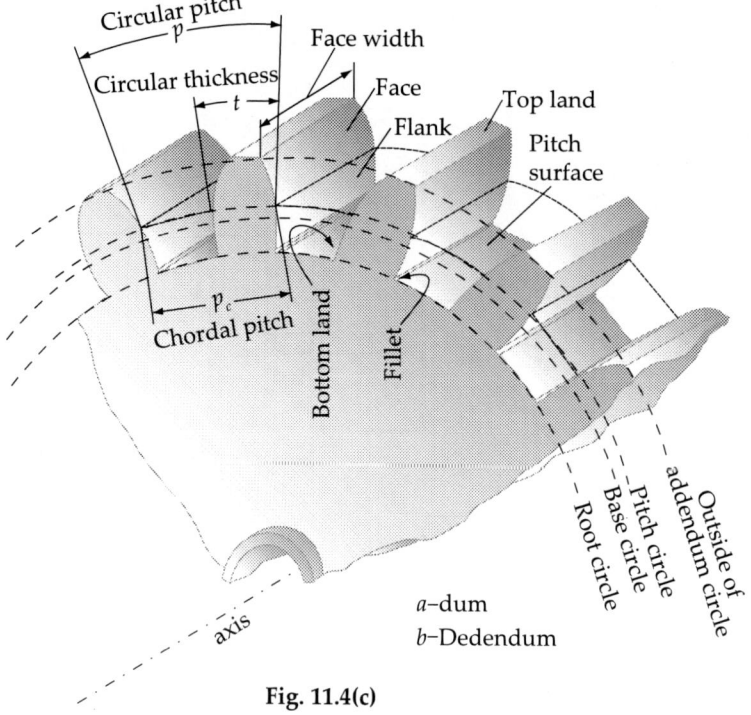

Fig. 11.4(c)

11.3.8 Flank

It is the surface of the gear tooth below the pitch circle.

11.3.9 Face Width

It is the width of the gear tooth measured parallel to the axis.

11.3.10 Top Land

The top land is the top surface of the tooth.

11.3.11 Tooth Profile

The curve forming the face and flank is called the tooth profile of the tooth.

11.3.12 Circular Pitch (P_c)

The circular pitch is the distance measured on the circumference of the pitch circle from a point on one tooth to a corresponding point on the adjacent tooth. It is denoted by P_c.

Circular pitch, $P_c = \dfrac{\pi D}{T}$, where D is the diameter of pitch circle and T is the number of teeth.

Diameter of pitch circle $(D) = \dfrac{P_c T}{\pi}$.

Two or more gears can mesh together correctly if the wheels have the same circular pitch. Then

$$P_c = \pi \frac{D_1}{T_1} = \pi \frac{D_2}{T_2}$$

$$\frac{D_1}{D_2} = \frac{T_1}{T_2}$$

11.3.13 Pitch Point (P)

It is the common point of contact of two pitch circles.

11.3.14 Diametral Pitch (P_d)

The diametral pitch is the number of teeth on the gear per millimetre of pitch diameter or it is the ratio of the number teeth T and the pitch circle diameter D in mm. The units of diametral pitch are the reciprocal of millimetres.

Hence, diametral pitch (P_d)

$$P_d = \frac{T}{D} = \frac{\pi}{P_c}$$

since $P_c = \dfrac{\pi D}{T}$. Hence, $P_d \times P_c = \pi$

11.3.15 Module (m)

The module 'm' is the ratio of the pitch circle diameter and the number of teeth, i.e. $m = \dfrac{D}{T}$. The unit of module is millimetre. The module is the index of tooth size in SI units. Addendum 'a' is equal to module 'm'. Dedendum 'd' is equal to 1.157 m. Clearance is equal to 0.157 m.

11.3.16 Pressure Angle or Obliquity (ψ)

It is the angle between the common normal to two gear teeth at the point of contact and the common tangent at the pitch point. The standard pressure angles are $14\frac{1}{2}°$ and $20°$.

11.3.17 Path of Contact

It is the path traced by the point of contact of two teeth from the beginning to end of engagement.

11.3.18 Length of Path of Contact

It is the length of the common normal cutoff by the addendum circles of the gear wheel and the pinion.

11.3.19 Arc of Contact

It is the path traced by a point on the pitch circle from the beginning to the end of engagement of a given pair of teeth. This consists of two parts:

(a) **Arc of approach:** It is the portion of the path of contact from the beginning of the engagement to the pitch point of a pair of teeth.
(b) **Arc of recess:** It is the portion of the path of contact from the pitch point to the end of the engagement of a pair of teeth.

11.4 CONDITION FOR CONSTANT VELOCITY RATIO OR LAW OF GEARING

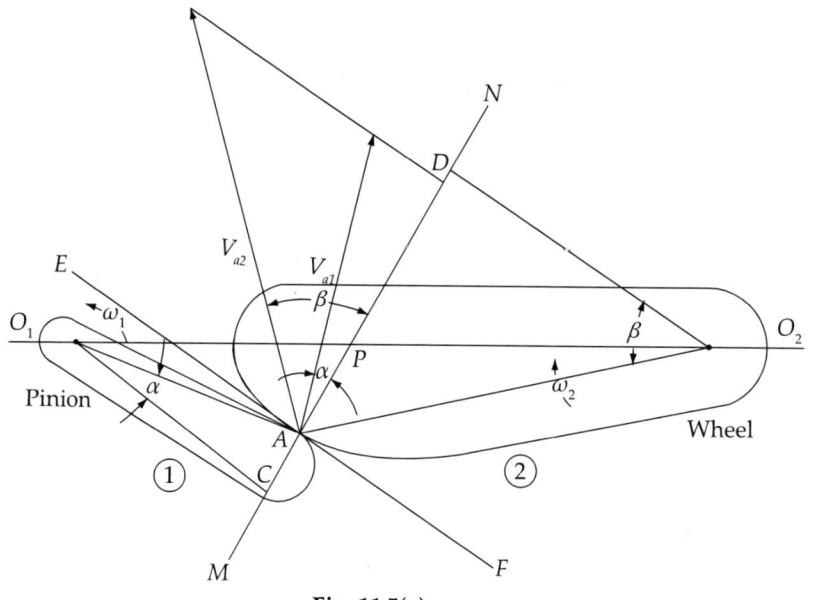

Fig. 11.5(a)

The law of gearing states the condition which must be satisfied by the gear tooth profiles to maintain a constant angular velocity ratio between two gears. Figure 11.5(a) shows pinion represented by 1 and wheel represented by 2. O_1 and O_2 represent the centre of rotation of pinion and wheel. A is the point of contact and MN is the common normal to the teeth profiles at A. The point P is the point of intersection of MN with the line joining the centres O_1 and O_2.

Let v_{a1} and v_{a2} be the velocities of point A regarded as on pinion and wheel equal to $\omega_1 \cdot O_1 A$ and $\omega_2 \cdot O_2 A$, where ω_1 and ω_2 are the angular velocities of pinion and wheel respectively. Their components along MN are $v_{a1} \cos \alpha$ and $v_{a2} \cos \beta$ respectively.

For the two surfaces to remain in contact, these two components must be equal, i.e.

$$v_{a1} \cos \alpha = v_{a2} \cos \beta \quad \text{or} \quad \omega_1 O_1 A \cdot \cos \alpha = \omega_2 O_2 A \cdot \cos \beta$$

But $O_1 A \cdot \cos \alpha = O_1 C$ and $O_2 A \cdot \cos \beta = O_2 D$ from Fig. 11.5(a), therefore

$$\omega_1 O_1 C = \omega_2 O_2 D \quad \text{or} \quad \frac{\omega_1}{\omega_2} = \frac{O_2 D}{O_1 C}$$

From the similar triangles, $O_1 CP$ and $O_2 DP$, the ratio $\dfrac{O_2 D}{O_1 C} = \dfrac{O_2 P}{O_1 P}$ and therefore

$$\frac{\omega_1}{\omega_2} = \frac{O_2 P}{O_1 P}$$

Hence, the ratio of the angular velocities of pinion (1) and wheel (2) is inversely proportional to the distance of point P from the centres O_1 and O_2.

The velocity ratio remains constant provided the common normal (MN), drawn at the point of contact (A) to the surfaces intersects the line of centres always at a fixed point (P) called the pitch point, i.e. the instantaneous centre of pinion and wheel I_{12}. This is the fundamental condition which must be satisfied by the profiles adopted for the teeth of gear wheels.

Law of gearing

The common normal at the point of contact between a pair of teeth must always pass through the pitch point.

If D_1 and D_2 represent the pitch circle diameters of pinion and wheel having teeth T_1 and T_2 respectively, then the velocity ratio

$$\frac{\omega_1}{\omega_2} = \frac{D_2}{D_1} = \frac{T_2}{T_1} = \frac{r_2}{r_1}$$

This equation is frequently used to define the law of gearing, which states that the pitch point must remain fixed on the line of centres. This means that all the lines of action for every instantaneous point of contact must pass through the pitch point. Thus, the shape of the teeth profiles meshing must satisfy the law of gearing. Therefore, the tooth profile selected must be reproduced economically. The teeth having involute profiles satisfy these requirements.

The velocity of sliding is the velocity of one tooth relative to its mating tooth along the common tangent EF at the point of contact A. In other words, it is the relative velocity between the points on the two teeth, say A_1 (i.e., point A on 1) and A_2 (i.e., point A on 2) along the common tangent to the teeth. The components of the velocities v_{a1} and v_{a2} along the line parallel to the common tangent EF at the point of contact A are $v_{a1} \sin \alpha$ and $v_{a2} \sin \alpha$ respectively.

The velocity of sliding (v_S) of point A_2 relative to point A_1 at the point of contact A,

$$v_S = v_{a2} \sin \beta - v_{a1} \sin \alpha$$
$$= \omega_2 O_2 A \sin \beta - \omega_1 O_1 A . \sin \alpha = \omega_2 AD - \omega_1 AC$$

since $O_2 A \sin \beta$ and $O_1 A \sin \alpha$ are equal to AD and AC. But $AD = AP + PD$ and $AC = CP - PA$, hence

$$v_S = \omega_2(AP + PD) - \omega_1(CP - PA)$$
$$= (\omega_1 + \omega_2)AP + \omega_2 PD - \omega_1 CP$$

Since the triangles $O_1 CP$ and $O_2 DP$ are similar, $\dfrac{O_2 P}{O_1 P} = \dfrac{PD}{CP}$.

According to the law of gearing,

$$\frac{\omega_1}{\omega_2} = \frac{O_2 P}{O_1 P} = \frac{PD}{CP} \quad \text{or} \quad \omega_1 CP = \omega_2 PD$$

Therefore, the velocity of sliding $v_S = (\omega_1 + \omega_2)AP$, i.e. the product of the sum of the angular velocities of the pinion and wheel and the distance of the point of contact on the mating teeth from the pitch point.

Figure 11.5(b) shows the line of action of transmission of motions at the point of contact. The most commonly used forms of teeth profiles which satisfy the law of gearing are involute and cycloidal profiles.

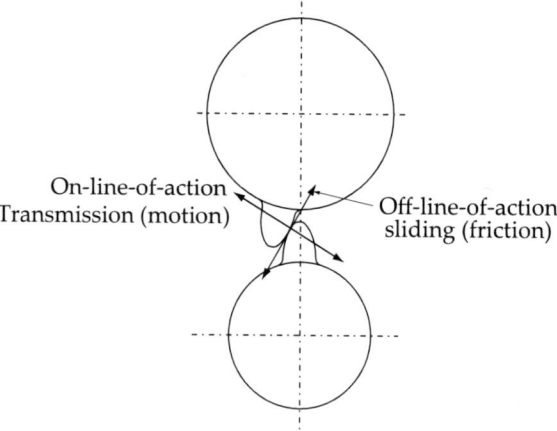

Fig. 11.5(b)

11.5 LENGTH OF THE ARC OF CONTACT

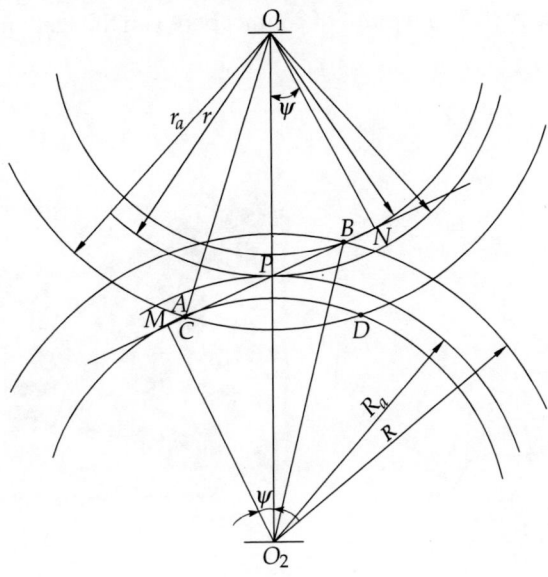

Fig. 11.6

The length of the arc of contact must be at least equal to the circular pitch of the teeth, so that a second pair of teeth will begin to engage before engagement between the preceding pair ends. As shown in Fig. 11.6, the addendum circles cut the common tangent MN at A and B points.

1. The length of path of contact

The distance between A and B is called the length of path of contact. But $AB = AP + PB$.

Since
$$AP = AN - PN$$
$$= \left[\sqrt{(O_1A^2 - O_1N^2)}\right] - O_1P \sin \psi$$
$$= \left[\sqrt{(O_1A^2 - O_1P^2 \cos^2 \psi)}\right] - O_1P \sin \psi$$

Similarly,
$$PB = BM - PM = \left[\sqrt{(O_2B^2 - O_2M^2)}\right] - O_2P \sin \psi$$
$$= \left[\sqrt{(O_2B^2 - O_2P^2 \cos^2 \psi)}\right] - O_2P \sin \psi$$

Substituting for $O_1A = r_a$; $O_1P = r$; $O_2B = R_a$; $O_2P = R$ for pinion and wheel.

The length of path of contact

$$AB = \sqrt{(r_a^2 - r^2 \cos^2 \psi)} + \sqrt{(R_a^2 - R^2 \cos^2 \psi)} - (r + R) \sin \psi \qquad (1)$$

2. The length of arc of contact

Picture 11.4 shows how the path of the point of contact between the teeth moves. Picture 11.5 shows two gear wheels in mesh.

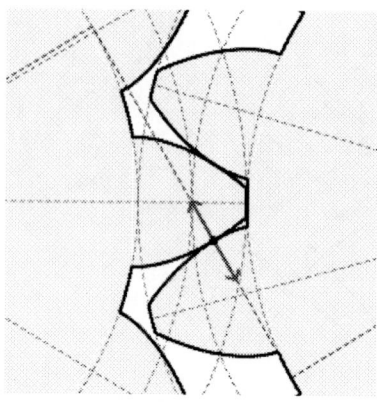

Pic. 11.4

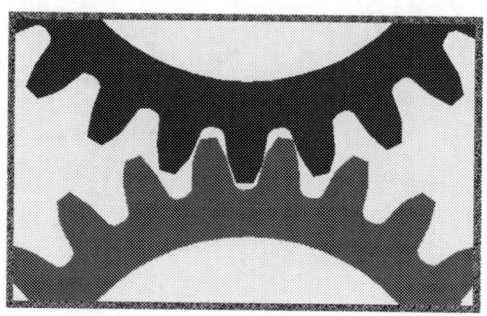

Pic. 11.5

The length of the arc of contact must be at least equal to the circular pitch of the teeth, so that a second pair of teeth will begin to engage before engagement between the preceding pair ends. Hence, the definition of the length of arc of contact is the path traced by a point on the pitch circle from the beginning to the end of a given pair of teeth.

From Fig. 11.6, pitch circle radius is equal to base circle radius divided by $\cos \psi$. Hence, the pitch circle radius of pinion $O_1P = \dfrac{O_1N}{\cos \psi}$. But the arc CD is equal to the length of path of contact AB.

$$\text{Therefore, arc } CD \text{ of contact} = \frac{\text{length of path of contact}}{\cos \psi}$$

$$= \frac{AB}{\cos \psi}$$

$$\text{Hence, the maximum length of arc of contact} = \frac{\text{maximum length of path of contact}}{\cos \psi}$$

$$= \frac{MN}{\cos \psi}$$

$$= \frac{(MP + PN)}{\cos \psi}$$

$$= \frac{(R + r) \sin \psi}{\cos \psi}$$

$$= (R + r) \tan \psi$$

since $MP = R \sin \psi$ and $PN = r \sin \psi$ as shown at Fig. 11.6.

3. Distance between the centres O_1O_2

$$O_1O_2 = O_1P + O_2P$$
$$= \left(\frac{r}{\cos\psi}\right) + \left(\frac{R}{\cos\psi}\right) = \frac{(r+R)}{\cos\psi}$$

since $O_1P = \dfrac{r}{\cos\psi}$ and $O_2P = \dfrac{R}{\cos\psi}$

4. Contact ratio or number of pairs of teeth in contact

This is defined as the ratio of the length of the arc of contact to the circular pitch (P_c). Theoretically the minimum value for the contact ratio is one, i.e. there must be at least one pair of teeth in contact for continuous action.

Note: Larger the contact ratio, more quickly the gears will operate.

W E 11.1: Two involute gears of 20° pressure angle are in mesh. The number of teeth on pinion is 20 and the gear ratio is 2. If the pitch expressed in module is 5 mm and pitch line speed is 1.2 m/sec, assuming addendum as standard and equal to one module.
Find
  (a) The angle turned through by pinion when one pair of teeth is in mesh
  (b) The maximum velocity of sliding
  (c) The number of pairs of teeth in contact
  (d) The angle turned through by the pinion and the gear wheel when one pair of teeth is in contact.
  (e) The ratio of sliding to rolling motion when the tip of a tooth on the larger wheel (i) is just making contact, (ii) is just leaving contact with its mating tooth, and (iii) is at the pitch point.

Given:
$\psi = 20°$;   Teeth on pinion $T_1 = 20$;   Teeth on gear be T_2;   Gear ratio $(G) = \dfrac{T_2}{T_1} = 2$
Module $m = 5$ mm;   Velocity $v = 1.2$ m/sec;   Addendum $a = 1$

(a) Angle turned through by pinion when one pair of teeth is in mesh

Pitch circle radius of pinion
$$r = m \cdot \frac{T_1}{2} = 5 \times \frac{20}{2} = 50 \text{ mm}$$

and pitch circle radius of wheel
$$R = \frac{mT_2}{2} = \frac{mGT_1}{2}$$
$$= 5 \times 2 \times \frac{20}{2} = 100 \text{ mm}$$

Therefore, radius of addendum circle of pinion
$$r_a = r + \text{addendum} = 50 + 5 = 55 \text{ mm}$$

and radius of addendum circle of wheel $R_a = R + \text{addendum} = 100 + 5 = 105$ mm

Length of the path of approach, i.e. the path of contact when engagement occurs

$$AP = \left[\sqrt{(R_a)^2 - R^2 \cos^2 \psi}\right] - R \sin \psi$$
$$= \left[\sqrt{(105)^2 - 100^2 \cos^2 20°}\right] - 100 \sin 20°$$
$$= 46.85 - 34.2 = 12.65 \text{ mm}$$

and length of path of recess (i.e., the path of contact when disengagement occurs).

$$PB = \left[\sqrt{(r_a)^2 - r^2 \cos^2 \psi}\right] - r \sin \psi$$
$$= \left[\sqrt{(55)^2 - 50^2 \cos^2 20°}\right] - 50 \sin 20°$$
$$= 28.0 - 17.1 = 11.5 \text{ mm}$$

Therefore, length of the path of contact,

$$AB = AP + PB$$
$$= 12.65 + 11.5 = 24.15 \text{ mm}$$

$$\text{Length of the arc of contact} = \frac{\text{length of the path of contact}}{\cos \psi}$$
$$= \frac{24.15}{\cos 20°} = 25.7 \text{ mm}$$

$$\text{Angle turned through by pinion} = \frac{(\text{length of arc of contact} \times 360°)}{\text{circumference of pinion}}$$
$$= \frac{25.7 \times 360°}{(2\pi \times 50)} = 29.45°$$

(b) Maximum velocity of sliding

Let ω_1 be the angular velocity of pinion and ω_2 be the angular velocity of wheel. Then the pitch line velocity $v = \omega_1 r = \omega_2 R$.

Therefore,

$$\omega_1 = \frac{v}{r} = \frac{120}{5} = 24 \text{ rad/sec} \quad \text{and} \quad \omega_2 = \frac{v}{R} = \frac{120}{10} = 12 \text{ rad/sec}$$

Therefore, maximum velocity of sliding

$$v_S = (\omega_1 + \omega_2) \cdot AP$$
$$= (24 + 12)12.65 = 455.4 \text{ mm/sec}$$

(c) Number of pairs of teeth in contact

Length of arc of contact/circular pitch.
Length of arc of contact = 25.7 mm

The circular pitch $p_c = \pi \times m = \pi \times 5 = 15.72$ mm.

Therefore, number of pairs of teeth in contact (or contact ratio) $= \dfrac{25.7}{15.72} = 1.64$ say 2.

(d) Angle turned through by the pinion and gear wheel when one pair of teeth is in contact

$$\text{Angle turned through by the pinion} = \frac{\text{length of arc of contact} \times 360°}{\text{circumference of pinion}}$$

$$= \frac{25.7 \times 360°}{(2\pi \times 50)} = 29.44°$$

$$\text{Angle turned through by the gear wheel} = \frac{\text{length of arc of contact} \times 360°}{\text{circumference of gear}}$$

$$= \frac{25.7 \times 360°}{(2\pi \times 100)} = 14.72°$$

(e) Ratio of sliding to rolling motion

Let ω_1 be the angular velocity of pinion and ω_2 be the angular velocity of wheel.

Then
$$\frac{\omega_1}{\omega_2} = \frac{T_2}{T_1}$$

$$\omega_2 = \omega_1 \times \frac{T_1}{T_2} = \omega_1 \times \frac{20}{40} = 0.5\omega_1$$

and rolling velocity
$$v_R = \omega_1 r = \omega_2 R = \omega_1 \times 50$$
$$= 50\omega_1 \text{ mm/sec}$$

(i) At the instant when the tip of a tooth on the larger wheel is just making contact with its mating teeth, i.e. when the engagement commences, the sliding velocity

$$v_S = (\omega_1 + \omega_2)AP = (\omega_1 + 0.5\omega_1)12.65$$
$$= 18.97\omega_1 \text{ mm/sec}$$

Therefore, ratio of sliding velocity to rolling velocity

$$\frac{v_S}{v_R} = \frac{18.97\omega_1}{50\omega_1} = 0.3795$$

(ii) At the instant when the tip of a tooth on the larger wheel is just leaving contact with its mating teeth, i.e. when the engagement terminates, the sliding velocity

$$v_S = (\omega_1 + \omega_2)PB = (\omega_1 + 0.5\omega_1)11.5$$
$$= 17.25\omega_1 \text{ mm/sec}$$

Therefore, ratio of sliding velocity to rolling velocity

$$\frac{v_S}{v_R} = \frac{17.25\omega_1}{50\omega_1} = 0.345$$

(iii) Since at the pitch point, the sliding velocity is zero, therefore, the ratio of sliding velocity to rolling velocity is zero.

11.6 MINIMUM NUMBER OF TEETH ON THE PINION TO AVOID INTERFERENCE

In order to avoid interference, the addendum circles of the two mating gears must cut the common tangent to the base circles between the points of tangency. The limiting condition reaches when the addendum circles of pinion and wheel pass through points N and M respectively as shown in Fig. 11.7. Let T_1 and T_2 represent the teeth on the pinion and wheel, m is the module of the teeth, r is the pitch circle radius of the pinion, G is the gear ratio $= \dfrac{T_2}{T_1} = \dfrac{R}{r}$ and ψ is the pressure angle or angle of obliquity.

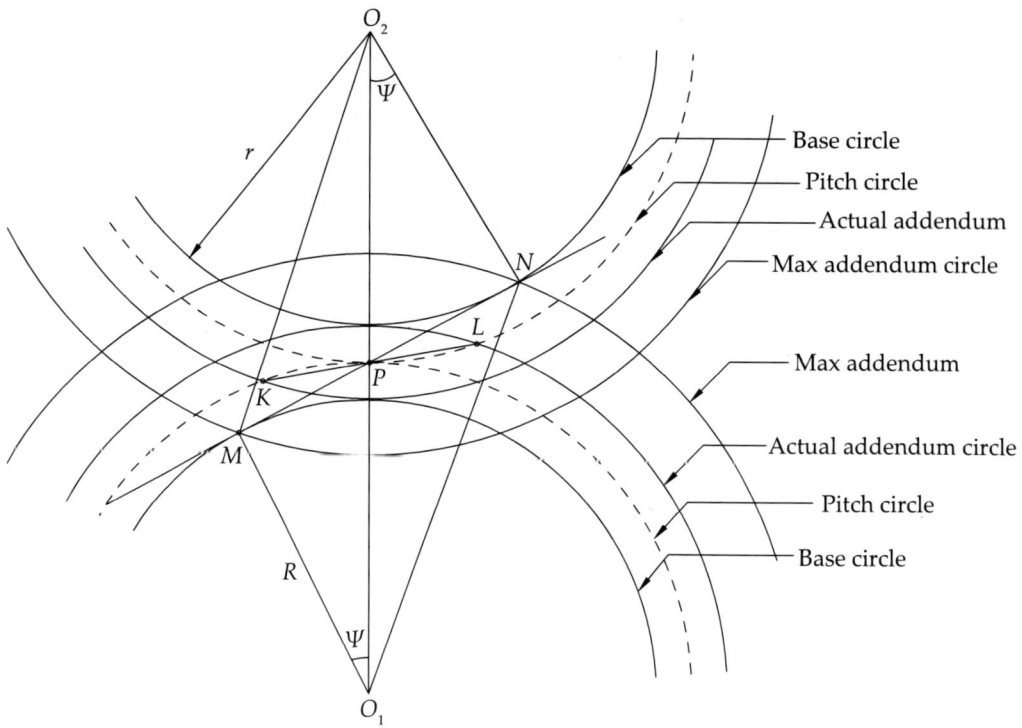

Fig. 11.7

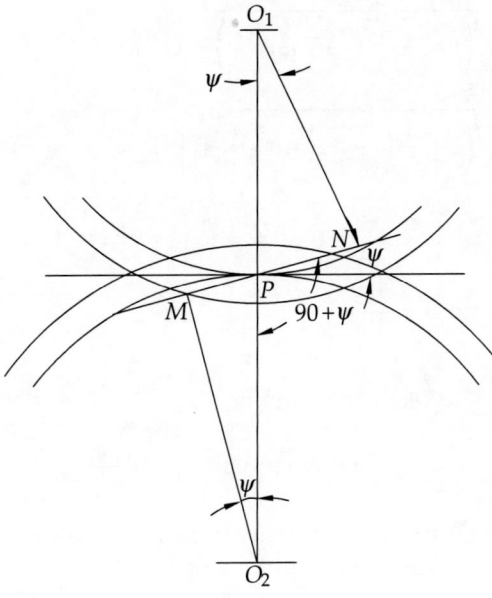

Fig. 11.8(a)

From Fig. 11.8(a), the triangle O_2NP,

$$O_2N^2 = O_2P^2 + PN^2 - 2O_2P \cdot PN \cos O_2PN$$
$$= O_2P^2 + PN^2 - 2O_2P \cdot PN \cos(90° + \psi)$$
$$= O_2P^2 + PN^2 + 2O_2P \cdot PN \sin\psi; \quad \text{But } NP = O_1P \sin\psi;$$
$$O_2N = \sqrt{[O_2P^2 + O_1P^2 \sin^2\psi + 2O_2P \cdot O_1P \sin^2\psi]}$$
$$= O_2P\sqrt{1 + \left(\frac{O_1P}{O_2P}\right)\left[\left(\frac{O_1P}{O_2P}\right) + 2\right]\sin^2\psi}$$

But the addendum of the wheel $= O_2N - O_2P$

$$= O_2P\left\{\sqrt{\left[1 + \left(\frac{O_1P}{O_2P}\right)\left[\left(\frac{O_1P}{O_2P}\right) + 2\right]\sin^2\psi\right]}\right\} - 1 \quad (1)$$

Let the addendum of the gear wheel be $a_w \cdot m$ where a_w is the fraction and m is the module. Then $\dfrac{O_1P}{O_2P} = \dfrac{T_1}{T_2} = \dfrac{1}{G}$, $O_2P = \dfrac{mT_2}{2}$ and $O_1P = \dfrac{mT_1}{2}$. Substituting in (1),

$$a_w m = \frac{T_2 m}{2}\left[\sqrt{\left(1 + \frac{T_1}{T_2}\left(\frac{T_1}{T_2} + 2\right)\sin^2\psi\right)} - 1\right]$$
$$= \frac{T_2 m}{2}\left[\sqrt{\left(1 + \frac{1}{G}\left(\frac{1}{G} + 2\right)\sin^2\psi\right)} - 1\right]$$

$$a_w = \frac{T_2}{2}\left[\sqrt{1 + \frac{T_1}{T_2}\left(\frac{T_1}{T_2} + 2\right)\sin^2\psi} - 1\right]$$

$$= \frac{T_2}{2}\left[\sqrt{1 + \frac{1}{G}\left(\frac{1}{G} + 2\right)\sin^2\psi} - 1\right]$$

$$= \frac{T_2}{2} Z \qquad \text{where } Z = \sqrt{\left[1 + \frac{1}{G}\left(\frac{1}{G} + 2\right)\sin^2\psi\right]} - 1$$

which depends on the gear ratio $G = \dfrac{T_2}{T_1}$

$$\text{Teeth on gear wheel } T_2 = \frac{2 \cdot a_w}{Z}$$

$$\text{Teeth on pinion } T_1 = \frac{2 \cdot a_w}{GZ} \qquad (2)$$

Equation (2) gives the minimum number of teeth required on the pinion in order to avoid interference between the flanks of the pinion and the tips of the gear wheel teeth.

For equal wheels $G = 1$, i.e. pinion = wheel,

Number of teeth on pinion

$$T_1 = \frac{2 a_w}{\left[\sqrt{(1 + 3\sin^2\psi)} - 1\right]}$$

Figure 11.8(b) shows a pinion gearing with a rack. PN is the tangent from the pitch point to the base circle and in order to avoid under-cutting of the flanks of the pinion the addendum of the rack must not exceed NV.

But $NV = PN \sin\psi$ and $PN = O_1P \sin\psi$.

Therefore,

$$NV = O_1P \cdot \sin^2\psi$$

But $NV = a_r \cdot m$ and $O_1P = \dfrac{T_1 \cdot m}{2}$

so that

$$a_r = \frac{T_1 \sin^2\psi}{2}$$

Note:

1. Normal to involute profile is a tangent to base circle.
2. At the point of contact, normal to tooth profiles for involutes become tangents to respective base circles.
3. Locus of the point of contact will be a common tangent.

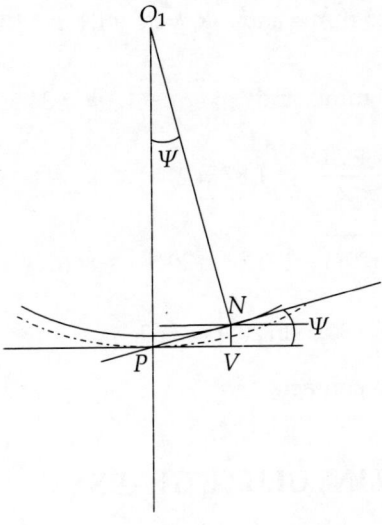

Fig. 11.8(b)

W E 11.2: Determine the minimum number of teeth required on each wheel, in order to avoid interference between two 20° involute gears mesh externally and give a velocity ratio of 3. Module is 3 mm and the addendum is equal to 1.1 module. If the pinion rotates at 120 r.p.m., determine also the number of pairs of teeth in contact.

Given:
$\psi = 20°$; Velocity ratio = 3; Module $m = 3$ mm; Speed of pinion, $N_1 = 120$ r.p.m.; Addendum = 1.1 m; $a_w = 1.1$; Gear ratio $G = 3$.

Let r and T_1 represent the radius and number of teeth on pinion; R and T_2 represent for the wheel.

Minimum number of teeth required on the wheel:

$$T_2 = \frac{2a_w}{Z} = \frac{2 \times 1.1}{\sqrt{\left[1 + \frac{1}{3}\left(\frac{1}{3} + 2\right)\sin^2 20°\right]} - 1} = 68.79$$

Taking the higher whole number divisible by the velocity ratio, i.e.

$$T_2 = 69 \quad \text{and} \quad T_1 = \frac{69}{3} = 23$$

Number of pairs of teeth in contact,

$$n = \frac{\text{Arc of contact}}{\text{Circular pitch}}$$

$$= \frac{\text{Path of contact}}{\cos \psi} \cdot \frac{1}{\pi m} = \frac{X + Y}{\cos \psi \times \pi m}$$

where $X = \left[\sqrt{(R_a^2 - R^2 \cos^2 \psi)} - R \sin \psi\right]$ and $Y = \left[\sqrt{(r_a^2 - r^2 \cos^2 \psi)} - r \sin \psi\right]$.

$$R = \frac{mT_2}{2} = \frac{3 \times 69}{2} = 103.5 \text{ mm} \quad \text{and} \quad R_a = R + 1.1m = 103.5 + 1.1 \times 3 = 106.8 \text{ mm}$$

$$r = \frac{mT_1}{2} = \frac{3 \times 23}{2} = 34.5 \text{ mm} \quad \text{and} \quad r_a = r + 1.1m = 34.5 + 1.1 \times 3 = 37.8 \text{ mm}$$

$$n = \frac{X + Y}{\cos 20° \pi \times 3} = \frac{8.866 + 7.69}{0.94 \pi \times 3} = 1.87 \simeq 2$$

where $X = \left[\sqrt{(106.8^2 - 103.5^2 \cos^2 20°)} - 103.5 \sin 20° \right] = 8.866$

and $Y = \left[\sqrt{(37.8^2 - 34.5^2 \cos^2 20°)} - 34.5 \sin 20° \right] = 7.69$.

Thus, two pairs of teeth will be in contact.

11.7 INTERFERENCE IN INVOLUTE GEARS

The tip of the tooth under-cutting the root on its mating gear is known as interference. MN is the common tangent to base circles. Points M and N are called interference points. To avoid interference, the path of contact should not extend beyond these interference points. Therefore, the limiting radius of the addendum circle of the pinion is O_1N and the gear wheel is O_2M.

Interference may only be prevented, when the addendum circles of the two mating gears cut the common tangent to the base circles within the interference points like A and B in Fig. 11.6.

11.8 METHODS OF AVOIDING INTERFERENCE

There are three ways by which interference or under-cutting of the flanks of pinions with small numbers of teeth may be avoided.

1. The part of the flank of the pinion tooth which lies within the base circle and the part of the face of the gear tooth which engages with it may be made cycloidal instead of involutes in shape.
2. The addenda of the teeth on the wheel and pinion may be modified, the addenda of the wheel being reduced by the amount necessary to avoid interference and of the pinion being correspondingly increased.
3. The centre distance for two mating gears may be made larger than the standard centre distance which has the effect of increasing the pressure angle and so avoiding interference.

11.9 FORMS OF TEETH

Generally, two types of teeth are commonly used. They are

1. Cycloidal
2. Involute.

They are explained below.

11.9.1 Cycloidal Teeth

A cycloid is the curve traced by a point on the circumference of a circle which rolls without slipping on a fixed straight line. When the circle rolls without slipping on the outside of a fixed circle, the curve traced by a point on the circumference of a circle is known as epi-cycloid. On the other hand, if a circle rolls without slipping on the inside of a fixed circle, then the curve traced by a point on the circumference of a circle is called hypo-cycloid.

The generation of cycloidal tooth profiles of a gear is as shown in Fig. 11.9(a) and (b). The circle C is rolled without slipping on the outside of the pitch circle and the point P on the circle C traces epi-cycloid PA representing the face of the cycloidal tooth. The circle D is rolled on the inside of pitch circle and the point P on the circle D traces hypo-cycloid PB representing the flank of the tooth profile. Thus, the profile BPA is one side of the cycloidal tooth. Similarly, the opposite side tooth.

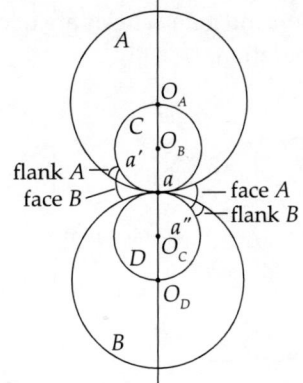

Fig. 11.9(a) Generating cycloidal tooth profiles.

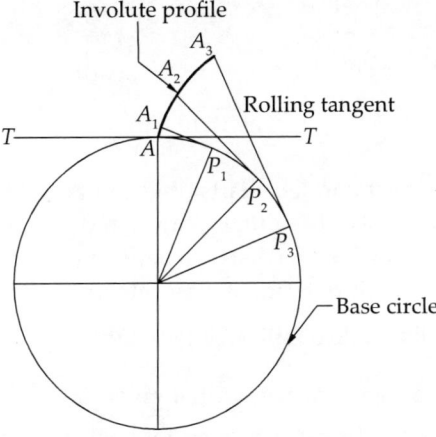

Fig. 11.9(b)

11.9.2 Involute Tooth

The involute of a circle is a plane curve generated by a point on a tangent, which rolls on the circle without slipping or by a point on a taught string which is unwrapped from a reel as shown in Fig. 11.9(b). In both the gears, the circle is called base circle.

The involute is traced as follows. Let A be the starting point of the involute. The base circle is divided into say three parts and named P_1, P_2, P_3 as shown. Then draw the tangents at those points and mark lengths P_1A_1, P_2A_2, P_3A_3 equal to arc lengths AP_1, AP_2, AP_3. Then join the points A, A_1, A_2, A_3 to obtain the involute curve AR. As seen from the figure at any instant the tangent to base circle is the normal to the involute or the normal to involute is the tangent to the base circle.

11.10 HELICAL GEARS

A pair of helical gears is shown in Fig. 11.10. The teeth of helical gears are cut so that each one forms part of a helix. The teeth of pinion and gear wheels are of opposite hand. These gears provide less noise and gives smoothness in operation.

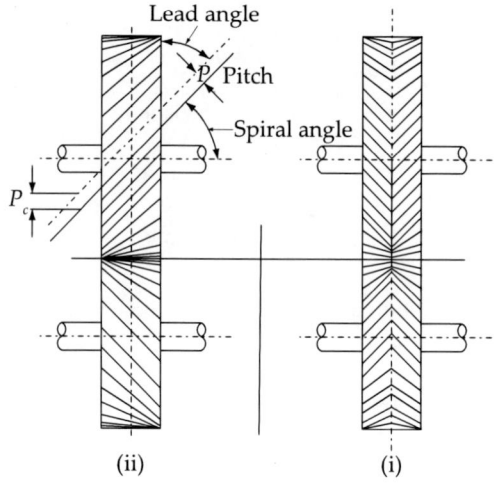

Fig. 11.10

Two gears can be used to connect parallel shafts (line contact). Helices will be right handed on one wheel and left handed on the other. The pitch is measured at right angles to the teeth is termed the normal pitch indicated by P_n which is the distance between similar faces of adjacent teeth along a helix. P_c is the circular pitch which determines the pitch circle diameters of the wheels and is equal to $\dfrac{P_n}{\cos \alpha}$ where α is the helix angle or spiral angle of the teeth. $P_c = \dfrac{P_n}{\cos \alpha}$.

The pitch circle diameter of a helical wheel with a given normal pitch is $\dfrac{1}{\cos \alpha}$ times the pitch circle diameter of a spur wheel with the same number of teeth and pitch.

Since the normal thrust between the teeth is inclined to the axis of rotation, there is an axial thrust on each shaft. To obviate this axial thrust, the teeth are more often cut in the form of a double helix as shown in Fig. 11.10(ii). Thereby equal and opposite thrusts are produced on each wheel and no axial thrust is transmitted to the shafts. The involute shape is invariably used for the profiles and the action of the teeth is essentially the same as for normal spur gears.

Advantages:

1. The chief advantage of these gears is gradual engagement of each pair of teeth, which contributes greatly to the smoothness and quietness of the operation and at the same time increases the load carrying capacity.
2. The contact between the engaging teeth is continuous and there is no sudden application of load.
3. Smooth drive with high efficiency of transmission.

Disadvantages:

Since the normal thrust between the teeth is inclined to the axis of rotation, there is an axial thrust on each shaft. In order to overcome this axial thrust, the teeth are more often cut in the form of a double helix as shown in Fig. 11.10(ii) instead of single helix as shown at Fig. 11.10(i), when equal and opposite thrust is transmitted to the shafts.

11.11 BEVEL GEARS

The bevel gears shown in Fig. 11.11 are used when the axes of the shafts intersect. The motion is transmitted by the pitch surfaces due to pure rolling action. The bevel gears are made from the frustum of the cones, whose apexes coincide with the point of intersection of the axes of the two shafts. The profiles of the teeth are made as in the case of spur gears so as to satisfy the fundamental condition for the transmission of uniform motion. Either cycloidal or involute profiles are used. As with spur gears, the teeth of bevel wheels are generally machined and involute profile is universally employed.

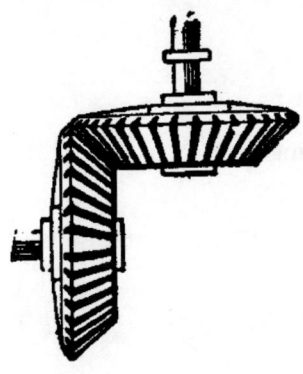

Fig. 11.11

11.12 SPIRAL GEARS

These are used to connect and transmit motion between two non-parallel and non-intersecting shafts. The pitch surfaces are cylindrical and the teeth have point contact. These are used for transmission of small power only. The shortest distance between the two shafts gives the centre distance for a pair of spiral gears.

Figure 11.12 shows two pairs of spiral gears in which the relative positions of the shafts are identical, but for the same direction of rotation of the lower shaft the direction of rotation of the upper shaft is reversed. The shaft angle is less than 90° for the left-hand pair and greater than 90° for the right-hand pair. The shaft angle is equal to the sum of the spiral angles of the teeth of the mating wheels.

Fig. 11.12

The normal pitch of the teeth must be the same for both wheels, but the circular pitches will be different unless the spiral angles for the teeth of both wheels are the same.

11.13 EXERCISE

11.13.1 Short Answer Questions

1. State and prove the law of gearing.
2. Explain the following:
   (a) circular pitch, (b) diametral pitch, (c) module, (d) pressure angle, (e) addendum and (f) dedendum.
3. Prove that velocity of sliding is proportional to the distance of the point of contact from the pitch point.
4. Derive an expression for the following:
   (a) length of the path of contact for two involutes profile gears in mesh, (b) length of arc of contact between two involute profile gears in mesh.

5. Derive an expression for minimum number of teeth in involute rack and pinion arrangements. How are these related to pressure angle?
6. Explain interference and how is this avoided.
7. Derive an expression for the centre distance of a pair of spiral gears.
8. Define the following terms related to gears: (i) diametral pitch (ii) dedendum.
9. What are the advantages and disadvantages of cycloidal teeth profile over involute teeth profile?
10. What are the different methods of eliminating interference in gear wheels?
11. Explain the terms related to gears module and addendum.
12. Define normal pitch and axial pitch relating to helical gears.
13. State and derive law of gearing. What is meant by interference in involute gears? Explain different methods to reduce or eliminate the interference.
14. What is path of contact? Derive the relation for its magnitude.
15. Deduce a relation between centre distances with base and pitch circle radii.
16. Discuss interference and under-cutting in gears.
17. What is arc of contact? Derive the relation for its magnitude.
18. Describe various types of gears used for connecting parallel shafts.
19. What is a higher pair? How are gears classified?
20. State and explain the law of gearing.

11.13.2 Problems

1. A pinion of 25 cm pitch circle diameter is cut with involute teeth of 12 mm module pitch and 25° angle of obliquity. If the addendum height is 12 mm in the pinion and also in 100 mm wheel with which it meshes, find the angles that the pinion turns through while any one pair of teeth continue to maintain contact.

2. A pair of gears has 16 and 22 teeth, module pitch 12.5 mm; addendum 12.5 mm; dedendum 14.25 mm and pressure angle 20°. The pinion drives in counterclockwise direction. Determine the following:
 (i) the pitch circle radii, (ii) the base circle radii, (iii) the circular pitch, (iv) the length of the path of contact, and (v) the angle of approach and recess.

3. The following particulars refer to 20° gears in mesh: Speed of pinion = 400 r.p.m.; Number of teeth on pinion = 36; Number of teeth on gear wheel = 42. Determine the addendum of the gears if the path of approach and recess is half velocity of sliding between the contacting teeth.

4. A pair of involute spur gears with 16° pressure angle and pitch of module 6 mm is in mesh. The number of teeth on pinion is 16 and its rotational speed is 240 r.p.m. When the gear ratio is 1.75, find in order that the interference is just avoided:

(i) the addenda on pinion and gear wheel
(ii) the length of path of contact
(iii) the maximum velocity of sliding to teeth on either side of the pitch point.

5. In a spiral gear drive connecting two shafts, the approximate centre distance is 400 mm and the speed ratio = 3. The angle between the two shafts is 500 and the normal pitch is 18 mm, the spiral angle for the driving and driven wheels are equal. Find
(i) the number of teeth on each wheel
(ii) exact centre distance
(iii) efficiency of the drive, if friction angle is 60°?

6. (a) Derive an expression for the minimum number of teeth required on the pinion in order to avoid interference in involute gear teeth.
(b) Two involute gears of 20° pressure angle are in mesh. The number of teeth on pinion is 20 and the gear ratio is 2. If the pitch expressed in module is 5 mm and the pitch line speed is 1.2 m/s, assuming addendum as standard and equal to one module, find:
(i) the angle turned through by pinion when one pair of teeth is in mesh
(ii) the maximum velocity of sliding.

7. Two 20° involute spur gears have a module of 10 mm, the addendum is one module. The larger gear has 50 teeth and the pinion 13 teeth. Does the interference occur? If it occurs, to what value should the pressure angle be changed to eliminate interference?

8. A pair of 200 full depth involute spur gears having 30 and 50 teeth respectively of module 4 mm is in mesh, the smaller gear rotates at 1000 r.p.m. Determine (i) sliding velocities at engagement and at disengagement of a pair of teeth (ii) the contact ratio. Take addendum = 1 module.

9. The velocity ratio of two spur gears in mesh is 0.4 and the centre distance 75 mm. For a module of 1.2 mm, find the number of teeth of the gears. What will be the pitch line velocity if the speed is 800 r.p.m.? Also find the speed of the gear wheel.

10. Two 20° involute spur gears having a velocity ratio of 2.5 mesh externally. Module is 4 mm and the addendum is equal to 1.23 module. Pinion rotates at 150 r.p.m. Find
(i) The minimum number of teeth on each wheel to avoid interference.
(ii) The number of pairs of teeth in contact.

11. Two gears in mesh have a module of 10 mm and a pressure angle of 250. The pinion has 20 teeth and the gear has 52. The addendum on both the gears is equal to one module. Determine
(i) The number of pairs of teeth in contact.
(ii) The angles of action of the pinion and the wheel.
(iii) The ratio of the sliding velocity to the rolling velocity at the pitch point and at the beginning and end of engagement.

12. An internal spur gear having 200 teeth meshes with a pinion having 40 teeth and a module of 2.5 mm. Determine
    (i) the velocity ratio if the pinion is the driver
    (ii) the centre distance
    (iii) if the centre distance is increased by 3 mm, find the resulting pressure angle.

11.13.3 Multiple Choice Questions

1. Two parallel shafts can be connected by _____
   (a) straight spur   (b) spiral   (c) cross helical   (d) straight bevel   [*Ans.* (a)]

2. Two intersecting shafts can be connected by _____ gears
   (a) straight spur   (b) spiral   (c) cross helical   (d) straight bevel   [*Ans.* (d)]

3. The product of diametral pitch and the module pitch is equal to
   (a) $p_c p_d = 2\pi$,   (b) $p_c p_d = \pi$,   (c) $p_c p_d = \dfrac{\pi}{2}$ (d) $p_c p_d = 1$ [*Ans.* (b)]

4. The product of diametral pitch and the module pitch is equal to
   (a) $p_d \times m = 2\pi$ (b) $p_d \times m = \pi$ (c) $p_d \times m = \dfrac{\pi}{2}$ (d) $p_d \times m = 1$ [*Ans.* (d)]

5. Two skew shafts can be connected by _____ gears
   (a) straight spur   (b) spiral   (c) cross helical   (d) straight bevel   [*Ans.* (c)]

6. The size of gears is usually specified by
   (a) circular pitch   (b) outside diameter
   (c) pitch circle diameter   (d) inside diameter   [*Ans.* (c)]

7. The circular pitch of spur gears is the ratio of the
   (a) number teeth to the pitch diameter
   (b) pitch diameter to the number of teeth
   (c) circumference of the pitch circle to the number of teeth
   (d) circumference of the pitch circle to the diameter of pitch circle   [*Ans.* (c)]

8. The module of spur gears is the ratio of the
   (a) number of teeth to the pitch diameter
   (b) pitch diameter to the number of teeth
   (c) circumference of the pitch circle to the number of teeth
   (d) circumference of the pitch circle to the diameter of pitch circle.   [*Ans.* (b)]

9. The pressure angle of spur gears is kept small
   (a) to reduce axial thrust
   (b) to increase the force for power transmission
   (c) for both a and b
   (d) none of (a) and (b)   [*Ans.* (c)]

10. The contact ratio of gears is always _____

    (a) more than one (b) one (c) less than one (d) zero. [Ans. (a)]

11. In case of involute gear teeth the pressure angle is _____

    (a) same at all points of contact
    (b) maximum at the engagement of teeth
    (c) minimum at the engagement of teeth
    (d) zero at the pitch point [Ans. (a)]

12. The ratio of circular pitch and the module is _____

    (a) π (b) $\dfrac{1}{\pi}$ (c) π^2 (d) $\dfrac{1}{\pi^2}$ [Ans. (a)]

13. The path of contact in involute tooth profile is a _____

    (a) parabola (b) circle (c) straight line (d) curve [Ans. (c)]

14. Interference occurs in case of _____

    (a) cycloid profile teeth (b) involute profile teeth
    (c) in both of them (d) none of them [Ans. (b)]

15. The minimum number of teeth in a rack and pinion for a 20° pair angle teeth is _____

    (a) 20 (b) 18 (c) 22 (d) 24 [Ans. (b)]

16. The normal circular pitch in helical gears is given by _____

    (a) $p \sin \psi$ (b) $\dfrac{p}{\sin \psi}$ (c) $p \cos \psi$ (d) $\dfrac{p}{\cos \psi}$ [Ans. (c)]

17. The maximum efficiency of spiral is given by _____

    (a) $\dfrac{[\cos(\theta - \phi) + 1]}{[\cos(\theta - \phi) - 1]}$ (b) $\dfrac{[\cos(\theta + \phi) - 1]}{[\cos(\theta - \phi) + 1]}$

    (c) $\dfrac{[\cos(\theta + \phi) + 1]}{[\cos(\theta - \phi) - 1]}$ (d) $\dfrac{[\cos(\theta + \phi) + 1]}{[\cos(\theta - \phi) + 1]}$ [Ans. (d)]

18. The maximum efficiency of a worm and worm wheel is given by _____

    (a) $\dfrac{(1 - \sin \phi)}{(1 + \sin \phi)}$ (b) $\dfrac{(1 + \sin \phi)}{(1 - \sin \phi)}$ (c) $\dfrac{(1 - \cos \phi)}{(1 + \sin \phi)}$ (d) $\dfrac{(1 - \sin \phi)}{(1 + \cos \phi)}$ [Ans. (a)]

Gear Trains

12

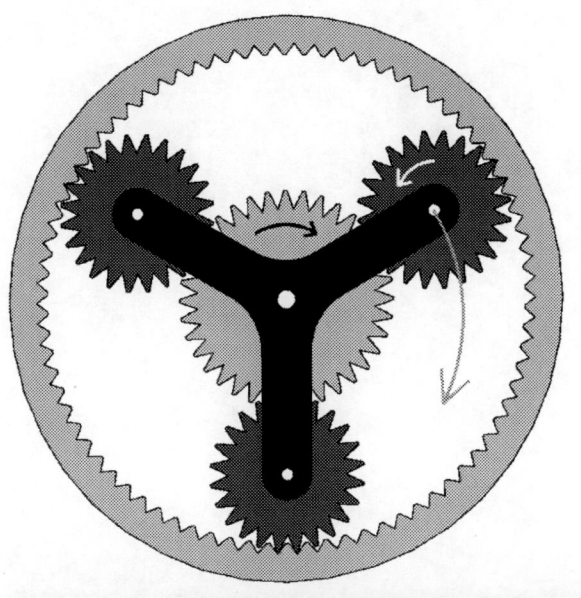

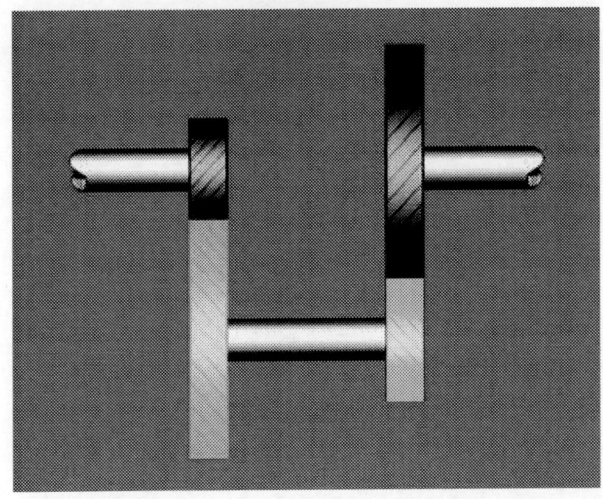

Gear Trains

12.1 INTRODUCTION

Slipping is a common phenomenon in the transmission of power by belt or rope as discussed in the previous chapter. The effect of slipping causes velocity ratio to reduce. It is possible to achieve a required definite velocity ratio especially in precision machines by means of gears. A gear drive has to be adopted when the distance between the shafts of driver and follower is small.

Advantages of gear drive

1. It transmits exact velocity ratio.
2. It has a high efficiency.
3. It has a compact layout.
4. It can transmit a large power.
5. It has a reliable service. The gear drive is called a positive drive as there won't be any slip.

Disadvantages of gear drive

1. Manufacturing of gears requires special equipment.
2. Any error in manufacturing causes vibrations and noise during operation.
3. Any defect in one wheel damages the whole setup.

Gear train

The group of gears used to transmit motion or power from one shaft to another shaft is called a gear train. The gear trains are classified broadly into four types:

1. Simple gear train.
2. Compound gear train.
3. Reverted gear train.
4. Epicyclic gear train.

12.2 SIMPLE GEAR TRAIN OR SIMPLE GEAR DRIVE

A simple gear train or drive consists of one gear wheel on each shaft, i.e. with one gear wheel on each shaft having similar teeth. The motion or power is transmitted between the driving and the driven (or follower) shafts through any number of intermediate wheels. The intermediate wheels act both as a driver and a follower.

The gears are usually represented by their pitch circles as shown in Fig. 12.1(a). Let each gear wheel be represented by either numbers 1, 2, etc. or by alphabets A, B, C etc. Let the angular velocity in radians per second, number of teeth and speed in revolutions per minute (r.p.m.) of any gear wheel be represented by ω, T and N followed by a subscript to indicate the wheel. For example, ω_1, T_1 and N_1 represent the respective values of wheel 1 or ω_A, T_A, and N_A of wheel A.

12.2.1 Speed Value or Speed Ratio or Velocity Ratio (VR)

The ratio of the speed of driving gear wheel to the speed of the driven gear wheel is defined as speed value. When there are only two gear wheels, it is $\dfrac{N_1}{N_2}$ and when there are total 5 gear wheels then it is $\dfrac{N_1}{N_5}$, i.e. $\dfrac{\text{speed of the first wheel}}{\text{speed of the last wheel}}$.

12.2.2 Train Value

The ratio of the speed of the driven gear wheel to the speed of the driving gear wheel is defined as train value. When there are only two gear wheels, it is $\dfrac{N_2}{N_1}$ and when there are total 5 gear wheels then it is $\dfrac{N_5}{N_1}$, i.e. $\dfrac{\text{speed of the last wheel}}{\text{speed of the first wheel}}$.

Then $\dfrac{N_2}{N_1}$ should be equal to $\dfrac{T_1}{T_2}$ as seen in the previous chapter. The direction of rotation of wheels 1 and 2 will be in the opposite directions as shown in the Fig. 12.1(a).

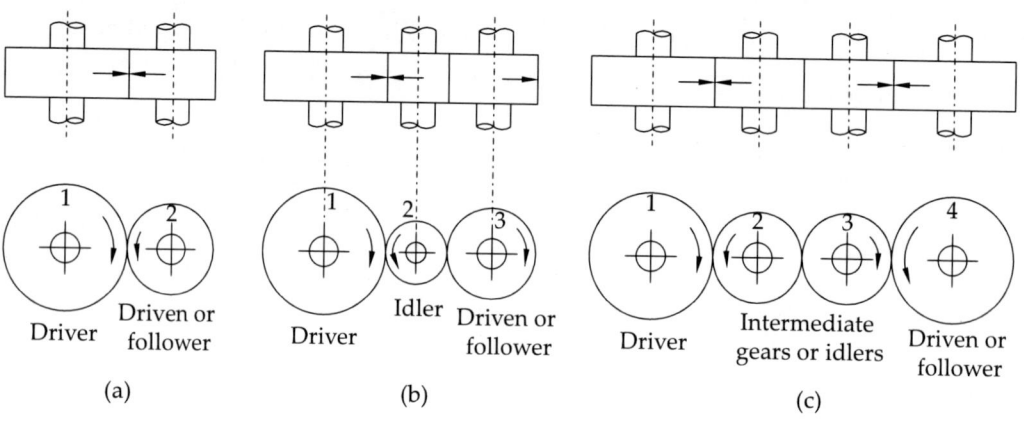

Fig. 12.1 Simple gear trains.

Figure 12.1(b) shows the simple gear train with total three gear wheels 1, 2 and 3 on three shafts. The middle shaft 2 is called an intermediate shaft and the wheel on it is called intermediate wheel or simply an idler. The idlers are also used when the distance between the shafts of driver and the driven is more.

The train value between 1 and 2, i.e. $\dfrac{N_2}{N_1} = \dfrac{T_1}{T_2}$ and between 2 and 3, $\dfrac{N_3}{N_2} = \dfrac{T_2}{T_3}$. Multiplying both gives the train value between shafts 1 and 3, $\dfrac{(N_2 N_3)}{(N_1 N_2)} = \dfrac{(T_1 T_2)}{(T_2 T_3)}$ giving $\dfrac{N_3}{N_1} = \dfrac{T_1}{T_3}$.

Hence, it is clear from this that the intermediate shaft has no influence on the train value. But the direction of rotation of 1 and 3 will be in the same direction as seen in Fig. 12.1(b).

Similarly, in Fig. 12.1(c), the train value between the four wheels is, $\dfrac{N_4}{N_1} = \dfrac{T_1}{T_4}$. But the direction of rotation will be opposite between the driving wheel and the driven wheel.

The number of intermediate gears or idlers used depends on the distance between the driving shaft and the driven shaft and also on the direction of rotation of the driven and the driving shaft, i.e. whether the driven wheel is to rotate in the same direction or in the opposite direction of the driving wheel.

Note:

1. If the total number of wheels is even, i.e. 2, 4, 6, etc., then the driver and the driven rotate in the opposite directions.
2. If the total number of wheels is odd, i.e. 3, 5, 7, etc., then the driver and the driven rotate in the same direction.
3. Any number of intermediate wheels will have no effect on the speed ratio.
4. Usually, the driver is called a wheel and the driven a pinion.

12.2.3 Power Transmitted by a Simple Gear Train

Let F be the tangential force in Newtons, exerted by the driver (also called pressure or force exerted by the teeth) and let v be the peripheral velocity in metres per second of the driving wheel at the pitch point. Then the power transmitted or work done by the driving wheel = Force × Peripheral velocity ($F \times v$) in Newton metre per second (N-m/s).

W E 12.1: *(Simple gear train)* In a simple gear train, the wheel has 120 teeth and the pinion has 40 teeth. What is the speed of pinion if the wheel rotates at 100 r.p.m.?

Given:
Let the driving wheel be called A and the pinion B. T_A represents the teeth on driver A = 120 and T_B represents the teeth on the pinion B = 40;

Let the speed of the wheel A be N_A = 100 r.p.m.;

Then the train value = $\dfrac{N_B}{N_A} = \dfrac{T_A}{T_B}$.

Speed of follower, i.e. Pinion (N_B)

$$N_B = N_A \left(\dfrac{T_A}{T_B}\right) = 100 \times \left(\dfrac{120}{40}\right) = 300 \text{ r.p.m.}$$

The direction of rotation of B will be in the opposite of A as there are only two wheels, i.e. even number.

W E 12.2: *(Simple gear train with idlers)* There are three gear wheels P, Q, R with 120, 40 and 60 teeth respectively. Gear P is the driver and gear R is the follower. The gear Q is an idler connecting the driver and the driven or follower. If the speed of driver P is 150 r.p.m. rotating in the clockwise direction, what is the speed of follower R and its direction of rotation?

Given:

The teeth on P, Q and R are given as $T_P = 120$; $T_Q = 40$; $T_R = 60$. Speed of the driver N_P is given as 150 r.p.m.

Gear P is the driver for gear Q and Q is the driver for gear R. The gear Q is an idler connecting the driver P with the follower or driven gear R.

Train value of P and Q is $\dfrac{N_Q}{N_P} = \dfrac{T_P}{T_Q}$ and the train value of Q and R is $\dfrac{N_R}{N_Q} = \dfrac{T_Q}{T_R}$.

Multiplying both gives

$$\frac{N_Q}{N_P} \times \frac{N_R}{N_Q} = \frac{T_P}{T_Q} \times \frac{T_Q}{T_R}$$

giving train value $\dfrac{N_R}{N_P} = \dfrac{T_P}{T_R}$. Speed of follower, i.e. Pinion (N_R)

$$N_R = \frac{N_P T_P}{T_R} = \frac{150 \times 120}{60} = 300 \text{ r.p.m.}$$

When P rotates clockwise, Q rotates anticlockwise and when Q rotates anticlockwise, R rotates clockwise. Therefore, the direction of R is same as the driver P. Hence, the single idler makes the follower and driver to rotate in the same direction. Since the total number of wheels are three, i.e. odd number, the driver and driven rotate in the same direction.

The train value between P and R without idler is $N_R = N_P \times \dfrac{T_P}{T_R} = 150 \times \dfrac{120}{60} = 300$ r.p.m. But the direction of rotation of R will be opposite of the driver as there are two wheels only, i.e. even number.

Note:

1. The odd number of idlers, i.e. 1, 3, 5, etc., make the follower and driver to rotate in the same directions.
2. The even number of idlers, i.e. 2, 4, 6, etc., make the follower and driver to rotate in the opposite directions.
3. Idlers have no effect on the train value or speed value or speed ratio.

12.3 COMPOUND GEAR TRAIN

The motion transmitted between the driving and the driven (or follower) shafts through an intermediate shaft having more than one gear wheel is called a compound gear train.

Figure 12.2 shows a compound gear train with two compound gears. The gear wheel 1 is the driving wheel, 2-3 and 4-5 are compound gears on two intermediate shafts and 6 is the driven gear. The gear wheel 1 is the driver for gear 2, 3 is the driver for 4 and 5 is the driver for 6.

The train value between 1 and 2 is $\dfrac{N_2}{N_1} = \dfrac{T_1}{T_2}$, between 3 and 4 is $\dfrac{N_4}{N_3} = \dfrac{T_3}{T_4}$ and between 5 and 6 is $\dfrac{N_6}{N_5} = \dfrac{T_5}{T_6}$. But the speeds of wheels 2 and 3 are equal as both are fixed on the same shaft, so $N_2 = N_3$ and similarly $N_4 = N_5$. Therefore, on multiplying all the three gives the speed ratio $\dfrac{N_6}{N_1} = \dfrac{T_1 T_3 T_5}{T_2 T_4 T_6}$.

Hence, the train value is the product of the number of teeth on the drivers divided by the product of the number of teeth on the driven wheels or followers.

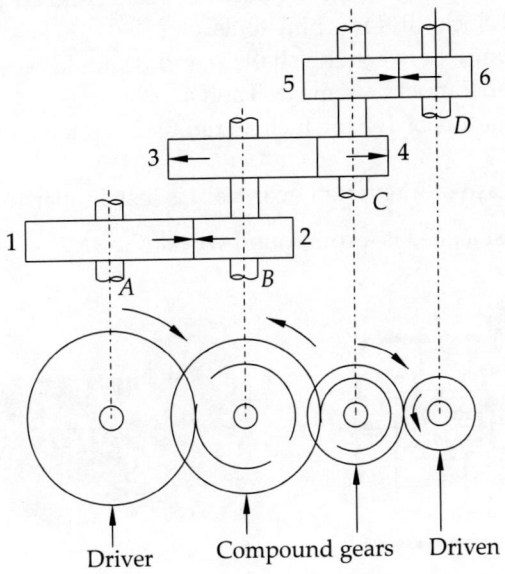

Fig. 12.2 Compound gear train.

12.4 REVERTED GEAR TRAIN

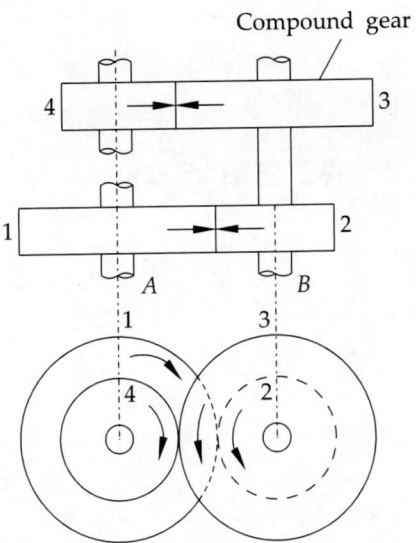

Fig. 12.3 Reverted gear train.

A reverted gear train is one in which the axis of the first wheel, i.e. the driver and the axis of the last gear wheel must be coaxial. That means the gear wheels 1 and 4 must have the common axis of rotation as shown in Fig. 12.3. Such an arrangement of gears is called the reverted gear train or as speed reducers adopted in clocks and machine tools, etc.

From Fig. 12.3, the distance between the shafts A and B should be same. Therefore, the sum of the pitch circle radii of 1 and 2 must be same as 3 and 4.

Hence, the pitch circle radius of 1 + pitch circle radius 2 = pitch circle radius of 3 + pitch circle radius 4.

The speed ratio in the case of a reverted gear train is just similar to compound gear train seen above and it is $\dfrac{N_1}{N_4} = \dfrac{T_2 T_4}{T_1 T_3}$ since 2-3 is a compound wheel $N_2 = N_3$.

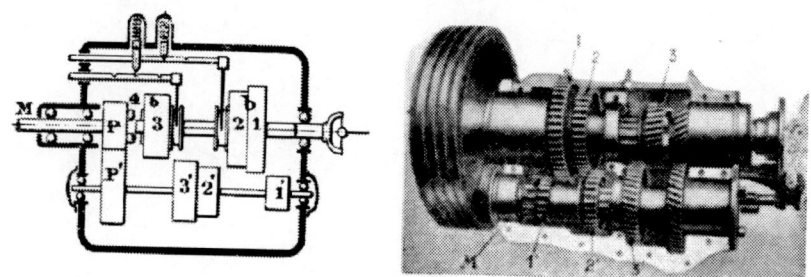

Pic. 12.1 Gearbox.

Picture 12.1 shows the gearbox used in automobiles to change speeds. This is also a reverted gear train. All reverted gear trains need not have a compound gear train. Picture 12.2 shows the three-dimensional view of the simple reverted gear train.

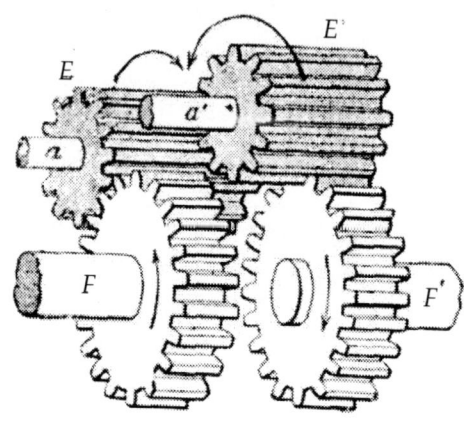

Pic. 12.2

Note: Reverted gear train is also a compound gear train but all compound gear trains are not reverted gear trains.

W E 12.3: *(Reverted gear train)* In a reverted gear train shown in Fig. 12.4, two shafts A and B are in the same straight line and are geared together through an intermediate parallel shaft C. The gears connecting the shafts A and C have a module of 2 mm and those connecting the shafts C and B have a module of 2.5 mm. The speed ratio of reverted gear train should not be less than 15 and the ratio at each reduction must be same. Find suitable number of teeth for all the four gears 1 to 4. The minimum number of teeth of each gear is to be not less than 16. Also find the exact velocity ratio. Take the distance of shaft C from A and B as 200 mm.

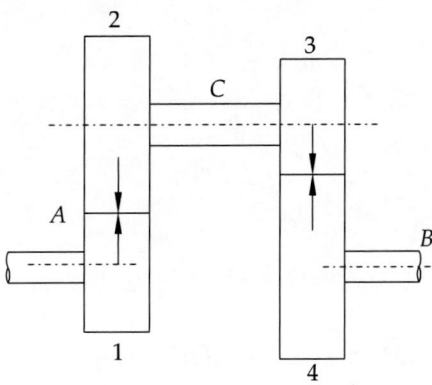

Fig. 12.4

Given:
Module of A and C means modules of wheels 1 and 2, i.e. m_1 and m_2 = 2 mm and module of C and B means modules of wheels 3 and 4, i.e. m_3 and m_4 = 2.5 mm.

The speed ratio $\dfrac{N_1}{N_2} = \dfrac{T_2}{T_1}$ and $\dfrac{N_3}{N_4} = \dfrac{T_4}{T_3}$

$$\dfrac{N_1}{N_2} \times \dfrac{N_3}{N_4} = \dfrac{T_2}{T_1} \times \dfrac{T_4}{T_3}$$

Since 2 and 3 are on the same shaft, their speeds will be equal, i.e. $N_2 = N_3$ hence the speed ratio is

$$\dfrac{N_1}{N_4} = \dfrac{T_2}{T_1} \times \dfrac{T_4}{T_3}$$

Given also that the total speed ratio of the train as 16, i.e. $\dfrac{N_1}{N_4} = 16$. Also mentioned in the problem that the speed ratio between the gears 1 and 2 and between the gears 3 and 4 are to be equal, therefore $\dfrac{N_1}{N_2} = \dfrac{N_3}{N_4}$ must be equal to $\sqrt{16}$, i.e. 4.

Speed ratio of any pair of gears in mesh is equal to the inverse of their number of teeth ratio, therefore

$$\dfrac{T_2}{T_1} = \dfrac{T_4}{T_3} = 4 \qquad (1)$$

The distance between the shafts of 1 and 2 must be same as the distance between the shafts of 3 and 4, i.e. sum of the radii of 1 and 2 must be equal to sum of the radii of 3 and 4.

Let x be the distance which is given as 200 mm. Hence,

$$x = r_1 + r_2 = r_3 + r_4 = 200 \text{ mm} \tag{2}$$

Module (m) of a gear wheel means a ratio of its pitch circle diameter and the number of teeth on that wheel $\dfrac{D}{T}$

i.e.
$$m = \frac{D}{T} \Rightarrow D = m \cdot T$$

Therefore, radius of pitch circle,

$$r = \frac{D}{2} = \frac{m.T}{2}$$

Rewriting equation (2) gives

$$\frac{m_1 T_1}{2} + \frac{m_2 T_2}{2} = \frac{m_3 T_3}{2} + \frac{m_4 T_4}{2} = 200 \tag{3}$$

Since $m_1 = m_2 = 2$ and $m_3 = m_4 = 2.5$, the equation (3) becomes

$$2(T_1 + T_2) = 2.5(T_3 + T_4) = 400$$

Hence,
$$T_1 + T_2 = \frac{400}{2} = 200 \tag{4}$$

$$T_3 + T_4 = \frac{400}{2.5} = 160 \tag{5}$$

From equation (1), $T_2 = 4T_1$. Substituting this value of T_2 in (4) gives $5T_1 = 200$ or $T_1 = 40$ and $T_2 = 200 - 40 = 160$.

Again from equation (1) $T_4 = 4T_3$. Substituting this value of T_4 in (5) gives $5T_3 = 160$ or $T_3 = 32$ and $T_4 = 160 - 32 = 128$.

Speed ratio of the reverted gear train with the calculated values of number of teeth on each gear

$$\frac{N_1}{N_4} = \frac{T_2}{T_1} \times \frac{T_4}{T_3} = \frac{160}{40} \times \frac{128}{32} = 4 \times 4 = 16$$

Hence, the condition is satisfied.

12.5 EPICYCLIC GEAR TRAIN

In all the gear trains discussed above, the gears rotate about the fixed axes only. But in the epicyclic gear train, the axes of some of the gears also rotate as shown in Fig. 12.5. The gear wheel B rotates about the axis O_2, which again rotates about the fixed axis O_1 of wheel A. Axes of the gears A and B are connected by an arm C.

Gear Trains 363

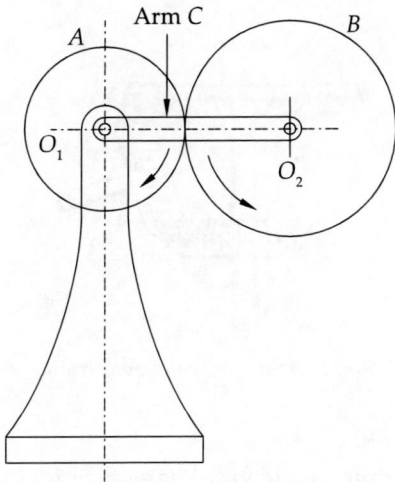

Fig. 12.5

In the epicyclic gear train showed in Fig. 12.5, the wheel A is fixed to the frame, while the arm C revolves carrying along with it the wheel B. This is a simple example of an epicyclic gear train. Such an arrangement is also possible with the compound wheels. Epicyclic gear trains are also called Sun and Planetary gear train as shown below.

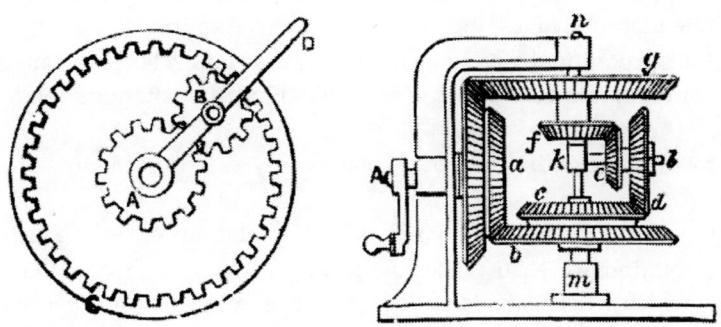

Pic. 12.3 Epicyclic gear trains.

Picture 12.3 shows the epicyclic gear trains with spur gears and bevel gears. The main advantage of the epicyclic gear trains is to have high velocity ratios. Picture 12.4 shows the simple bevel gear train.

These trains are used especially in the back gears of lathes, watches, automobiles, hoists and pulley blocks, etc.

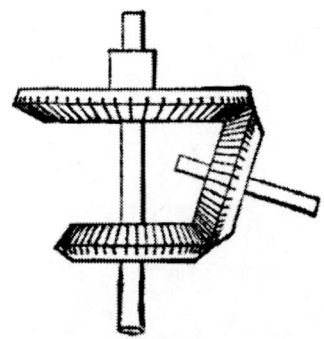

Pic. 12.4 Simple bevel gear train.

Velocity ratio of epicyclic gear train

The velocity ratio cannot be determined like in the case of simple and compound gear trains. The velocity ratio is to be determined in a different way. The most convenient and easy method is by using the tabulation or tabular method. Different types of problems are explained below using this method.

Tabular method:

1. Follow some convention to indicate direction of rotation, i.e. for clockwise rotation (CW) as positive sign (+) and for anticlockwise or counterclockwise (ACW or CCW) as negative sign (−).
2. Convert the epicyclic train into simple or compound gear train by fixing the arm and making the wheels rotate their own axes, thus making the axis rotating stationary.
3. Give one clockwise rotation to any one gear in a simple epicyclic gear train or to a compound gear wheel in a compound epicyclic gear train. Find the rotations made by all the other wheels.
4. Multiply the rotations of each wheel by a constant say x. In other words, make each wheel to rotate by x times.
5. Add some rotations say y rotations to each wheel of the train including arm.
6. Find the total rotations of each wheel by adding each column and then apply the given conditions, i.e. which wheel is fixed and which wheel rotates or number of revolutions given along with their directions to determine the constants x and y first.
7. Knowing the x and y values, determine the rotations of the required gear wheels along with the directions.

W E 12.4: *(Simple epicyclic gear train with two wheels)* In an epicyclic gear train showed in Fig. 12.6, the arm C carries two gears A and B having 40 and 60 teeth respectively. If the arm rotates at 150 r.p.m. in the anticlockwise direction about the centre of the gear A which is fixed, determine the speed of gear B. If the gear A instead of being fixed makes 300 r.p.m. in the clockwise direction, what will be the speed of gear B?

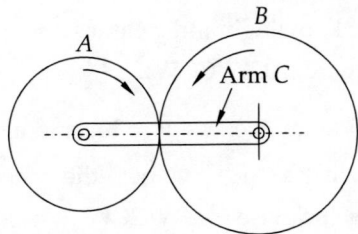

Fig. 12.6 Simple epicyclic gear train.

Given:
The speed of arm C, $N_C = -150$ r.p.m. as given anticlockwise (ACC or CCW), i.e. $(-ve)$

1. Finding N_B, when wheel A is fixed, i.e. $N_A = 0$
2. Finding N_B, when speed of wheel A is 300 r.p.m. clockwise (CW), i.e. $N_A = +300$.

The various steps are as follows:

1. First draw the figure and indicate the names given.
2. To start with assume the arm C as fixed, i.e. its speed or rotation is zero. (N_{arm} or $N_C = 0$).
3. Give initially one clockwise rotation to one of the two wheels, i.e. to either A or B.

Note: Any wheel can be chosen in simple epicyclic gear trains. Remember the result will be same irrespective of the wheel chosen as shown below:

(i) Wheel A is given one clockwise rotation

Given:
$T_A = 40$; $T_B = 60$; $NC = -150$ r.p.m.

Step	Condition of Motion	Revolutions of Wheels (N)		
		Arm C (N_C)	Gear A (N_A)	Gear B (N_B)
1.	Fix arm C and rotate gear A by + 1 revolution (CW)	0	+1	$-(T_A/T_B)$
2.	Multiply with +x revolutions throughout	0	+x	$-x(T_A/T_B)$
3.	Add +y revolutions to all	+y	+y	+y
4.	Total of each element	+y	x + y	$y - x(T_A/T_B)$

Substituting in the total, the given number of teeth, revolutions of respective elements and the direction of rotation and then solve for x and y values as follows:

1. Finding N_B taking the wheel A as fixed, i.e. $N_A = 0$. Therefore, equate total of element of A,

$$x + y = 0 \quad \text{or} \quad x = -y$$

and similarly the total of wheel C given, i.e. $N_C = y = -150$.

Hence, $x = +150$ and therefore knowing x and y values, the speed of $N_B = -150 - \left(40 \times \dfrac{150}{60}\right) = -250$, i.e. 250 r.p.m. anticlockwise (ACW/CCW).

2. Finding N_B taking $N_A = +300$ r.p.m., i.e. $x + y = 300$ and also given $N_C = y = -150$. Hence, $x = +450$ and therefore knowing x and y values, the speed of $N_B = -150 - \left(40 \times \dfrac{450}{60}\right) = -150 - 300$, i.e. 450 r.p.m. anticlockwise (ACW/CCW).

(ii) Wheel B is given one clockwise rotation

Step	Condition of Motion	Arm C (N_C)	Gear A (N_A)	Gear B (N_B)
1.	Fix arm C and rotate gear B by +1 revolution (CW)	0	$-(T_B/T_A)$	+1
2.	Multiply with $+x$ revolutions throughout	0	$-x(T_B/T_A)$	$+x$
3.	Add $+y$ revolutions to all	$+y$	$+y$	$+y$
4.	Total of each element	$+y$	$y - x(T_B/T_A)$	$x + y$

Substituting in the total, the given number of teeth, revolutions of respective elements and the direction of rotation and then solve for x and y values as follows:

1. Finding N_B taking wheel A as fixed, i.e. $N_A = 0$.

   Therefore, total of element of A,
   $$y - x\left(\dfrac{T_B}{T_A}\right) = 0 \quad \text{or} \quad y = x\left(\dfrac{T_B}{T_A}\right)$$

   and the total of element C is given as -150, i.e. $N_C = y = -150$. Hence, $y = -150 = x\left(\dfrac{60}{40}\right)$ and therefore $x = -150 \times 40/60 = -100$.

   Hence, $N_B = x + y = -100 - 150 = -250$, i.e. 250 r.p.m. anticlockwise (ACW/CCW).

2. Finding N_B taking $N_A = +300$ r.p.m i.e. $y - x\left(\dfrac{T_B}{T_A}\right) = 300$ and also given $N_C = y = -150$

   Hence, $x = -\dfrac{(300 + 150)\,40}{60}$ and therefore knowing x and y values, s the speed of $N_B = x + y = -300 - 150 = -450$, i.e. 450 r.p.m. anticlockwise (ACW/CCW).

   Hence, any wheel can be taken to start with for giving one rotation.

Try: Figure 12.6 shows a simple epicyclic gear train. The wheel A is stationary. What is the speed of C, when B rotates about its own axis at 100 r.p.m.? The number of teeth on A and B are 45 and 120 respectively.

W E 12.5: *(Simple epicyclic gear train with three wheels)* In an epicyclic gear train, as shown in Fig. 12.7(a), the number of teeth on wheels A, B, and C are 48, 24 and 50 respectively. If the arm rotates at 400 r.p.m., clockwise, find: (1) Speed of wheel C when A is fixed and (2) Speed of wheel A when C is fixed.

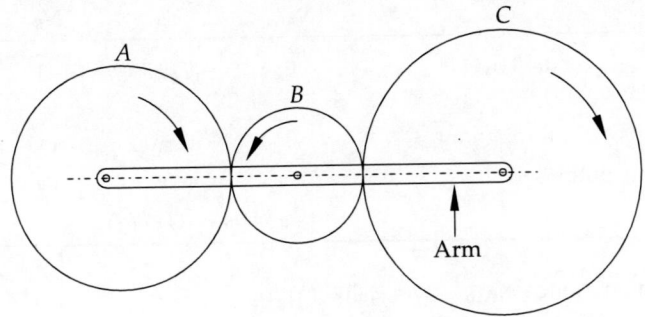

Fig. 12.7(a)

Given:
$T_A = 48$; $T_B = 24$; $T_C = 50$ $N_{Arm} = +400$ r.p.m. given as clockwise (CW)

(i) Here wheel A is considered to give one clockwise rotation

Step	Condition of Motion	Revolutions of Elements N			
		N_{Arm}	N_A	N_B	N_C
1.	Fix arm and rotate A by +1 revolutions (CW)	0	+1	$-(T_A/T_B)$	(T_A/T_C)
2.	Multiply with x throughout	0	+x	$-x(T_A/T_B)$	$x(T_A/T_C)$
3.	Add y revolutions to all	+y	+y	+y	+y
4.	Total	+y	x + y	$y - x(T_A/T_B)$	$y + x(T_A/T_C)$

The speed of arm is given, hence $N_{Arm} = y = +400$ r.p.m.

I. Speed of C when A is fixed, i.e. $N_A = x + y = 0$.

Therefore, $x = -y = -400$; therefore $N_C = 400 - 400\left(\dfrac{48}{50}\right) = 16$ r.p.m. (CW)

II. Speed of A when C is fixed, i.e. $N_C = 0$.

Therefore, $y + x\dfrac{48}{50} = 0$. Since $y = +400$, then

$$x = -y\dfrac{50}{48} = -400\dfrac{50}{48} = -\dfrac{5000}{12}$$

Therefore, speed of A, $N_A = x + y = -\dfrac{5000}{12} + 400 = -16.67$ r.p.m. (ACW or CCW)

(ii) Now the middle wheel B is considered to give one clockwise rotation

Step	Condition of Motion	Revolutions of Elements N			
		N_{Arm}	N_A	N_B	N_C
1.	Fix arm and rotate B by +1 revolutions (CW)	0	$-(T_B/T_A)$	+1	$-(T_B/T_C)$
2.	Multiply with x throughout	0	$-x(T_B/T_A)$	$+x$	$-x(T_B/T_C)$
3.	Add y revolutions to all	$+y$	$+y$	$+y$	$+y$
4.	Total	$+y$	$y - x(T_B/T_A)$	$y + x$	$y - x(T_B/T_C)$

The speed of arm is given, hence $N_{Arm} = y = +400$ r.p.m.

I. Speed of C when A is fixed, i.e. $N_A = y - x\dfrac{T_B}{T_A} = 0$

Therefore, $400 - x\dfrac{24}{48} = 0; x = 800.$

$$N_C = 400 - 800\dfrac{24}{50} = 400 - 384 = 16 \text{ r.p.m. (CW)}$$

II. Speed of A when C is fixed, i.e. $N_C = 0$

Therefore, $y - x\dfrac{24}{50} = 0$ or $y = x\dfrac{12}{25}$. Since $y = +400$, then $x = y\dfrac{25}{12} = 400\dfrac{25}{12} = \dfrac{2500}{3}$.

Therefore, speed of A,

$$N_A = y - x\dfrac{T_B}{T_A} = 400 - \dfrac{2500}{3}\dfrac{24}{48}$$
$$= 400 - 416.67 = -16.67 \text{ r.p.m. (ACW or CCW).}$$

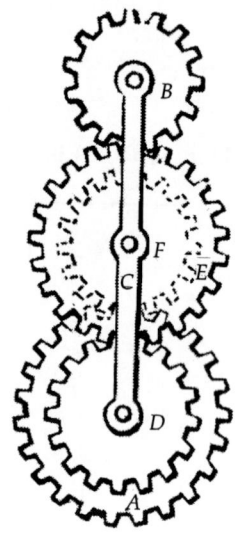

Fig. 12.7(b)

Similarly, with wheel C can be tried.

Note: In fact, the selection of middle wheel is more convenient.

Try: Figure 12.7 (b) shows an epicyclic gear train with gears $E - F$ and $A - D$ are coaxial connected to simple gear B by an arm C. Assume the teeth on B as 24, teeth on $E - F$ as 50-48 and teeth on $A - D$ also as 50-48. Note E & D and A & F are of same size. If the arm C rotates at 400 r.p.m. clockwise, find: (1) Speed of wheel D when B is fixed and (2) Speed of wheel B when D is fixed.

12.6 TORQUE IN EPICYCLIC GEAR TRAINS

Power (P) = Torque (T) × Angular velocity (ω). There will be total three different torques acting in any gear train to be in equilibrium. They are (1) Input torque T_I, (2) Output torque T_O, and (3) Holding or braking torque, T_H.

Net torque applied in a gear train is zero

$$\text{Net torque} = T_I + T_O + T_H = 0 \tag{1}$$

i.e. $F_I.r_I + F_O.r_O + F_H.r_H = 0$ where F is the force and r is the radius.

Total power or kinetic energy in a gear train

$$\text{Kinetic energy} = T_I.\omega_I + T_O.\omega_O + T_H.\omega_H \tag{2}$$

But $\omega_H = 0$ because it is fixed.

Therefore,

$$T_I.\omega_I = -T_O.\omega_O \Rightarrow T_O = -T_I \frac{\omega_I}{\omega_O}$$

and from (1),

$$T_H = -(T_I + T_O) = -\left(T_I - \frac{T_I \omega_I}{\omega_O}\right)$$

Holding torque

$$T_H = T_I \left[\frac{\omega_I}{\omega_O} - 1\right] \quad \text{or} \quad T_I \left[\frac{N_I}{N_O} - 1\right] \tag{3}$$

W E 12.6: *(Simple reverse epicyclic gear train with one internal gear)* An epicyclic gear consists of three gears A, B and C as shown in Fig. 12.8.

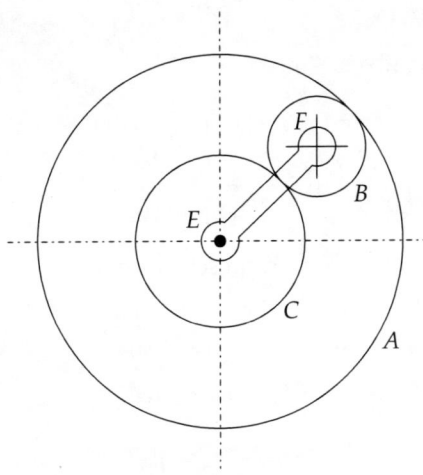

Fig. 12.8 Simple reverse epicyclic gear train with one internal gear.

Theory of Machines

The gear A has 120 internal teeth and gear C has 80 external teeth. The gear B meshes with both A and C and is carried on an arm EF which rotates about the centre of A at 200 r.p.m. If the arm EF receives 7.5 kW and A is driven in the same direction at 100 r.p.m., determine the speed of gear C and torque on its shaft. This is a simple reverse epicyclic gear train as the axes of wheels C and A are co-axial.

Given:
$T_A = 120$; $T_C = 80$; Arm EF speed, i.e. $N_{EF} = 200$ r.p.m.; Determine the speed of C and torque on it.

Diameter of A = (Diameter of C + Twice the diameter of B), similarly the number of teeth on A = (Number of teeth on C + Twice the number of teeth on B), i.e. $T_A = T_C + 2.T_B$

Therefore, teeth on B,
$$T_B = \frac{(T_A - T_C)}{2} = \frac{(120 - 80)}{2} = 20$$

Middle wheel B is chosen to give one clockwise rotation as it will be easy.

Step	Condition of Motion	Revolutions of Elements N			
		N_{EF}	N_B	N_C	N_A
1.	Fix arm EF and rotate B by +1 revolution (CW)	0	+1	$-(T_B/T_C)$	(T_B/T_A)
2.	Multiply with +x throughout	0	+x	$-x(T_B/T_C)$	$x(T_B/T_A)$
3.	Add +y revolutions to all	+y	+y	+y	+y
4.	Total	+y	x + y	$y - x(T_B/T_C)$	$y + x(T_B/T_A)$

Speed of arm EF, $N_{EF} = 200 = +y$ (CW) and speed of gear A is given as 100,
$$N_A = 100 = y + x\frac{T_B}{T_A}$$

On substituting for y, T_B and T_A, gives
$$x = (100 - 200)\frac{120}{20} = -600$$

Speed of $C = y - x\frac{T_B}{T_C} = 200 - \left(-600 \times \frac{20}{80}\right) = 350$ r.p.m. (CW);

Let the torque be represented by Tr and Tr_A represent torque on A, etc.
$$\frac{Tr_A}{Tr_C} = \frac{N_C}{N_A} = \frac{350}{100} = 3.5$$

$$Tr_A = 3.5 \times Tr_C \quad (1)$$

Reaction torque on arm EF,
$$Tr_{EF} = Tr_A - Tr_C \quad (2)$$

Power receiving on arm $EF = 7.5$ kW

$$= 2 \times \pi \times N_{EF} \times \frac{Tr_{EF}}{60} \text{ (or)}$$

$$Tr_{EF} = \frac{60 \times 7.5 \times 1000}{(2 \times \pi \times 200)}$$

$$= 358 N-m = Tr_A - Tr_C \qquad (3)$$

Solving (1) and (3) gives torque on C, $Tr_C = 143.2$ N-m

12.7 COMPOUND EPICYCLIC GEAR TRAIN

Epicyclic gear trains using compound gears are called compound epicyclic gear train. If the driver and the driven are coaxial, then that compound epicyclic gear train is called a reversed epicyclic gear train. The working principle of this can be seen from the following worked examples:

Note: Choose the compound wheel to give one revolution instead of simple gear.

W E 12.7: *(Reversed compound epicyclic gear train)* In a reversed epicyclic gear train shown in Fig. 12.9, the arm A carries two gears B and C and a compound gear D-E. The gear B meshes with gear E and gear C meshes with gear D. The number of teeth on gears B, C and D are 75, 30 and 90 respectively. Find the speed and direction of rotation of gear C when gear B is fixed and the arm A makes 100 r.p.m. clockwise (CW).

Given:
The teeth on B, $T_B = 75$; teeth on C, $T_C = 30$ and teeth on D, $T_D = 90$;

To determine the speed of C, N_C when the speed of B is zero, i.e. $N_B = 0$ and the speed of A, $N_A = +100$ r.p.m. since given as clockwise (CW).

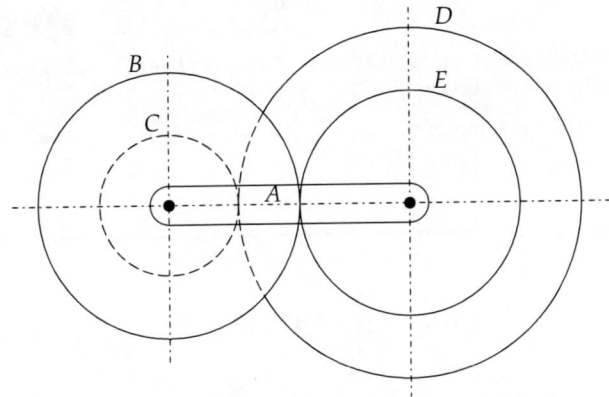

Fig. 12.9 Reversed epicyclic compound gear train.

From Fig. 12.9, it is clear that the sum of the teeth on B and on E = sum of the teeth on C and on D as the centre distances between the shafts are to be same, i.e. $r_B + r_E = r_C + r_D$. Since the number of teeth on each gear for the same module are proportional to their pitch circle diameters, therefore, $T_B + T_E = T_C + T_D$. Hence, $T_E = T_C + T_D - T_B = 30 + 90 - 75 = 45$

The problem can be solved by two ways:

1. Taking wheel C to give one rotation
2. Taking compound wheel $D - E$ to give one rotation.

(1) Single gear wheel C is considered to give one clockwise rotation

Step	Condition of Motion	Revolutions of Elements N			
		N_A	N_C	N_{DE}	N_B
1.	Fix arm A and rotate C by +1 revolution (CW)	0	+1	$-(T_C/T_D)$	$(T_C/T_D)(T_E/T_B)$
2.	Multiply with x throughout	0	$+x$	$-x(T_C/T_D)$	$x(T_C/T_D)(T_E/T_B)$
3.	Add y revolutions to all	$+y$	$+y$	$+y$	$+y$
4.	Total	$+y$	$x+y$	$y - x(T_C/T_D)$	$y + x.(T_C/T_D)(T_E/T_B)$

Given:

$N_A = y = 100$; $N_B = 0 = 100 + x\left(\dfrac{30 \times 45}{90 \times 75}\right)$ or $-100 = \dfrac{x}{5}$

Therefore, $x = -500$ and $N_C = x + y = 100 - 500 = -400$ or 400 anticlockwise (ACW/CCW)

(2) Compound gear wheel D-E is considered to give one clockwise rotation

Step	Condition of Motion	Revolutions of Elements N			
		N_A	N_C	N_{DE}	N_B
1.	Fix arm A and rotate compound wheel DE by +1 revolution (CW)	0	$-(T_D/T_C)$	+1	$-(T_E/T_B)$
2.	Multiply with x throughout	0	$-x(T_D/T_C)$	$+x$	$-x(T_E/T_B)$
3.	Add y revolutions to all	$+y$	$+y$	$+y$	$+y$
4.	Total	$+y$	$y - x(T_D/T_C)$	$y + x$	$y - x(T_E/T_B)$

Given:

$N_A = y = 100$; $N_B = 0 = y - x\dfrac{T_E}{T_B} = 100 - x\dfrac{45}{75}$ or $100 = \dfrac{3x}{5}$

Therefore, $x = \dfrac{500}{3}$ and therefore,

$$y - x\dfrac{T_D}{T_C} = 100 - \dfrac{500}{3} \times \dfrac{90}{30}$$
$$= 100 - 500 = -400 \text{ or } 400 \text{ anticlockwise (ACW/CCW)}$$

Note: The result is same by both ways but the latter one is easier compared to the former and so hereafter only compound wheel is selected to give initial one revolution.

W E 12.8: *Compound epicyclic reversed gear train with internal gears.* Figure 12.10 shows the compound epicyclic gear train. Wheels A, D and E are free to rotate independently on spindle O, while B and C compound wheel rotates on a spindle P, on the end of arm OP. All the teeth on different wheels have the same module. A has 12 teeth, B has 30 teeth and C has 14 teeth cut externally. Find the number of teeth on wheels D and E which are cut internally.

If the wheel A is driven clockwise at 1 r.p.s., while D is driven counterclockwise at 5 r.p.s., determine the magnitudes and direction of the angular velocities of arm OP and wheel E. Find the torque exerted on the output shaft of D if the input shaft of A develops a torque of 100 N-m.

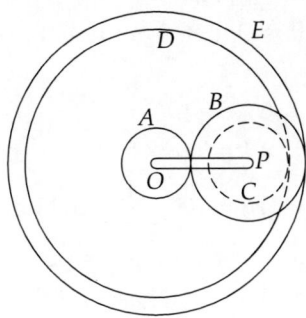

Fig. 12.10

Given:
Teeth on $A = 12 (T_A)$; Teeth on $B = 30 (T_B)$; Teeth on $C = 14 (T_C)$

To find speed of B and speed of C knowing that $T_E = T_A + 2.T_B = 60 + 12 = 72$;

Therefore,
$$T_D = T_E - (T_B - T_C) = [72 - (30 - 14)] = 56$$

Step	Condition	Arm OP	Compound B-C	N_A	N_D	N_E
1.	Fix arm and give +1 to BC	0	+1	$-T_B/T_A$	$+T_C/T_D$	$+T_B/T_E$
2.	Multiply with x	0	$+x$	$-xT_B/T_A$	$+xT_C/T_D$	$+xT_B/T_E$
3.	Add y to all	$+y$	$+y$	$+y$	$+y$	$+y$
4.	Total	$+y$	$x + y$	$y - xT_B/T_A$	$y + xT_C/T_D$	$y + xT_B/T_E$

Given:
$N_A = 1$ r.p.s. or 60 r.p.m. (CW).

Therefore,
$$N_A = y - x\frac{T_B}{T_A} = y - x\frac{30}{12} = 60 \tag{1}$$

$N_D = 300$ r.p.m. (ACW).

Therefore,
$$N_D = y + x\frac{T_C}{T_D} = y + x\frac{14}{56} = -300 \qquad (2)$$

Solving (1) and (2) gives $x = -130.90$ and $y = -267.28$.

Therefore, the speed of arm

$$P = -267.28 \text{ r.p.m.} \text{ or } -4.455 \text{ r.p.s.} = 4.455 \text{ r.p.s. anticlockwise (ACW)}$$

Speed of $E = y + x\dfrac{T_B}{T_E} = -267.28 + (-130.90)\dfrac{30}{72} = -321.82$ r.p.m. or 5.364 r.p.s. *(ACW)*.

Here input shaft is A and output shaft is D. Let Tr represent torque.

Therefore,
$$Tr_A N_A = Tr_D N_D$$
$$100 \times 1 = Tr_D \times 5$$
$$Tr_D = 100 \times \frac{1}{5} = 20 \text{ N-m}.$$

W E 12.9: Two shafts A and B are coaxial. Gear C with 25 teeth is rigidly mounted on the shaft A. A compound gear D-E gears with C and an internal gear G. D has 20 teeth and gears with C. E has 30 teeth and gears with an internal gear G. The gear C is fixed. The compound gear D-E is mounted on a pin which projects from an arm which is keyed to the shaft B as shown at Fig. 12.11. (i) Find the number of teeth on the internal gear G. Assume that all gears have the same module. If the shaft B (arm) rotates at 1000 r.p.m., (ii) find the speed of wheel G, (iii) the torque on the fixed gear T_C.

Given:
$N_C = 0$; $N_{arm} = 1000$ r.p.m.; $T_E = 30$; $T_D = 20$; $T_C = 25$; Compound wheel D-E is given one revolution.

Step No.	Condition of Motion	Revolutions of Elements N			
		Arm	Speed of D-E	Speed of G	Speed of C
1.	Fix arm A and rotate D-E by +1 revolution (CW)	0	+1	(T_E/T_G)	$-(T_D/T_C)$
2.	Multiply with x throughout	0	$+x$	$+x(T_E/T_G)$	$-x(T_D/T_C)$
3.	Add y revolutions to all	$+y$	$+y$	$+y$	$+y$
4.	Total	$+y$	$x + y$	$y + x(T_E/T_G)$	$y - x(T_D/T_C)$

1. Teeth on G: From figure,
$$\frac{T_C}{2} + \frac{T_D}{2} = \frac{T_G}{2} - \frac{T_E}{2}$$
on substituting the respective teeth, $T_G = 75$

2. As $N_C = 0 = y - x \cdot \dfrac{20}{25}$; and $N_{arm} = 1000 = y$, solving gives, $x = 1250$,

   Therefore, speed of shaft

   $$G = y + x \dfrac{T_E}{T_G} = 1000 + 1250 \dfrac{30}{75} = 1500 \text{ r.p.m. (CW)}$$

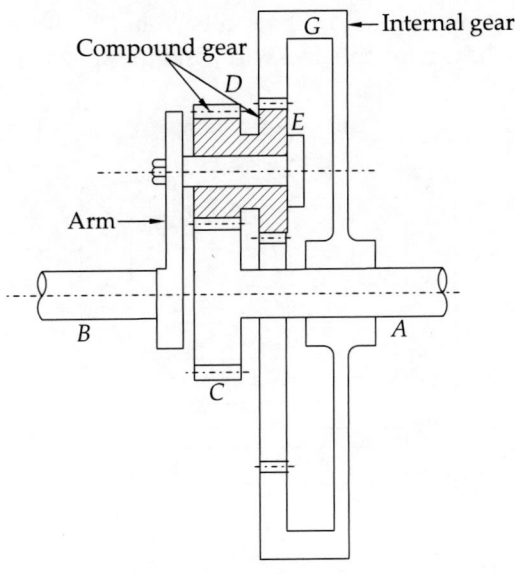

Fig. 12.11

3. Power transmitted through arm = 7.5 kW = 7500 W; Knowing the speed of arm, the torque on arm, $Tr_{arm} = 71.59$ N-m (input torque). As the power output is same as power input. Output torque on G,

   $$Tr_G = 71.59 \times \dfrac{1000}{1500} = 47.73 \text{ N-m}$$

   Sum of the torques must be zero, i.e.

   $$T_I + T_O + T_H = 0, \quad T_H = 71.59 - 47.73 = 23.86 \text{ N-m}$$

   Therefore, torque on the fixed wheel C is 23.86 N-m.

12.8 EPICYCLIC GEAR TRAINS WITH BEVEL GEARS

For more compact, epicyclic system with bevel gears are used as they provide a very high-speed reduction with few gears. For example: Humpages speed reduction gear, differential gear of an automobile, etc.

Note: Axes of rotation of the bevel gears will not be parallel like spur gears. Hence, ± sign is to be used for non-parallel axes. Between two gears which rotate about a common axis or axes parallel, +ve sign is to be used for same direction and −ve sign for opposite direction of rotation. This sign convention is to be followed carefully for bevel gear trains. Observe the following examples.

W E 12.10: Two bevel gears A and B having 40 teeth and 30 teeth respectively are rigidly mounted on two coaxial shafts X and Y as shown in Fig. 12.12(a). A bevel gear C having 50 teeth meshes with A and B and rotates freely on one end of an arm. At the other end of the arm is welded a sleeve and the sleeve is riding freely on the axes of the shafts X and Y. If the shaft X makes 100 r.p.m. clockwise and arm rotates at 100 r.p.m. anticlockwise, find the speed of shaft Y.

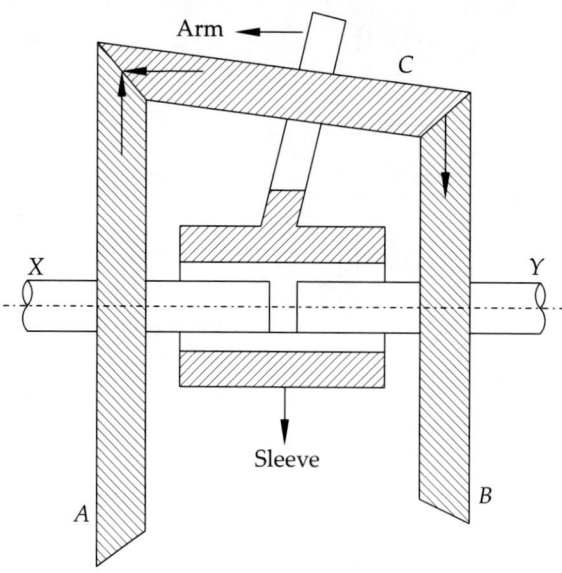

Fig. 12.12(a)

Given:
$T_A = 40$, $T_B = 30$, $T_C = 50$ $N_A = 100$ r.p.m. (CW);
$N_{arm} = 100$ r.p.m. (ACW); $N_B = ?$

Step No.	Condition of Motion	Revolutions of Elements N			
		Arm	N_A	N_C	N_B
1.	Fix arm and rotate gear A by +1 revolution (CW)	0	+1	$\pm(T_A/T_C)$	$-(T_A/T_C)/(T_CT_B) = -T_A/T_B$
2.	Multiply with x throughout	0	$+x$	$\pm x(T_A/T_C)$	$-x(T_A/T_B)$
3.	Add y revolutions to all	$+y$	$+y$	$+y$	$+y$
4.	Total	$+y$	$x+y$	$y \pm (T_A/T_C)$	$y - x(T_A/T_B)$

Given speed of $N_A = x + y = 100$ and $N_{arm} = y = -100$. Because anticlockwise.

So $x = 200$, therefore, $N_B = -100 - 200\dfrac{40}{30} = -366.7$ r.p.m. or 366.7 r.p.m. (ACW/CCW) or in the opposite direction of rotation of wheel A.

Try: In Fig. 12.12(b), a similar gear train as above is shown. Two bevel gears C and D of equal size having 40 teeth on each wheel, are rigidly mounted on two coaxial shafts A and A as shown. A bevel gear B having 30 teeth meshes with C and D, rotates freely on one end of an arm G. At the other end of the arm is welded a sleeve F and the sleeve F can ride freely on the axes of the shafts A and A. If the shaft A connected to D makes 100 r.p.m. clockwise and arm rotates at 100 r.p.m. anticlockwise, find the speed of the other shaft A connected to C.

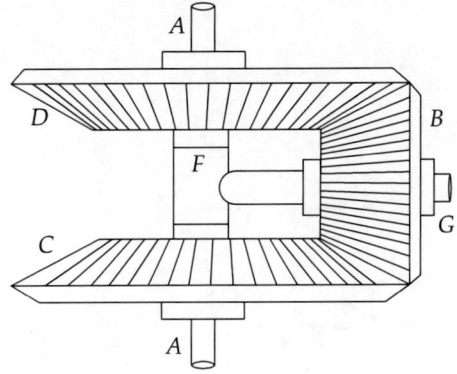

Fig. 12.12(b)

Differential mechanism: The purpose of the differential mechanism is to enable the back wheels to revolve at different speeds when the car is rounding a corner. The bevel wheels are used. Two equal wheels C and D shown in Fig. 2.13 and in Pic. 12.5 are keyed to the halves of the rear axle and the bevel wheels E and F revolve on pins which are carried by the arms. The drive is transmitted

Pic. 12.5

from the propeller shaft to the arms through the bevel wheels A and B. The bevel pinions E and F gear respectively with the wheels C and D, so that if the arms are stationary the wheels C and D revolve in opposite directions. It is easily seen that if the vehicle is moving along a straight path the wheels C, D and also the arms will revolve at the same speed and the bevel pinions E and F will remain stationary relative to the arms.

W E 12.11: *(Differential gear)* Figure 12.13 shows a differential gear used in a motor car. The pinion A on the propeller shaft has 12 teeth and gears with the crown gear B which has 60 teeth. The shafts P and Q form the rear axes to which the road wheels are attached. If the propeller rotates at 1000 r.p.m. and the road wheel attached to axle Q has a speed of 210 r.p.m. while taking a turn, find the speed of road wheel attached to axle P.

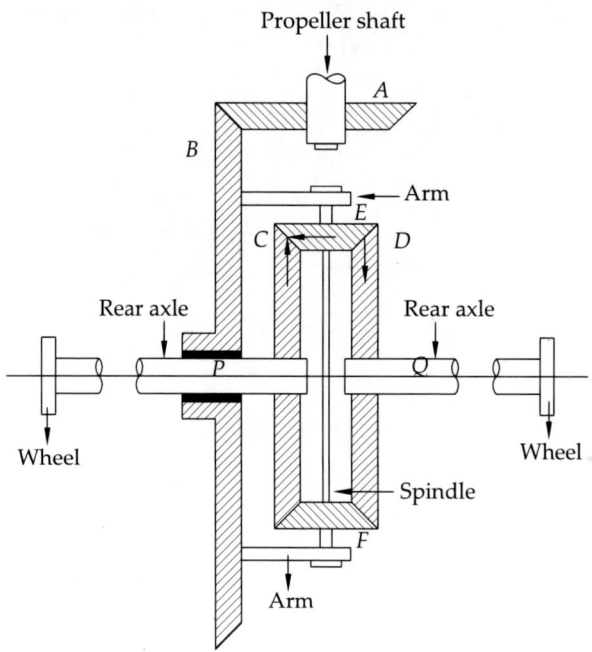

Fig. 12.13

Step No.	Condition of Motion	Revolutions of Elements			
		Arm gear B, N_B	N_C	N_E	N_D
1.	Fix arm or gear B and rotate C by +1 revolution (CW)	0	+1	$\pm(T_C/T_E)$	$-(T_C/T_D) = -1$ since $T_C = T_D$
2.	Multiply with x throughout	0	+x	$\pm x(T_C/T_E)$	$-x$
3.	Add y revolutions to all	+y	+y	+y	+y
4.	Total	+y	x + y	$y \pm (T_C/T_E)$	y − x

Given:

$T_A = 12$; $T_B = 60$; $N_A = 1000$ r.p.m.

Therefore,

$$N_B = N_A \times \frac{T_A}{T_B} = 1000 \times \frac{12}{60} = 200 \text{ r.p.m. (CW)}$$

Hence, arm gear B speed = $y = 200$ r.p.m. (CW). Given the speed of D, $N_D = 210$ r.p.m. (CW);

Therefore, $y - x = 210$. Hence $x = -10$.

Therefore, the speed of wheel C, $N_C = x + y = -10 + 200 = 190$ r.p.m. in the same direction as gear wheel B.

W E 12.12: A shaft Y is driven by a coaxial shaft X by means of an epicyclic gear train, as shown in Fig. 12.14. The wheel A is keyed to X and E to Y. The wheels B and D are compound and carried on an arm F which can turn freely on the common axes of X and Y. If the number of teeth on A, B, C, D and E are respectively 18, 54, 72, 22 and 44 and the shaft X makes (N_A) 200 r.p.m., determine (i) the speed in r.p.m. and sense of rotation of the shaft Y (N_E) when wheel C is fixed, (ii) when wheel C rotates at 30 r.p.m. opposite to wheel A.

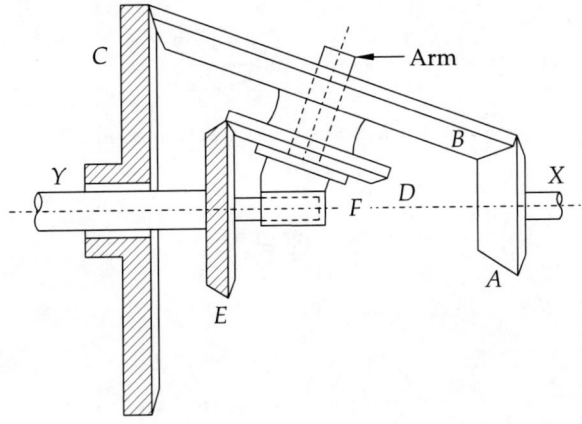

Fig. 12.14

Given:

The number of teeth on A, $T_A = 18$,

on B, $T_B = 54$,

on C, $T_C = 72$,

on D, $T_D = 22$,

on E, $T_E = 44$.

Step No.	Condition of Motion	Revolutions of Elements				
		Arm F, N_F	N_A	N_{B-D}	N_C	N_E
1.	Fix arm F and rotate A by +1 revolution (CW)	0	+1	$\pm(T_A/T_B)$	$-(T_A/T_C)$	$-(T_A/T_B)\cdot(T_D/T_E)$
2.	Multiply with x throughout	0	$+x$	$\pm(T_A/T_B)$	$-x(T_A/T_C)$	$-x(T_A T_D/T_B T_E)$
3.	Add y revolutions to all	$+y$	$+y$	$+y$	$+y$	$+y$
4.	Total	$+y$	$x+y$	$y \pm x(T_A/T_B)$	$y - x(T_A/T_C)$	$y - x(T_A T_D/T_B T_E)$

Speed of A, i.e. $x + y = 200$, $N_C = 0 = y - x\dfrac{T_A}{T_C} = y - x \cdot \dfrac{18}{72}$.

Solving gives, $x = 160$ and $y = 40$.

(i) The speed of E

$$N_E = y - x\frac{T_A T_D}{T_B T_E} = 40 - 160\frac{18 \times 22}{54 \times 44}$$
$$= \frac{40}{3} \text{ r.p.m. in the same direction as shaft } X.$$

(ii) Speed of A

$N_A = x + y = 200$ and $N_C = -30 = y - x\dfrac{18}{72}$. Solving gives

$x = \dfrac{920}{5}$ and $y = \dfrac{80}{5}$

Therefore, the speed of E,

$$N_E = y - x\frac{T_A T_D}{T_B T_E} = 16 - \frac{920}{5}\left(\frac{18 \times 22}{54 \times 44}\right)$$
$$= -\frac{44}{3} = 14.67 \text{ r.p.m. (ACW)}$$

12.9 EXERCISE

12.9.1 Short Answer Questions

1. Explain the advantages of gear drive over the belt and rope drives.
2. Distinguish between the terms: pitch circle, addendum circle and dedendum circle.
3. What are the various types of gears? Explain each one of them with sketches.
4. Explain the procedure adopted for designing the spur wheels.

5. Explain with a neat sketch the sun and planet wheel.
6. Explain the working of an epicyclic gear train with bevel wheels.
7. How will you find the velocity ratio of an epicyclic gear train by tabular method?
8. Design the spur gears having a velocity ratio of 4 and the pitch of teeth as 7.5 mm. The approximate distance between the two shafts is 300 mm.
   [**Ans.** $T_1 = 200; T_2 = 50; d_1 = 477.3$ mm; $d_2 = 119.3$ mm; $x = 298.3$ mm]
9. Two spurs A and B of an epicyclic gear train as shown in Fig. 12.5 have 24 and 30 teeth respectively. The arm rotates at 100 r.p.m. in the clockwise direction. Find the speed of the gear B on its own, when the gear A is fixed. [**Ans.** 180 r.p.m.].
   If in the above example, the wheel A rotates at 200 r.p.m. in the anticlockwise direction, what will be the speed of the gear B. [**Ans.** 260 r.p.m.]
10. Give a neat sketch of a reverted gear train.
11. What is differential gear? What are its applications? What is an epicyclic gear train?
12. Explain the use of a gear train. What is the speed reduction of a simple train of 16, 18, 20 and 30 toothed gears? The first mentioned gear is the driver and all are external spur gears. In what respect does a compound gear train differ from a simple train?
13. Discuss in brief the points to be taken into consideration while designing a gear drive for a particular service.
14. Using a 3-dimensional sketch, derive an equation relating the pressure angle in a plane normal to the teeth; pressure angle in the plane of rotation and the helix angle.
15. A bevel gear is required to transmit 32 kW with a velocity ratio of 6:1 from a bevel pinion with the following shaft angles (i) 75° (ii) 90°. Determine the semi-pitch cone angles for pinion and gear.
16. State the most important advantages and disadvantages of worm gearing compared to other types of gearing.
17. Derive an expression for the centre distance of a pair of spiral gears.
18. Explain epicyclic gear train. What are its merits and demerits as compared to reversed gear?
19. In what respects does a compound gear train differ from a simple train and epicyclic gear train?
20. Define gear train and why is it necessary.
21. Explain with a neat sketch the 'sun and planet wheel'.
22. What is gear train? What are its main types?
23. Explain briefly the differences between simple, compound and epicyclic gear trains.
24. What is reverted gear train? Where is it used?
25. What is differential gear of an automobile? How does it function?

12.9.2 Problems

1. What is the speed reduction of a simple train of 16, 18, 20 and 30 toothed gears? The first mentioned gear is the driver and all the other are external spur gears.

2. Layout a pair of involutes spur gears to the following specifications: Distance between shaft centres 30 cm, driving shaft turns at 200 r.p.m., module pitch = 6.25 mm, black lash = 0. Angle of obliquity can be taken as 20° degrees.

3. An epicyclic gear train shown has a sun wheel S of 30 teeth and two planet wheels p-p of 50 teeth. The planet wheels mesh with the internal teeth of fixed annulas A. The driving shaft carrying the sun wheel transmits 4 kW at 300 r.p.m. The driven shaft is connected to an arm which carries the planet wheels. Determine the speed of the driven shaft and the torque transmitted, if the overall efficiency is 95%.

4. Explain the use of a gear train. What is the speed reduction of a simple train of 16, 18, 20 and 30 toothed gears? The first mentioned gear is the driver and all are external spur gears. In what respect does a compound gear train differ from a simple train?

5. Find the speed and direction of gear 8 in Fig. 12.15. What is the speed ratio of the train?

[Ans. N_8 = 68.2 r.p.m. CW; speed ratio = $-\dfrac{5}{88}$]

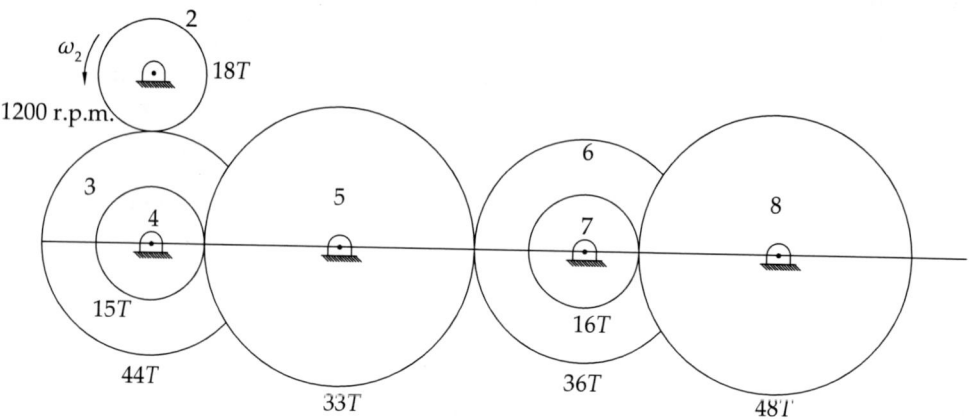

Fig. 12.15

6. The internal gear 7 shown in Fig. 12.16 turns at 60 r.p.m. What are the speeds and directions of arm 3? If the arm 3 rotates CCW at 300 r.p.m., find the speed and direction of rotation of internal gear 7?

[Ans. 231 r.p.m. CCW]

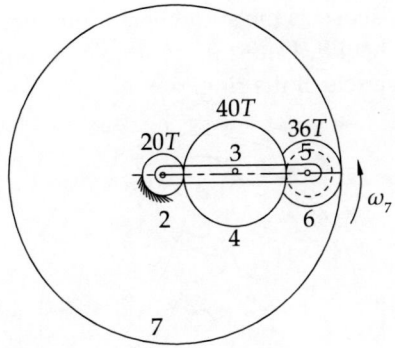

Fig. 12.16

7. In Fig. 12.17 shaft C is stationary. If gear 2 rotates at 800 r.p.m. CCW, what are the speed and direction of rotation of shaft B?

   (a) Consider shaft B as stationary. If shaft C is driven at 380 r.p.m. CCW, what are speed and direction of rotation of shaft A? [*Ans.* 645 r.p.m. CW]

   (b) Determine the speed and direction of rotation of shaft C if (a) shaft A and B both rotate at 360 r.p.m. CCW and (b) shaft A rotates at 360 r.p.m. CCW and shaft B rotates at 360 r.p.m. CCW.

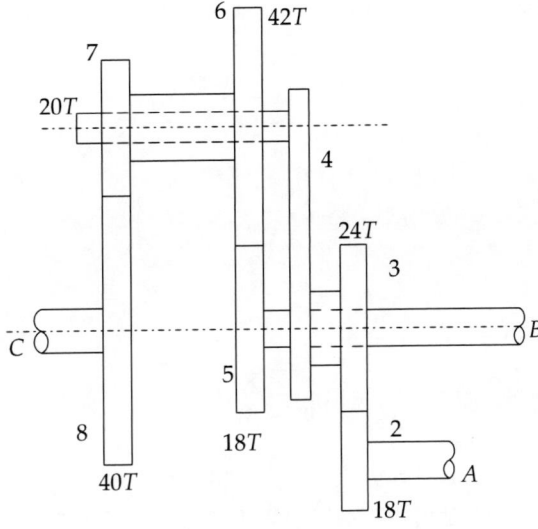

Fig. 12.17

8. In an epicyclic gear train of the 'sun and planet type' as shown in Fig. 12.18 the pitch circle diameter of the internally toothed ring D is to be 216 mm and the module 4 mm. When the ring D is stationary, the spider A, which carries three planet wheels C of equal size, is to make

one revolution in the same sense as the sun wheel B for every five revolutions of the driving spindle carrying the sun wheel B. Determine suitable number of teeth for all the wheels and the exact diameter of pitch circle of the ring.

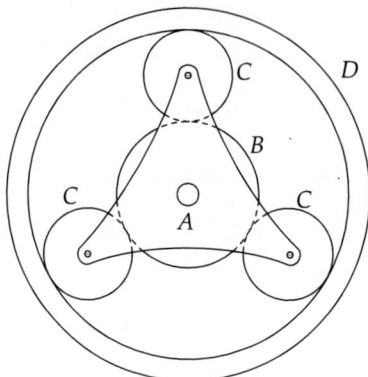

Fig. 12.18

9. An epicyclic gear train, as shown in Fig. 12.19 is composed of a fixed annular wheel A having 150 teeth. The wheel A is meshing with wheel B which drives wheel D through an idle wheel C, D being concentric with A. The wheels B and C are carried on an arm which revolves clockwise at 100 r.p.m. abut the axis of A and D. If the wheels B and D have 25 teeth and 40 teeth respectively, find the number of teeth on C and the speed and sense of rotation of C.

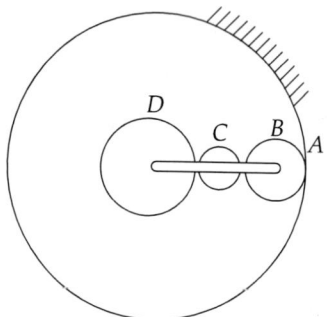

Fig. 12.19

10. A compound epicyclic gear is shown in Fig. 12.20. The gears A, D and E are free to rotate on the axis P. The compound gears B and C rotate together on the axis Q at the end of arm F. All the gears have equal pitch. The number of external teeth on the gears A, B and C are 18, 45 and 21 respectively. The gears D and E are annular gears. Gear A rotates at 100 r.p.m. in the anticlockwise direction and the gear D rotates at 450 r.p.m. clockwise. Find the speed and direction of the arm and the gear E.

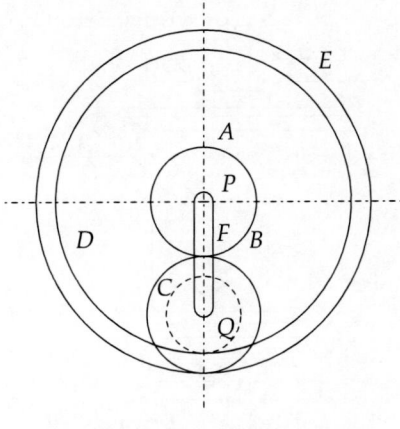

Fig. 12.20

11. Bevel gear 2 shown in Fig. 12.21 is driven by the engine in the reduction unit shown. Bevel planets 3 mesh with crown gear 4 and are pivoted on the spider (arm), which is connected to propeller shaft B. Find the per cent speed reduction. [*Ans.* 41%]

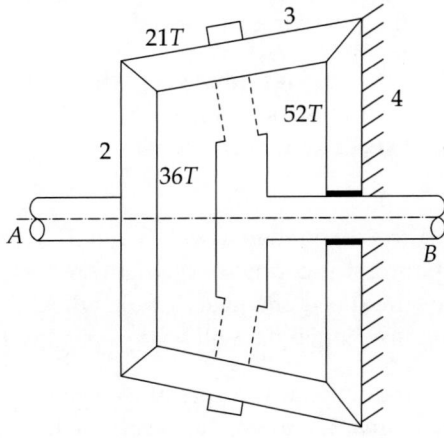

Fig. 12.21

12. An epicyclic bevel gear train shown in Fig. 12.22 has fixed gear B meshing with pinion C. The gear E on the driven shaft meshes with the pinion D. The pinions C and D are keyed to a shaft, which revolves in bearings on the arm A. The arm A is keyed to the driving shaft. The number of teeth: $T_B = 75$, $T_C = 20$, $T_D = 18$ and $T_E = 70$. Find the speed of the driven shaft, if (1) the driven shaft makes 1000 r.p.m. and (2) the gear B turns in the same sense as the driving shaft at 400 r.p.m., the driving shaft still making 1000 r.p.m.

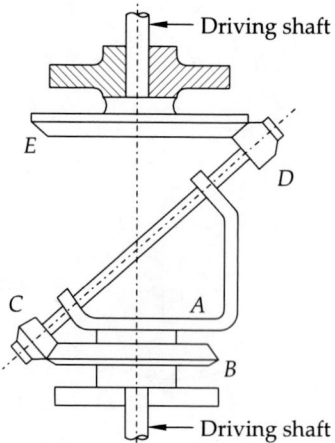

Fig. 12.22

13. The speed ratio of reverted gear train is to be 12. The module pitch of gears A and B is 3.125 mm and of gears C and D is 2.5 mm. Calculate the suitable number of teeth for the gears. No gear is to have less than 24 teeth. B-C is the compound gear and A and D are the driver and driven with their axes being the same or coaxial.

14. In an epicyclic gear train, an arm carries three gear wheels A, B and C having 48, 24 and 50 teeth respectively. The wheel A meshes with B and B meshes with C. If the arm rotates at 400 r.p.m. clockwise, find (i) speed of wheel C when A is fixed (ii) speed of wheel A when C is fixed.

15. In an epicyclic gear train, an arm carries two wheels A and B having 36 and 45 teeth respectively. If the arm rotates at 150 r.p.m. in the counterclockwise direction about the centre of the wheel A which is fixed, determine the speed of wheel B, if the wheel A instead of being fixed, makes 300 r.p.m. in the clockwise direction, what will be the speed of B?

16. In a reverted epicyclic gear train, the arm A carries two gears B and C and a compound gear D-E. The gear B meshes with gear E and gear C meshes with gear D. The number of teeth on gears B, C and D are 75, 30 and 90 respectively. Find the speed and direction of gear C when gear B is fixed and the arm A makes 100 r.p.m. clockwise.

12.9.3 Multiple Choice Questions

1. Which of the following circle is an imaginary circle in the study of toothed gears?
   (a) Pitch circle (b) Addendum circle (c) Dedendum circle (d) All of them. [**Ans.** (a)]

2. The pitch of a toothed wheel is _____
   (a) $\dfrac{\pi d}{T}$; (b) $\dfrac{\pi T}{d}$; (c) $\dfrac{\pi}{Td}$; (d) $\dfrac{T}{\pi d}$ where d is the diameter of the pitch circle and T is the number of teeth on the wheel. [**Ans.** (a)]

3. Which of the following statements is correct?
   (a) The gears are used for transmission of power
   (b) The gears are used where exact velocity ratio is required
   (c) In a single train of wheels, motion of both the driver and follower is like
   (d) All the above. [*Ans.* (d)]

4. In a compound train of wheels, we provide idle wheels, which do not affect the velocity ratio of the system.
   (a) agree   (b) disagree [*Ans.* (b)]

5. In an epicyclic gear train, one of the wheels is fixed and the other makes revolutions about the fixed wheel.
   (a) True   (b) False [*Ans.* (a)]

6. In a simple gear train, there is an odd number of idlers. The direction of rotation of the driver and the driven gears will be _____.
   (a) opposite   (b) same   (c) depends upon number of teeth of the gears [*Ans.* (b)]

7. In a reverted gear train the axes of the first and last gear are _____.
   (a) parallel   (b) coaxial   (c) neither parallel nor coaxial [*Ans.* (b)]

8. If the axes of the first gear and the last gear of a compound gear train are coaxial, the gear train is known as _____.
   (a) simple   (b) epicyclic   (c) reverted   (d) compound [*Ans.* (c)]

9. In a gear train, the train value is given by _____.
   (a) $\dfrac{N_1}{N_n}$ (b) $\dfrac{N_n}{N_1}$ (c) $N_1 \times N_n$ (d) $N_n - N_1$ [*Ans.* (b)]

10. The speed ratio of a gear train is _____.
    (a) equal to the train value   (b) reciprocal of the train value [*Ans.* (b)]

11. A gear train in which axes of gears have motion are called _____ gear trains.
    (a) epicyclic   (b) simple   (c) compound   (d) reverted [*Ans.* (a)]

12. In a clock mechanism, the hours and minute hands are connected by _____ gear train.
    (a) simple   (b) epicyclic   (c) compound   (d) reverted [*Ans.* (d)]

13. A differential uses _____ gear train.
    (a) simple   (b) epicyclic   (c) reverted   (d) compound [*Ans.* (b)]

Balancing of Rotating Masses

13

Balancing of Rotating Masses

13.1 INTRODUCTION

Balancing is the technique of correcting or eliminating unwanted inertia forces and moments in rotating machinery. Thus, determining the unbalance and the application of corrections is the principal problem in the study of balancing. Rotating elements such as different wheels of vehicles, pulleys, gear wheels and flywheels, etc., shown in Pic. 13.1 have certain masses. Due to casting and manufacturing defects such as blow holes and hard spots, mass distribution will not be uniform. This causes the centre of gravity of these elements not coinciding with the axis of rotation. The mass which is acting at a distance from the axis of rotation produces an unbalanced dynamic force also called a disturbing force. The magnitude of these dynamic forces will be constant but their directions vary as they are rotating elements. The rotating elements which are supported on bearings will be subjected to dynamic forces that are proportional to the square of the speed and directly proportional to the distance (eccentricity) from the axis.

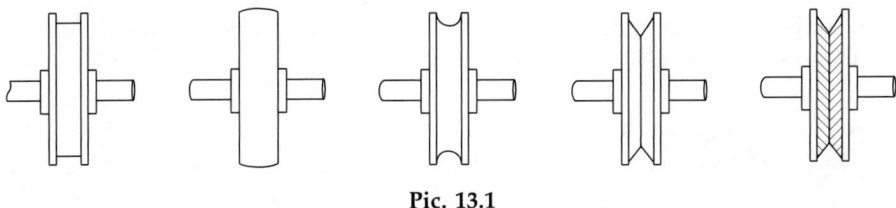

Pic. 13.1

These forces tend to bend the shaft and produce vibrations. These dynamic forces not only subject bearings to repeated loads which may cause parts to fail prematurely by fatigue but also cause stresses to develop in the various members. It is, therefore, very essential for all rotating elements or bodies to be completely balanced.

13.2 CHECKING OF A ROTATING ELEMENT

The arrangement shown in Fig. 13.1 is a simple experiment to determine whether the rotating elements are statically balanced or not as follows. Two wheel discs A and B are of same size with numbers marked as in watches from 1 to 12 are mounted on two shafts to a frame so that they can rotate freely without friction. Take one wheel at a time and make it revolve gently by hand and observe where they stop. Repeat this for four to five times. You may observe one wheel disc say A always stopping at the same position whereas the other wheel disc B, stopping each time at different positions with reference to an arrow marked on the frame below each wheel. Then you can understand from this that the wheel disc A is statically unbalanced which means that the axis of the shaft and the centre of mass of the disc are not coincident or due to excess mass away from the axis of rotation. The wheel disc B which stops each time at different places is a statically balanced wheel because its mass is uniformly distributed and the centre of mass of the disc coinciding with the axis of rotation of the shaft.

The static unbalance can be corrected by drilling out material or by adding mass on the diametrically opposite side. There are static balancing machines for testing and balancing gears, fans, impellers and wheels of vehicles, etc. The machine measures the unbalance by indicating its magnitude and location.

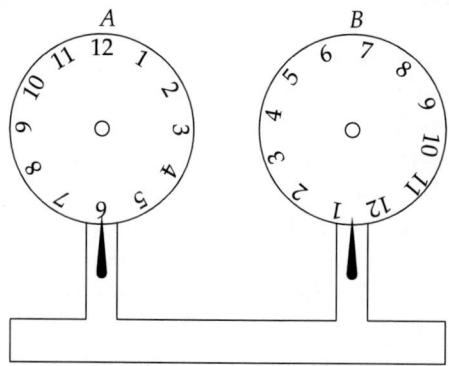

Fig. 13.1 Testing of wheels.

In order to eliminate the centrifugal force due to the unbalanced mass, another mass is attached on the opposite side of the axis of rotation such that it produces a centrifugal force equal to the force due to the unbalanced mass. In other words, the centrifugal forces due to the unbalanced and balanced masses are made equal and opposite. This process of providing the balancing mass, in order to counteract the effect of the centrifugal force due to unbalanced mass is called the balancing.

13.3 TYPES OF BALANCING OF ROTATING ELEMENTS

Each rotating element has to be checked whether it is perfectly balanced or not. Each element may have a single or several unbalanced masses. There may be a single or several rotating elements on a single shaft. Balancing can be taken up each element separately or all elements together simultaneously. In other words, each element can be taken up separately or all elements taken up together which are on a single shaft. Hence, balancing of rotating elements is studied in the following order for complete understanding of the balancing of rotating masses.

13.3.1 Balancing of a Single Unbalanced Rotating Mass

(a) By a single balancing mass in the same plane of rotation.
(b) By two balancing masses in two different planes.

13.3.2 Balancing of Several Unbalanced Rotating Masses

Picture 13.2 shows the several rotors mounted on a single shaft.

Pic. 13.2

(a) By a single balancing mass in the same plane of rotation when several masses are in the same planes.
(b) By two balancing masses in two different planes when several masses are in several planes.

Let the unbalanced masses be represented by m, m_1, m_2, etc. at radii r, r_1, r_2, etc. and the balancing masses be represented by m_b, m_{b1}, m_{b2}, etc. at radii r_b, r_{b1}, r_{b2}, etc.

Let the centrifugal force due to the unbalanced mass be represented by F_C equal to $m\omega^2 r$ and the centrifugal force due to the balancing mass be represented by F_{Cb} equal to $m_b\omega^2 r_b$.

13.4 BALANCING OF A SINGLE UNBALANCED ROTATING MASS

The balancing of a single unbalanced mass can be done in two ways as follows:

13.4.1 By a Single Balancing Mass Rotating in the Same Plane

Figure 13.2 shows the rotating element such as a flywheel or gear wheel or pulley with a single unbalanced mass m concentrated at a distance r from the centre of rotation. This mass when rotates at ω radians per second produces a centrifugal force F_C equal to $m\omega^2 r$ which causes the shaft to bend. In order to reduce this effect, the wheel has to be balanced by adding a single balancing mass m_b on the diametrically opposite side at such a radius r_b so that it produces a centrifugal force F_{Cb} of $m_b\omega^2 r_b$ equal to F_C.

As the centrifugal forces are equal, $F_C = F_{Cb}$ or $m\omega^2 r = m_b\omega^2 r_b$ or $mr = m_b r_b$. The centrifugal force is proportional to the product of mass and its distance from the centre of rotation because ω^2 is common on both sides. The product $m_b r_b$ can be split up conveniently by choosing either m_b or r_b depending upon the working conditions.

W E 13.1: A wheel has an unbalanced mass 10 kg, with 250 mm from the axis of rotation is to be balanced by a balancing mass of 4 kg. Find the radius at which the centre of gravity of balancing mass should be placed.

Given:
Unbalanced mass $(m) = 10$ kg and its radius of rotation $(r) = 250$ mm;
Balancing mass $(m_b) = 4$ kg and let r_b be its radius or distance.
Then for balancing $m.r = m_b.r_b$
Therefore,
$$r_b = \frac{m.r}{m_b}$$
$$= 10 \times \frac{250}{4} = 625 \text{ mm}$$

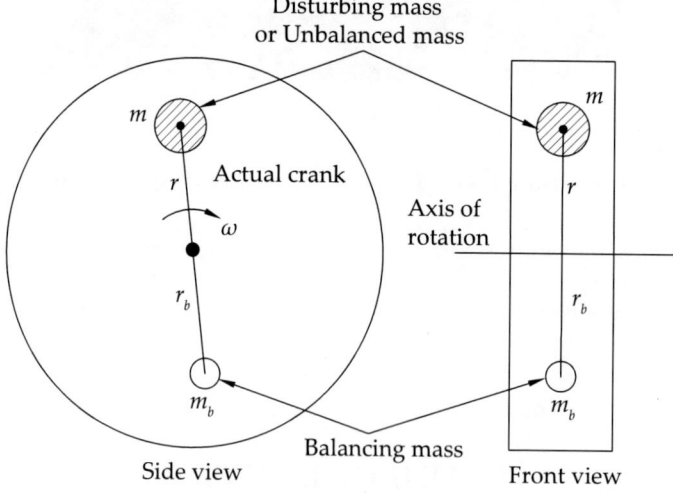

Fig. 13.2 Unbalanced wheel.

13.4.2 By Two Balancing Masses in Two Different Planes

Always it is not possible to add the balancing mass on the diametrically opposite side in the same plane. In such a case, to balance a single unbalanced mass, two balancing masses in two different planes parallel to the plane of rotation of the unbalanced mass are essential. All the three masses in three different planes must be so arranged such that the resultant of the dynamic forces and couples acting on the shaft must be zero separately. Hence, they should satisfy the following two conditions for a dynamic balancing:

1. Resultant dynamic force on the shaft is zero.
2. Resultant dynamic couple on the shaft is zero.

The two balancing masses in two balancing planes can be considered in two different ways as follows:

There are total three planes. Let the plane D represents the plane of the unbalanced mass m, and planes B_1 and B_2 represent the two planes for two balancing masses m_{b1} and m_{b2}.

Let the distance L represents the distance between the planes B_1 and B_2 having balancing masses. Let the distances of the balancing planes B_1 and B_2 from the plane D of the unbalanced mass be L_1 and L_2 respectively as shown. Let the three masses be at radii of r, r_{b1} and r_{b2} respectively.

Here there are two options in choosing the planes of the balancing masses depending upon the requirement or convenience as follows:

(A) Planes of the balancing masses on either side of the plane D:

Figure 13.3 shows the plane D in between the two planes B_1 and B_2 having balancing masses.

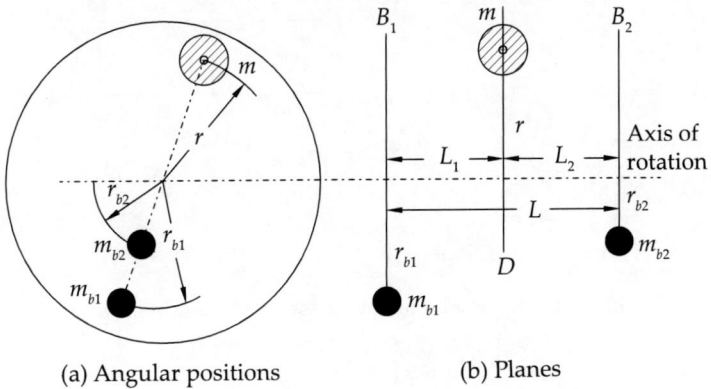

(a) Angular positions (b) Planes

Fig. 13.3 Planes of balancing masses on either side.

Note: The distance between the balancing planes $L = L_1 + L_2$.

The first condition to be satisfied is: *Sum of the upward centrifugal forces must be equal to the sum of the downward centrifugal forces.*

Hence, from the figure

$$m\omega^2 r = m_{b1}\omega^2 r_{b1} + m_{b2}\omega^2 r_{b2}$$

Cancelling ω^2 common on both sides gives,

$$mr = m_{b1}r_{b1} + m_{b2}r_{b2} \tag{1}$$

The second condition to be satisfied is: *Sum of the moments taken about the balancing planes must be zero.*

Take moments about the plane B_1,

$$L \cdot m_{b2} r_{b2} = mr \cdot L_1$$

$$m_{b2} r_{b2} = mr \frac{L_1}{L} \tag{2}$$

Then take moments about the plane B_2,

$$L \cdot m_{b1} r_{b1} = mr \cdot L_2$$

$$m_{b1} r_{b1} = mr \frac{L_2}{L} \tag{3}$$

From the above three conditions, distances L_1, L_2 or radii r_{b1}, r_{b2} or balancing masses m_{b1}, m_{b2} from the known, the unknown can be determined.

(B) Both planes of balancing masses on the same side:

Figure 13.4 shows the plane D on one side of the two planes B_1 and B_2 having balancing masses.

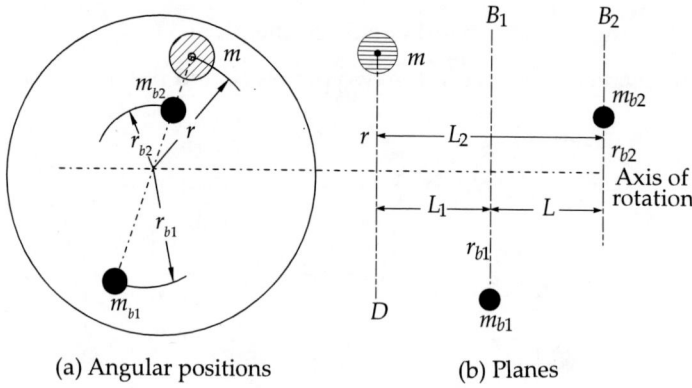

(a) Angular positions   (b) Planes

Fig. 13.4 Planes of balancing masses on the same side.

Note: Distance between the balancing planes $L = L_1 - L_2$

The first condition to be satisfied is: Sum of the upward centrifugal forces must be equal to the sum of the downward centrifugal forces.

Hence, from the figure,

$$m\omega^2 r + m_{b2}\omega^2 r_{b2} = m_{b1}\omega^2 r_{b1}$$

cancelling ω^2 common on both sides gives,

$$mr + m_{b2}r_{b2} = m_{b1}r_{b1} \tag{1}$$

The second condition to be satisfied is: Sum of the moments taken about the balancing planes must be zero.

Take moments about the plane B_1,

$$L.m_{b2}r_{b2} = mr.L_1$$

$$m_{b2}r_{b2} = mr.\frac{L_1}{L} \tag{2}$$

Then take moments about the plane B_2,

$$L.m_{b1}r_{b1} = mr.L_2$$

$$m_{b1}r_{b1} = \frac{mr.L_2}{L} \tag{3}$$

From the above conditions, distances L_1, L_2 or radii r_{b1}, r_{b2} or balancing masses m_{b1}, m_{b2} from the known, the unknown can be determined. This can be understood from the following example.

Note: Assuming L_1, L_2, and L, the product of $m_{b1}r_{b1}$ or $m_{b2}r_{b2}$ can be found.

W E 13.2: A wheel has an unbalanced mass of 10 kg at a radius of 40 mm. Determine the balancing mass required (i) in the same plane opposite at a radial distance of 50 mm, (ii) balancing masses in two planes at 30 mm and 45 mm radii with (a) on either side (b) on the same side with the distance between the balancing planes be equal to 1.2 m.

Given:
$m = 10$ kg; $r = 40$ mm; $r_b = 50$ mm; $r_{b1} = 30$ mm and $r_{b2} = 45$ mm and $L = 1.2$ m.

To determine

(i) In the same plane: m_b
(ii) (a) on either side: m_{b1} and m_{b2}; at L_1 and L_2
 (b) on the same side: m_{b1} and m_{b2}; at L_1 and L_2

(i) In the same plane: The centrifugal force due to the unbalanced mass F_C is proportional to

$$m \cdot r = 10 \times 40$$
$$= 400 \text{ kg.mm}$$
$$= 0.4 \text{ kg-m} \qquad (1)$$

Let m_b be the balancing mass at 50 mm radius. Then the centrifugal force F_{Cb} due to the balancing mass will be proportional to

$$m_b \cdot r_b = m_b.50 \text{ kg-mm}$$
$$= 0.05 m_b \text{ kg-m} \qquad (2)$$

For balancing equate both (1) and (2), so 0.4 kg-m = $0.05 m_b$ kg-m,

Hence, $m_b = \dfrac{0.4}{0.05} = 8$ kg

(ii) In two different planes:

(a) On either side: The distance $L = L_1 + L_2$. Given $L = 1.2$ m and assuming $L_1 = L_2 = 0.5L = 0.6$ m.

Let the centrifugal forces due to the two balancing masses be F_{Cb1} and F_{Cb2} in B_1 and B_2 planes at L metres apart are proportional to $m_{b1} r_{b1}$ and $m_{b2} r_{b2}$ kg-mm.
Sum of the upward and downward forces must be equal,

$$mr = m_{b1} r_{b1} + m_{b2} r_{b2} = 0.4 \text{ kg-m} \qquad (1)$$

Taking moments about B_1 plane

$$m \cdot r \cdot L_1 = m_{b2} r_{b2} L \qquad (2)$$

Taking moments about B_2 plane

$$m \cdot r \cdot L_2 = m_{b1} r_{b1} L \qquad (3)$$

From (2) and (3)

$$m_{b2} r_{b2} = m \cdot r \cdot \dfrac{L_1}{L} = 0.4 \times \dfrac{0.6}{1.2} = 0.2 \text{ kg-m}$$

and

$$m_{b1}r_{b1} = m \cdot r \cdot \frac{L_2}{L} = 0.4 \times \frac{0.6}{1.2} = 0.2 \text{ kg-m}$$

(b) On the same side: The distance $L = L_1 - L_2$. Given $L = 1.2$ m and assume $L_1 = 0.6$ m and $L_2 = 1.8$ m. Let the centrifugal forces due to the two balancing masses be F_{Cb1} and F_{Cb2} in B_1 and B_2 planes at L metres apart are proportional to $m_{b1}r_{b1}$ and $m_{b2}r_{b2}$ kg-mm.

Sum of the upward and downward forces must be equal

$$m_{b1}r_{b1} = mr + m_{b2}r_{b2} \tag{4}$$

Taking moments about B_1 plane,

$$m.r.L_1 = m_{b2}r_{b2}L \tag{5}$$

Therefore,

$$m_{b2}r_{b2} = 0.4 \times \frac{0.6}{1.2} = 0.2 \text{ kg-m}$$

Taking moments about B_2 plane,

$$m.r.L_2 = m_{b1}r_{b1}L \tag{6}$$

Therefore

$$m_{b1}r_{b1} = 0.4 \times \frac{1.8}{1.2} = 0.6 \text{ kg-m}$$

Assume either balancing mass or balancing radius. Generally, radius will be assumed and mass will be determined.

W E 13.3: A 40 kg mass (*A*) mounted on an axle at a distance of 1 m is to be balanced by two masses (*B*) and (*C*). The balancing masses are to be mounted in the planes at distances 1 m and 2 m on either sides of 40 kg mass at radii 1 m and 2 m respectively from the axis of rotation. Find the magnitudes of the balancing masses.

Given:
Unbalanced mass of the body *A*, $m_A = 40$ kg; Radius r_A of rotation of m_A is 1 m; the centrifugal force is proportional to $m_A r_A$, i.e. 40 kg-m.

Distance between the planes of *A* and *B* (L_1) = 1 m; Distance between the planes of *A* and *C* (L_2) = 2 m; Radius r_B of rotation of mass m_B is 1 m and Radius r_C of rotation of m_C is 2 m; Let m_B and m_C be the balancing masses in planes *B* and *C* in kg.

Balancing planes on either side: Sum of the forces $m_A r_A = m_B r_B + m_C r_C$

$$40 \times 1 = m_B \times 1 + m_C \times 2$$
$$m_B + 2m_C = 40 \tag{1}$$

and taking moments about plane A,

$$m_B r_B L_1 = m_C r_C L_2$$
$$m_B \times 1 \times 1 = m_C \times 2 \times 2$$
$$m_B = 4 m_C \qquad (2)$$

Substituting for m_B in equation (1) gives

$$4 m_C + 2 m_C = 40;\ 6 m_C = 40\ \text{or}\ m_C = \frac{40}{6} = 6.67\ \text{kg}$$

$$m_B = 4 m_C = 4 \times 6.67 = 26.67\ \text{kg}$$

13.5 BALANCING OF SEVERAL UNBALANCED MASSES ROTATING IN THE SAME PLANE

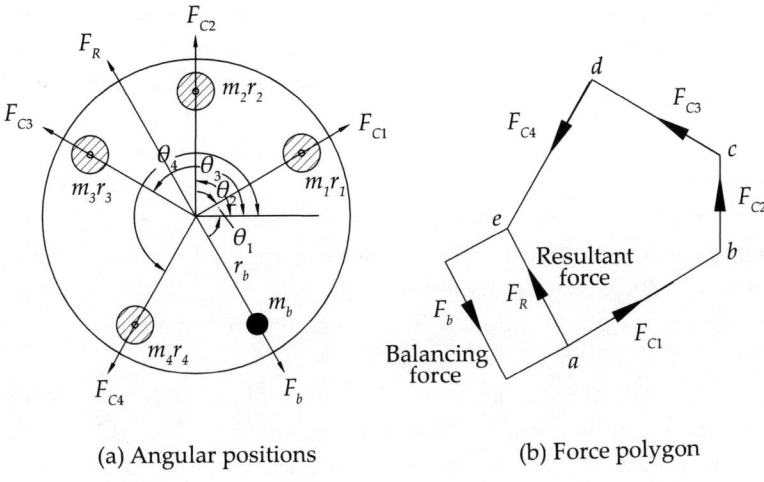

(a) Angular positions   (b) Force polygon

Fig. 13.5 Several unbalanced masses in the same plane.

Sometimes a single rotating wheel may have several unbalanced masses at different radii. All the several masses can be balanced by a single balancing mass in the same plane. As shown in Fig. 13.5, consider that there are four unbalanced masses m_1, m_2, m_3, and m_4, at radii of r_1, r_2, r_3 and r_4 from the axis of the rotation of the shaft. Let θ_1, θ_2, θ_3 and θ_4 be the angles of these masses with respect to the horizontal line as shown. Let these masses rotate in a single plane representing the plane of the paper with a constant angular velocity of ω radians per second.

The magnitude m_b and radial distance r_b of the balancing mass may be found out analytically or graphically as follows:

13.5.1 Analytical Method

Find the centrifugal force due to each mass and resolve the forces horizontally and vertically. Find their sum as given below:

Sum of the horizontal components $\Sigma_H = m_1 r_1 \cos\theta_1 + m_2 r_2 \cos\theta_2 + m_3 r_3 \cos\theta_3 + m_4 r_4 \cos\theta_4$

Sum of the vertical components $\Sigma_V = m_1 r_1 \sin\theta_1 + m_2 r_2 \sin\theta_2 + m_3 r_3 \sin\theta_3 + m_4 r_4 \sin\theta_4$. Magnitude of the resultant centrifugal force $F_C = \sqrt{[(\Sigma H)^2 + (\Sigma V)^2]}$ and if θ is the angle made by the resultant force with the horizontal, then $\tan\theta = \tan^{-1}\left(\dfrac{\Sigma V}{\Sigma H}\right)$. The balancing force F_{Cb} must be equal and opposite of the resultant force F_C. Find the magnitude of the balancing mass m_b and its radial distance r_b knowing $F_{Cb} = (m_b \cdot r_b)$, choosing either balancing mass m_b or its radius r_b.

13.5.2 Graphical Method

The various steps involved are as follows:

1. First draw the angular position diagram as shown at Fig. 13.5(a).
2. Find out the centrifugal forces which are proportional to the product of the mass and its radial distance of rotation, exerted by each mass on the rotating shaft.
3. Now draw force polygon, parallel to the corresponding masses in the space diagram by choosing a suitable scale. Side ab of the force polygon represents the centrifugal force due to m_1, side bc due to m_2, side cd due to m_3 and side de due to m_4 respectively as shown in Fig. 13.5(b).
4. As per the law of force, the closing side ae represents the resultant force in magnitude and direction as shown.
5. The balancing force is equal and opposite to the resultant force, i.e. ea.
6. Find the magnitude of the balancing mass m_b at the given radius r_b such that $m_b r_b$ equal to the resultant centrifugal force represented by ae.
7. Measure the angle made by the side ea with respect to horizontal line.

W E 13.4: There are four unbalanced masses m_1, m_2, m_3 and m_4 of 200 kg; 300 kg, 240 kg and 260 kg respectively rotating in the same plane at radii of rotation 0.2 m, 0.15 m, 0.25 m and 0.3 m respectively. The corresponding angles between the successive masses are 45°, 75°, 120°. Find the position and magnitude of the balancing mass m_b required, if its radius r_b of rotation is 0.2 m.

Given:

Masses m_1, m_2, m_3 and m_4 as 200 kg; 300 kg, 240 kg and 260 kg; and radii r_1, r_2, r_3 and r_4 as 0.2 m, 0.15 m, 0.25 m and 0.3 m. The angles between the consecutive masses $\theta_1, \theta_2, \theta_3$ and θ_4 as 0°, 45°, 120°, 240°; with respect to mass m_1.

Let m_b be the balancing mass and θ_b be the angle which the balancing mass makes with m_1. The magnitudes of centrifugal forces are proportional to the product of each mass and its radius.

Therefore,

$m_1 r_1 = 200 \times 0.2 = 40$ kg; $m_2 r_2 = 300 \times 0.15 = 45$ kg;
$m_3 r_3 = 240 \times 0.25 = 60$ kg; $m_4 r_4 = 260 \times 0.3 = 78$ kg;

Now the problem can be solved analytically or graphically as follows:

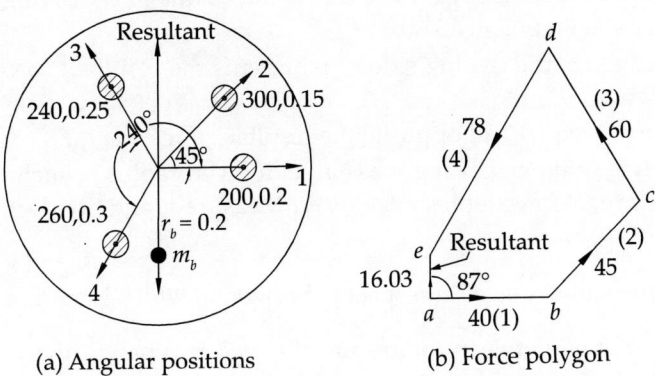

(a) Angular positions   (b) Force polygon

Fig. 13.6

Analytical Method: The angular position diagram is shown in Fig. 13.6(a). Resolving the centrifugal forces horizontally and vertically,

$$\Sigma F_H = m_1 r_1 \cos\theta_1 + m_2 r_2 \cos\theta_2 + m_3 r_3 \cos\theta_3 + m_4 r_4 \cos\theta_4$$
$$= 40\cos 0° + 45\cos 45° + 60\cos 120° + 78\cos 240°$$
$$= 40 + 31.8 - 30 - 39 = 2.8 \text{ kg-m}$$

$$\Sigma F_V = m_1 r_1 \sin\theta_1 + m_2 r_2 \sin\theta_2 + m_3 r_3 \sin\theta_3 + m_4 r_4 \sin\theta_4$$
$$= 40\sin 0° + 45\sin 45° + 60\sin 120° + 78\sin 240°$$
$$= 0 + 31.8 + 52 - 67.59 = 16.21 \text{ kg-m}$$

$$\text{Resultant force} = \sqrt{[(\Sigma F_H)^2 + (\Sigma F_V)^2]}$$
$$= \sqrt{[(2.8)^2 + (16.21)^2]} = 16.45 \text{ kg-m}$$

We know that
$$m_b r_b = \text{Resultant force} = 16.45$$
$$m_b = \frac{16.45}{r_b} = \frac{16.45}{0.2} = 82.25 \text{ kg}$$

and
$$\tan\theta = \frac{\Sigma F_V}{\Sigma F_H} = \frac{16.21}{2.8} = 5.786$$
$$\theta = 80.20°$$

Graphical method–The various steps are as follows:

1. First draw the angular position diagram as shown in Fig. 13.6(a).
2. Find out the centrifugal forces exerted by each mass which are proportional to the product of the mass and its radius of rotation on the rotating shaft.

3. Now draw force polygon by drawing parallel to corresponding masses, choosing a suitable scale ab representing the centrifugal force due to m_1, bc due to m_2, cd due to m_3 and de due to m_4 respectively as shown in Fig. 13.6(b).
4. As per the law of force, the closing side ae represents the resultant force in magnitude and direction as shown.
5. The balancing force is equal and opposite the resultant force, i.e. ea.
6. Find the magnitude of the balancing mass m_b at a given radius r_b such that $m_b r_b$ is equal to the resultant centrifugal force represented by ae. Measure ae and divide it with 0.2. Therefore,
$$m_b = \frac{16.03}{0.2} = 80.15 \text{ kg}.$$
7. Measure the angle made by ae with respect to horizontal which is equal to 87°.

Note: There is little variation between the results obtained by analytical and graphical methods because of errors while drawing force polygon and the scale adopted.

13.6 BALANCING OF SEVERAL UNBALANCED MASSES ROTATING IN SEVERAL PLANES

The problem of unbalance due to several masses rotating in several planes can be taken up in two ways:

(i) Balancing each unbalanced mass in the same plane on the diametrically opposite side as explained in the case of single unbalanced mass. Hence, the number of balancing masses will be equal to the number of unbalanced masses. But this is a tedious procedure to balance individually.

(ii) Take two planes convenient for finding the two balancing masses to balance all the unbalanced masses in several planes. The principle involved is very simple as explained below:

The principle to be followed here is

1. to fix two planes conveniently and
2. then shifting the unbalanced forces acting in different planes to one of the balancing planes.

The shifting of force from one place to other gives rise to, same force acting at the new place together with a couple equal to that force multiplied by the distance moved as shown in the Fig. 13.7. It is shown at Fig. 13.7(a) that there is one force F acting at 1. Two equal and opposite forces F are seen added at 2 in Fig. 13.7(b). One upward force at 1 and one downward force at 2 forms a couple C equal to product of the force F and the distance shifted say L, leaving one upward force F at 2 as shown at Fig. 13.7(c) respectively. Hence, it is to be understood that when a force is shifted, it results in one couple (equal to product of force and the distance shifted) and the same force at the shifted place. This concept is adopted here in the case of several unbalanced masses rotating in several planes.

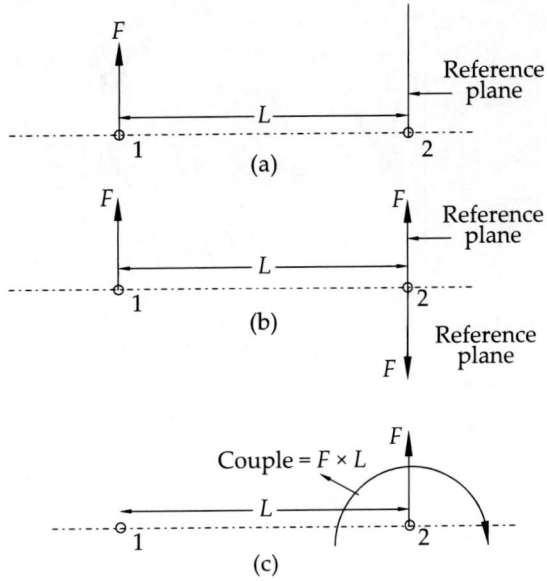

Fig. 13.7 Effecting of shifting the force.

Reference planes are the planes parallel to the plane of rotation of the unbalanced masses and perpendicular to the axis of rotation at some points on the shaft.

Hence, in order to have a complete balance of the several revolving unbalanced masses in different planes, the following two conditions must be satisfied:

1. All the forces must balance, i.e. the resultant force must be zero including the forces due to the balancing masses in two balancing planes.
2. All the couples must balance, i.e. the resultant couple must be zero including the couples due to the balancing masses.

Note:

1. Choose the planes for balancing masses at the specified or convenient places.
2. Follow the sign conventions for distances from the reference planes on either side. The usual practice is to consider the distance on the right-hand side of the reference plane as positive and negative for left-hand side of the reference plane.
3. The couples are represented by vectors as per right-hand thumb rule. Consider clockwise couples as +ve and anticlockwise couples as −ve.

Figure 13.8 shows four unbalanced masses m_1, m_2, m_3, m_4 regarded as revolving in four planes 1, 2, 3 and 4 at the distances indicated respectively as shown at (a) and with their relative angular positions as shown at (b). L and M represent the planes specified or selected for applying balancing masses. Let m_{bL} and m_{bM} represent the balancing masses in planes L and M which are to be determined as follows:

404 Theory of Machines

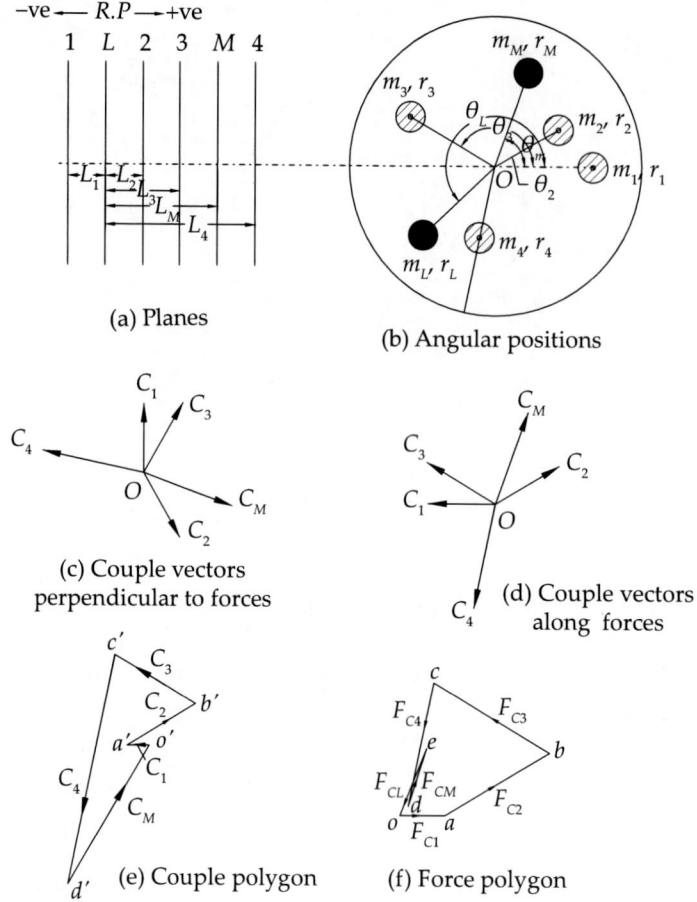

Fig. 13.8 Several unbalanced masses in several planes (a, b, c, d, e and f).

1. Choose plane L as reference (R.P.). The distances of all other planes from L are to be indicated as per the sign conventions given above in the note.
2. Tabulate the data as shown below by taking the planes in the same order from left to right.

Plane	Mass m	Radius r	Force/ω^2 (F_C) mr	Distance from (Ref. P) L	Couple/ω^2 (C) mrL
1	m_1	r_1	$(F_{C1}) m_1 r_1$	$-L$	$-(C_1) m_1 r_1 L_1$
L(Ref. P)	m_{bL}	r_{bL}	$(F_{CL}) m_{bL} r_{bL}$	0	$(C_L) 0$
2	m_2	r_2	$(F_{C2}) m_2 r_2$	$+L_2$	$+(C_2) m_2 r_2 L_2$
3	m_3	r_3	$(F_{C3}) m_3 r_3$	L_3	$+(C_3) m_3 r_3 L_3$
M	m_{bM}	r_{bM}	$(F_{CM}) m_{bM} r_{bM}$	$+L_M$	$+(C_M) m_{bM} r_{bM} L_M$
4	m_4	r_4	$(F_{C4}) m_4 r_4$	L_4	$+(C_4) m_4 r_4 L_4$

3. Represent the couple by a vector drawn perpendicular to the plane of the couple as per right-hand thumb rule. The couple C_1 is due to transfer of force $(F_1)m_1r_1$ from plane 1 to the reference plane L and it is proportional to $m_1r_1L_1$ and acts perpendicular to the plane in which the centrifugal force of m_1 acts. Follow some convention for negative and positive couples. Here assumed upward for negative couples, thus through O, draw OC_1 upwards as shown Fig. 13.2(c). Similarly, draw the vectors OC_2, OC_3 and OC_4 perpendiculars to planes of centrifugals due to masses m_2, m_3 and m_4 respectively as shown downwards being positive.

4. Turn all the couple vectors drawn at (c), counterclockwise through 90°. Now it can be observed that the couple vectors are along centrifugal forces (for positive away from the centre and for negative towards centre) as shown at (d). This is more convenient for drawing couple polygon as shown in Fig. 13.8(e). The relative position of masses remains unchanged as shown.

5. Now draw the couple polygon as shown at Fig. 13.8(e) and the closing side $d'o'$ represents the balanced couple. Since the balanced couple CM is proportional to $m_{bM}r_{bM}L_M$, the balancing mass m_{bM} = vector $\dfrac{o'd'}{(r_{bM} \cdot L_M)}$ and its angle of inclination θ_M can be measured from Fig. 13.8(b) by drawing a radial line parallel to $d'o'$.

6. Now draw the force polygon as shown at Fig. 13.8(f). The closing side vector eo (in the direction from e to o) represents the balancing force. Since the balancing force is proportional to $m_{bL}r_{bL}$, therefore $m_{bL}r_{bL}$ = vector eo or m_{bL} = vector $\dfrac{eo}{r_{bL}}$. From this expression, the value of the balancing mass m_{bL} in the plane L can be obtained and then measure the angle of inclination θ_L of this mass with the horizontal as shown in Fig. 13.8(b).

W E 13.5: A shaft of 3 m span carries two masses A and B of magnitude 5 kg and 10 kg respectively and revolving at radii 450 mm and 600 mm respectively. The planes in which the masses rotate are 1.2 m and 2.4 m respectively from the left support and the angles between the cranks is 60°. If the speed of rotation is 100 r.p.m., find the displacing force on the two supports of the machine shaft. The balancing masses are to be placed in planes X and Y as shown in Fig. 13.9(a). The distances between the planes X and A is 0.9 m and between X and Y is 2.4 m. If the balancing masses resolve at a radius of 300 mm, find their magnitudes and angular positions.

Given:
PQ be the shaft supported as shown at Fig. 13.9(a). Masses are, $m_A = 5$ kg; $m_B = 10$ kg; and radii are $r_A = 450$ mm = 0.45 m. $r_B = 600$ mm = 0.06 m. Take either X or Y as reference plane. Here X is taken as reference plane and the radii of the balancing masses are $r_X = r_Y = 300$ mm = 0.3 m.

Let m_X and m_Y be the balancing masses in the X and Y planes. The angular position of the mass B, assuming the mass A as in the horizontal position is as shown in the Fig. 13.9(b). Assume X plane as reference plane and indicate it by Ref. P. The distances of the other planes to the right side of the X plane are taken as +ve. Tabulate the data in a tabular form as indicated below:

Plane	Mass m	Radius r	Force/ω^2 mr	Distance from (Ref)L L	Couple/ω^2 LmrL
X (Ref. P)	m_X	0.3	$0.3m_X$	0	0
A	5	0.45	2.25	0.9	2.025
B	10	0.6	6.0	2.1	12.60
Y	m_Y	0.3	$0.3m_Y$	2.4	$0.72m_Y$

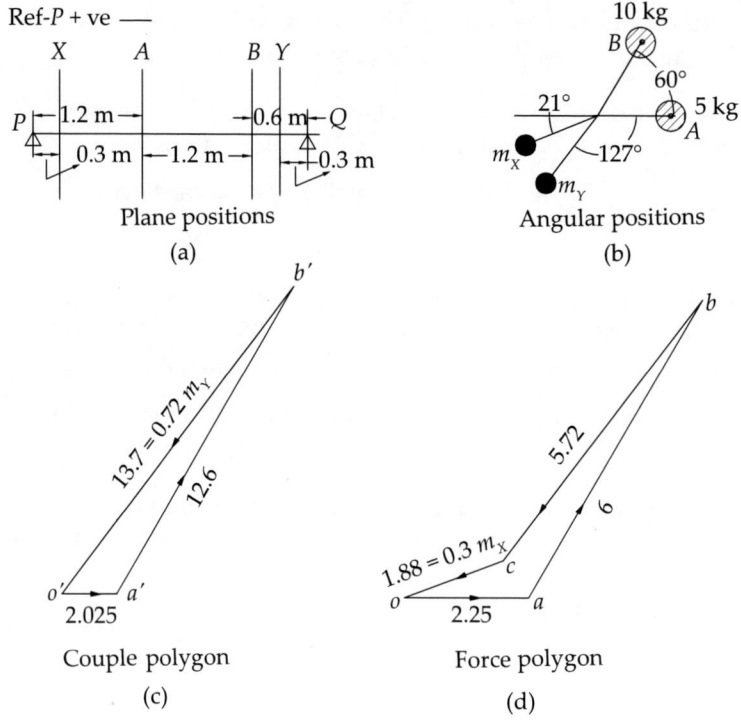

Fig. 13.9

Determine the balancing masses m_X and m_Y and their angular positions graphically as below:

1. First draw the couple polygon from the above table as shown in Fig. 13.9(c) to some suitable scale. The resultant couple represented by vector $b'o'$ gives the balanced couple equal to $0.72m_Y$, therefore, $0.72m_Y = b'o' = 13.7$ kg-m² or $m_Y = 19.1$ kg.
Obtain the angular position of the mass m_Y by drawing as shown in Fig. 13.9, parallel to vector bo. By measurement, the angular position of mass m_Y, $\theta_Y = 127°$ in the clockwise direction from mass m_A.

2. Now draw the force polygon from the data given in column 4 of table as shown in Fig. 13.9(d) to some suitable scale. The vector co represents the balancing force proportional to $0.3m_X$, therefore by measurement, $0.3m_X = co = 1.875$ kg-m or $m_X = 6.25$ kg.

Obtain the angular position of the mass m_X by drawing as shown in Fig. 13.9(b), parallel to vector co. By measurement, the angular position of m_X is $\theta_X = 201°$ in the anticlockwise direction from mass m_A.

W E 13.6: Four masses A, B, C and D as shown below are to be completely balanced.

	A	B	C	D
Mass (kg)	-	25	40	35
Radius (mm)	150	200	100	180

The planes containing masses B and C are 250 mm apart. Masses C and D make angles of 90° and 195° respectively with B in the same sense. Find the magnitude and the angular position of mass A and the position of planes A and D.

Given:
$r_A = 150$ mm $= 0.15$ m; $\quad m_B = 25$ kg; $\quad r_B = 200$ mm $= 0.20$ m; $\quad m_C = 40$ kg; $\quad r_C = 100$ mm $= 0.1$ m; $m_D = 35$ kg; $\quad r_D = 180$ mm $= 0.18$ m; $\quad \angle BOC = 90°$; $\quad \angle BOD = 195°$.

Assume the planes A and D as shown in Fig. 13.10(a).

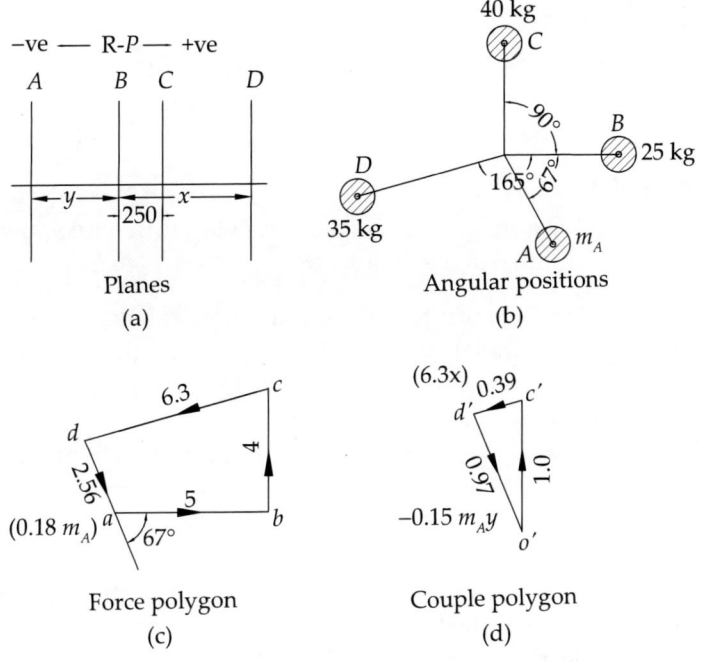

Fig. 13.10

1. The magnitude and the angular position of mass A

Let m_A be the magnitude of mass A, X and Y be the distances of planes A and D from B as shown in Fig. 13.10. The angular positions of the masses are as shown in Fig. 13.10(b) respectively.

Assume the plane B as the reference plane (Ref. B) and the mass B (m_B) as along the horizontal line. The data is tabulated as below.

Plane	Mass m	Radius r	Force/ω^2 mr	Distance from (Ref)L L	Couple/ω^2 $LmrL$
A	m_A	0.15	$0.15 m_A$	$-Y$	$-0.15 m_A . Y$
B (Ref. P)	25	0.2	5.0	0	0
C	40	0.1	4.0	0.25	1.0
D	35	0.18	6.3	X	6.3X

The magnitude and angular position of mass A has to be determined by drawing the force polygon first from the data given in the column 4 of the table as shown in Fig. 13.10, to some suitable scale. Since the masses are to be completely balanced, therefore the force polygon must be a closed figure. The closing side (i.e., vector do) is proportional to $0.15 m_A$.

By measurement $0.15 m_A$ = vector do = 2.56 kg-m or m_A = 17.3 kg.

In order to find the angular position of mass A, draw OA in Fig. 13.10 parallel to vector do and measure the angular position of mass A from mass B in the anticlockwise direction giving $\angle AOB = 293°$ or clockwise giving $\angle AOB = 67°$.

2. Position of planes A and D

The positions of planes A and D are to be obtained by drawing the couple polygon, as shown in Fig. 13.10 taking the data in column 6 of the table:

(a) Draw vector $o'c'$ parallel to OC equal to 1.0 kg-m², to some suitable scale.
(b) From c' and o', draw lines parallel to OD and OA, such that they intersect at point d'. By measurement, you find $6.3X$ = vector $c'd'$ = 0.39 kg-m² or X = 0.062 m, i.e. on the positive side of the reference line as assumed.
Therefore, the plane of mass D is at 0.062 m or 62 mm towards right of the reference plane B.
(c) Again from couple polygon, closing side $d'o'$, by measurement is 0.97 which represents $-0.15 m_A.Y$. Substituting for m_A as 17.3 kg, gives Y = -0.386 m or -0.386 m, i.e. opposite of the assumed direction.
The negative sign indicates that the plane A is 0.386 m or 386 mm towards right of plane B, but not towards left of B, as assumed.
Therefore, the plane of mass A is at 0.386 m or 386 mm towards right of the reference plane B.

W E 13.7: A, B, C and D are four masses carried by a rotating shaft at radii 100, 125, 200 and 150 mm respectively. The planes in which the masses rotate are spaced 600 mm apart and the masses of B,

C and D are 10 kg, 5 kg, and 4 kg respectively. Find the required mass A and the relative angular position of the four masses so that the shaft shall be in complete balance.

Given:

$r_A = 100$ mm $= 0.1$ m; $r_B = 125$ mm $= 0.125$ m; $r_C = 200$ mm $= 0.2$ m; $r_D = 150$ mm $= 0.15$ m; $m_B = 10$ kg; $m_C = 5$ kg; $m_D = 4$ kg

The positions of the planes in the front view are shown in Fig. 13.11. Assuming the plane of mass A as reference, the data is tabulated as follows:

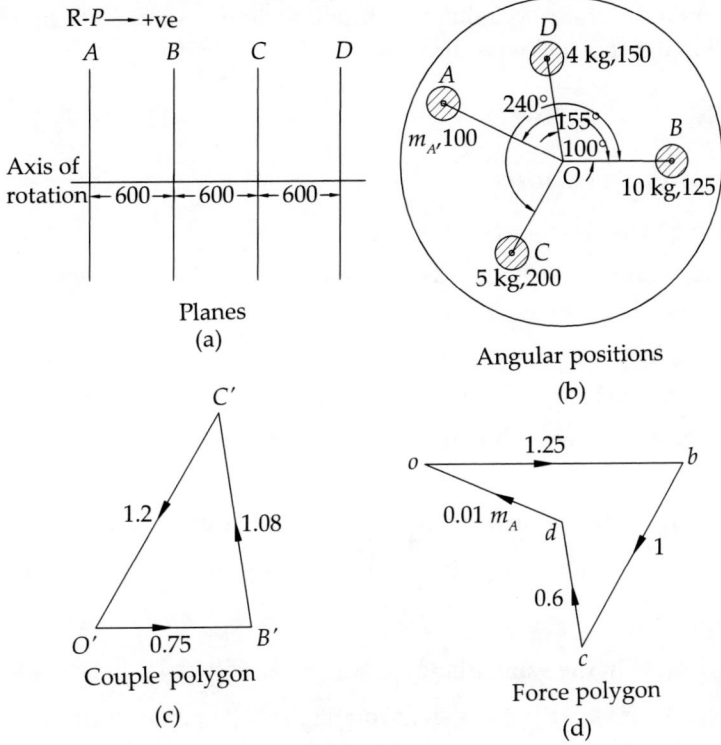

Fig. 13.11

Plane	Mass m	Radius r	Force/ω^2 mr	Distance from (Ref)L L	Couple/ω^2 $LmrL$
A(Ref)	m_A	0.1	$0.1 m_A$	0	0
B	10	0.125	1.25	0.6	0.75
C	5	0.2	1	1.2	1.2
D	4	0.15	0.6	1.8	1.08

First draw the couple polygon as shown at Fig. 13.11 from the above data by assuming the position of mass B in the horizontal direction OB as shown in Fig. 13.11 with O'B' horizontal to some suitable scale to represent 0.75 kg-m². From O' and B' draw vectors $O'C'$ and $B'C'$ equal to 1.2 and 1.08 kg-m² respectively which intersect at C'. Now draw in Fig. 13.11(d) OC parallel to $O'C'$ and OD parallel to $B'C'$. By measurement the angular settings of mass C from mass B in the anticlockwise direction, i.e. $\angle BOC = 240°$, and that of D from B in the anticlockwise direction, i.e. $\angle BOD = 100°$.

In order to find the required mass A (m_A) and its angular settings, draw the force polygon to some suitable scale, as shown at Fig. 13.10 from the data given in table column (4). The closing side vector do 0.7 is equal to 0.1 m_A, therefore $m_A = 7$ kg.

Now draw OA in Fig. 13.10(b), parallel to vector do. By measurement, the angular setting of mass A from mass B in the anticlockwise direction, i.e. $BOA = 155°$.

13.7 EXERCISE

13.7.1 Short Answer Questions

1. What is balancing? Discuss its advantages.
2. Describe the procedure for the balancing of rotating bodies. Why is balancing of rotating parts necessary for high-speed engines?
3. State clearly the difference between (i) balancing of a single rotating body by another body in the same plane, and (ii) balancing of a single rotating by two bodies in two different planes.
4. How will you balance several bodies rotating in one plane by a body in the same plane analytically?
5. Explain the graphical method of balancing several bodies in one plane by a body in the same plane.

13.7.2 Problems

I Balancing of single mass in the same plane

1. A body of mass 30 kg is attached to a shaft rotating at 300 r.p.m. at a distance of 500 mm from its axis. The body is to be balanced by mass, which has to be attached at a distance of 300 mm from the axis of the shaft. Find the magnitude of the balancing mass. (*Ans.* 50 kg)

II Balancing of single mass in two other planes

1. A body of mass 10 kg is attached to a rotating shaft at a radius of 500 mm from its axis of rotation. It is to be balanced by two masses with their centre of gravity in the same plane in such a way that one of the mass is 200 mm from 10 kg mass and the other 300 mm on the opposite side. Find the masses of the balancing masses, if the radii are 400 mm from the axis of the rotating shaft. (*Ans.* 7.5 kg)

III Several masses in the same plane

1. Three masses A, B and C of 20 kg, 18 kg and 32 kg respectively rotate at radii of 0.4, 0.5 and 0.2 m respectively, in one plane. The angular positions of B and C are 60° and 135° respectively from A. Find the magnitude and position of mass D on a radius of 0.6. (*Ans.* 24.5 kg; 212.9°)

2. In a mechanism, four masses m_1, m_2, m_3 and m_4 are 20 kg, 30 kg, 24 kg and 26 kg respectively. The corresponding radii of rotation are 200 mm, 150 mm, 250 mm and 300 mm respectively. The angles between the successive masses are 45°, 75° and 135° respectively. Estimate the position and magnitude of the mass, which when attached at a radius of 200 mm in the same plane of rotation will balance the system. **(Ans. 248.70; 11.6 kg)**

3. Two weights of 8 kg and 16 kg rotate in the same plane at radii of 1.5 and 2.25 m respectively. The radii of these weights are 60° apart. Find the position of the third weight of the magnitude of 12 kg in the same plane which can produce static balance of the system.

IV Several masses in several planes

1. A, B, C, D are the four masses carried by a rotating shaft at 100 mm, 150 mm, 150 mm and 200 mm radius respectively. The planes in which the masses rotate are spaced at 500 mm apart and the magnitudes of the masses B, C and D are 9 kg, 5 kg and 4 kg respectively. Find the required mass A and the relative angular setting of the four masses so that the shaft shall be in complete balance.

2. A shaft carries four masses A, B, C and D of 12, 20, 30 and 16 kg respectively spaced 18 cm apart. Measuring angle anticlockwise from A, B is 240°, C is 135° and D is 270°. The radii are 15 cm, 12 cm, 6 cm and 18 cm and the speed of the shaft is 120 r.p.m. Find the magnitude and direction relative to A of the resultant moment at a plane midway between A and B.

3. Four masses m_1, m_2, m_3 and m_4 having 100, 175, 200 and 25 kg are fixed to cranks of 20 cm radius and revolve in places 1, 2, 3 and 4. The angular position of the cranks in planes 2, 3 and 4 with respect to the crank in plane 1 are 75°, 135° and 200° taken in the same sense. The distance of planes 2, 3 and 4 from plane 1 are 60 cm, 186 cm and 240 cm respectively. Determine the position and magnitude of the balance mass at a radius of 60 cm in plane L and M located at middle of the plane 1 and 2 and the middle of the planes 3 and 4 respectively.

4. A shaft 3 m span between the bearings carries two masses of 5 kg and 10 kg acting at the extremities of the arms 0.45 m and 0.6 m long respectively. The planes in which the masses rotate are 1.2 m and 2.4 m respectively from the left-hand bearing and the angle between the arms is 600. If the speed of rotation is 100 r.p.m., find the displacing force on the two bearings of the machine. If the masses are balanced by two additional masses acting at a radius 0.3 m and placed 0.3 from each bearing, estimate the magnitude of the two balanced masses and angles at which they may be set with respect to the two arms.

5. A, B, C, D are the four masses carried by a rotating shaft at radii of 100 mm, 125 mm, 200 mm and 150 mm respectively. The planes in which the masses rotate are spaced at 600 mm apart and the magnitude of the masses B, C and D are 10 kg, 5 kg and 4 kg respectively. Find the required mass A and the relative angular setting of the four masses so that the shaft shall be in complete balance. (Assume the plane of the mass as the reference plane.)

6. Four masses A, B, C and D are to be completely balanced. The planes centering masses B and C are 30 cm apart. The angle between planes containing masses B and C is 90°. C and B

make angles of 120° and 210° respectively with D in the same sense. Find: (a) The weight and angular position of mass A. (b) The position of planes A and D.

Details of rotating masses:

Mass	Weight (kg)	Radius (cm)
A	W	18
B	30	24
C	50	12
D	40	15

13.7.3 Multiple Choice Questions

1. In order to balance a rotating body, another body of the same mass is attached to the rotating body on its opposite side.

   (a) Yes (b) No [*Ans.* (b)]

2. The principle involved in balancing a system of balancing masses is that the resultant centrifugal force of the rotating bodies should be equal and opposite to that of the balancing body.

   (a) Agree (b) Disagree [*Ans.* (a)]

3. In order to balance a system of rotating bodies, we must know the angular velocity of the rotating shaft.

   (a) True (b) False [*Ans.* (b)]

4. A single body can be balanced only by another body of _____.

   (a) smaller mass (b) small mass (c) bigger mass (d) any one of them [*Ans.* (d)]

5. Static balancing involves balancing of _____.

   (a) forces (b) couples (c) both (d) masses [*Ans.* (a)]

6. In case of rotating masses the magnitude of the balancing mass is _____ when the speed of the shaft is doubled.

   (a) doubled (b) halved (c) unaffected (d) quadrupled [*Ans.* (c)]

7. For complete dynamic balance, at least _____ mass/masses are necessary.

   (a) two (b) three (c) four (d) one [*Ans.* (a)]

8. If rotating system is dynamically balanced, it is statically _____.

   (a) balanced (b) unbalanced (c) partially balanced [*Ans.* (a)]

Balancing of Reciprocating Masses

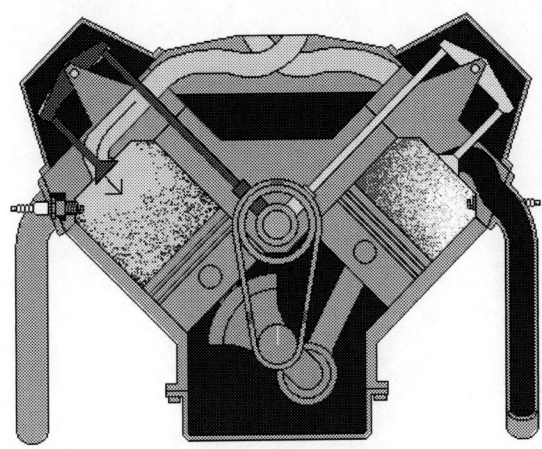

Balancing of Reciprocating Masses

14.1 INTRODUCTION

In the previous chapter, the balancing of rotating parts like cranks, pulleys, gears and flywheels, etc. were seen. But there are elements which reciprocate such as pistons in I.C. engines, cross head in a steam engine and there are some parts like connecting rod which oscillate. These elements also have certain mass and when they undergo acceleration give forces. These forces act on the crankshaft causing bending stress and produce vibrations. Hence, the study of balancing of these forces due to reciprocating masses is dealt with in this chapter. The various forces acting on the reciprocating parts were discussed in the chapter on inertia forces due to masses having reciprocating motions. The resultant of all these various forces acting on the body of the engine is known as an unbalanced force. The purpose of balancing the reciprocating masses is to reduce or minimise the unbalanced forces. The Pic. 14.1 and Fig. 14.1 shows the reciprocating mechanism, where m_R is the mass of the reciprocating parts, L is the length of the connecting rod, r is the radius of the crank, θ is the angle made by the crank with the line of stroke, w is the angular velocity of the crank and n is the ratio of $\dfrac{L}{r}$.

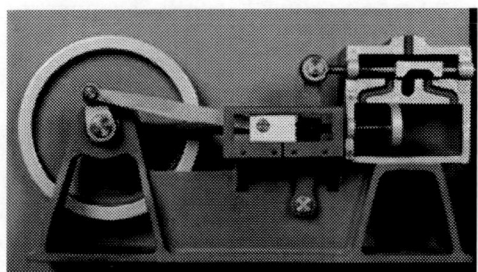

Pic. 14.1

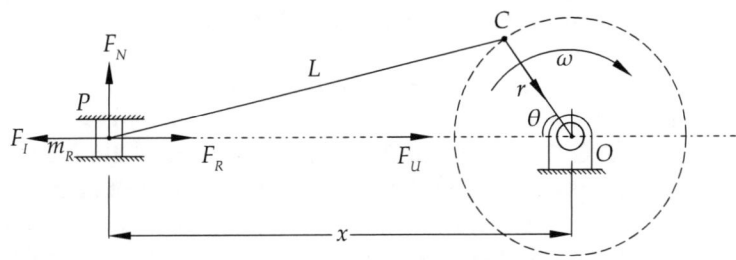

Fig. 14.1 Reciprocating mechanism.

The analytical expression for the acceleration a_R of the reciprocating parts is $\omega^2 r[\cos\theta + \dfrac{\cos 2\theta}{n}]$ and the inertia force F_I due to the reciprocating parts having mass m_R is equal and opposite to the accelerating force F_R of the reciprocating parts. The expression for F_R is $m_R.\omega^2 r[\cos\theta + \dfrac{\cos 2\theta}{n}]$. This is the unbalanced force F_U due to reciprocating masses acting on the crankshaft which can be written as sum of two components $m_R\omega^2 r\cos\theta + m_R\omega^2 r\dfrac{\cos 2\theta}{n}$. The first term $m_R\omega^2 r\cos\theta$ is called the primary unbalanced force F_P and the second term $m_R\omega^2 r\dfrac{\cos 2\theta}{n}$ is called the secondary unbalanced force F_S. Both these forces F_P and F_S depend on the crank angle θ and their maximum and minimum values are as follows:

F_P is maximum at $\theta = 0°$ or $180°$ and it is $F_{P\max} = mR\omega^2 r$ occurs twice per revolution.

F_S is maximum at $\theta = 0°, 90°, 180°, 360°$ and it is $F_{S\max} = \dfrac{m_R\omega^2 r}{n}$ occurs four times per revolution.

The maximum primary force is n times secondary force. Based on the value of n and at moderate speeds, secondary force is negligible. Therefore, the study of balancing of primary unbalanced forces is important.

Note: The unbalanced forces due to the reciprocating masses vary in magnitude depending upon the value of θ but they are constant in direction. But the unbalanced forces due to the rotating masses vary in the direction depending on the value of θ but they are constant in magnitude.

Let m_R represent the reciprocating mass; F_P primary force; F_I inertia force; F_U unbalanced force; F_S secondary force; F_R accelerating force due to reciprocating masses; θ crank angle; N speed in r.p.m.; ω speed in radians per sec; r crank radius; L length of connecting rod; n ratio of $\dfrac{L}{r}$.

14.2 PARTIAL BALANCING

Partial balancing of unbalanced primary forces in a reciprocating engine using a revolving balancing mass m_b: Imagine for the time being, the primary force as the horizontal component of the centrifugal force due to a rotating mass equal to m_R assumed at a crank radius r and rotating at ω radians per sec as shown in the Fig. 14.2. Assume as shown in the figure, a balancing mass m_b attached opposite the crank at a radius r_b and rotating at the same angular velocity as crank, i.e. ω rad/sec. This gives a centrifugal force of $m_b\omega^2 r_b$ and its horizontal component is $m_b\omega^2 r_b \cos\theta$. This is equal to the primary force F_P provided the product of rotating balancing mass m_b and its radius r_b is equal to the product of reciprocating mass m_R and its crank radius r as both are rotating at an angular velocity ω.

Then the two components $m_b\omega^2 r_b \cos\theta$ and $m_R\omega^2 r\cos\theta$ which are equal and opposite cancel, leaving a perpendicular component $m_b\omega^2 r_b \sin\theta$ as the unbalanced force. This again gives a maximum value of $m_b\omega^2 r_b$ when θ is $90°$ and $270°$. Therefore, now the unbalanced force due to the reciprocating mass is perpendicular to the line of stroke instead of along the line of stroke. This method of balancing the reciprocating masses gives only a change in the direction of the unbalanced force hence this is not a correct solution. It is only to change the problem by changing the direction of the unbalanced force.

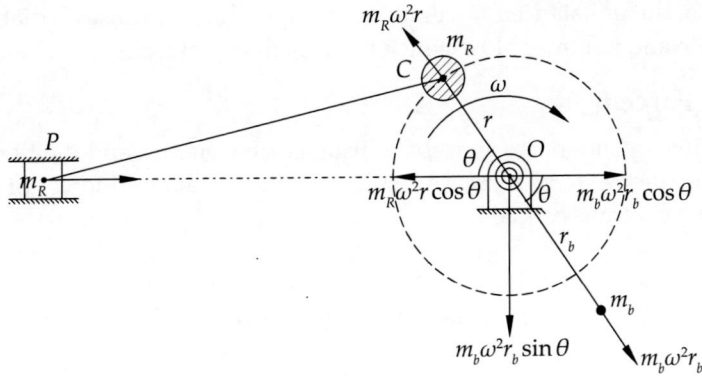

Fig. 14.2 Balancing of reciprocating mass.

The method of reducing the magnitude and direction of the unbalanced force is called partial balancing. That means trying to reduce the magnitude of the unbalanced force by assuming the balancing mass as a fraction of the reciprocating masses, i.e. ($m_b = cm_R$) where c is a constant having value less than one. Now there will be two components both along the horizontal direction F_H and along the vertical direction F_V as follows:

Therefore, the net horizontal component F_H along the line of stroke = $(1-c)m_R\omega^2 r \cos\theta$ and vertical component $F_V = cm_R\omega^2 r \sin\theta$.

The resultant of these two components gives the unbalanced force at any instant

$$\sqrt{[(1-c)m_R\omega^2 r \cos\theta]^2 + [cm_R\omega^2 r \sin\theta]^2} = m_R\omega^2 r \sqrt{[(1-c)^2 \cos^2\theta + c^2 \sin^2\theta]}$$

By plotting curves for different values of c, the magnitude of the unbalanced force on the engine frame can be observed having minimum value for c equals to 0.5. The resultant unbalanced force for $c = 0.5$, becomes $\dfrac{m_R\omega^2 r}{2}$. This method of balancing is called the partial balancing as it has not only reduced the magnitude but also changed the direction of the unbalanced force.

Applications of reciprocating engines

1. Locomotives (two cylinders)
2. Multi-cylinders in line engines (automobiles and marine)
3. V-engines and radial engines (air compressors and aircraft).

14.3 EFFECT OF PARTIAL BALANCING IN TWO-CYLINDER LOCOMOTIVES

The effect of partial balancing of the reciprocating parts produces unbalanced primary forces both along the line of the stroke and also perpendicular to the line of stroke. In a two-cylinder locomotive, the two cranks of the two cylinders will be at 90° in phase difference. If θ is the crank angle of cylinder (1), then the other crank angle of cylinder (2) is $(90° + \theta)$ as shown in Fig. 14.3(a). The effect of partial balancing of the unbalanced forces along the line of stroke causes variation in the tractive force, swaying couple and hammer blow which are explained below.

14.3.1 Tractive Force (F_T)

In a locomotive the two cylinders are apart by a distance of 'a' metres and their cranks are at 90° in phase difference. The tractive force F_T is the resultant of the unbalanced forces of the two cylinders acting along their line of strokes equal to

$$F_T = (1-c)m_R\omega^2 r[\cos\theta + \cos(90° + \theta)]$$
$$= (1-c)m_R\omega^2 r[\cos\theta - \sin\theta] \tag{1}$$

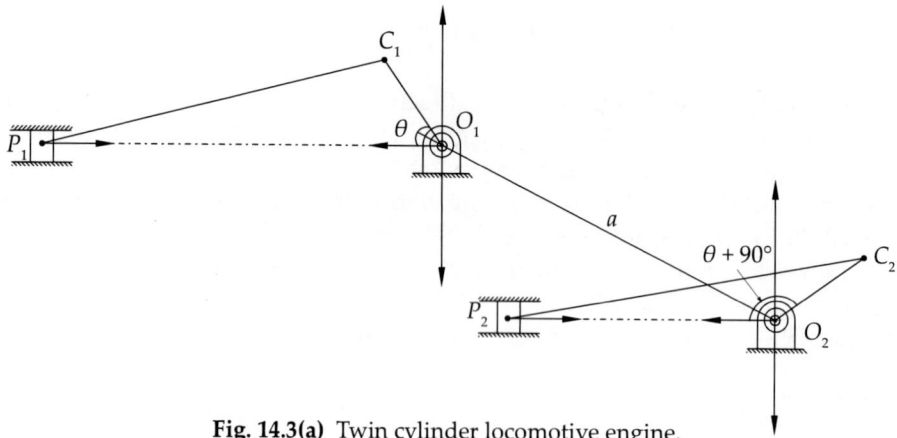

Fig. 14.3(a) Twin cylinder locomotive engine.

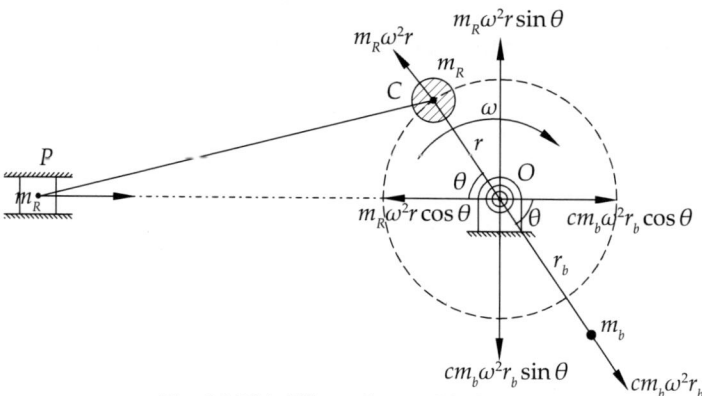

Fig. 14.3(b) Effect of partial balancing.

The tractive force F_T is maximum or minimum when the value of $\dfrac{d(\cos\theta - \sin\theta)}{d\theta}$ is zero or the value of $(-\sin\theta - \cos\theta)$ is zero or $\tan\theta = -1$. Hence, the value of θ should be 135° or 315°. Therefore, the maximum or minimum tractive force

$$F_T = \pm(1-c)m_R\omega^2 r(\cos 135° - \sin 135°) = \pm\sqrt{2}(1-c)m_R\omega^2 r$$

14.3.2 Swaying Couple

The two cylinders are at a distance of 'a' metres as shown in Fig. 14.4(a). Taking moments about the centre line of the two cylinders gives the

$$\text{Swaying couple} = \left[(1-c)m_R\omega^2 r\cos\theta \cdot \frac{a}{2}\right] - \left[(1-c)m_R\omega^2 r\cos(90°+\theta) \cdot \frac{a}{2}\right]$$

$$= (1-c)m_R\omega^2 r(\cos\theta + \sin\theta)\frac{a}{2}.$$

This will be maximum or minimum when $\dfrac{d(\cos\theta + \sin\theta)}{d\theta}$ is equal to zero. That is, $-\sin\theta + \cos\theta = 0$ or $\tan\theta = 1$, i.e. $\theta = 45°$ or $225°$.

Therefore the maximum or minimum swaying couple $= \pm(1-c)m_R\omega^2 r(\cos 45° + \sin 45°)\cdot\dfrac{a}{2}$

$$= \pm(1-c)m_R\omega^2 r \cdot \frac{a}{\sqrt{2}}$$

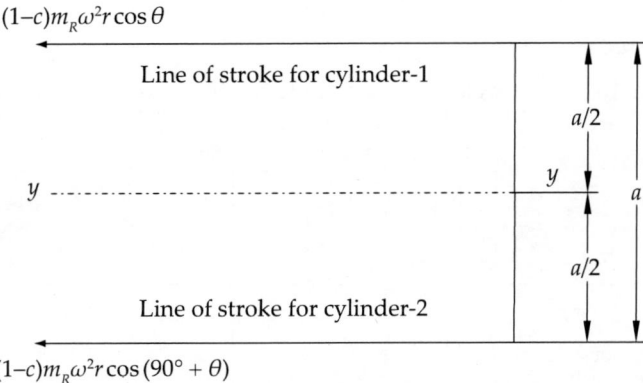

Fig. 14.4(a) Swaying couple.

14.3.3 Hammer Blow

The effect of variation in the magnitude of unbalanced force perpendicular to the line of stroke is called the Hammer blow. So Hammer blow is $m_b\omega^2 r_b \sin\theta$. The maximum value of Hammer blow is $m_b\omega^2 r_b$ when θ is 90° or 270° as shown in Fig. 14.4(b).

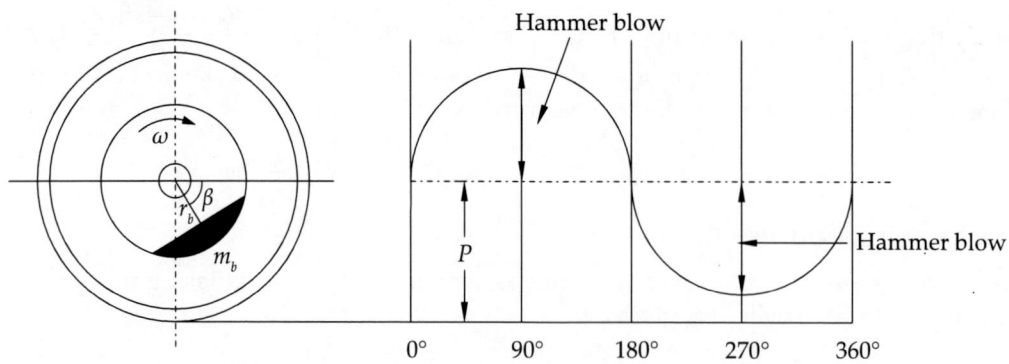

Fig. 14.4(b) Hammer blows.

Let P be the static load acting downwards on the rails. Therefore, the net pressure between the wheel and the rail is $(P \pm m_b\omega^2 r_b)$. If $(P \pm m_b\omega^2 r_b)$ is negative, then the wheel will be lifting from the rails or the wheels lose contact with rails. Hence, the limiting value for P should be $m_b\omega^2 r_b$ or the limiting speed or angular velocity $\omega = \sqrt{\dfrac{P}{m_b r_b}}$. If the speed exceeds this limiting speed, then the wheel lifts from the rails and causes accidents.

Note:
Let m_R and m_{RO} represent the reciprocating and rotating masses respectively, then the total balancing mass m_b on the opposite side of the crank is $(m_{RO} + cm_R)$.

14.3.4 Types of Locomotives

The difference between the crank angles of two cylinders is 90° in order to produce uniform turning moment or torque or effort (T).

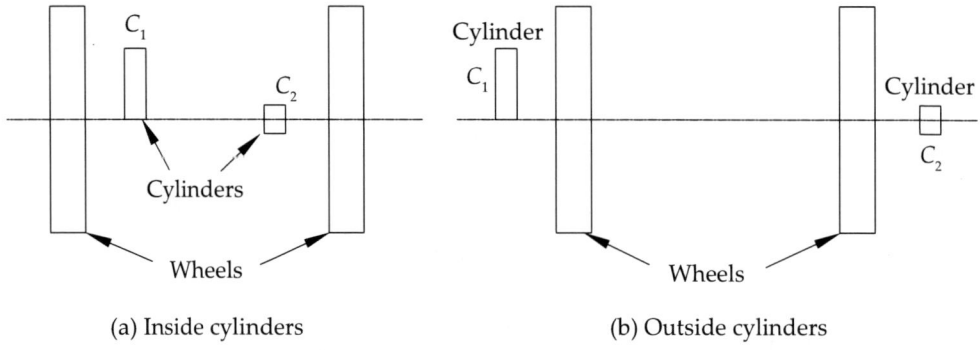

Fig. 14.5 Types of locomotives.

There are two types of locomotives:

1. Uncoupled: Here the effort is transmitted to one pair of the driving wheels only.

2. Coupled: The driving wheels are connected to the leading and trailing wheels by outside coupling rods. There are again two types with cylinders inside and wheels outside or cylinders outside and wheels inside as shown in Fig. 14.5 (a) and (b) below. These are explained by the worked examples.

W E 14.1: Inside cylinder: An inside cylinder locomotive has its cylinder centre lines 0.7 m apart and has a stroke of 0.6 m. The rotating masses per cylinder are equivalent to 150 kg at the crankpin, and the reciprocating masses per cylinder are 180 kg. The wheel centre lines are 1.5 m apart. The cranks are at right angles. The whole of the rotating masses and $\frac{2}{3}$ of the reciprocating masses are to be balanced by masses placed at a radius of 0.6 m.

Find (1) the magnitude and direction of the balancing masses, (2) the fluctuation in the rail pressure under one wheel, (3) the variation in the tractive effort, and (4) the magnitude of the swaying couple at a crank speed of 300 r.p.m.

Figure 14.6(a) shows the space diagram of the planes of wheels and cylinders at the given distances with each plane shown as a reference plane. Figure 14.6(b) shows the crank positions of two cylinders at 90°.

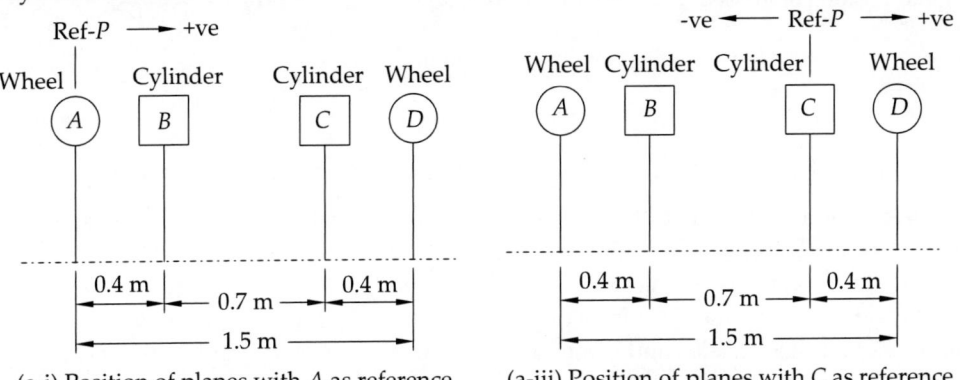

(a-i) Position of planes with A as reference

(a-iii) Position of planes with C as reference

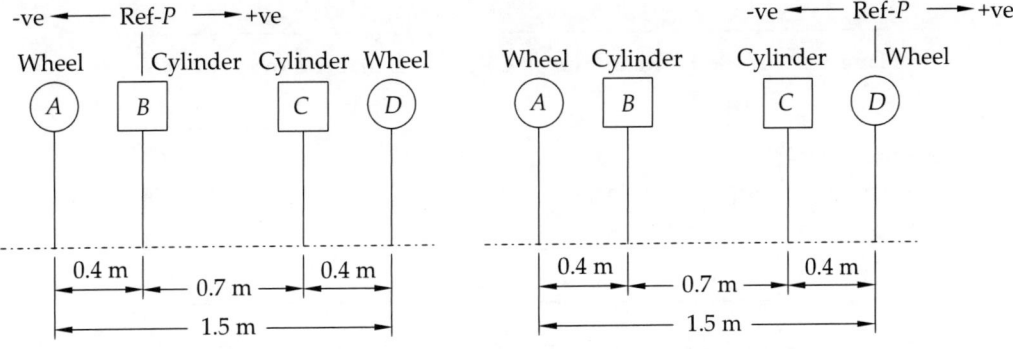

(a-ii) Position of planes with B as reference

(a-iv) Position of planes with D as reference

(a) Planes of cylinders and wheels

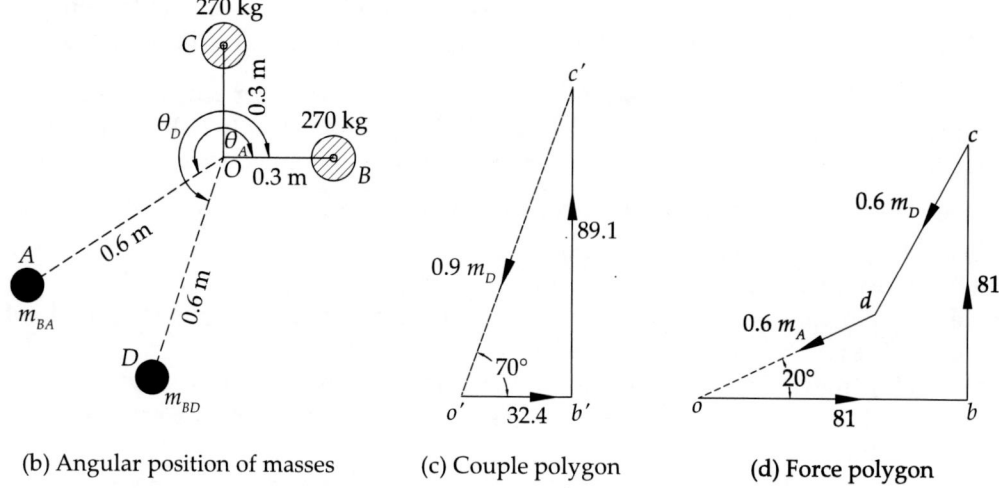

(b) Angular position of masses (c) Couple polygon (d) Force polygon

Fig. 14.6 Inside cylinder.

Given:

The total rotating masses of the cylinders = $\left(m_{RO} + \dfrac{2m_R}{3}\right) = \left(150 + 2 \times \dfrac{180}{3}\right) = 270$ kg and reciprocating masses $m_R = 180$ kg. Radius of the cranks $r = 0.3$ m as stroke is 0.6 m. The radius of the balancing mass is 0.6 m. The speed N is 300 r.p.m. or $\omega = 31.42$ rad/sec.

The equivalent balancing mass in the cylinder planes B and C be m_B and m_C equal to $(m_{RO} + cm_R)$ = 270 kg. Let m_A and m_D be the masses in the planes of the wheels A and D and θ_A and θ_D be their angular positions. Their magnitudes and directions are to be determined graphically as follows:

(i) Open a table as below and assume wheel A as the reference plane (Ref. P) as shown at Fig. 14.6(a-i).

Plane	Mass (kg) m	Radius (m) r	C.F./ω^2 m.r.	Distance L (m) from Ref. P	Couple/ω^2 m.r.L.
A (Ref. P)	m_A	0.6	$0.6 m_A$	0	0
B	270	0.3	81	0.4	32.4
C	270	0.3	81	1.1	89.1
D	m_D	0.6	$0.6 m_D$	1.5	$0.9 m_D$

Note: C.F. means centrifugal force and Ref. P means reference plane.

2. Looking at the centrifugal forces and the couples, two unknowns are in the forces and one unknown in the couples. Hence, first draw the couple polygon by choosing a proper scale as shown at Fig. 14.6(c), then equate the closing side $c'o'$ to $0.9 m_D$. Then $m_D = \dfrac{c'o'}{0.9}$ giving 105 kg.

Balancing of Reciprocating Masses 423

3. Draw a line OD parallel to $c'o'$ in the direction from c' towards o' in Fig. 14.6(b) and measure the angle θ_D with respect to B as shown.

4. Determine the other unknown force $0.6\,m_A$ by drawing a force polygon as shown at Fig. 14.6(d). The closing side do gives the magnitude and direction of m_A. The length do equals to $0.6\,m_A$. Then $m_A = \dfrac{do}{0.6} = 105$ kg.

5. Now draw a line OA parallel to do in Fig. 14.6(b) and measure the angle θ_A with respect to B as shown.

6. The balancing mass determined for wheels in the planes A and D, 105 kg represents the combined balancing mass of both rotating masses and reciprocating masses together. But the balancing mass for only rotating masses of 150 kg is $150 \times \dfrac{105}{270} = 58.5$ kg and balancing mass for only reciprocating masses of 120 kg is $120 \times \dfrac{105}{270} = 46.5$ kg. The total of both equals to 105 kg.

Hammer blow:

The balancing mass 46.5 kg for the reciprocating mass alone gives rise a centrifugal force which causes fluctuations in the pressure applied on the rails called hammer blow = $46.5(31.42)^2 0.6$ = 27602 N.

Variation in tractive force:

The maximum variation in the tractive force or effort = $\pm \sqrt{2}(1-c)mR\omega^2 r$

$$= \pm \sqrt{2}\left(1 - \dfrac{2}{3}\right) 180 \times 31.42^2 \times 0.3$$

$$= \pm 25127 \text{ N}$$

$$\text{Swaying couple} = a(1-c)m_R\omega^2 r \dfrac{1}{\sqrt{2}}$$

$$= 0.7\left(1 - \dfrac{2}{3}\right) 180 \times 31.42^2 \dfrac{0.3}{\sqrt{2}}$$

$$= 8797 \text{ N-m}$$

Note: *Choice of reference plane:*

1. Sometimes the reference plane will be specified in the problem then there is no doubt. Simply take it and solve the problem.
2. When it is not mentioned, the reference plane has to be selected.
3. Taking the plane A as reference plane, the problem was solved as above.
4. But one has to think of why not planes B or C or D be taken as reference planes? It is shown below taking one after another for better understanding and explained whether it works out or not.

Theory of Machines

(ii) Taking B as reference plane as shown at Fig. 14.6 (a-ii) and prepare the table as follows:

Plane	Mass (kg) m	Radius (m) r	C.F./ω^2 m.r.	Distance L (m) from Ref. P	Couple/ω^2 m.r.L.
A	m_A	0.6	0.6m_A	−0.4	−0.24m_A
B (Ref. P)	270	0.3	81	0	0
C	270	0.3	81	0.7	56.7
D	m_D	0.6	0.6m_D	1.1	0.66m_D

Look at the couples and forces there are two unknowns in both. So it is not possible to determine the unknowns. Hence, one should not take B as reference plane.

(iii) Take plane C as reference plane as shown at Fig. 14.6(a-iii) and fill the table as follows:

Plane	Mass (kg) m	Radius (m) r	C.F./ω^2 m.r.	Distance L (m) from Ref. P	Couple/ω^2 m.r.L.
A	m_A	0.6	0.6m_A	−1.1	−6.6m_A
B	270	0.3	81	−0.7	−56.7
C (Ref. P)	270	0.3	81	0	0
D	m_D	0.6	0.6m_D	0.4	0.24m_D

By looking at the couple and force columns, there are two unknowns here again in both forces and couples. So it is also not possible to determine the unknowns. Hence, no one should take C as reference plane.

(iv) Take plane D as reference as shown at Fig. 14.6 (a-iv) and fill the table as follows:

Plane	Mass (kg) m	Radius (m) r	C.F./ω^2 m.r.	Distance L (m) from Ref. P	Couple/ω^2 m.r.L.
A	m_A	0.6	0.6m_A	−1.5	−0.9m_A
B	270	0.3	81	−1.1	−89.1
C	270	0.3	81	−0.4	−32.4
D (Ref. P)	m_D	0.6	0.6m_D	0	0

By looking at the couples and forces, there is one unknown in the couples and two unknowns in the forces. So just as Ref.P A, it is possible to determine the unknowns here in the same manner. Try and see whether same results are obtained or not.

Note: So one can take either A or D planes as reference plane but not B and C planes.

W E 14.2: Outside cylinder: An outside cylinder locomotive has its cylinder centre lines 1.75 m apart and has a crank radius of 0.3 m. The rotating masses per cylinder are equivalent to 300 kg at the crankpin, and the reciprocating masses per cylinder are 300 kg. The wheel centre lines are 1.45 m apart. The cranks are at right angles. The whole of the rotating masses and $\frac{2}{3}$ of the reciprocating masses are to be balanced in the planes of the driving wheels. Find, taking the radius of balance masses as 0.75 m.
1. The magnitude and angular positions of the balancing masses
2. Find the speed in kilometres per hour at which the wheel will lift off the rails when the load on each driving wheel is 30 kN and the diameter of tread of driving wheel is 1.8 m
3. The magnitude of the swaying couple at the speed arrived at in (2). (Fig. 14.7 a,b,c,d)

Figure 14.7(a) shows the space diagram of the planes of the cylinders and wheels at the given distances as indicated. Figure 14.7(b) shows the crank positions of A and D at 90°.

Given that rotating masses of A and D as 360 kg, i.e. m_{RoA} and m_{RoD} = 360 kg and their reciprocating masses as m_{RA} and m_{RD} = 300 kg. Radii of cranks B and C are r_B and r_C given as 0.3 m. The radii of the wheels r_A and r_D are given as 0.75 m. The angular velocity ω or speed N is to be determined.

The equivalent balancing masses m_b in the planes A and D be m_A and m_D equal to $m_{RoA} + cm_{RA}$ and $m_{RoD} + cm_{RD}$, so $360 + 2 \times \frac{300}{3} = 360 + 200 = 560$ kg. Let m_B and m_C are the balancing masses in B and C planes, θ_B and θ_C be their angular positions. Their magnitudes and directions are to be determined graphically as follows:

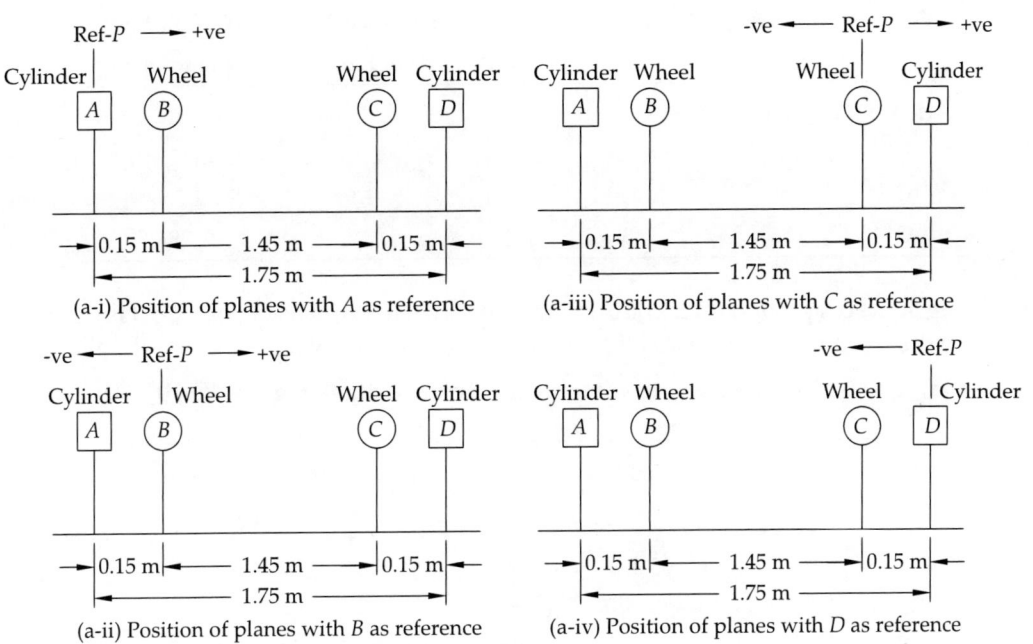

Fig. 14.7(a)

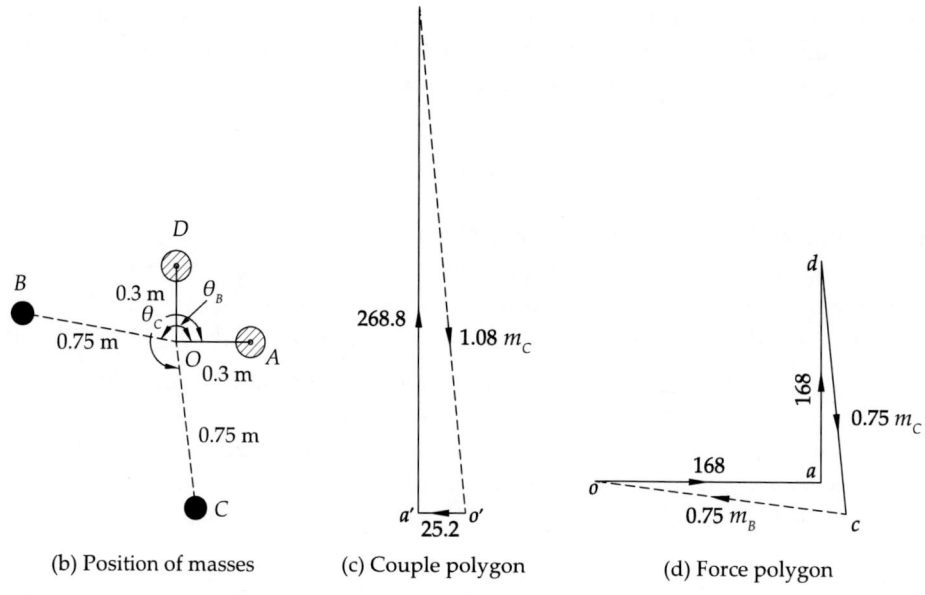

(b) Position of masses   (c) Couple polygon   (d) Force polygon

Fig. 14.7 Outside cylinders.

(I) Open a table as below and assume wheel A as reference plane Fig. 14.7(a-i)

Plane	Mass (kg) m	Radius (m) r	C.F./ω^2 m.r.	Distance L (m) from Ref. P A	Couple/ω^2 kg.m.m² m.r.L.
A (Ref. P)	560	0.3	168	0	0
B	m_B	0.75	0.75m_B	0.15	0.1125m_B
C	m_C	0.75	0.75m_C	1.6	1.2m_C
D	560	0.3	168	1.75	294

By looking at the couple and force columns, there are two unknowns in both. So it is not possible to determine the unknowns. So we should not take A as reference plane.

(II) Open a table as below and assume wheel B as reference plane Fig. 14.7(a-ii)

Plane	Mass (kg) m	Radius (m) r	C.F./ω^2 m.r.	Distance L (m) from Ref. P A	Couple/ω^2 kg.m.m² m.r.L.
A	560	0.3	168	−0.15	−25.2
B Ref. P	m_B	0.75	0.75m_B	0	0
C	m_C	0.75	0.75m_C	1.45	1.0875m_C
D	560	0.3	168	1.60	268.8

From the table and by looking at centrifugal force, there are two unknowns and from couples there is one unknown respectively. Hence, first draw a couple polygon choosing a scale as shown, Fig. 14.7(c), the closing side $d'o'$ gives both magnitude and direction. The magnitude, i.e. its length $d'o'$ is equal to 1.0875 m_C. Hence, $m_C = \dfrac{d'o'}{1.0875}$ giving 249 kg. Draw a line OC parallel to $d'o'$ in the direction from d to o in Fig. 14.7 (b) shown.

Now only unknown force is $0.75m_B$, which can be found by drawing a force polygon as shown at Fig. 14.7(d). The closing side co gives magnitude and direction of m_B. The length co is equal to $0.75m_B$. Therefore, $m_B = \dfrac{co}{0.75}$ giving $m_B = 249$ kg and draw a line OB parallel to co as in Fig. 14.7(b) and measure angles θ_B and θ_C from OA.

Balancing mass for reciprocating mass is $cm_R \times \dfrac{249}{560} = 89$ kg.

We know that
$$\omega_A = \sqrt{\dfrac{P}{m_B r_B}} = \sqrt{\dfrac{30 \times 1000}{89 \times 0.75}} = 21.2 \text{ rad/sec}$$

and
$$v = \omega \times \dfrac{D}{2} = 21.2 \times \dfrac{1.8}{2} = 19.08 \text{ m/s} = 19.08 \times \dfrac{3600}{1000} = 68.7 \text{ km/hr}$$

Swaying couple:

$$\text{The swaying couple} = a(1-c)\, m_R \omega^2 \dfrac{r}{\sqrt{2}}$$

$$= 1.75\left(1 - \dfrac{2}{3}\right) 300 \times 21.2^2 \dfrac{0.3}{\sqrt{2}} = 16.687 \text{ kN-m}$$

(III) Take plane C as reference and fill the table as follows Fig. 14.7(a-iii)

Plane	Mass (kg) m	Radius (m) r	C.F./ω^2 m.r.	Distance L (m) from Ref. P A	Couple/ω^2 kg.m.m^2 m.r.L
A	560	0.3	168	−1.6	−268.8
B	m_B	0.75	$0.75m_B$	−1.45	−1.0875m_B
C Ref. P	m_C	0.75	$0.75m_C$	0	0
D	560	0.3	168	0.15	25.2

By looking at the couple and force columns, there is one unknown in the couples and two unknowns in the forces. So just as Ref. $P\,B$, it is possible to determine the unknowns in the same manner. So we can take either B or C planes as reference plane.

(IV) Take plane D as reference and fill the table as follows Fig. 14.7(a-iv)

By looking at the couple and force columns, there are two unknowns in the couples and two unknowns in the forces. So just as Ref. $P\,A$, it is not possible to determine the unknowns in the same manner. So we should not take either A or D planes as reference planes.

428 Theory of Machines

Plane	Mass (kg) m	Radius (m) r	C.F./ω^2 m.r.	Distance L (m) from Ref. P A	Couple/ω^2 kg.m.m² m.r.L.
A	560	0.3	$0.3m_A$	−1.75	$-0.9m_A$
B	m_B	0.75	$0.75m_B$	−1.6	−89.1
C	m_C	0.75	$0.75m_C$	−0.15	$-32.4m_C$
D (Ref. P)	560	0.3	$0.3m_D$	0	0

Note: Try by taking the reference plane as C and see whether same results are obtained or not.

14.4 MULTI-CYLINDER IN-LINE ENGINES

Picture 14.2 shows the multi-cylinder engine with four cylinders having a common crankshaft.

Pic. 14.2

The crankshaft is common for all the cylinders in multi-cylinder in-line engines and the line of stroke of all cylinders are parallel. But the piston, connecting rod and crank of each cylinder move in separate parallel planes. The balancing of multi-cylinder in-line engines is done by using the primary and the secondary crank method as follows:

Primary and Secondary crank method

In this method, the primary and secondary components are assumed due to two separate primary and secondary cranks. The primary crank (C_P) is an actual crank assumed as carrying a rotating mass equal to the reciprocating mass (m_R) at radius r and rotating at ω. But the secondary crank (C_S) is an imaginary crank assumed as carrying a rotating mass equal to reciprocating mass (m_R) at radius $\dfrac{r}{4n}$ and rotating at 2ω radians per second as shown in Fig. 14.8.

For complete balancing of multi-cylinder in-line engines, all the following four conditions for primary and secondary forces and couples must be satisfied:

I. Balancing of the primary forces

1. Algebraic sum of all the primary forces of all cylinders must be zero, i.e.

$$\Sigma m_R \omega^2 r \cos\theta = 0 \quad (1)$$

2. Algebraic sum of all the primary couples must be zero, i.e.

$$\Sigma a . m_R \omega^2 r \cos\theta = 0 \quad (2)$$

where a is the distance between the cylinders.

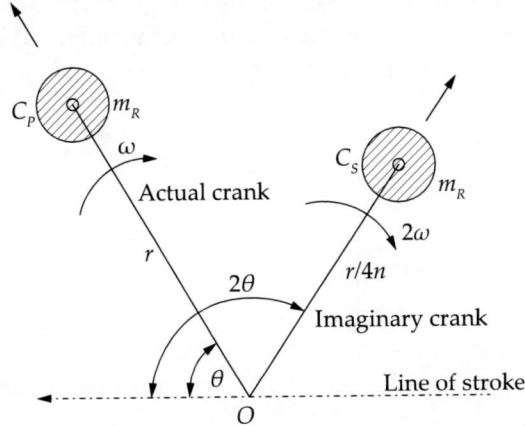

Fig. 14.8 Actual (primary) and imaginary (secondary) cranks.

II. Balancing of the secondary forces

1. Algebraic sum of all the secondary forces of all cylinders must be zero, i.e.

$$\Sigma m_R (2\omega)^2 \frac{r}{4n} \cos 2\theta = 0 \quad (1)$$

2. Algebraic sum of the secondary couples must be zero, i.e.

$$\Sigma a . m_R (2\omega)^2 \frac{r}{4n} \cos 2\theta = 0 \quad (2)$$

where a is the distance between the cylinders.

The detailed procedure is explained through the following example:

430 Theory of Machines

W E 14.3: A four crank engine has two outer cranks set at 120° to each other, and their reciprocating masses are each 400 kg. The distances between the planes of rotation of adjacent cranks are 450 mm, 750 mm and 600 mm. If the engine is to be in complete primary balance, find the reciprocating masses and their relative angular position of each of the inner cranks. If the length of each crank is 300 mm, the length of each connecting rod is 1.2 m and the speed of rotation is 240 r.p.m., what is the maximum secondary unbalanced force and couple? Let m_1, m_2, m_3, m_4 represent the four reciprocating masses and $\theta_1, \theta_2, \theta_3$ and θ_4 the crank angles of the four cylinders. Let r be the radius of the cranks.

Given:

$m_1 = m_4 = 400$ kg; $\quad r = 300$ mm $= 0.3$ m; $\quad L = 1.2$ m; $\quad N = 240$ r.p.m. or $\omega = \dfrac{2\pi 240}{60}$
$= 25.14$ rad/sec.

The masses and their angles of m_2, θ_2 and m_3, θ_3 with respect to 1 are to be determined. The positions of the planes of rotation of the cranks and their angular settings are shown in Fig. 14.9(a) and (b) assuming the crank of cylinder 1 as in the horizontal position. The data is tabulated taking the plane of cylinder 2 as the reference plane as below:

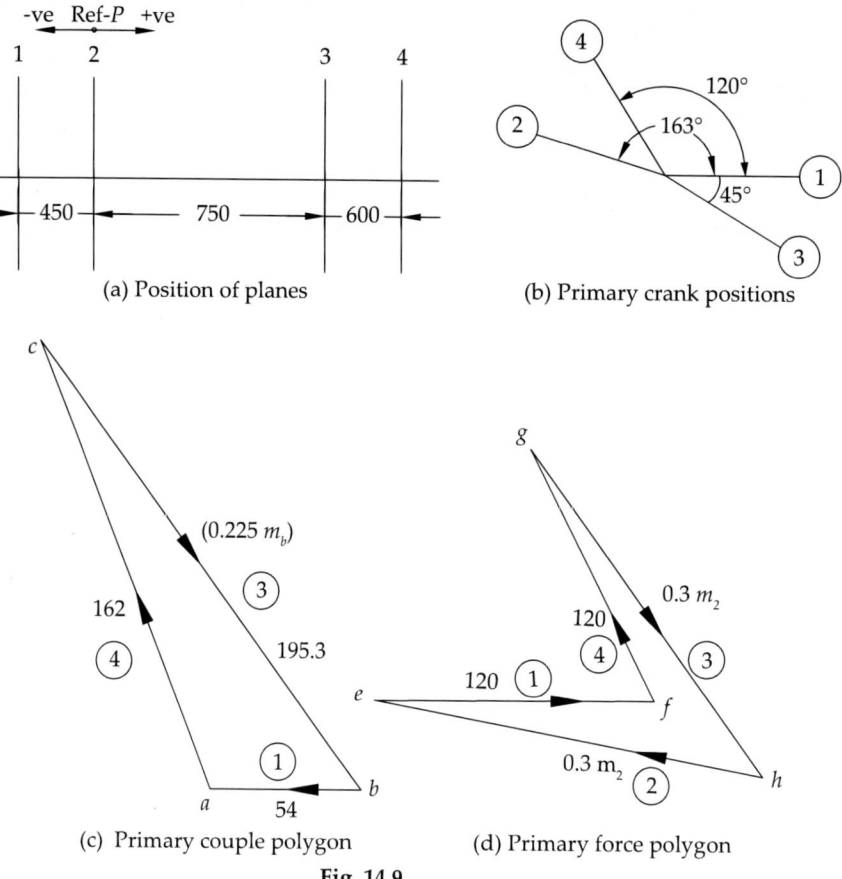

Fig. 14.9

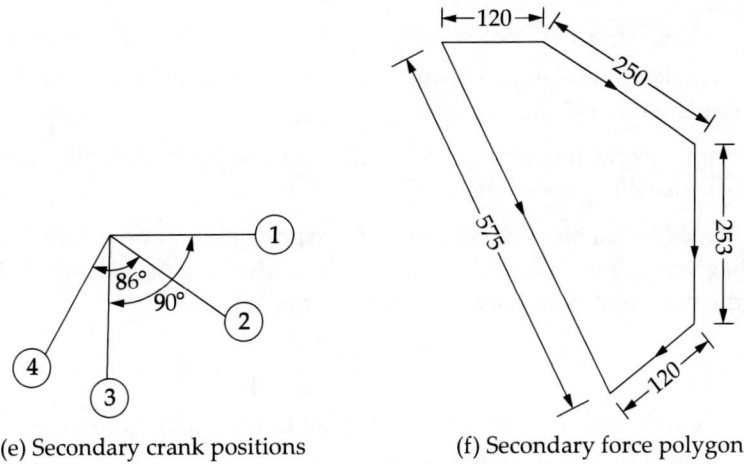

(e) Secondary crank positions   (f) Secondary force polygon

Fig. 14.9 Four cylinders in-line engine.

First take up balancing of primary couples and forces and then secondary couples and forces as indicated below:

Plane	Mass (kg) m	Radius (m) r	C.F./ω^2 m.r.	Distance L (m) from Ref. P 2	Couple/ω^2 kg.m.m^2 m.r.L.
1	400	0.3	120 (ef)	−0.45	−54 (ab)
2(Ref. P)	m_2	0.3	0.3m_2 (he)	0	0
3	m_3	0.3	0.3m_3 (gh)	0.75	0.225m_3 (bc)
4	400	0.3	120 (fg)	1.35	162 (ca)

In the force column there are two unknown forces whereas the couples column only one unknown couple is there. Hence, start first the couple polygon.

Primary Balancing: For the engine to be in complete primary balance, the primary couple polygon and the primary force polygon must close.

First draw the primary couple polygon as shown in Fig. 14.9(c) taking suitable scale for the data given in the last column of the above table in order to determine the reciprocating mass of cylinder 3.

Measure the closing side of the couple polygon according to the scale which gives 195.3 kg. This is equal to 0.225m_3 and hence $m_3 = \dfrac{195.3}{0.225} = 868$ kg and draw a line parallel to this to get angular position of crank 3 in the figure (b) and measure its angular position θ_3 with respect to crank 1 in the anticlockwise direction giving $\theta_3 = 315°$.

Now take the primary forces in order to find the reciprocating mass of cylinder 2. Draw the primary force polygon as shown at Fig. 14.9(d) to some suitable scale from the data given in the fourth column of the above table. Now measure the closing side according to the scale which gives

253 kg. This is equal to $0.3m_2$ and hence $m_2 = \dfrac{253}{0.3} = 843$ kg and draw a line parallel to this to get angular position of crank 2 in the figure (b) and measure its angular position with respect to crank 1 in the anticlockwise direction giving $\theta_2 = 163°$.

Secondary Balancing: Obtain the secondary crank positions by rotating the primary cranks by twice in the anticlockwise direction as shown in Fig. 14.9(e).

Secondary unbalanced force: Draw the secondary force polygon as shown in Fig. 14.9(f) to suitable scale from the data given in the column 4 of above table. Measure the closing side of the polygon which represents the maximum secondary unbalanced force.

$$\text{Maximum secondary unbalanced force} = \dfrac{575\omega^2}{n} = \dfrac{575(25.14)^2}{(1.2/0.3)} = 90854 \text{ N} = 90.854 \text{ kN}.$$

Try: The same problem do again by taking the plane of the cylinder 3 as reference as follows and see whether you get the same results or not.

Plane	Mass (kg) m	Radius (m) r	C.F./ω^2 $m.r.$	Distance L (m) from Ref. P 3	Couple/ω^2 kg.m.m² $m.r.L.$
1	400	0.3	120	−1.2	−144
2	m_2	0.3	$0.3m_2$	−0.75	$-0.225m_2$
3(Ref. P)	m_3	0.3	$0.3m_3$	0	0
4	400	0.3	120	0.60	72

W E 14.4: The cranks and connecting rods of a 4-cylinder in-line engine running at 1800 r.p.m. are 600 mm and 2400 mm each respectively and the cylinders are spaced 150 mm apart. If the cylinders are numbered 1 to 4 in sequence from one end, the cranks appear at intervals of 90° in end view in the order 1-4-2-3. The reciprocating mass corresponding to each cylinder is 1.5 kg. Determine: 1. Unbalanced primary and secondary forces, if any, and 2. Unbalanced primary and secondary couples with reference to central plane of the engine.

Note: Here the reference plane is specified clearly and hence no need to assume.

Given:
$N = 1800$ r.p.m. or $\omega = 188.52$ rad/s; $r = 600$ mm = 0.6 m; $L = 2400$ mm = 2.4 m; mass $m = 1.5$ kg

The position of the cylinder planes and their cranks are shown in Fig. 14.10(a) and (b) respectively. The reference plane is mentioned in the problem as in between the cylinders 2 and 3, hence tabulate the data accordingly as below:

Take up primary and secondary forces first and then primary and secondary couples for convenience.

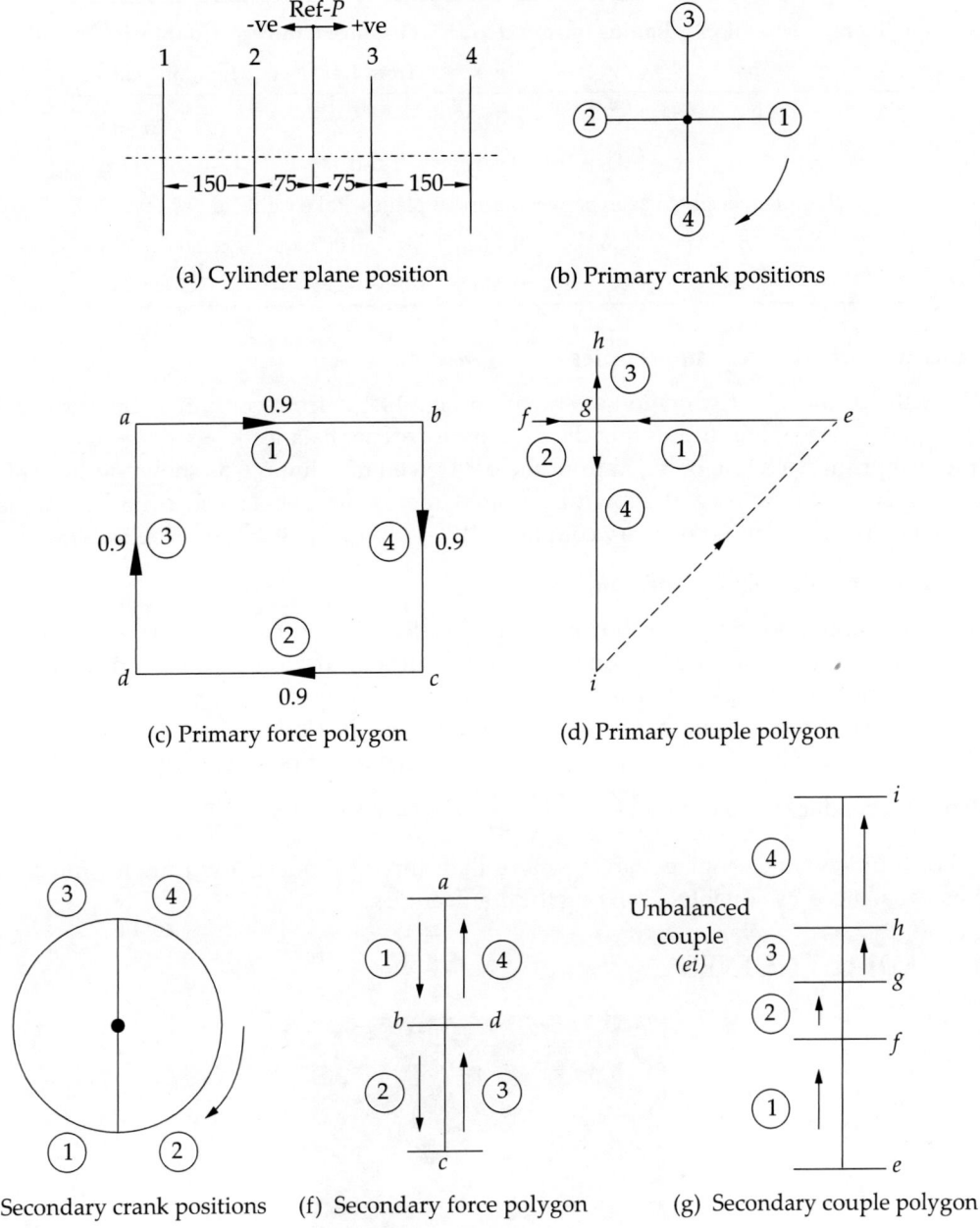

Fig. 14.10 Four cylinder engine.

Plane	Mass (kg) m	Radius (m) r	C.F./ω^2 $m.r.$	Distance L (m) from Ref. P A	Couple/ω^2 kg.m.m² $m.r.L.$
1	1.5	0.6	0.9 (ab)	−0.225	−0.2025 (ef)
2	1.5	0.6	0.9 (bc)	−0.075	−0.0675 (fg)
	Ref. plane is given as in between 2 and 3 planes. Take right side as +ve				
3	1.5	0.6	0.9 (cd)	+0.075	+0.0675 (gh)
4	1.5	0.6	0.9 (da)	+0.225	+0.2025 (hi)

1. Balance of primary forces and couples

First draw the primary force polygon as shown in Fig. 14.1(c) from column 4. Since the primary force polygon is closed, it indicates that the primary forces are balanced.

Draw the primary couple polygon from the data given in column 6 as shown in Fig. 14.10(d). The closing side of the polygon shown by dotted represents unbalanced primary couple. By measurement, the unbalanced primary couple = $0.19\omega^2 = 0.19 \times (188.52)^2 = 6752.56$ Nm.

2. Balance of secondary forces and couples

The secondary crank positions are shown in Fig. 14.10(e). Draw the secondary force polygon as shown in Fig. 14.10(f). There are two upward and two downward forces so they cancel each other, hence the secondary forces are balanced.

Draw the secondary couple polygon as shown in Fig. 14.10(g). The unbalanced secondary couple is the sum of all couples as they are all in the same direction as shown. By adding all, the unbalanced secondary couple = $\dfrac{0.54\omega^2}{n} = \dfrac{0.54(188.52)^2}{(2.4/0.6)} = 4797.8$ N m.

Note: For multi-cylinder engines having more than three cylinders, the primary forces can be completely balanced by suitably arranging the crank angles.

14.5 RADIAL ENGINES

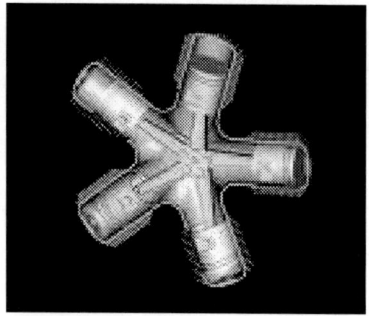

Pic. 14.3 Five cylinder radial engine.

The radial engines will have two or more than two cylinders arranged with their line of strokes in radial direction. Two cylinder radial engines are called V-engines. When there are more than two cylinders, then those engines are called multi-cylinder radial engines. Picture 14.3 shows the five cylinder radial engine and Pic. 14.4 shows the line diagram of the seven-cylinder radial engine.

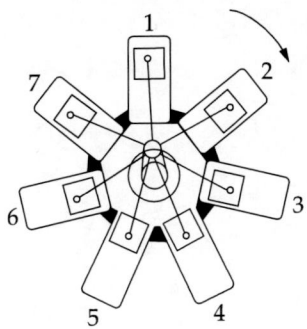

Pic. 14.4 Seven cylinder radial engine.

The angle between the lines of strokes of two adjacent cylinders is equal to $\dfrac{360°}{n}$ where n is the number of cylinders. There will be only one crank common to all cylinders. The connecting rod of each cylinder is connected to the same crank. All the cylinders work in the same plane so all the primary and secondary forces act in a single plane unlike in the multi-cylinder in-line engines. Hence, the unbalanced forces won't give any couples in radial engines. Therefore, only the balancing of primary and secondary forces has to be considered. The balancing of radial engines can be done either by the graphical means using the direct and reverse crank method or by the analytical method. The direct and reverse crank method is generally adopted because of its simplicity and easy for analysis.

14.5.1 Direct and Reverse Crank Method

The unbalanced force consists of both primary and secondary components as seen earlier. The primary and secondary components can be calculated using primary and secondary cranks. Here in the direct and reverse crank method, the primary and secondary cranks are again replaced by two imaginary cranks called direct and reverse cranks. Direct crank rotates in the clockwise and the reverse crank in the anticlockwise direction as shown in the Fig. 14.11 carrying half the reciprocating mass $\dfrac{m_R}{2}$ at the crank ends. As seen earlier the radius of the primary crank is r and rotates at ω rad/sec whereas the radius of the secondary crank is $\dfrac{r}{4n}$ and rotates at 2ω rad/sec.

The centrifugal force due to $\dfrac{m_R}{2}$ in each crank gives one horizontal and one vertical component. The horizontal components get added up whereas the vertical components get cancelled. The total horizontal force due to primary direct and reverse cranks is $2\left[\dfrac{m_R}{2}\omega^2 r \cos\theta\right]$ which is equal to the primary component $m_R \omega^2 r \cos\theta$. Due to secondary direct and reverse cranks is $2\left[\dfrac{m_R}{2}(2\omega)^2 \dfrac{r}{4n} \cos 2\theta\right]$ which is equal to the secondary component $m_R(2\omega)^2 \dfrac{r}{4n} \cos 2\theta$. Figure 14.11(a) shows the primary direct and reverse cranks and (b) shows the secondary direct and reverse cranks.

The position of the cranks of the primary and secondary cranks has to be drawn carefully.

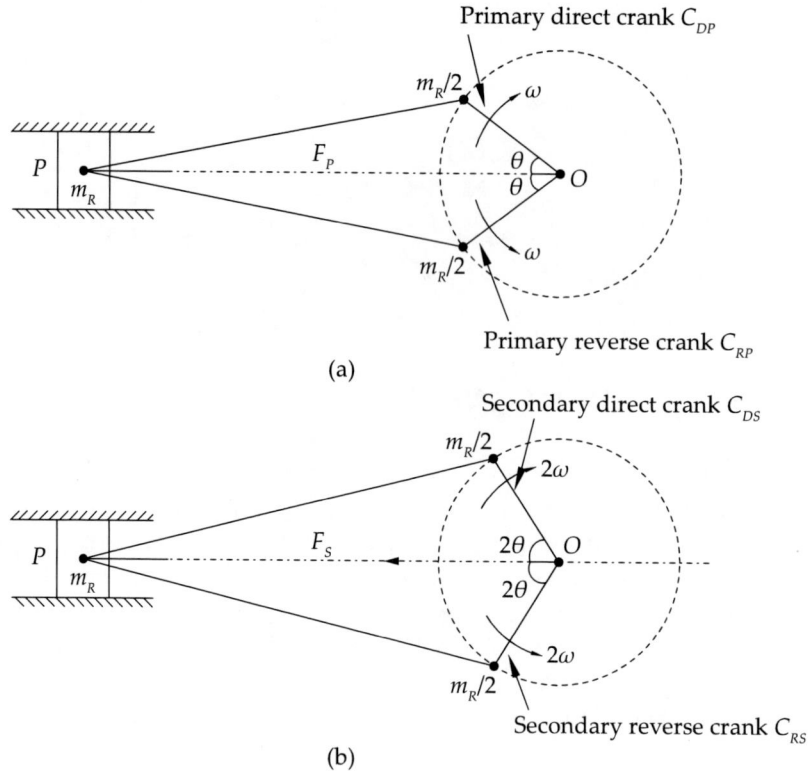

Fig. 14.11 Direct and reverse cranks.

14.5.2 Analytical Method

Each cylinder of the radial engine gives both primary and direct component ($F_P + F_S$) along the line of stroke of the respective cylinders. Take separately the primary and the secondary components and resolve horizontally and vertically to determine the resultant unbalanced primary and secondary forces just as balancing of several revolving masses in the same plane as discussed in the earlier chapter.

The concept of the direct and reverse crank method is applied to the following worked example:

W E 14.5: The three cylinders of an air compressor have their axes 120° to one another and their connecting rods are coupled to a single crank. The stroke is 100 mm and the length of each connecting rod is 150 mm. The mass of the reciprocating parts per cylinder is 1.5 kg. Find the maximum primary and secondary forces acting on the frame of the reciprocating compressor when running at 3000 r.p.m. Describe clearly whether both primary and secondary forces are balanced or not.

Balancing of Reciprocating Masses 437

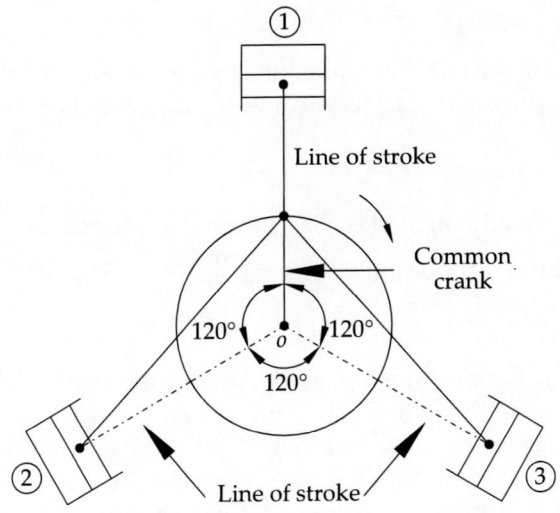

(a) Radial engine

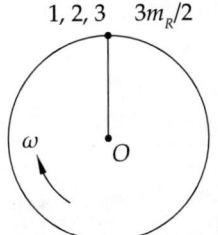

(b) Primary cranks

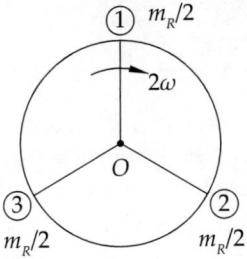 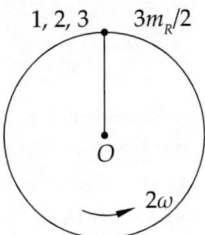

(c) Secondary cranks

Fig. 14.12 Radial compressor.

Direct and Reverse Crank Method

First draw the line diagram of the radial compressor as shown at Fig. 14.12(a). Draw the direct and reverse primary cranks as shown at (b). Next draw the direct and reverse secondary cranks as shown at (c).

Given:

Connecting rod $L = 150$ mm $= 0.15$ m; Stroke $= 100$ mm and so the crank radius $r = \dfrac{\text{stroke}}{2} = \dfrac{100}{2} = 50$ mm $= 0.05$ m. Mass of the reciprocating parts $m_R = 1.5$ kg; $N = 3000$ r.p.m. $= 314.2$ rad/sec, $n = \dfrac{L}{r} = \dfrac{150}{50} = 3$.

Figure 14.12(a) shows the three cylinders at 120°. Assume the crank angle of the common crank OC as zero ($\theta = 0°$) with respect to cylinder (1) and assume the direction of rotation of crank as clockwise. The direct cranks rotate in the clockwise and reverse cranks in the counterclockwise directions.

The primary direct and reverse crank positions are shown in Fig. 14.12(b) and they have to be drawn at the following angles with respect to the crank of the cylinder 1.

Primary crank angles	Cylinders		
	1	2	3
θ (Direct)	0°	120°	240°
θ (Reverse)	0°	−120°	−240°

The secondary direct and reverse crank positions are shown in Fig. 14.12(c) and they have to be drawn at the following angles with respect to the crank of the cylinder 1.

Secondary crank angles	Cylinders		
	1	2	3
θ (Direct)	0°	240°	480°
θ (Reverse)	0°	−240°	−480°

Draw four circles of some diameter, two for primary and two for secondary cranks as shown. In all the 4 circles, draw 3 radial lines at 120° with cylinder 1 in the vertical position as shown.

Indicate clockwise direction for direct cranks and counterclockwise or anticlockwise direction for reverse cranks of both primary and secondary cranks respectively.

Indicate crank positions of the cylinders as 1, 2 and 3 from their respective line of strokes clockwise by the angles indicated above at Fig. 14.12(b). All the three direct cranks can be seen in the vertical position and the three reverse primary cranks in three radial directions. The primary direct cranks give an unbalanced force whereas the reverse cranks give no unbalanced force. Hence, the maximum unbalanced force due to the primary direct cranks is

$$3 m_R \omega^2 \dfrac{r}{2} = 3 \times 1.5 \times (314.2)^2 \times \dfrac{0.05}{2} = 11112.6 \text{ N or } 11.1126 \text{ kN}$$

This can be balanced by attaching a rotating mass on the diametrically opposite side of the crank and rotating with the crank of magnitude m_{bP1} at radius r_{bP1} such that their product

$$m_{bP1} r_{bP1} = 3\left(\frac{m_R \cdot r}{2}\right) = 3 \times \frac{1.5 \times 0.05}{2} = 0.1125 \text{ kg-m}$$

In Fig. 14.12(c), the secondary direct cranks form a balanced system, as three components are along the three directions as shown. The force polygon closes and so there is no unbalanced force due to the secondary direct cranks. But the three secondary reverse cranks are seen in vertical position. The maximum unbalanced force due to the secondary reverse cranks is

$$3\left(\frac{m_R}{2}\right)(2\omega)^2\left(\frac{r}{4n}\right) = 3\left(\frac{1.5}{2}\right)(2 \times 314.2)^2 \times \frac{0.05}{4 \times 3}$$
$$= 3702 \text{ N} = 3.702 \text{ kN}$$

This can be balanced by a rotating mass attached on the diametrically opposite side of the crank and rotating with the crank of magnitude m_{bS1} at radius r_{bS1} such that their product

$$m_{bS1} r_{bS1} = 3\left(\frac{m_R \cdot r}{2}\right)\frac{1}{4n} = 3\left[\frac{1.5 \times 0.05}{2}\right]\frac{1}{4 \times 3} = 0.009375 \text{ kg-m}$$

14.6 V-ENGINES

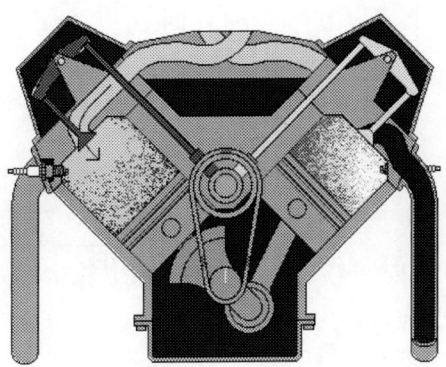

Pic. 14.5

Picture 14.5 shows the picture of the V-engine. V-engine can be called radial engine with two cylinders having the line of strokes at an angle of 2α as shown in Fig. 14.13. The common crank OC drives the two connecting rods CP_1 and CP_2 of the two cylinders indicated by (1) and (2). Each cylinder has unbalanced primary and secondary components $(F_{P1} + F_{S1})$ for cylinder (1) and $(F_{P2} + F_{S2})$ for cylinder (2) as shown along their respective line of strokes.

The V-engine can be analysed by using either direct and reverse crank method used for radial engines or by analytical method.

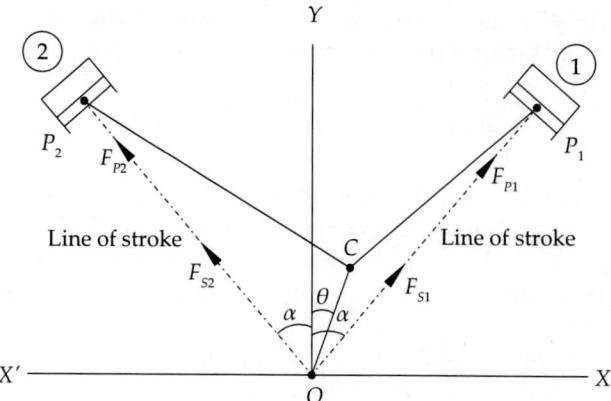

Fig. 14.13 V-engine.

14.6.1 Analytical Method

Take the crank OC inclined at θ to y axis as shown.

Let the line of stroke of each cylinder makes an angle α with the vertical on either side.

The crank angle of the two cylinders as per the diagram are $(\alpha - \theta)$ and $(\alpha + \theta)$ and accordingly the primary and secondary forces of the two cylinders are:

Cylinder (1) is $F_{P1} + F_{S1} = m_R \omega^2 r \left[\cos(\alpha - \theta) + \dfrac{\cos 2(\alpha - \theta)}{n} \right]$ and

Cylinder (2) is $F_{P2} + F_{S2} = m_R \omega^2 r \left[\cos(\alpha + \theta) + \dfrac{\cos 2(\alpha + \theta)}{n} \right]$

Resolve the two components of the two cylinders separately along the horizontal and vertical as follows:

The total of the primary components along the vertical OY is $F_{PV} = (F_{P1} + F_{P2}) \cos \alpha$ and along the horizontal $X'X$ is $F_{PH} = (F_{P1} - F_{P2}) \sin \alpha$.

Let F_P be the resultant of the two primary components $= \sqrt{[F_{PH}^2 + F_{PV}^2]}$.

Similarly, the total of the secondary components along the vertical OY is $F_{SV} = (F_{S1} + F_{S2}) \cos \alpha$ and along the horizontal $X'X$ is $F_{SH} = (F_{S1} - F_{S2}) \sin \alpha$.

Let F_S be the resultant of the two secondary components $= \sqrt{[F_{SH}^2 + F_{SV}^2]}$.

Substitute for F_{P1}, F_{P2}, F_{S1} and F_{S2} then the net primary force

$$F_P = 2m_R \omega^2 r \sqrt{\left[(\cos^2\alpha . \cos\theta)^2 + (\sin^2\alpha . \sin\theta)^2\right]} = 2m_R \omega^2 r \left(\cos^2\alpha\right) \text{ for } \theta = 0°$$

the net secondary force

$$F_S = 2m_R \omega^2 \dfrac{r}{n} \sqrt{\left[(\cos\alpha . \cos 2\alpha . \cos 2\theta)^2 + (\sin\alpha . \sin 2\alpha . \sin 2\theta)^2\right]} = 2m_R \omega^2 \dfrac{r}{n} (\cos\alpha . \cos 2\alpha) \text{ for } \theta = 0°$$

Balancing of Reciprocating Masses **441**

The maximum and minimum values $F_{P\,max}$, $F_{P\,min}$ and $F_{S\,max}$, $F_{S\,min}$ depend upon the value of the crank angle θ. The simplified expressions of F_P and F_S when θ is zero for three different V-angles are as given below:

	α	F_P	F_S
1	30°	$\dfrac{m_R}{2}\omega^2 r\left[\sqrt{(9\cos^2\theta + \sin^2\theta)}\right]$ $\dfrac{3}{2}m_R\omega^2 r$ for $\theta = 0$	$\dfrac{\sqrt{3}m_R}{2n}\omega^2 r$
2	45°	$m_R\omega^2 r$	$\dfrac{\sqrt{2}m_R}{n}\omega^2 r \sin^2\theta$ 0 for $\theta = 0$
3	60°	$\dfrac{m_R}{2}\omega^2 r\sqrt{(\cos^2\theta + 9\sin^2\theta)}$ $m_R\omega^2 \dfrac{r}{2}$ for $\theta = 0$	$\dfrac{m_R}{2n}\omega^2 r\left[\sqrt{(\cos^2 2\theta + 9\sin^2 2\theta)}\right]$ $m_R\omega^2 \dfrac{r}{2n}$ for $\theta = 0$

14.6.2 Direct and Reverse Crank Method

Figure 14.14 gives the direct (in clockwise) and reverse crank (in anticlockwise) positions of both primary and secondary cranks of the two cylinder V-engines for three different V-angles assuming θ as zero.

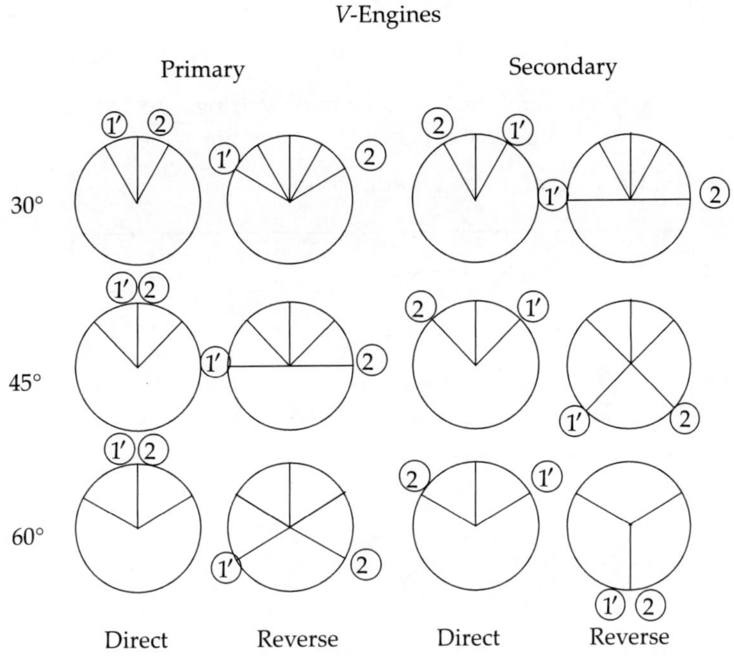

V-Engines

Fig. 14.14

The corresponding expressions for crank angle θ equal to 0 are given as follow in the Table-II:

	Angle α	Primary	Secondary
1.	30°	$\dfrac{3m_R}{2}\omega^2 r$	$\dfrac{\sqrt{3}m_R}{2n}\omega^2 r$
2.	45°	$m_R \omega^2 r$	0
3.	60°	$\dfrac{m_R}{2}\omega^2 r$	$\dfrac{m_R}{2n}\omega^2 r$

Note:

1. The analytical method is advisable to calculate unbalanced forces for any crank angle θ.
2. The direct and reverse crank method is easy to adopt for $\theta = 0°$. The balancing can be easily be decided by this method.

W E 14.6: Three V-twin engines have the cylinder axes at (i) 60° (ii) 90° and 120° angles and the connecting rods operate a common crank. The reciprocating mass per cylinder is 11.5 kg and the crank radius is 75 mm. The length of the connecting rod is 0.3 m. Determine the primary and secondary unbalanced components for all the three cases when the engine speed is 500 r.p.m. and for crank angle 0°.

Given:

Reciprocating mass per cylinder m_R = 11.5 kg; Radius of crank r = 75 mm = 0.075 m; Length of connecting rod L = 0.3 m and speed N = 500 r.p.m. or ω = 52.37 rad/sec, Ratio $\dfrac{L}{r} = \dfrac{0.3}{0.075} = 4$.

The problem can be solved by both analytical method and by using direct and reverse crank method. The expressions for $\theta = 0°$ for all the three types of V-engines are same.

Calculate the unbalanced forces in all three types as follows:

	α	F_P	F_S
1	30°	$\dfrac{3}{2}m_R\omega^2 r = 3548.26\text{ N}$	$\dfrac{\sqrt{3}}{2}m_R\dfrac{\omega^2 r}{n} = 512.15\text{ N}$
2	45°	$m_R \omega^2 r = 2365.5\text{ N}$	0
3	60°	$m_R \dfrac{\omega^2 r}{2} = 2365.51\text{ N}$	$\dfrac{m_R \omega^2 r}{2n} = 591.38\text{ N}$

14.7 EXERCISE

14.7.1 Short Answer Questions

1. Prove that the resultant unbalanced force is minimal, when half of the reciprocating masses are balanced by the rotating masses.

2. A machine weighing 680 N is mounted on springs of stiffness, 11 kN/cm with an assumed damping factor of 0.2. A piston within the machine weighing 20 N has a reciprocating motion with a stroke of 7.5 cm and a speed of 3,000 r.p.m. Compute:

    (a) the amplitude of the machine
    (b) its phase angle w.r.t. the exciting force and
    (c) the amplitude of the machine when damper is removed.

3. (a) Discuss, in detail, the effect of partial balancing of reciprocating parts of two cylinders.
   (b) Explain the effects of partial balancing of locomotives.

4. (a) Explain the method of direct and reverse cranks to determine the unbalance forces in radial engines.
   (b) What is hammer blow?

5. (a) What do you mean by primary and secondary unbalance in reciprocating engines?
   (b) Define 'Primary' and 'Secondary' balancing for reciprocating masses.

6. Deduce expressions for variation in tractive force, swaying couple and hammer blow for an uncoupled two-cylinder locomotive engine.

7. Find the magnitudes of the unbalanced primary and secondary forces in V-engines. Deduce the expressions when the lines of stroke of the two cylinders at 60° and 90° to each other.

14.7.2 Problems

Partial Balancing

1. A single cylinder reciprocating engine has a reciprocating mass of 60 kg. The crank rotates at 60 r.p.m. and the stroke is 320 mm. The mass of the revolving parts at 160 mm radius is 40 kg. If two-thirds of the reciprocating parts and the whole of the revolving parts are to be balanced, determine the (i) balance mass required at a radius of 350 mm. (ii) unbalanced force when the crank has turned 500 degrees from the top dead centre. [*Ans.* 36.57 kg; 209.9 N]

2. A single cylinder reciprocating engine has a speed of 240 r.p.m., stroke 300 mm, mass of reciprocating 50 kg, mass of revolving parts at 150 mm radius 37 kg. If two-thirds of the reciprocating parts and all the revolving parts are to be balanced, find: (a) the balance mass required at a radius of 400 mm and (b) the residual unbalanced force when the crank has rotated 60°, from top dead centre.

3. The following data relate to a single cylinder reciprocating engine. Mass of reciprocating parts = 40 k. Mass of revolving parts = 30 kg at 180 mm radius. Speed = 150 r.p.m. Stroke length = 350 mm. If 60% of the reciprocating parts and all the revolving parts to be balanced, determine the (a) balancing mass required at a radius of 320 mm. (b) The unbalanced force when the crank has turned 450 from the top dead centre.

Tractive force, swaying couple

1. A 2-cylinder uncoupled locomotive with cranks at 900 has a crank radius of 32.4 cm. The distance between centres of driving wheel is 150 cm. The pitch of cylinders is 60 cm. The diameter of treads of driving wheel is 180 cm. The radius of centre of gravity of balance weights is 65 cm. The pressure due to dead load on each wheel is 4 tonnes. The weight of reciprocating and rotating parts per cylinder are 330 kg and 300 kg respectively. The speed of locomotive is 60 kmph. Find: (a) The balance weights both in magnitude and position required to be placed in the planes of driving wheels to balance whole of the revolving and 2/3 of the reciprocating masses. (b) Swaying couple. (c) The variation of tractive force. (d) The maximum and minimum pressure in rails. What is the maximum speed at which it is possible to run the locomotive, in order that the wheels are not lifted from the rail?

Locomotive

1. *(Inside cylinder)* Using the given data, determine for inside cylinder uncoupled locomotive: (i) The magnitude and position of balance weights required at 90 cm radius in the plane of the wheels, (ii) the hammer blow (iii) the maximum variation of tractive effort and swaying couple when the cranks make 300 r.p.m.; Stroke = 75 cm, Revolving mass per cylinder = 240 kg, Reciprocating mass per cylinder = 280 kg, Distance between centre lines of cylinders = 60 cm, Distance between centres of wheels = 160 cm. It is required to balance whole of revolving and 3/4 th of reciprocating masses.

2. *Outside cylinder:* The following data apply to an outside cylinder unbalanced locomotive: Mass of the rotating parts per cylinder = 360 kg. Mass of reciprocating parts per cylinder = 300 kg. Angle between cranks = 90°. Crank radius = 300 mm. Cylinder centres = 1.75 m. Radius of balance masses = 750 mm. Wheel centres = 1.75 mm. If the whole of rotating and 2/3 of the reciprocating parts are to be balanced in planes of driving wheels, find: (a) Magnitude and angular position of balance masses. (b) Speed in kilometres per hour at which the wheel will lift off the rails when the load on each driving wheel is 50 kN and the diameter of tread of driving wheel is 1.8 m. (c) Swaying couple at the speed arrived in the (b) above.

3. *(Inside cylinder)* A four coupled-wheel locomotive with two inside cylinders has reciprocating and revolving parts per cylinder as 300 kgf and 250 kgf respectively. The distance between planes of driving wheels is 150 cm. The pitch of cylinders is 60 cm. Diameter of tread and driving wheels is 190 cm and the distance between planes of coupling rod cranks is 190 cm. The revolving parts for each coupling rod crank are 125 kgf. The angle made by coupling rod crank with adjustment crank is 180°. The distance of centre of gravity of balance weights in planes of driving wheels from centre is 75 cm. Crank radius is 32 cm. Determine: (a) The magnitude and position of balance weights required in leading and trailing wheels to balance

2/3 of reciprocating and whole of revolving parts if half of the required reciprocating parts are to be balanced in each pair of coupled wheels. (b) The maximum variation of tractive force and hammer blow when locomotive speed is 100 kmph.

Radial Engine:

1. The connecting rods of a three-cylinder air compressor are coupled to a single crank and the axes are at 120° to one another. Each connecting rod is 180 mm long and the stroke is 120 mm. The reciprocating parts have a mass of 1.8 kg per cylinder. Find the magnitude of the primary and secondary forces when the engine runs at 1200 r.p.m. [*Ans.* 2.558 kN, 852.7 N]

2. A three-cylinder radial engine has axes at 120° to one another and their connecting rods are coupled to a single common crank. The stroke length is 100 mm and length of each connecting rod is 150 mm. If the mass of reciprocating parts per cylinder is 1 kg, determine the primary and secondary force of the engine running at 2400 r.p.m.

V-Engine:

1. An engine has two cylinders in the form of V, the centre lines of the cylinders being in one plane and inclined at 45° on either side of a central vertical. The two connecting rods work on the same crank. The mass of the reciprocating parts for each cylinder is 0.5 kg. The crank radius is 4.5 cm and connecting rod length is 16.75 cm. Show that the vertical force on this engine due to secondary inertia forces zero and that of a suitable balance masses are attached to the crankshaft, the primary inertia forces can be reduced to zero. For this value of the balance mass, find the greatest out of balance force acting on the engine in the horizontal direction when the speed is 3000 r.p.m. [*Ans.* 834 N]

14.7.3 Multiple Choice Questions

1. The magnitude of the secondary force is _____ the primary force.

    (a) more than   (b) less than   (c) equal to. [*Ans.* (b)]

2. In reciprocating engines the primary unbalanced force

    (a) cannot be balanced   (b) can be fully balanced   (c) can be partially balanced. [*Ans.* (c)]

3. The primary unbalanced force is maximum when the angle of crank with the line of stroke is _____

    (a) 45°   (b) 90°   (c) 135°   (d) 180° [*Ans.* (d)]

4. If the ratio of the length of connecting rod to the crank increases

    (a) primary unbalanced forces increase   (b) primary unbalanced forces decrease
    (c) secondary unbalanced forces increase   (d) secondary unbalanced decrease. [*Ans.* (d)]

5. The resultant unbalanced force is minimum in reciprocating engines

    (a) when one-third of the reciprocating masses are balanced by rotating masses
    (b) when half the reciprocating masses are balanced by the rotating masses

(c) when three-fourths of the reciprocating masses are balanced by the rotating masses

(d) none of the above. [*Ans.* (b)]

6. The frequency of secondary force as compared to primary force for ratio of connecting rod length radius of 4 is

   (a) half   (b) twice   (c) four times   (d) sixteen times. [*Ans.* (b)]

7. In order to balance the reciprocating mass of a reciprocating engine

   (a) primary forces must balance   (b) primary couples must balance
   (c) secondary forces must balance   (d) all of the above. [*Ans.* (d)]

8. The balancing masses are introduced in the plane parallel to the plane of rotation of the disturbing mass. To obtain complete dynamic balance, a minimum number of balance masses to be introduced is

   (a) one   (b) two   (c) three   (d) four. [*Ans.* (b)]

Longitudinal and Transverse Vibrations

15

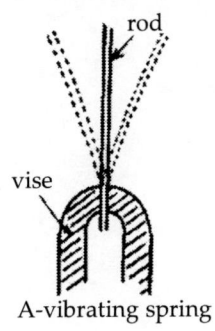

A-vibrating spring

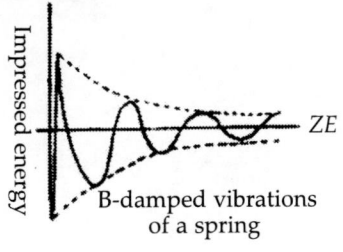

B-damped vibrations of a spring

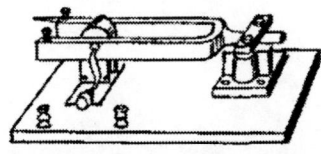

C-electric driven tuning fork

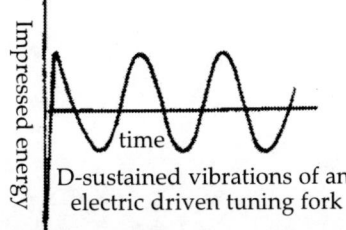

D-sustained vibrations of an electric driven tuning fork

Longitudinal and Transverse Vibrations

15.1 INTRODUCTION

Any motion that exactly repeats itself after a certain interval of time is a periodic motion and is called a vibration. The unwanted noise, wear and frequent, premature failure of one or more of the parts in any mechanical system is due to vibrations. All moving parts of all machines produce vibrations. Hence, while designing one has to anticipate the vibration problems and try to minimise its undesirable effects. Any element is said to vibrate if it undergoes to and fro motion in either rectilinear or angular direction. All bodies possessing mass and elasticity are capable of vibration. So any element made of elastic materials when get displaced from their equilibrium position, work is done on the elastic constraints of the body, thus the energy is stored in the form of strain energy. When the body is released, the internal forces cause the body to move towards its equilibrium position. Examples: Shafts, springs or beams, etc. If the motion is frictionless, the strain energy stored in the body is converted into kinetic energy during the period the body reaches the equilibrium position. When the body passes through the equilibrium or mean position, the kinetic energy is utilised to overcome the elastic forces and that is stored in the form of strain energy, and so on.

15.2 BASIC ELEMENTS OF ANY VIBRATORY SYSTEM

Mathematical model of any vibratory system must be represented by the inertial element, restoring elements and damping elements:

15.2.1 Inertial Element or Mass

These are represented by lumped masses (m) for rectilinear motion and by lumped mass moment of inertia (I) for angular motion. From Newton's law of motion, the product of the mass and its acceleration is equal to the force applied to the mass and the acceleration takes place in the direction in which the force acts. The work is force times displacement in the direction of the force. The kinetic energy increases if work is positive and decreases if work is negative. If the motion is periodic, the system repeats its motion at equal time or regular interval of time. A motion that does not repeat itself at equal time intervals is called an aperiodic motion.

15.2.2 Restoring Element or Spring

Linear or torsional springs of negligible mass to represent restoring elements for rectilinear (stiffness 's') and torsional (torsional stiffness 'q') motions respectively. A spring force exists if the spring is deformed and the work done in deforming a spring is transformed into potential energy, i.e. the strain energy stored in the spring.

15.2.3 Damping Elements or Damper

The symbol for damper is the piston in a cylinder of negligible mass and elasticity. Damping force exists only if there is relative motion between the two ends of the damper. The work or the energy input to a damper is converted into heat. Viscous damping in which the damping force is proportional to the velocity is called linear damping.

15.3 VARIOUS TERMS USED IN VIBRATION AND THEIR MEANINGS

15.3.1 Period

The minimum time required for the system to repeat its motion is called a period and is measured in (t_p) seconds.

15.3.2 Cycle

The motion completed during one time period.

15.3.3 Frequency

Frequency is the number of times the motion repeats itself per unit time. It is expressed in hertz (Hz) and is equal to one cycle per second.

15.3.4 Resonance

When the frequency of the external force is same as that of the natural frequency, resonance takes place, i.e. amplitude or deformation or displacement will reach to maximum at resonance and the system will fail due to breakdown.

15.4 TYPES OF VIBRATIONS

The vibrations can be classified into various categories as follows:

15.4.1 Free or Natural Vibrations

When the body vibrates just due to some sudden disturbance, then the body is said to be under free or natural vibrations. Sometimes when a body is initially displaced from its equilibrium position and released, then it will vibrate just like a simple pendulum.

15.4.2 Forced Vibrations

When the body vibrates under the influence of external forces acting on it, then the body is said to be under forced vibrations. The vibrations exist as long as the external forces are acting on it.

15.4.3 Damped Vibrations

When the energy of a vibrating system is gradually dissipated to overcome friction or through other resistances, there won't be vibrations after a certain time or the vibrations die down. Such vibrations are called damped vibrations. There will be a reduction in amplitude for every cycle of vibration.

15.5 TYPES OF VIBRATIONS BASED ON THE DEFLECTION

Figure 15.1 shows a massless shaft, one end of which is fixed and the other end carrying a heavy disc. The system can execute the following three types of vibrations based on the deflection of the shaft longitudinal, transverse and torsional as follows:

15.5.1 Longitudinal Vibrations
When the particles of the shaft or disc move along the axis of the shaft to and fro in the longitudinal direction, then the vibrations are known as longitudinal vibrations as shown at Fig. 15.1(a).

15.5.2 Transverse Vibrations
When the particles of the shaft move perpendicular to the axis of the shaft on either side in the transverse direction, then the vibrations are known as transverse vibrations as shown at Fig. 15.1(b).

15.5.3 Torsional Vibrations
When the particles of the shaft or disc move in an angular or circular form, i.e. alternately twisting and untwisting the shaft about its axis are known as torsional vibrations as shown at Fig. 15.1(c).

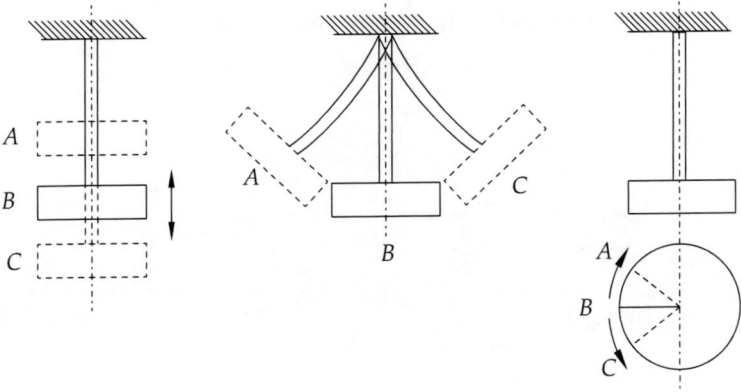

Fig. 15.1 Types of vibrations.

15.6 NATURAL FREQUENCY OF FREE LONGITUDINAL VIBRATIONS

There are three methods by which the natural frequency of free longitudinal vibrations are calculated. Generally, any vibratory motion is considered the simple harmonic motion (SMS).

Stress is proportional to strain. If the limit of proportionality does not exceed in all the above cases, then the restoring force is directly proportional to the displacement of the disc from its equilibrium or mean position. Hence, it follows that the acceleration towards the equilibrium position is directly proportional to the displacement from that position.

Let the displacement be taken as a function of sine or cosine function, i.e. simple harmonic motion. Then,

$$\text{Displacement } x = X \sin \omega t$$

$$\text{Velocity } v = \frac{dx}{dt} = \omega X \cos \omega t$$

$$\text{Acceleration } a = \frac{dv}{dt} = \frac{d^2x}{dt^2}$$

$$= -\omega^2 X \sin \omega t = -\omega^2 x$$

Hence, acceleration is proportional to displacement and opposite in direction of displacement.

15.6.1 Equilibrium Method

Let s be the stiffness of the element; m is the mass of the element attached and let δ be the static deflection as shown at Fig. 15.2.

At Fig. 15.2(i) one end of the spring is fixed and the other end is left free as shown. At (ii) a mass m is attached to the spring. Then the spring gets elongated by δ called static deflection. The free body diagram of the mass is showed under equilibrium position with one upward force of the spring and the other due to gravitational force. Spring force is equal to gravitational force, i.e. $s\delta = mg$ (or) $\dfrac{s}{m} = \dfrac{g}{\delta}$.

The equation of motion of the mass m after a time interval of t seconds is

$$m\frac{d^2x}{dt^2} = -sx$$

as shown in Fig. 15.2(iii).

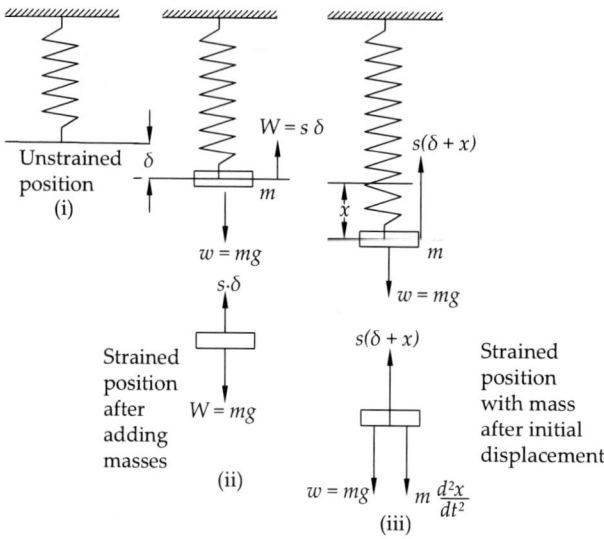

Fig. 15.2 Spring mass system.

$$m\frac{d^2x}{dt^2} + sx = 0$$

$$\frac{d^2x}{dt^2} + \frac{s}{m}x = 0 \qquad (1)$$

According to fundamental equation of SHM seen above,

$$\frac{d^2x}{dt^2} + \omega^2 x = 0 \qquad (2)$$

Comparing equations (1) and (2),

$$\omega^2 = \frac{s}{m} \quad \text{or} \quad \omega = \sqrt{\frac{s}{m}} \qquad (3)$$

$$\text{Time period } t_P = \frac{2\pi}{\omega} = \frac{2\pi}{\sqrt{s/m}}$$

$$\text{Natural frequency of vibration } f_n = \frac{1}{t_P} = \frac{1}{2\pi}\sqrt{\frac{s}{m}}$$

since $\dfrac{s}{m} = \dfrac{g}{\delta}$

Natural frequency

$$f_n = \frac{1}{t_P} = \frac{1}{2\pi}\sqrt{\frac{g}{\delta}} = \frac{1}{2\pi}\sqrt{\frac{9.81}{\delta}} = \frac{0.4985}{\sqrt{\delta}} \text{ Hz} \qquad (4)$$

where deflection δ must be in metres.

15.6.2 Energy Method

There are two kinds of energies (1) Kinetic energy (KE) which is due to the motion of the body and (2) Potential energy (PE) which is due to the position of the body raised or lowered from the mean position. Instead of potential energy, strain energy (SE) exists when a body under goes strain.

At mean position of the body, the potential energy $PE = 0$ whereas the kinetic energy KE is maximum. At extreme positions, the potential energy PE is maximum and kinetic energy $KE = 0$. The best example is the simple pendulum.

In the free vibrations, no energy is transmitted into the system or from the system to outside. Therefore sum total of two energies is constant, i.e.

$$(KE + PE) = \text{a constant} \qquad (1)$$

Hence, if it is differentiated with respect to time,

$$\frac{d(KE + PE)}{dt} = 0 \qquad (2)$$

The expression for $KE = \dfrac{mv^2}{2} = m\left(\dfrac{dx}{dt}\right)^2 \cdot \dfrac{1}{2}$

The expression for $PE = x\left(\dfrac{0+sx}{2}\right) = \dfrac{sx^2}{2}$ or the strain energy. Substituting in equation (2) and differentiating gives the equation of motion of the body as follows.

$$\dfrac{d}{dt}\left[m\left(\dfrac{dx}{dt}\right)^2 \dfrac{1}{2} + \dfrac{sx^2}{2}\right] = m\dfrac{d^2x}{dt^2} + sx = 0$$

$$\dfrac{d^2x}{dt^2} + \dfrac{s}{m}x = 0 \tag{3}$$

Comparing (3) with the fundamental equation of SHM, $\dfrac{d^2x}{dt^2} + \omega^2 x = 0$ gives $\omega = \sqrt{\dfrac{s}{m}}$;

Time period, $t_P = \dfrac{2\pi}{\omega} = \dfrac{2\pi}{\sqrt{s/m}} = \dfrac{2\pi}{\sqrt{g/\delta}}$

Natural frequency

$$f_n = \dfrac{1}{2\pi}\sqrt{\dfrac{g}{\delta}} = \dfrac{1}{2\pi}\sqrt{\dfrac{9.81}{\delta}} = \dfrac{0.4985}{\sqrt{\delta}} \text{ Hz} \tag{4}$$

where deflection δ must be in metres.

15.6.3 Rayleigh's Method

This method states that in a system the maximum kinetic energy (KE_{max}) is equal to the maximum potential energy (PE_{max}) or maximum strain energy (SE_{max}).

Assuming SHM, the displacement $x = X\sin\omega t$; the maximum displacement $x_{max} = X$; the velocity $v = \dfrac{dx}{dt} = \omega X \cdot \cos\omega t$; the maximum velocity $v_{max} = \omega X$.

Therefore, the maximum kinetic energy

$$KE_{max} = \dfrac{mv_{max}^2}{2} = \dfrac{m\omega^2 X^2}{2} \tag{1}$$

and the maximum potential or strain energy

$$PE_{max} = \dfrac{sx_{max}^2}{2} = \dfrac{sX^2}{2} \tag{2}$$

Then equating (1) and (2),

$$\dfrac{m\omega^2 X^2}{2} = \dfrac{sX^2}{2} \quad \text{or} \quad \omega^2 = \dfrac{s}{m}$$

Angular velocity $\omega = \sqrt{\dfrac{s}{m}}$

Time period $t_P = \dfrac{2\pi}{\omega} = \dfrac{2\pi}{\sqrt{s/m}}$

$$\text{Natural frequency } f_n = \frac{1}{t_p} = \frac{1}{2\pi}\sqrt{\frac{s}{m}} \qquad (3)$$

From (3), natural frequency

$$f_n = \frac{1}{2\pi}\sqrt{\frac{g}{\delta}} = \frac{1}{2\pi}\sqrt{\frac{9.81}{\delta}} = \frac{0.4985}{\sqrt{\delta}} \text{ Hz} \qquad (4)$$

where deflection δ must be in metres.

Note: ω is called the natural circular frequency denoted by ω_n.

W E 15.1: Determine the equivalent spring constant for the following spring arrangements as shown below at Fig. 15.2(a) and (b).

1. Two springs of stiffness s_1 and s_2 are connected in series as shown at (a). The springs will stretch by $\frac{1}{s_1}$ and $\frac{1}{s_2}$ when one end of the springs is fixed at O and at the other free end P a unit force F is applied. The total stretch $= \frac{1}{s_1} + \frac{1}{s_2} = \frac{(s_1 + s_2)}{s_1 s_2}$.

Therefore, the stiffness s_o of the equivalent spring which will stretch by the same amount is given by

$$\frac{1}{s_o} = \frac{1}{s_1} + \frac{1}{s_2} = \frac{(s_1 + s_2)}{s_1 s_2} \quad \text{or} \quad s_o = \frac{s_1 s_2}{(s_1 + s_2)}$$

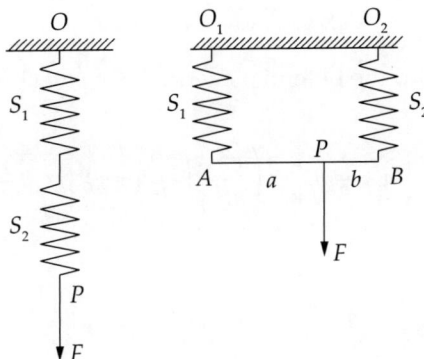

Fig. 15.2(a, b)

2. Two springs of stiffness s_1 and s_2 are connected in parallel as shown at (b). One end of each spring is fixed at O_1 and O_2 and the other ends are connected by a bar of negligible mass AB. An unit force F is applied at a point P on the bar at a and b distances from the other ends A and B of the springs as shown. The unit force will be shared by the two springs and they stretch by: $\dfrac{b}{s_1(a+b)}$ and $\dfrac{a}{s_2(a+b)}$.

When a and b are equal to 1, that means the unit force is applied at the midpoint, then the stretch of the springs: $\dfrac{1}{2s_1}$ and $\dfrac{1}{2s_2}$ and the stiffness of the equivalent spring $s_o = \dfrac{4}{\left[\dfrac{1}{s_1} + \dfrac{1}{s_2}\right]}$.

W E 15.2: Determine the natural frequency of the simple pendulum of Fig. 15.2(c) by the energy method if (a) the mass of rod L is negligible and (b) if the mass of the rod is not negligible. Solve the problem also by Newton's method.

Energy method

1. *Mass of the rod negligible:* For rotation about the pivot O. The mass moment of inertia of the rod about O, $I_o = mL^2$ where L is the length of the rod.

The kinetic energy $KE = \dfrac{1}{2}mL^2\left(\dfrac{d\theta}{dt}\right)^2 = \dfrac{1}{2}mL^2\omega^2$ and the potential energy $PE = mgL(1 - \cos\theta)$.

Thus,
$$\dfrac{d}{dt}[KE + PE] = mL^2\left(\dfrac{d^2\theta}{dt^2}\right)\left(\dfrac{d\theta}{dt}\right) + mgL\left[\dfrac{d\theta}{dt}\right] = 0$$

$$= \dfrac{d^2\theta}{dt^2} + \dfrac{g}{L}\sin\theta = 0$$

But for small oscillations, $\sin\theta = \theta$ or $\dfrac{d^2\theta}{dt^2} + \dfrac{g}{L}\theta = 0$ and natural frequency $f_n = \dfrac{1}{2\pi}\sqrt{\dfrac{g}{L}}$ Hz.

2. *Mass of the rod not negligible:* Let m_r be the mass of the rod of uniform cross section, then $I_o = mL^2 + \dfrac{1}{3}m_rL^2$.

The kinetic energy $KE = \dfrac{1}{2}I_o\omega^2$ and the potential energy $PE = mgL(1 - \cos\theta) + \dfrac{1}{2}m_rgL(1 - \cos\theta)$.

Thus,
$$\dfrac{d}{dt}[KE + PE] = \left[m + \dfrac{1}{3}m_r\right]L^2\left(\dfrac{d^2\theta}{dt^2}\right)\left(\dfrac{d\theta}{dt}\right) + gL\left[m + \dfrac{1}{2}m_r\right]\sin\theta\left(\dfrac{d\theta}{dt}\right) = 0$$

$$\left(\dfrac{d^2\theta}{dt^2}\right) + \dfrac{x}{y}\left(\dfrac{g}{L}\right)\sin\theta = 0$$

where $x = \left[m + \dfrac{1}{2}m_r\right]$ and $y = \left[m + \dfrac{1}{3}m_r\right]$.

But for small oscillations, $\sin\theta = \theta$ or $\left(\dfrac{d^2\theta}{dt^2}\right) + \dfrac{x}{y}\left(\dfrac{g}{L}\right)\theta = 0$ and natural frequency $f_n = \dfrac{1}{2\pi}\sqrt{\left(\dfrac{g}{L}\right)\left[\dfrac{x}{y}\right]}$ Hz.

W E 15.3: A U-tube as shown in Fig. 15.2(d) contains a liquid of density r. Determine the equation of motion of the liquid by the energy method if the liquid is set into motion. Solve the problem also by Newton's method.

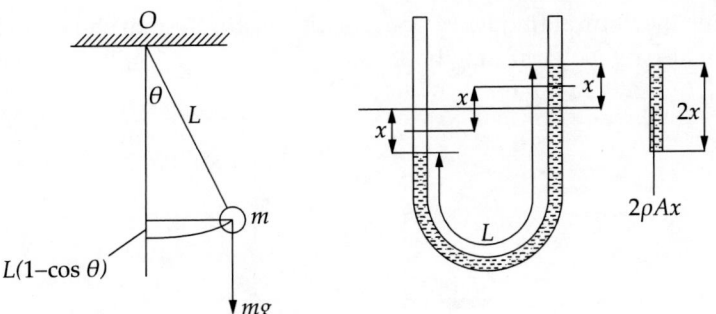

Fig. 15.2(c, d)

Assume the tube is of uniform cross section of area A and density of the fluid as ρ.

Energy method

Total mass of liquid be $m = AL\rho$ kg. Kinetic energy $KE = \frac{1}{2}AL\rho\left(\frac{dx}{dt}\right)^2$.

The potential energy PE is equal to the work done on the fluid to disturb the fluid level or work required to transfer the fluid column of length x from the left to the right side of the tube without disturbing the remaining fluid. The work done is result of a constant force ($mg = \rho Agx$) acting through the distance x.

Hence, $PE = mgh = Ax\rho g x = A\rho g x^2$ where $m = Ax\rho$ and $h = x$ in this case, and

$$\frac{d}{dt}[KE + PE] = AL\rho\left(\frac{dx}{dt}\right)\left(\frac{d^2x}{dt^2}\right) + 2\rho gAx\left[\frac{dx}{dt}\right] = 0$$

on taking out common terms $\left(\frac{d^2x}{dt^2}\right) + \frac{2g}{L}x = 0$ and the natural frequency $f_n = \frac{1}{2\pi}\sqrt{\left(\frac{2g}{L}\right)}$ Hz.

Newton's method

Column of liquid of length $2x$ on the right side exerts a restoring force on the remaining liquid. Total mass of the liquid $= AL\rho$ kg.

Newton's law: $m\left(\frac{d^2x}{dt^2}\right)$ = sum of the forces in x direction, i.e.

$$AL\rho\left(\frac{d^2x}{dt^2}\right) = -2xA\rho g$$

$$\left(\frac{d^2x}{dt^2}\right) + \left(\frac{2g}{L}\right)x = 0$$

and $f_n = \frac{1}{2\pi}\sqrt{\left(\frac{2g}{L}\right)}$ Hz.

W E 15.4: Determine the natural frequency of oscillation of the system shown in Fig. 15.2(e) by the energy method and also by Newton's method.

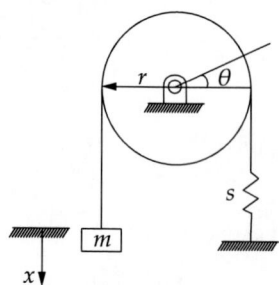

Fig. 15.2(e)

Energy method
Distance moved by mass $m = x = r\theta$. Using θ as coordinate

$$KE = \left(\frac{1}{2}\right)m\left(\frac{dx}{dt}\right)^2 + \left(\frac{1}{2}\right)I_o\theta^2 = \left(\frac{1}{2}\right)mr^2\omega^2 + \left(\frac{1}{2}\right)I_o\omega^2$$

$$PE = \left(\frac{1}{2}\right)sx^2 = \left(\frac{1}{2}\right)sr^2\theta^2$$

$$\frac{d}{dt}(KE + PE) = \frac{d\omega}{dt} + \left[\frac{sr^2}{(I_o + mr^2)}\right]\theta = 0$$

$$f_n = \frac{1}{2\pi}\sqrt{\left(\frac{sr^2}{(I_o + mr^2)}\right)} \text{ Hz}$$

Newton's method
Distance moved by mass $m = x = r\theta$. Consider the motion of the disc with θ as coordinate. For the mass, m

$$m\frac{d^2x}{dt^2} = F \tag{1}$$

and for the disc

$$I_o\frac{d^2\theta}{dt^2} = -F.r - sr^2\theta \tag{2}$$

Substituting (1) in (2)

$$I_o\frac{d^2\theta}{dt^2} = -m.r\left(\frac{d^2x}{dt^2}\right) - sr^2\theta$$

but $\dfrac{d^2x}{dt^2} = r\left(\dfrac{d^2\theta}{dt^2}\right)$, so

$$I_o\frac{d^2\theta}{dt^2} = -m.r^2\left(\frac{d^2\theta}{dt^2}\right) - sr^2\theta = 0$$

$$\frac{d^2\theta}{dt^2} + \frac{sr^2}{(I_o + mr^2)}\theta = 0$$

and $f_n = \dfrac{1}{2\pi}\sqrt{\left(\dfrac{sr^2}{(I_o + mr^2)}\right)}$ Hz.

15.7 NATURAL FREQUENCY OF FREE TRANSVERSE VIBRATIONS

First the expressions for the deflections due to the concentric loads and distributed loads on shafts for various end conditions such as simply supported and fixed ends, etc. studied in the strength of materials must be remembered here in the transverse vibrations as they will be used for determining the natural frequencies of free transverse vibrations as follows:

From Fig. 15.3, the accelerating force is mass times acceleration = $m\dfrac{d^2x}{dt^2}$ and the restoring force is stiffness times the displacement = $-sx$ and the equation of motion is $m\dfrac{d^2x}{dt^2} + sx = 0$ (or) $\dfrac{d^2x}{dt^2} + \dfrac{s}{m}x = 0$.

This shows that the natural frequency of the transverse vibration is same as longitudinal vibration, $f_n = \dfrac{1}{t_p} = \dfrac{1}{2\pi}\sqrt{\dfrac{s}{m}}$. Replacing $\dfrac{s}{m}$ by $\dfrac{g}{\delta}$, the frequency $f_n = \dfrac{1}{2\pi}\sqrt{\dfrac{g}{\delta}} = \dfrac{0.4985}{\sqrt{\delta}}$ Hz.

The expressions to determine the deflection (δ) for various types of loading and end conditions are given below for ready reference in the following table followed by figures:

Table 15.1 Types of loading

S. No.	Type of Beam	Deflection (δ)	
1	Cantilever beam with a point load W at the free end	$WL^3/3EI$ (at the free end)	Fig. T15.1
2	Cantilever beam with a uniformly distributed load of w/length	$wL^4/8EI$ (at the free end)	Fig. T15.2
3	Simply supported beam with an eccentric point load W	$Wa^2b^2/3EIL$ (at the point end)	Fig. T15.3
4	Simply supported beam with a central point load W	$WL^3/48EI$ (at the centre)	Fig. T15.4
5	Simply supported beam with a uniformly distributed load of w/length.	$5wL^4/384EI$ (at the centre)	Fig. T15.5
6	Fixed beam with an eccentric point load W	$Wa^3b^3/3EI^3$ (at the point load)	Fig. T15.6
7	Fixed beam with a central point load W	$WL^3/192EI$ (at the centre)	Fig. T15.7
8	Fixed beam with a uniformly distributed load of w/length	$wL^4/384EI$ (at the centre)	Fig. T15.8

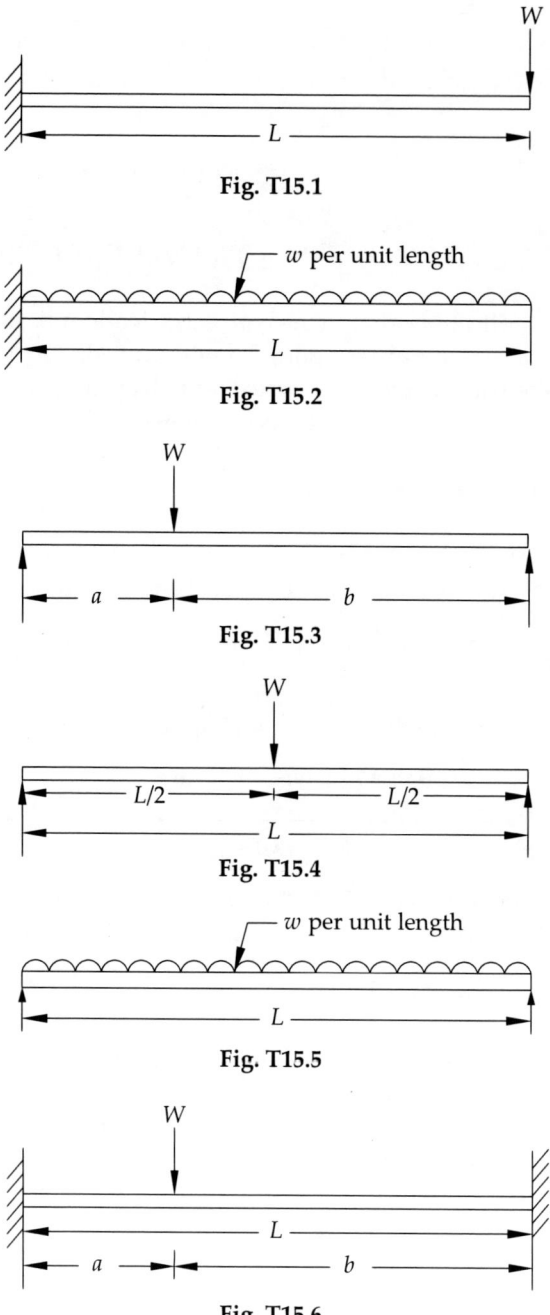

Fig. T15.1

Fig. T15.2

Fig. T15.3

Fig. T15.4

Fig. T15.5

Fig. T15.6

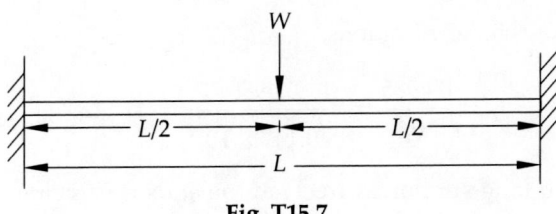

Fig. T15.7

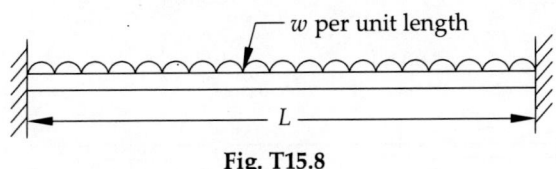

Fig. T15.8

W E 15.5: A cantilever shaft of 50 mm diameter and 300 mm long has a disc of mass 100 kg at its free end. The Young's modulus of the shaft material is 200×10^9 N/m². Determine the longitudinal and transverse vibrations of the shaft.

Given:
Diameter $d = 50$ mm $= 0.05$ m; Length of the shaft $L = 300$ mm $= 0.3$ m; Mass acting at the end of the cantilever $m = 100$ kg; Young's modulus $E = 200 \times 10^9$ N/m²;

Area of cross section

$$A = \frac{\pi d^2}{4} = \frac{\pi (0.05)^2}{4} = 1.96 \times 10^{-3} \text{ m}^2$$

and moment of inertia

$$I = \frac{\pi d^4}{64} = \frac{\pi (0.05)^4}{64} = 0.3 \times 10^{-6} \text{ m}^4$$

Longitudinal vibrations of the shaft

The longitudinal deflection of the shaft due to load W,

$$\delta = \frac{WL}{AE} = \frac{mgL}{AE} = \frac{100 \times 9.81 \times 0.3}{1.96 \times 10^{-3} \times 200 \times 10^9} = 0.751 \times 10^{-6} \text{ m}$$

The natural frequency of longitudinal vibrations,

$$f_n = \frac{0.4985}{\sqrt{\delta}} = \frac{0.4985}{\sqrt{0.751 \times 10^{-6}}} = 575 \text{ Hz}$$

Transverse vibrations of the shaft

The static deflection of the shaft due to transverse load W,

$$\delta = \frac{WL^3}{3EI} = \frac{mgL^3}{3EI} = \frac{100 \times 9.81 \times 0.3^3}{3 \times 200 \times 10^9 \times 0.3 \times 10^{-6}} = 0.147 \times 10^{-3} \text{ m}.$$

The natural frequency of transverse vibrations,

$$f_n = \frac{0.4985}{\sqrt{\delta}} = \frac{0.4985}{\sqrt{0.147 \times 10^{-3}}} = 41 \text{ Hz}$$

W E 15.6: A shaft of 0.75 m long supported freely at the ends is carrying a body of mass 90 kg at 0.25 m from one end. Find the natural frequency of transverse vibration. Assume $E = 200 \times 10^9$ N/m² and shaft diameter = 50 mm.

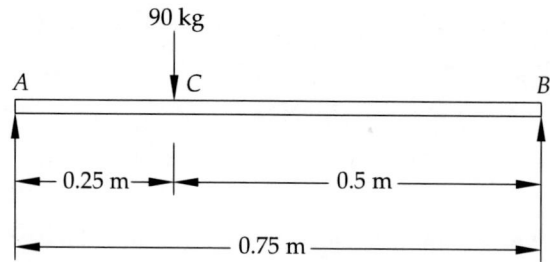

Fig. 15.3 Simply supported with one concentrated load.

Given:
Length of the shaft $L = 0.75$ m;
Mass of the concentrated load, $m = 90$ kg acting at distance of = 0.25 m from one end;
Young's modulus $E = 200 \times 10^9$ N/m²; Diameter of shaft $d = 50$ mm = 0.05 m.

Moment of inertia of the shaft

$$I = \frac{\pi d^4}{64} = \frac{\pi (0.05)^4}{64} = 0.307 \text{ m}^4$$

and the static deflection δ at the load point C as shown in Fig. T15.3,

$$\delta = \frac{Wa^2b^2}{3EIL} = \frac{90 \times 9.81 \times 0.25^2 \times 0.05^2}{3 \times 200 \times 10^9 \times 0.307 \times 10^{-6} \times 0.75} = 0.1 \times 10^{-3} \text{ m}$$

The natural frequency of transverse vibration,

$$f_n = \frac{0.4985}{\sqrt{\delta}} = \frac{0.4985}{\sqrt{0.1 \times 10^{-3}}} = 49.85 \text{ Hz}$$

W E 15.7: Calculate the whirling speed of a shaft 20 mm diameter and 0.6 m long carrying a mass of 1 kg at its midpoint. The density of the shaft material is 40 Mg/m³, and Young's modulus is 200 GN/m². Assume that the shaft is simply supported.

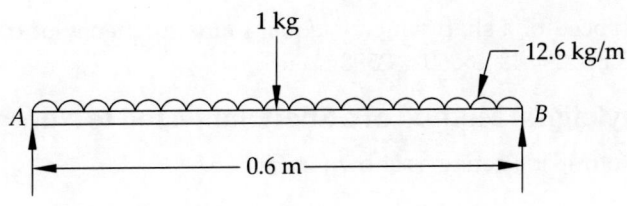

Fig. 15.4

Given:
The diameter of shaft $d = 20$ mm $= 0.02$ m; Length of the shaft $L = 0.6$ m;
Concentrated load of mass $m = 1$ kg; density of the shaft material $\rho = 40$ Mg/m^2 = 40×10^6 g/m^3 = 40×10^3 kg/m^3; $E = 200$ GN/m^2 = 200×10^9 N/m^2.

The shaft is shown in Fig. 15.4. The moment of inertia

$$I = \frac{\pi d^4}{64} = \frac{\pi (0.02)^4}{64} = 7.855 \times 10^{-9} \text{ m}^4$$

Since the density of shaft material is 40×10^3 kg/m^3, therefore mass of the shaft per unit length,

$$m_s = \text{area} \times \text{length} \times \text{density}$$

$$= \frac{\pi 0.02^2}{4} \times 1 \times 40 \times 10^3 = 12.6 \text{ kg/m}$$

The static deflection due to 1 kg of mass at the centre

$$\delta = \frac{WL^3}{48EI} = \frac{1 \times 9.81 \times 0.6^3 \times 0.05^2}{48 \times 200 \times 10^9 \times 7.855 \times 10^{-9}} = 28 \times 10^{-6} \text{ m}$$

Static deflection due to mass of the shaft,

$$\delta_s = \frac{5wL^4}{384EI} = \frac{5 \times 9.81 \times 12.6 \times 0.6^4}{384 \times 200 \times 10^9 \times 7.885 \times 10^{-9}} = 0.133 \times 10^{-3} \text{ m}$$

Frequency of transverse vibration

$$f_n = \frac{0.4985}{\sqrt{\delta + \frac{\delta_s}{1.27}}}$$

$$= \frac{0.4985}{\sqrt{28 \times 10^{-6} + \left(\frac{0.133 \times 10^{-3}}{1.27}\right)}}$$

$$= \frac{0.4985}{\left(11.52 \times 10^{-3}\right)} = 43.3 \text{ Hz}$$

Let N_c be the whirling speed of a shaft which is equal to the frequency of transverse vibration in Hz, therefore, N_c = 43.3 r.p.s. = 43.3 × 60 = 2598 r.p.m.

15.7.1 Energy (Rayleigh's) Method of a Shaft Subjected to Number of Point Loads

Assume that there are four loads acting as shown in Fig. 15.5.

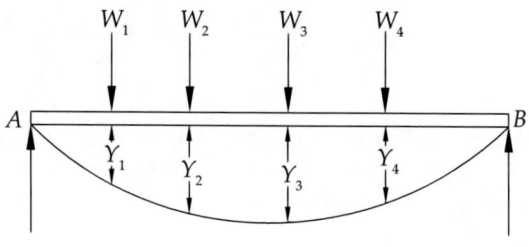

Fig. 15.5

The potential energy is proportional to mass and deflection. The maximum potential energy (PE_{max}) depends on maximum deflections at each load and it is given by

$$PE_{max} = \frac{m_1 g y_1}{2} + \frac{m_2 g y_2}{2} + \frac{m_3 g y_3}{2} + \frac{m_4 g y_4}{2} = \frac{g}{2}\Sigma my \quad (1)$$

The kinetic energy is proportional to mass and square of the velocity which again depends on displacement. The maximum kinetic energy depends on maximum velocity and it is given by

$$KE_{max} = \frac{m_1(\omega y_1)^2}{2} + \frac{m_2(\omega y_2)^2}{2} + \frac{m_3(\omega y_3)^2}{2} + \frac{m_4(\omega y_4)^4}{2} = \frac{\omega^2}{2}\Sigma my^2 \quad (2)$$

Hence, first determine the values of y_1, y_2, y_3, y_4 using the expression $\frac{Wa^2b^2}{3EIL}$ where a and b are the distances of the load from the ends A and B.

Since $PE_{max} = KE_{max}$ so from (1) and (2),

$$\frac{g}{2}\Sigma my = \frac{\omega^2}{2}\Sigma my^2$$

Hence,

$$\omega^2 = \frac{\Sigma mgy}{\Sigma my^2} \quad \text{or} \quad \omega = \sqrt{\frac{\Sigma mgy}{\Sigma my^2}}$$

Therefore,

$$\text{Natural frequency } f_n = \frac{\omega}{2\pi} \quad (3)$$

15.7.2 Dunkerley's Method for a Shaft Subjected to a Number of Point Loads

Assume three concentrated loads and the mass of the shaft as uniformly distributed load as shown in Fig. 15.6.

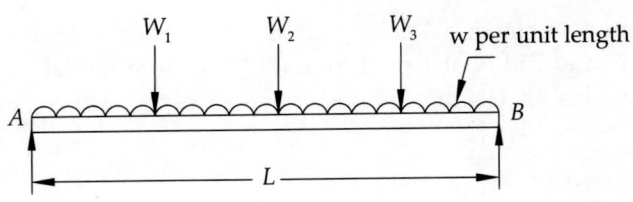

Fig. 15.6

Let f_1, f_2, f_3 and f_u are the natural frequencies due to each individual load of W_1, W_2, W_3 and W_u uniformly distributed load represented.

The expression for the natural frequencies due to concentric loads are: Due to load W_1, $f_1 = \dfrac{0.4985}{\sqrt{y_1}}$, due to W_2, $f_2 = \dfrac{0.4985}{\sqrt{y_2}}$, due to W_3, $f_3 = \dfrac{0.4985}{\sqrt{y_3}}$, and due to uniformly distributed load W_u, $f_u = \dfrac{0.5615}{\sqrt{y_u}}$ where y_1, y_2, y_3 and y_u are the deflections at W_1 due to W_1, at W_2 due to W_2, at W_3 due to W_3 and at the centre due to uniformly distributed load w respectively.

Dunkerley's empirical formula for the natural frequency (f_n) of the whole system is given as a function of individual frequencies as given below:

$$\frac{1}{f_n^2} = \frac{1}{f_1^2} + \frac{1}{f_2^2} + \frac{1}{f_3^2} + \frac{1}{f_u^2}$$

$$= \frac{y_1}{0.4985^2} + \frac{y_2}{0.4985^2} + \frac{y_3}{0.4985^2} + \frac{y_u}{0.5615^2}$$

$$= \left[\frac{1}{0.4985^2}\right]\left[y_1 + y_2 + y_3 + \frac{y_u}{1.27}\right]$$

The natural frequency of vibration $f_n = \dfrac{0.4985}{\sqrt{\left[y_1 + y_2 + y_3 + \dfrac{y_u}{1.27}\right]}}$

Note: When the mass of the shaft m_s is considered, then the natural frequency is as follows:

1. Longitudinal Vibration: $f_n = \dfrac{1}{2\pi}\sqrt{\dfrac{s}{m + \dfrac{m_s}{3}}}$ for cantilever

2. Transverse Vibration: $f_n = \dfrac{1}{2\pi}\sqrt{\dfrac{s}{m + \left(\dfrac{33m_s}{140}\right)}}$ for cantilever

3. For Both ends fixed: $f_n = \dfrac{1}{2\pi}\sqrt{\dfrac{s}{m + \left(\dfrac{13m_s}{35}\right)}}$

4. For Simply supported: $f_n = \dfrac{1}{2\pi}\sqrt{\dfrac{s}{m + \left(\dfrac{17m_s}{35}\right)}}$

W E 15.8: A shaft 50 mm diameter and 3 m long is simply supported at the ends and carries three loads of 1000 N, 1000 N and 750 N at 1 m, 2 m, and 2.5 m from the left support. The Young's modulus for shaft material is 200 GN/m²; find the frequency of transverse vibration.

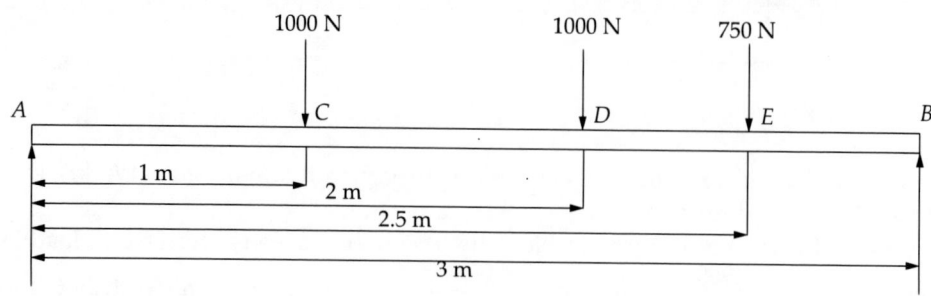

Fig. 15.7

Given:
Diameter of shaft $d = 50$ mm $= 0.05$ m; Length of shaft $L = 3$ m; Loads $W_1 = 1000$; $W_2 = 1000$; $W_3 = 750$ N; Young's modulus $E = 200$ GN/m² $= 200 \times 10^9$ N/m².

The shaft carrying loads W_1, W_2, W_3 is shown in Fig. 15.7,

Moment of inertia of the shaft

$$I = \left(\frac{\pi}{64}\right)d^4 = \frac{\pi}{64}(0.05)^4 = 0.307 \times 10^{-6} \text{ m}^4$$

Using the expression $\delta = \dfrac{Wa^2b^2}{(3EIL)}$, for the static deflection due to a point load W at distances a and b from the ends.

The static deflection due to $W_1 = 1000$ N; with $a = 1$ m, $b = 2$ m and $L = 3$ m is

$$\delta_1 = \left(\frac{1000 \times 1^2 \times 2^2}{3 \times 200 \times 10^9 \times 0.307 \times 10^{-6} \times 3}\right) = 7.24 \times 10^{-3} \text{ m}$$

Due to $W_2 = 1000$ N, $a = 2$ m, $b = 1$ m and $L = 3$ m is

$$\delta_2 = \left(\frac{1000 \times 2^2 \times 1^2}{3 \times 200 \times 10^9 \times 0.307 \times 10^{-6} \times 3}\right) = 7.24 \times 10^{-3} \text{ m}$$

Due to $W_3 = 750$ N, $a = 2.5$ m, $b = 0.5$ m and $L = 3$ m,

$$\delta_3 = \left(\frac{750 \times 2.5^2 \times 0.5^2}{3 \times 200 \times 10^9 \times 0.307 \times 10^{-6} \times 3}\right) = 2.12 \times 10^{-3} \text{ m}$$

The frequency of transverse vibration

$$f_n = \frac{0.4985}{\sqrt{[\delta_1 + \delta_2 + \delta_3]}} = \frac{0.4985}{\sqrt{[7.24 + 7.24 + 2.12] \times 10^{-3}}} = 3.867 \text{ Hz}$$

W E 15.9: A shaft 1.5 m long supported in flexible bearings at the ends carries two wheels each of mass 50 kg. One wheel is situated at the centre of the shaft and the other at a distance of 375 mm from the centre towards left. The shaft is hollow of external diameter 75 mm and internal diameter 40 mm. The density of the shaft material is 7700 kg/m³ and its modulus of elasticity is 200 GN/m². Find the frequency of transverse vibration.

Given:
Length of shaft $L = 1.5$ m;   Masses, $m_1 = m_2 = 50$ kg; $d_1 = 75$ mm $= 0.075$ m;
$d_2 = 40$ mm $= 0.04$ m; $\rho = 7700$ kg/m³; $E = 200$ GN/m² $= 200 \times 10^9$ N/m².

The shaft is shown in Fig. 15.8.

The M.I. of the shaft, $I = \left(\dfrac{\pi}{64}\right)\left(d_1^4 - d_2^4\right)$

$= \dfrac{\pi}{64}\left(0.075^4 - 0.04^4\right) = 1.4 \times 10^{-6}$ m⁴

since the density of shaft material is 7700 kg/m³.

The mass of the shaft per metre length,

m_s = area of cross section of the shaft × density

$= \dfrac{\pi}{4}[(0.075)^2 - (0.04)^2] \times 7700 = 24.34$ kg/m

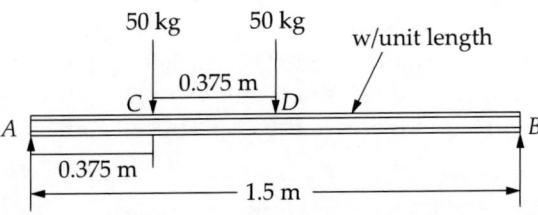

Fig. 15.8

Using the expression for the static deflection due to load W,

$\dfrac{Wa^2b^2}{3EIL} = \dfrac{mga^2b^2}{3EIL} = ma^2b^2\left(\dfrac{g}{3EIL}\right)$

$= ma^2b^2\left(\dfrac{9.81}{3 \times 200 \times 10^9 \times 1.4 \times 10^{-6} \times 1.5}\right)$

$= \dfrac{ma^2b^2}{(0.7785 \times 10^{-5})}$

Therefore, static deflection due to $m_1 = 50$ kg at C, $\delta_1 = \dfrac{m_1 a_1^2 b_1^2}{(0.7785 \times 10^{-5})}$ where $a_1 = 0.375$ m, and $b_1 = 1.125$ m.

$\delta_1 = \left(\dfrac{50 \times 0.375^2 \times 1.125^2}{0.7785 \times 10^{-5}}\right) = 69.29 \times 10^{-6}$ m

Similarly, the static deflection due to $m_2 = 50$ kg at D, $\delta_2 = \dfrac{m_2 a_2^2 b_2^2}{(0.7785 \times 10^{-5})}$ where $a_2 = b_2 = 0.75$ m.

$$\delta_2 = \left(\frac{50 \times 0.75^2 \times 0.75^2}{0.7785 \times 10^{-5}}\right) = 123.2 \times 10^{-6} \text{ m}$$

The static deflection due to uniformly distributed load or mass of the shaft

$$\delta_s = \frac{5 W_u L^4}{384 EI} \quad \text{where } W_u = m_u g$$

$$= \frac{5 \times 24.34 \times 9.81 \times 1.5^4}{384 \times 200 \times 10^9 \times 1.4 \times 10^{-6}} = 56.21 \times 10^{-6} \text{ m}$$

The frequency of transverse vibration

$$f_n = \frac{0.4985}{\sqrt{\left[\delta_1 + \delta_2 + \dfrac{\delta_s}{1.27}\right]}} = \frac{0.4985}{\sqrt{\left[69.29 + 123.2 + \dfrac{56.21}{1.27}\right] \times 10^{-6}}} = 32.4 \text{ Hz}$$

15.8 CRITICAL SPEED OR WHIRLING SPEED OF A SHAFT

When the gears or pulleys are put on the shaft, the centre of gravity (G) of the pulley or gear does not coincide with the centre line of the bearings or axis of the shaft, when the shaft is stationary. Due to this distance of G from the axis of rotation called eccentricity (e) the shaft is subjected to centrifugal force (C.F.). This force will bend the shaft which in turn increases the distance of G from the axis of rotation. This will again increase the C.F. Thus, the effect will be cumulative and ultimately the shaft fails.

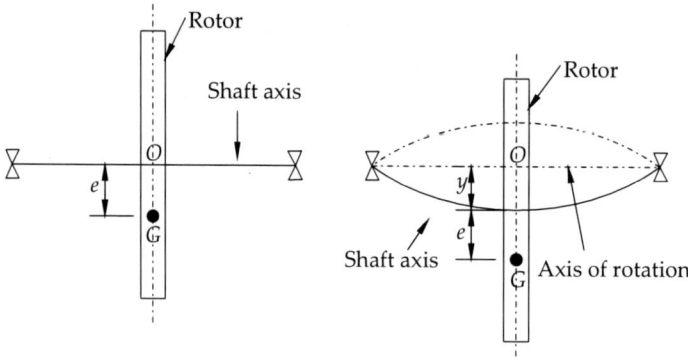

Fig. 15.9 Whirling of shaft.

Hence, the bending of shaft depends not only on the eccentricity (or distance of G) but also depends upon the speed at which the shaft rotates as centrifugal force C.F. is proportional to square of the speed (ω^2).

Definition: The speed at which the shaft runs so that the additional deflection of the shaft from the axis of rotation becomes infinite is known as critical or whirling.

Let the mass of the disc be m, e be the eccentricity and y be the deflection. s be the stiffness of shaft.

Centrifugal force $F_C = m\omega^2(y + e)$ and resisting force $= s \cdot y$.

Therefore,
$$y = \frac{m\omega^2 e}{s - m\omega^2} = \frac{\omega^2 e}{\dfrac{s}{m} - \omega^2}$$

$$y = \frac{\omega^2 e}{\left(\omega_n^2 - \omega^2\right)} = \frac{e}{\left[\dfrac{\omega_n^2}{\omega^2} - 1\right]}$$

The value y becomes infinite, when the circular frequency ω is equal to the natural frequency ω_n, i.e. ($\omega = \omega_n$).

This particular speed is called the critical or whirling speed

$$\omega_c = \omega_n = \sqrt{\frac{s}{m}} = \sqrt{\frac{g}{\delta}} \text{ Hz}$$

If N_C is the critical speed or whirling speed in r.p.s. then

$$2\pi N_C = \omega_c = \sqrt{\frac{g}{\delta}} \quad \text{and} \quad N_C = \frac{1}{2\pi}\sqrt{\frac{g}{\delta}} = \frac{0.4895}{\sqrt{\delta}} \text{ r.p.s.}$$

where δ is the static deflection.

Hence, the critical or whirling speed is the same as the natural frequency of transverse vibration but its unit will be in r.p.s.

W E 15.10: Determine the critical speed of the shaft of W E 15.9 loaded in the same way.

The lowest whirling speed of the shaft given in W E 15.9. (N_c) in r.p.s. is equal to the frequency of transverse vibration in Hz, therefore $N_c = 32.4$ r.p.s. $= 32.4 \times 60 = 1944$ r.p.m.

W E 15.11: A vertical shaft of 14 mm diameter is 1.2 m long and is supported in long bearings at its ends. A disc of mass 16 kg is attached to the centre of the shaft. Neglecting any increase in stiffness due to the attachment of the disc to the shaft, find the critical speed of rotation and the range of speed over which it is unsafe to run the shaft. Assume the shaft to be mass less. The centre of the disc is 0.4 mm from the geometric axis of the shaft. Permissible stress in shaft material is 70×10^6 N/m² and modulus of elasticity of shaft material E is 200 GN/m².

Given:
Diameter of shaft = 14 mm;   Length $L = 1.2$ m;   mass $m = 16$ kg, so weight $W = m \times 9.81$; $E = 200 \times 10^9$;   Permissible stress $f = 70 \times 10^6$ N/m².

Moment of inertia of shaft

$$I = \left(\frac{\pi d^4}{64}\right) = \frac{\pi}{64}\left(0.014^4\right)$$

Assuming fixed ends, deflection

$$\delta = \left(\frac{WL^3}{192EI}\right)$$

$$= \frac{16 \times 9.81 \times 1.2^3}{192 \times 200 \times 10^9 \times \pi \times \frac{0.014^4}{64}} = 0.00375 \text{ m}$$

Critical speed of rotation

$$N_C = \frac{0.4985}{\sqrt{3.75 \times 10^{-3}}} = 8.145 \text{ r.p.s.}$$

The speed at which the shaft is rotating $N_C = 489$ r.p.m.

Maximum bending stress (s): Bending equation is given by

$$\frac{M}{I} = \frac{f}{y}$$

$$M = I\frac{f}{y} \quad (1)$$

where $y = \frac{d}{2}$.

But bending moment due to the additional dynamic load, W_1

$$M = \frac{W_1 L}{8} \quad (2)$$

Equating (1) and (2),

$$W_1 = \frac{f \times I \times 8}{0.5d \times L} = 125.7 \text{ N}$$

Additional deflection due to W_1 is

$$\frac{W_1}{W}\delta = \frac{125.7 \times 0.00375}{16 \times 9.81} = 0.03 \text{ m} \quad (3)$$

But the additional deflection is also equal to

$$\pm \frac{e}{\left[\frac{\omega_C^2}{\omega^2} - 1\right]} = \pm \frac{e}{\left[\left(\frac{N_C}{N}\right)^2 - 1\right]} \quad (4)$$

Equating (3) and (4),

$$0.003 = \pm \frac{0.0004}{\left[\left(\frac{489}{N}\right)^2 - 1\right]}$$

$$\left[\left(\frac{489}{N}\right)^2 - 1\right] = \pm 0.1333$$

giving $N = 459$ and 525.

Thus, the range of unsafe speed is from 459 r.p.m. to 525 r.p.m.

15.9 FREQUENCY OF FREE DAMPED VIBRATIONS (VISCOUS DAMPING)

Pic. 15.1

The free vibrations reduce gradually and become zero due to resistance to motion called friction or damping. The Pic. 15.1 and Pic. 15.2 show the shock absorbers used in two-wheel and four-wheel vehicles to reduce vibrations.

Hence, it is ideal to consider vibrations without friction. Resistance to motion is called damping and it is proportional to velocity. Any vibrating system must consider, apart from mass and stiffness, another important factor called damping which is indicated by a piston in a cylinder as shown in Fig. 15.10. Hence, the periodic vibrations of decreasing amplitude are called damped vibrations.

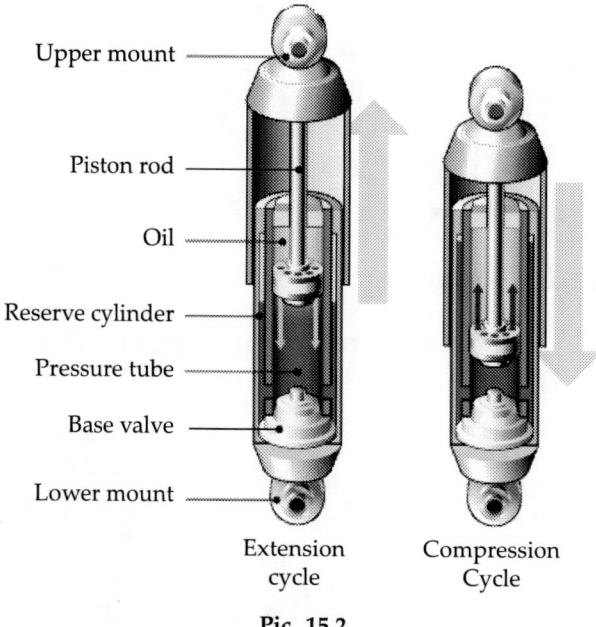

Pic. 15.2

The equation of motion consists of another term $c\dfrac{dx}{dt}$ due to damping and it is

$$m\frac{d^2x}{dt^2} + c\frac{dx}{dt} + sx = 0 \qquad (1)$$

where c is the damping or the damping force per unit velocity.

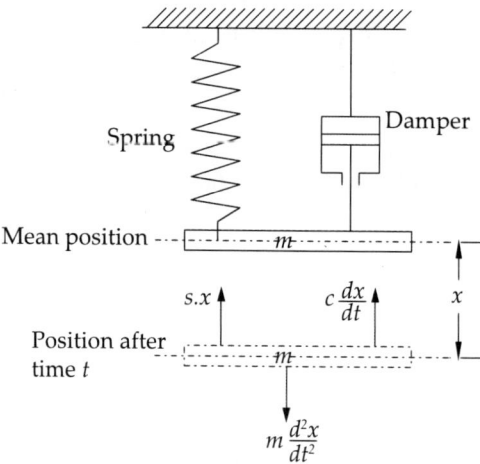

Fig. 15.10 Free damped vibrations.

Damping provided by fluid resistance is known as Viscous Damping. Dividing (1) by mass m gives

$$\frac{d^2x}{dt^2} + \left(\frac{c}{m}\right)\frac{dx}{dt} + \left(\frac{s}{m}\right)x = 0 \tag{2}$$

This is a differential equation of the second order.

Assume the solution of the form $x = e^{kt}$ where k is a constant to be determined. Substituting for x in (2) gives,

$$k^2 e^{kt} + \left(\frac{c}{m}\right)k e^{kt} + \left(\frac{s}{m}\right)e^{kt} = 0$$

$$k^2 + \left(\frac{c}{m}\right)k + \left(\frac{s}{m}\right) = 0 \tag{3}$$

The two roots of the quadratic equation are, k_1 and $k_2 = -\left(\frac{c}{2m}\right) \pm \sqrt{\left[\left(\frac{c}{2m}\right)^2 - \frac{s}{m}\right]}$.

Taking the solution of the equation (2) as $x = C_1 e^{k_1 t} + C_2 e^{k_2 t}$ where C_1 and C_2 are two arbitrary constants determined from the initial conditions.

The two roots k_1 and k_2 may be real or equal or complex conjugate (imaginary).

15.9.1 When the Roots are Real (Overdamping or Large Damping)

The two roots will be real and negative if $\left(\frac{c}{2m}\right)^2 > \frac{s}{m}$. Then the solution of the equation is $x = C_1 e^{k_1 t} + C_2 e^{k_2 t}$. This won't give vibratory motions and so it is called an aperiodic motion as the motion will be on one side of the equilibrium position.

15.9.2 When the Roots are Equal (Critical Damping)

If $\left(\frac{c}{2m}\right)^2 = \frac{s}{m}$, then the radical becomes zero. The two roots k_1 and k_2 will be equal. This case is called critical damping (C_c) where the mass moves back rapidly to its equilibrium position in the shortest possible time. Therefore,

$$x = (C_1 + C_2)e^{-\left(\frac{c}{2m}\right)t} = (C_1 + C_2)e^{-\omega_n t} \text{ since } \frac{c}{2m} = \sqrt{\frac{s}{m}} = \omega_n$$

This is also an aperiodic motion, i.e. not a vibratory motion.

Damping factor (ζ) is the ratio of damping to critical damping, $\zeta = \dfrac{c}{C_c} = \dfrac{c}{2m\omega_n}$

15.9.3 When the Roots are Complex Conjugate (Underdamping or Small Damping)

The two roots will be complex conjugate when $\dfrac{s}{m} > \left(\dfrac{c}{2m}\right)^2$.

Then the two roots k_1 and $k_2 = -\left(\dfrac{c}{2m}\right) \pm i\sqrt{\left[\dfrac{s}{m}\right] - \left[\dfrac{c}{2m}\right]^2}$ where, $i = \sqrt{-1}$.

Let us take $\dfrac{c}{2m} = a$ and $\dfrac{s}{m} = (\omega_n)^2$.

The damped circular frequency $\omega_d = \sqrt{[(\omega_n)^2 - a^2]}$.

Therefore, the two roots are $k_1 = -a + i\omega_d$ and $k_2 = -a - i\omega_d$.

The equation of motion,

$$x = e^{(-k_1 t)} + C_2 e^{(-k_2 t)} = e^{(-at)} \left[C_1 e^{i\omega_d t} + C_2 e^{-i\omega_d t} \right]$$

Using Euler's theorem $e^{i\theta} = \cos\theta + i\sin\theta$ and $e^{-i\theta} = \cos\theta - i\sin\theta$ giving the equation of motion as $x = e^{(-at)}(A\cos\omega_d t + B\sin\omega_d t)$ or $x = Ce^{(-at)}\cos(\omega_d t)$ where $\omega_d = \sqrt{\left[\dfrac{s}{m} - \left(\dfrac{c}{2m}\right)^2\right]}$.

$$\text{Time period } t_p = \dfrac{2\pi}{\omega_d}$$

and damped frequency

$$f_d = \dfrac{1}{t_p} = \dfrac{\omega_d}{2\pi} = \dfrac{1}{2\pi}\sqrt{\left[\dfrac{s}{m} - \left(\dfrac{c}{2m}\right)^2\right]}$$

If $c = 0$, free undamped vibrations occur, then

$$f_d = f_n = \dfrac{1}{2\pi}\sqrt{\left(\dfrac{s}{m}\right)}$$

15.9.4 Logarithmic Decrement

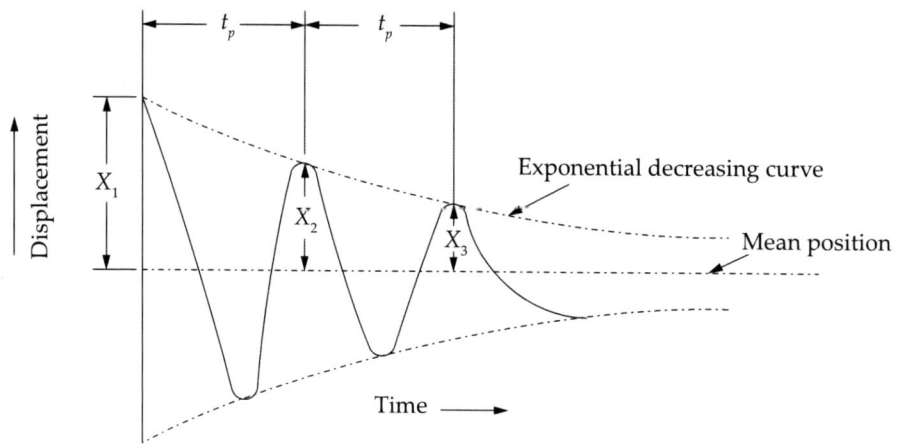

Fig. 15.11 Logarithmic decrement.

The effect of damping on the vibrations is to reduce the amplitude. Figure 15.11 shows how the amplitude reduces gradually. This reduction of amplitude is expressed as the natural logarithm of the ratio of any two successive amplitudes on the same side of the mean position and also called amplitude reduction factor.

From Fig. 15.11, $\dfrac{x_1}{x_2} = \dfrac{Ce^{(-at)}}{Ce^{[-a(t+t_p)]}} = e^{at_p} = \text{constant}$.

Taking logarithms to the base on both sides gives

$$\delta = \log_e \dfrac{x_1}{x_2} = \log_e \left(e^{at_p}\right) = at_p = a\left(\dfrac{2\pi}{\omega_d}\right)$$

$$= \left(\dfrac{c}{2m}\right) \dfrac{2\pi}{\sqrt{\left[(\omega_n)^2 - \left(\dfrac{c}{2m}\right)^2\right]}}$$

$$= \dfrac{2\pi c}{\sqrt{\left[C_c^2 - c^2\right]}} \text{ where, } C_c = 2m\omega_n$$

W E 15.12: A vibrating system consists of a mass of 100 kg, a spring of stiffness 75 N/mm and a damper with damping coefficient of 1000 N/m/s, determine the frequency of vibration of the system.

Given:
$m = 100$ kg; $s = 75$ N/mm $= 75 \times 10^3$ N/m; $c = 1000$ N/m/s

The circular frequency of undamped free vibrations

$$\omega_n = \sqrt{\left(\dfrac{s}{m}\right)} = \sqrt{\left(\dfrac{75 \times 10^3}{100}\right)} = 27.4 \text{ rad/s}$$

The circular frequency of damped vibrations $\omega_d = \sqrt{(\omega_n^2 - a^2)}$ where $a = \dfrac{c}{(2m)}$

$$\omega_d = \sqrt{\left[(27.4)^2 - \dfrac{1000}{(2 \times 100)^2}\right]} = 27.39 \text{ rad/s}$$

Frequency of vibration of the system

$$f_n = \dfrac{\omega_d}{2\pi} = \dfrac{27.39}{2\pi} = 4.35 \text{ Hz}$$

W E 15.13: The following data refer to a vibratory system with viscous damping. Mass 2.5 kg; spring constant 3 N/mm and the amplitude decreases to 0.25 of the initial value after five consecutive cycles. Determine the damping coefficient of the damper in the system.

Given:
$m = 2.5$ kg; $s = 3$ N/mm $= 3000$ N/m; $x_6 = 0.25 x_1$

Let c be the damping coefficient of the damper in N/m/s; x_1 be the initial amplitude and x_6 be the final amplitude after five consecutive cycles, i.e. $x_6 = 0.25 x_1$

The amplitude ratio of any consecutive amplitude is

$$\frac{x_1}{x_2} = \frac{x_2}{x_3} = \frac{x_3}{x_4} = \frac{x_4}{x_5} = \frac{x_5}{x_6} = e^\delta$$

Therefore

$$\frac{x_1}{x_6} = \frac{x_1}{x_2}\frac{x_2}{x_3}\frac{x_3}{x_4}\frac{x_4}{x_5}\frac{x_5}{x_6} = \left(\frac{x_1}{x_2}\right)^5 = e^{5\delta} = \frac{1}{0.25} = 4$$

or $\delta = \dfrac{1}{5}\log(4) = 0.278$

Also δ (Logarithmic decrement) $= \dfrac{2\pi\zeta}{\sqrt{[1-\zeta^2]}} \approx 2\pi\zeta$.

First solve it by approximate method, if the value comes greater than 0.3, the exact relationship may be used.

Hence,
$$0.278 = 2\pi\zeta$$
$$\zeta = 1.278/(2\pi) = 0.0442$$

Also, the damping coefficient is given by

$$c = 2\zeta\sqrt{(sm)} = 2 \times 0.0442\sqrt{(3000 \times 2.5)}$$
$$= 7.665 \text{ N-sec/m} = 0.07656 \text{ N-sec/cm}$$

15.10 FREQUENCY OF FORCED DAMPED VIBRATION

Here in this system, a periodic external disturbing force acts which is a function of sine or cosine say $F_x = F\cos\omega_t$, where F is the static force and ω is the angular velocity of the periodic disturbing force as shown in the Fig. 15.12.

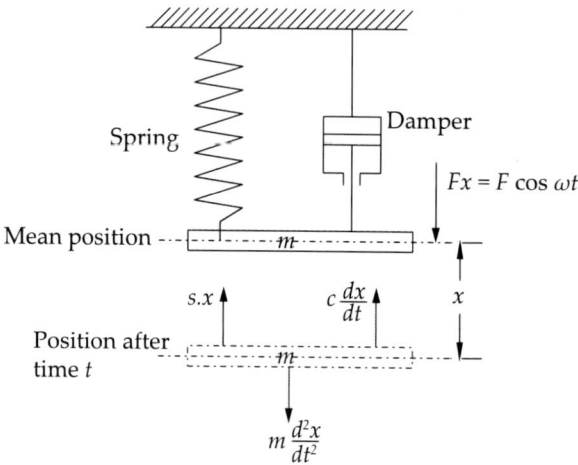

Fig. 15.12 Forced damped vibration.

The equation of motion is similar to equation of motion with free damped vibrations. But the right-hand side of the equation instead of zero equals to the external force as shown below.

$$\frac{md^2x}{dt^2} + \frac{cdx}{dt} + sx = F\cos\omega t \quad (1)$$

This can easily be solved by a graphical method by taking

Displacement $x = A\cos(\omega t - \phi)$ where A is the amplitude of vibration.

Velocity $v = \dfrac{dx}{dt} = \omega A \cos[90° + (\omega t - \phi)]$ and

Acceleration $a = \dfrac{d^2x}{dt^2} = \omega^2 A \cos[180 + (\omega t - \phi)]$

The external force F is represented by OS in the vector diagram shown in Fig. 15.13(i) and the spring force, damping force and accelerating force are respectively as below:

The spring force

$$sx = sA\cos(\omega t - \phi) \quad (1)$$

represented by OP in Fig. 15.13(i)

Damping force

$$c.v = \frac{cdx}{dt} = c\omega A \cos[90° + (\omega t - \phi)] \quad (2)$$

represented by OQ.

Inertia force is the force required to accelerate the mass 'm'

$$= \frac{md^2x}{dt^2} = m\omega^2 A \cos[180° + (\omega t - \phi)] \quad (3)$$

represented by OR.

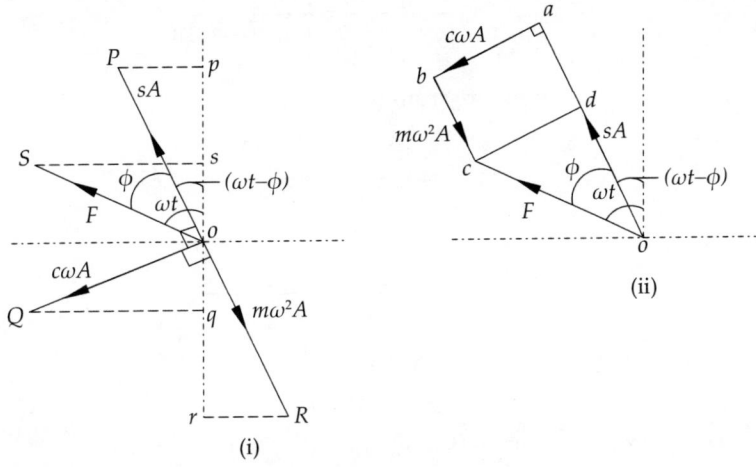

Fig. 15.13 Vector diagram.

Figure 15.13(i) represents the vector diagram of the forces. Then draw the force polygon as shown at Fig. 15.13(ii). From the force polygon, *oc* represents the force F.

$$F = \sqrt{[(od)^2 + (dc)^2]} = \sqrt{[(oa - ad)^2 + (dc)^2]} = A\sqrt{[(s - m\omega^2)^2 + c^2\omega^2]}.$$

Therefore, amplitude A or maximum amplitude $x_{max} = \dfrac{F}{\sqrt{[(s - m\omega^2)^2 + c^2\omega^2]}}$ and

$$\tan\phi = \frac{cd}{od} = \frac{c\omega A}{(sA - m\omega^2 A)} = \frac{c\omega}{(s - m\omega^2)}.$$

15.10.1 Magnification Factor or Dynamic Magnifier (D)

It is the ratio of maximum displacement (x_{max}) of the forced vibration and the deflection x_0 due to the static force

$$D = \frac{x_{max}}{x_0} = \frac{x_{max}}{\left(\dfrac{F}{s}\right)}$$

But,

$$x_{max} = \frac{\left(\dfrac{F}{s}\right)}{\sqrt{\left[\left(\dfrac{c\omega}{s}\right)^2 + \left(\dfrac{s - m\omega^2}{s}\right)^2\right]}}$$

$$= \frac{x_0}{\sqrt{\left[\left(\dfrac{c\omega}{s}\right)^2 + \left(\dfrac{s - m\omega^2}{s}\right)^2\right]}} \qquad (1)$$

1. When damping $c = 0$

$$x_{max} = x_0 \frac{\dfrac{s}{m}}{(\omega_n^2 - \omega^2)} = \frac{F}{[m(\omega_n^2 - \omega^2)]}$$

2. At resonance $\omega = \omega_n = \sqrt{\dfrac{s}{m}}$, therefore from (1)

$$x_{max} = x_0 \frac{s}{(c\omega_n)} = \frac{F}{[c\omega_n]}$$

Magnification factor

$$D = \frac{x_{max}}{x_0} = \frac{1}{\sqrt{\left(\dfrac{2c\omega}{C_c\omega_n}\right)^2 + \left(1 - \left(\dfrac{\omega}{\omega_n}\right)^2\right)}} \qquad (2)$$

we know that $\dfrac{c\omega}{s} = \dfrac{2c\omega}{2ms/m} = \dfrac{2c\omega}{2m\omega_n^2} = \dfrac{2c\omega}{(C_c\omega_n)}.$

Hence, $x_{max} = x_0 \cdot D$

3. When $c = 0$; $D = \dfrac{x_{max}}{x_0} = \dfrac{\omega_n^2}{(\omega_n^2 - \omega^2)}$

4. At resonance,
$$\omega = \omega_n; \quad D = \dfrac{x_{max}}{x_0} = \dfrac{s}{(c\omega_n)}$$

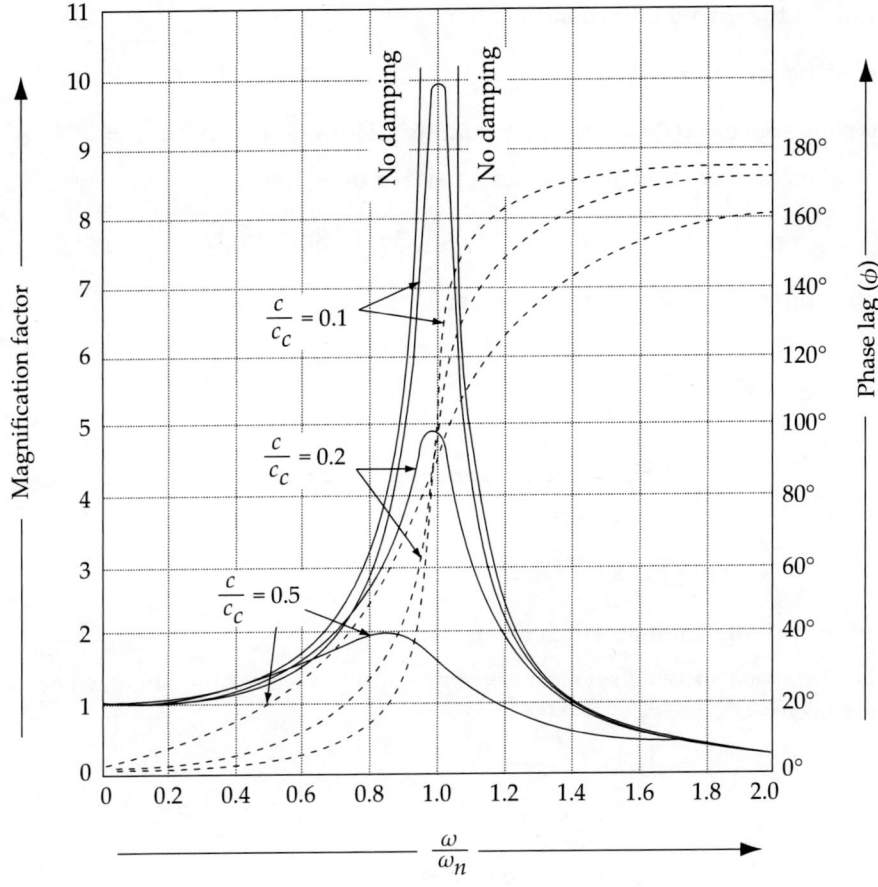

Fig. 15.14

The Fig. 15.14 shows the variation of D as $\dfrac{\omega}{\omega_n}$ varies for different values of $\dfrac{C}{C_c}$ or ζ.

W E 15.14: A single cylinder vertical petrol engine of total mass 400 kg is mounted upon a steel chassis frame and causes a vertical static deflection of 2.4 mm. The reciprocating parts of the engine have a mass of 18 kg and move through a vertical stroke of 160 mm with simple harmonic motion. A dashpot is provided whose damping resistance is directly proportional to the velocity and amounts to 2 N/mm/s.

Considering that the steady state of vibration has reached, determine: (1) The amplitude of forced vibrations, when the driving shaft of the engine rotates at 500 r.p.m. and (2) The speed of the driving shaft at which the resonance will occur.

Given:

$m = 400$ kg; $\delta = 2.4$ mm $= 2.4 \times 10^{-3}$ m; $m_E = 20$ kg; $L = 160$ mm $= 0.16$ m; $c = 2$ N/mm/s $= 2000$ N/m/s; $N = 500$ r.p.m. or $\omega = 2\pi \times \dfrac{500}{60} = 52.36$ rad/s.

1. **Amplitude of the forced vibrations**

Stiffness of the frame $s = \dfrac{mg}{\delta} = \dfrac{400 \times 9.81}{2.4 \times 10^{-3}} = 1.635 \times 10^6$ N/m

Length of the stroke = 160 mm and so the radius of the crank $r = 80$ mm $= 0.08$ m.

Centrifugal forces due to the reciprocating parts or the static force

$$F = m_E \omega^2 r = 18(52.36)^2 0.08 = 3948 \text{ N}$$

Maximum amplitude due to forced vibrations

$$x_{max} = \dfrac{F}{\sqrt{\left[(s - m\omega^2)^2 + c^2\omega^2\right]}}$$

$$= \dfrac{3948}{\sqrt{\left[\left(1.635 \times 10^6 - 400 \times 52.36^2\right)^2 + 2000^2 \times 52.36^2\right]}}$$

$$= 7.2 \times 10^{-3} \text{ m or 7.2 mm}$$

2. **Speed of the driving shaft at which resonance occurs**

Let N be the speed at which resonance occurs in r.p.m. and the angular speed at which resonance occurs is

$$\omega = \omega_n = \sqrt{\left(\dfrac{s}{m}\right)} = \sqrt{\left(\dfrac{1.635 \times 10^6}{400}\right)} = 63.93 \text{ rad/sec or}$$

$\dfrac{2\pi \times N}{60} = 63.93$; hence $N = 610.5$ r.p.m.

15.11 EXERCISE

15.11.1 Short Answer Questions

1. Explain the concept of overdamped, critically damped and underdamped system. Also show that in a free underdamped vibrating system, the amplitude of vibration decays exponentially.
2. Explain the Rayleigh method for computing natural frequency of multi-degree of free down vibratory system.

3. Develop an equation for natural frequency of transverse vibration explaining each term.
4. What is meant by vibration? How are they caused?
5. What are free, damped and forced vibrations? Explain.
6. What type of vibrations can be executed by a massless shaft, one end of which is fixed and the other end carries a heavy disc?
7. Distinguish between longitudinal, transverse and torsional vibrations.
8. What are the basic elements of a vibratory system? What is the degree of freedom?
9. Show that the ratio of two successive amplitudes of oscillations is constant in a damped vibratory system.
10. What is a logarithmic decrement? Or define logarithmic decrement. Derive the relation for the same.
11. What is Dunkerley's method?
12. Define whirling speed and discuss its importance.
13. Determine natural frequency of the pendulum system.
14. Define: (i) Free vibrations (ii) Forced vibrations (iii) Damping.
15. Describe Dunkerley's method and how it is used to find the natural frequency of a shaft carrying several loads.

15.11.2 Problems

Longitudinal vibrations:

1. In a spring mass vibrating system, the natural frequency of vibration is reduced to half the value when a second spring is added to the first spring is series. Determine the stiffness of the second spring in terms of that of the first spring. [**Ans.** $s^2 = \frac{s^1}{3}$]

2. In a spring-mass vibrating system, the natural frequency of vibration is 3.56 Hz. When the amount of the suspended mass is increased by 5 kg, the natural frequency is lowered to 2.9 Hz. Determine the original unknown mass and spring constant. [**Ans.** (10 kg; 5 N/mm)]

3. A spring mass system is excited by a force $F \sin \omega t$. On measuring the amplitude of vibration is found to be 12 mm at resonance. However, at a frequency 0.8 times the response frequency, the amplitude reduces to 8 mm. Determine the damping ratio of the system. [**Ans.** 0.142]

4. Determine the equation of motion and the natural frequency of the system in Fig. 15.15 by the energy method. Solve the same by Newton's method.

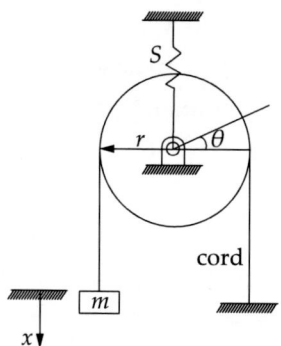

Fig. 15.15

5. A steel shaft 75 mm diameter and 900 mm long fixed at one end carries a flywheel of 1 tonne weight and radius of gyration 5 cm at its free end. Find the frequency of the free longitudinal and transverse vibration. $E = 2 \times 10^6$ kg/cm².

Damped vibrations:

1. In a single degree damped vibratory system, the suspended mass of 4 kg makes 24 oscillations in 20 seconds. The amplitude decreases to 0.3 of the initial value after 4 oscillations. Find the stiffness of the spring, the logarithmic decrement, the damping factor and damping coefficient.
   [*Ans.* 227 N/m; 0.3; 0.0478; 2.88 N/m/s]

2. The measurement on a mechanical vibrating system show that it has a mass of 8 kg and that the springs can be combined to give an equivalent spring of stiffness 5.4 N/mm. If the vibrating system have a dashpot attached which exerts a force of 40 N when the mass has a velocity of 1 m/s, find: (a) critical damping coefficient (b) damping factor (c) logarithmic decrement and (d) the ratio of two consecutive amplitudes.

3. A mass weighing 85 kgf is supported on springs which deflects 1.8 cm under the weight of the mass. The vibration of the mass are constrained to be linear and vertical and are damped by a dashpot which reduces the amplitude to one-quarter of its initial value in two complete oscillations, find (a) the magnitude of the damping force at unit speed and (b) periodic time of damped vibrations.

Transverse Vibrations:

1. A 22 mm wide and 45 mm deep steel bar is freely supported at two points that are 800 mm apart and carries a load of 180 kg midway between them. Determine the natural frequency of the transverse vibration, neglecting the weight of the bar. Also find the frequency of vibration if an additional load of 180 kg is distributed uniformly along the length of the shaft.
   [*Ans.* 23.5 Hz; 19.2 Hz]

2. A machine weighing 680 N is mounted on springs of stiffness, 11 kN/cm with an assumed damping factor of 0.2. A piston within the machine weighing 20 N has a reciprocating motion

with a stroke of 7.5 cm and a speed of 3000 r.p.m. Compute: (a) the amplitude of the machine (b) its phase angle w.r.t. the exciting force and (c) the amplitude of the machine when damper is removed.

3. A hollow shaft 1.8 m long is supported in flexible bearings at the ends. It carries two wheels each of 60 kg mass, one at the centre of the shaft and the other at 450 mm from the centre. The external and internal diameters of the shaft are 80 and 50 mm respectively. Determine the lowest whirling speed of the shaft. The density of the shaft material is 7500 kg/m^3 and the modulus of elasticity 210 GN/m^2. [*Ans.* 169.4 r.p.m.]

4. A shaft 50 mm diameter and 3 m long. It is simply supported at the ends and carries three masses 100 kg, 120 kg and 80 kg at 1.0 m, 1.75 m and 2.5 m respectively from the left support. Taking $E = 20$ N/m^2. Find the frequency of transverse vibrations using Rayleigh's method.

Whirling speed:

1. A vertical shaft 12.5 mm diameter rotates in long bearing and a disc weighing 16 kg is attached to the mid span of the shaft. The span of the shaft between the bearings is 50 cm. The mass centre of the disc is 0.05 cm from the axis of the shaft. Neglecting the mass of the shaft and taking the deflections as for a beam fixed at both ends, find the critical speed of rotation. Determine the range of speed over which the stress in the shaft due to the bending will exceed 1260 kg/cm^2. Take $E = 2 \times 10^6$ kg/cm^2.
Calculate the whirling speed of a shaft 20 mm diameter and 0.6 m long carrying a mass of 1 kg at its midpoint. The density of the shaft material is 40 Mg/m^3, and Young's modulus is 200 GN/m^2. Assume the shaft to be simply supported.

15.11.3 Multiple Choice Questions

1. The particles of a body move _____ its axis in longitudinal vibrations.
   (a) in a circle about   (b) parallel to   (c) perpendicular   (d) away from   [*Ans.* (b)]

2. The particles of a body move _____ its axis in torsional vibrations.
   (a) in a circle about   (b) parallel to   (c) perpendicular   (d) away from   [*Ans.* (a)]

3. In a spring mass system if the mass is halved and the spring stiffness is doubled the natural frequency is
   (a) halved   (b) doubled   (c) unchanged   (d) quadrupled   [*Ans.* (b)]

4. In free vibrations the velocity vector leads the displacement vector by
   (a) π (b) $\dfrac{\pi}{2}$ (c) $\dfrac{\pi}{3}$ (d) $\dfrac{2\pi}{3}$ [*Ans.* (b)]

5. In free vibrations the acceleration vector leads the displacement vector by
   (a) π (b) $\dfrac{\pi}{2}$ (c) $\dfrac{\pi}{3}$ (d) $\dfrac{\pi}{3}$ [*Ans.* (a)]

6. The amplitude ratio of two successive oscillations of a damped vibratory system is

   (a) more than one   (b) less than one   (c) equal to one   (d) variable   [*Ans.* (b)]

7. An overdamped system

   (a) does not vibrate at all
   (b) vibrates with frequency more than the natural frequency of system
   (c) vibrates with frequency less than the natural frequency of system
   (d) vibrates with frequency equal to the natural frequency of system   [*Ans.* (a)]

8. The ratio of the amplitude of the steady-state response of forced vibrations to the static deflection under the action of a static force is known as

   (a) damping ratio   (b) damping factor   (c) transmissibility   (d) magnification factor
   [*Ans.* (d)]

9. The frequency of damped vibrations is always _____ the natural frequency.

   (a) equal to   (b) more than   (c) less than   (d) double   [*Ans.* (c)]

10. If $\dfrac{\omega}{\omega_n}$ is more than $\sqrt{2}$ in a vibration isolation system, then for all values of the damping factor, the transmissibility

    (a) less than 2   (b) more than 2   (c) less than unity   (d) more than unity   [*Ans.* (c)]

11. Resonance is a phenomenon in which the frequency of the exciting force is _____ to the natural frequency of the system

    (a) double   (b) half   (c) equal   (d) thrice   [*Ans.* (c)]

12. At resonance the amplitude of vibration is

    (a) very large   (b) small   (c) zero   (d) depends upon frequency   [*Ans.* (a)]

13. At a certain speed, revolving shafts tend to vibrate violently in transverse directions

    (a) whirling speed   (b) critical speed   (c) whipping speed   (d) all of these   [*Ans.* (d)]

14. The critical speed of a rotating shaft with a mass at the centre is _____ the natural frequency of transverse vibrations of the system

    (a) equal   (b) less than   (c) more than   (d) dependent upon   [*Ans.* (a)]

15. A reduction in amplitude of successive oscillations indicate _____ vibrations.

    (a) free   (b) force   (c) damped   (d) natural   [*Ans.* (c)]

Torsional Vibrations

16

Torsional Vibrations

16.1 INTRODUCTION

When the particles of a shaft or disc move to and fro in a circular way about the axis of a shaft, then the vibrations are known as torsional vibrations. The shaft undergoes twists and untwists alternately, which causes torsional shear stresses induced in the shaft. Determinations of the torsional frequency of vibrations of various systems are dealt with in this chapter.

16.2 NATURAL FREQUENCY OF FREE TORSIONAL VIBRATIONS

Consider a shaft of negligible mass whose one end is fixed and the other end carrying a disc as shown in Fig. 16.1.

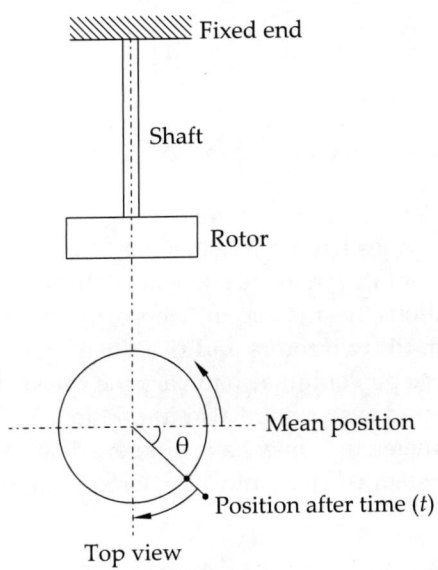

Fig. 16.1

Let the angular displacement of the shaft from the mean position after a time interval of t be θ in radians; m be the mass of the disc in kg. The mass moment of inertia of disc be $I = mk^2$ where k is the radius of gyration in metres; q be the torsional stiffness of the shaft in N-m.

$$\text{Restoring force} = q.\theta \quad (1)$$

$$\text{Accelerating force} = I\frac{d^2\theta}{dt^2} \quad (2)$$

From (1) and (2), the equation of motion is $\dfrac{I d^2\theta}{dt^2} + q\theta = 0$ or $\dfrac{d^2\theta}{dt^2} + \dfrac{q}{I}\theta = 0$.

Then comparing this with the fundamental equation of simple harmonic motion $\dfrac{d^2\theta}{dt^2} + \omega^2\theta = 0$;

Time period $t_p = \dfrac{2\pi}{\omega} = 2\pi\sqrt{\dfrac{I}{q}}$.

The natural frequency $f_n = \dfrac{1}{t_p} = \dfrac{1}{2\pi}\sqrt{\dfrac{q}{I}}$.

The torsional stiffness q may be obtained from the torsional equation $\dfrac{T}{J} = \dfrac{C\theta}{L}$ or $\dfrac{T}{\theta} = \dfrac{CJ}{L}$. Therefore, the torsional stiffness, i.e. the torque per unit twist $q = \dfrac{T}{\theta} = \dfrac{CJ}{L}$ where C is the modulus of rigidity of the shaft material. L is the length of the shaft of diameter d. The polar moment of inertia J of the shaft cross section is $\dfrac{\pi d^4}{32}$. By considering the mass moment of inertia of the shaft I_s, the natural frequency of torsional vibration is $f_n = \dfrac{1}{2\pi}\sqrt{\dfrac{q}{\left[I + \left(\dfrac{I_s}{3}\right)\right]}}$.

16.3 TORSIONAL VIBRATIONS OF A SHAFT WITH NUMBER OF ROTORS

The power is transmitted from one shaft to another using gears or pulleys, motors coupled to pumps, generators driven by engines having flywheel, etc. Each shaft carries a number of such wheels called rotors. The shafts vibrate torsionally, whenever the power is switched on or switched off or due to any sudden fluctuations that may occur. Hence, the study of the free natural vibrations of a shaft with (i) a single rotor, (ii) two rotors and (iii) three rotors is important. The gears are used not only for transmitting the power but also to vary the speeds from one shaft to other shaft. Hence, the vibrations of the geared systems are also important. The diameter of the shafts used instead of having a uniform diameter, they may have different diameters for different lengths along the shaft. The conversion of the stepped shaft into a torsionally equivalent uniform diameter shaft is also given below.

16.3.1 Free Torsional Vibrations of a Single Rotor System

The Pic. 16.1 is a single rotor in generators or motors mounted on a shaft supported in bearings. The shaft of length L and diameter d with one end fixed and other end having a heavy rotating mass called rotor is shown in Fig. 16.2 at the top and the variation of twist at the bottom. Let C be the modulus of rigidity of the shaft material and J be the polar moment of inertia of the shaft equal to $\dfrac{\pi d^4}{64}$. The natural frequency of a single rotor system

$$f_n = \dfrac{1}{2\pi}\sqrt{\dfrac{q}{I}} \text{ or } \dfrac{1}{2\pi}\sqrt{\dfrac{CJ}{IL}}$$

Pic. 16.1 Single rotor.

Another diagram shown below in Fig. 16.2 gives the twist along the length of the shaft from fixed end O to rotor end P. The twist at the fixed end O is zero, called node and at P having a certain twist represented by 'a'. The twist can be seen varying along the length of the shaft.

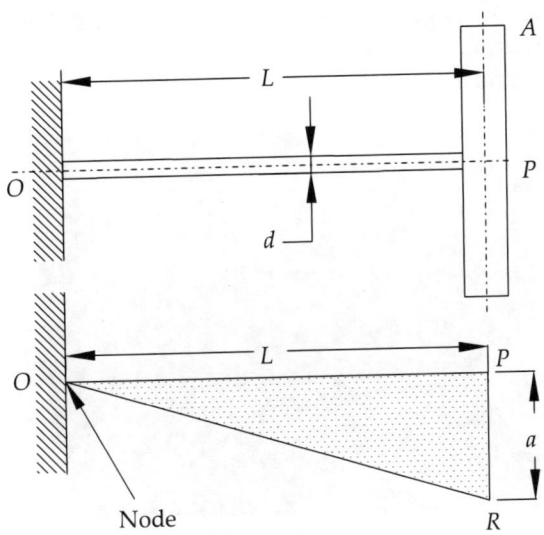

Fig. 16.2 Single rotor.

W E 16.1: A shaft of 100 mm diameter and 1 m long is fixed at one end and the other end carries a flywheel of mass 1 tonne. The radius of the flywheel is 0.5 m. Find the frequency of torsional vibrations, if the modulus of rigidity of the shaft material is 80 GN/m².

Given:
Diameter of the shaft $d = 100$ mm $= 0.1$ m; and length $L = 1$ m.
Modulus of rigidity $C = 80 \times 10^9$ N/m²; Radius of gyration of the flywheel $k = 0.5$ m;

Polar moment of inertia

$$J = \frac{\pi d^4}{32} = \frac{\pi (0.1)^4}{32} = 9.82 \times 10^{-6} \text{ m}^4$$

Torsional stiffness

$$q = \frac{CJ}{L} = \frac{80 \times 10^9 \times 9.82 \times 10^{-6}}{1} = 785.6 \times 10^3 \text{ Nm}$$

Moment of inertia

$$I = mk^2 = 1000(0.5)^2 = 250 \text{ kgm}^2$$

Therefore, natural frequency of vibration of a single rotor system

$$f_n = \frac{1}{2\pi}\sqrt{\frac{q}{I}} = \frac{1}{2\pi}\sqrt{\frac{CJ}{IL}}$$

$$= \frac{1}{2\pi}\sqrt{\frac{785.6 \times 10^3}{250}} = 8.9 \text{ Hz}$$

16.3.2 Free Torsional Vibrations of a Two-Rotor System

Here two different rotors are arranged as shown in Fig. 16.3 at the ends on a shaft. When the rotors move in opposite directions at the same instance alternately, then torsional vibrations occur. Picture 16.2 shows two rotors on a single shaft.

Pic. 16.2 Two rotors.

The point N shown in the other figure below in Fig. 16.3 is called the node where the twist is zero like at the fixed end of a cantilever. The lengths of the shafts on either side of the node, NP and NQ with a rotor at their ends represent like two single rotor systems seen above.

Let L be the total length of the shaft PQ. L_A be the length of the shaft from node N to P and L_B be the length of the shaft from node N to Q. The two rotors A and B are fixed at P and Q as shown. Let I_A and I_B be the moment of inertia of rotors A and B.

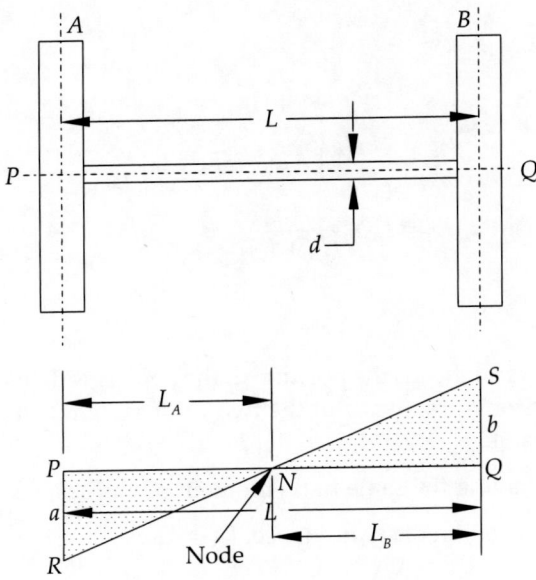

Fig. 16.3 Two-rotor system.

Treat the two-rotor system as two single rotor systems. Then the torsional frequency of the rotor A,

$$f_{nA} = \frac{1}{2\pi}\sqrt{\frac{q_A}{I_A}} = \frac{1}{2\pi}\sqrt{\frac{CJ}{I_A L_A}} \qquad (1)$$

where $q_A = \dfrac{CJ}{L_A}$

and the rotor B,

$$f_{nB} = \frac{1}{2\pi}\sqrt{\frac{q_B}{I_B}} = \frac{1}{2\pi}\sqrt{\frac{CJ}{I_B L_B}} \qquad (2)$$

where $q_B = \dfrac{CJ}{L_B}$

But the natural frequency of the two rotor system must be same whether calculated based on rotor A or B. Hence, the frequencies $f_{nA} = f_{nB}$.

Equating (1) and (2) gives

$$\frac{L_A}{L_B} = \frac{I_B}{I_A}$$

Hence,

$$L_A = \frac{I_B \times L_B}{I_A} \qquad (3)$$

and

$$L_B = \frac{I_A \times L_A}{I_B} \qquad (4)$$

and total shaft length
$$L = L_A + L_B \qquad (5)$$

Substituting (3) in (5)
$$L = \left(\frac{L_B I_B}{I_A}\right) + L_B = (I_A + I_B)\frac{L_B}{I_A}$$
$$L_B = I_A \times \frac{L}{(I_A + I_B)}$$

and so
$$L_A = L - L_B$$

Knowing the values of L_A and L_B means the position of the node N is determined. Then using either L_A or L_B, determine the torsional frequency of the two rotor system using either expression (1) or (2) as both give the same result.

Note: If the two rotors are having the same mass moment of inertias, i.e. $I_A = I_B$, then the node N will be at the midpoint of the length of shaft. Hence, $L_A = L_B = \dfrac{L}{2}$.

W E 16.2: A shaft of 100 mm diameter and 1 m long carries two flywheels of mass 1 tonne at its end. The radius of the flywheel is 0.5 m. Find the frequency of torsional vibrations, if the modulus of rigidity of the shaft material is 80 GN/m². If one of the flywheel is of 500 kg mass, what is the frequency of torsional vibration and where will be the node N?

Given:
Diameter of the shaft $d = 100$ mm $= 0.1$ m; and length $L = 1$ m.
Modulus of rigidity $C = 80 \times 10^9$ N/m²; Radius of gyration of the flywheel $k = 0.5$ m;
Polar moment of inertia
$$J = \frac{\pi d^4}{32} = \frac{\pi \times 0.1^4}{32} = 9.82 \times 10^{-6} \text{ m}^4$$
$$\text{Product of } CJ = 80 \times 10^9 \times 9.82 \times 10^{-6}$$
$$= 785.6 \times 10^3 \text{ Nm}^2;$$

(a) Both flywheels A and B are equal

Consider Fig. 16.3 with flywheels. Moment of inertia of each flywheel
$$I = I_A = I_B = mk^2 = 1000(0.5)^2 = 250 \text{ kgm}^2$$

$$I_A L_A = I_B L_B \text{ or } L_A = \frac{I_B L_B}{I_A}$$

$$L = L_A + L_B = \frac{I_B L_B}{I_A} + L_B = \frac{(I_B + I_A) L_B}{I_A} = 2L_B$$

Therefore, $L_B = \dfrac{L}{2} = 0.5$ m. Hence, node N is at the centre of shaft.
$$L_A = L - L_B = 1.0 - 0.5 = 0.5 \text{ m}$$

Therefore, natural frequency of vibration of two rotor system

$$f_n = \frac{1}{2\pi}\sqrt{\frac{q}{I}} = \frac{1}{2\pi}\sqrt{\frac{CJ}{I_A L_A}} = \frac{1}{2\pi}\sqrt{\frac{785.6 \times 10^3}{250 \times 0.5}} = 12.61 \text{ Hz}$$

(b) Both flywheels A and B are unequal

Moment of inertia of flywheel A,

$$I_A = 1000(0.5)^2 = 250 \text{ kg m}^2$$

Moment of inertia of flywheel B,

$$I_B = 500(0.5)^2 = 125 \text{ kgm}^2$$

$$I_A L_A = I_B L_B \text{ or } L_A = \frac{I_B L_B}{I_A}$$

$$L = L_A + L_B = \frac{I_B L_B}{I_A} + L_B$$

$$= \frac{(I_B + I_A)L_B}{I_A}$$

$$L = (250 + 125)\frac{L_B}{250} = \frac{375}{250}L_B$$

So $L_A = L - L_B = 1.0 - \frac{2}{3} = \frac{1}{3}$ m = 0.333 m.

Hence, node N is at one-third the length of the shaft from flywheel A.

Therefore, natural frequency of vibration of two rotor system

$$f_n = \frac{1}{2\pi}\sqrt{\frac{q}{I}} = \frac{1}{2\pi}\sqrt{\frac{CJ}{IL}} = \frac{1}{2\pi}\sqrt{\frac{785.6 \times 10^3}{250 \times 0.333}} = 15.455 \text{ Hz}$$

Try: Two flywheels with moment of inertia of 4.2 kg m² and 6.6 kg m² are separated by a uniform shaft 68.5 cm long. The stiffness of the shaft is 27×10^5 Nm/rad. Determine the position of the node and the natural frequency of torsional vibration. *[Ans. 43 cm from one flywheel, 163 Hz]*

16.3.3 Free Torsional Vibrations of a Three Rotor System

Here there are two possible ways in which the rotors can rotate. The first one is (a) the middle rotor rotating in the opposite direction of rotation of the end rotors and the other one is (b) the two adjacent rotors rotating in the opposite direction of rotation of the third rotor. The torsional vibrations of both the cases have to be studied as follows:

Let A, B and C be the names of the three rotors mounted on a shaft of diameter d and of total length L as shown in Fig. 16.4. The rotors A and C are attached at the ends of the shaft, whereas the rotor B is attached in between A and C. Let the distance between A and B rotors be L_1 and between B and C be L_2. Let I_A, I_B and I_C be the mass moment of inertia of rotors A, B and C. Let J be the polar moment of inertia of the shaft and C be the modulus of rigidity of the shaft material.

Figure 16.4(a) shows a three rotor system where the middle rotor rotates in the opposite direction of the end rotors. The Pics. 16.3 and 16.4 show three rotors on a shaft.

Pic. 16.3 Multi rotors.

Pic. 16.4

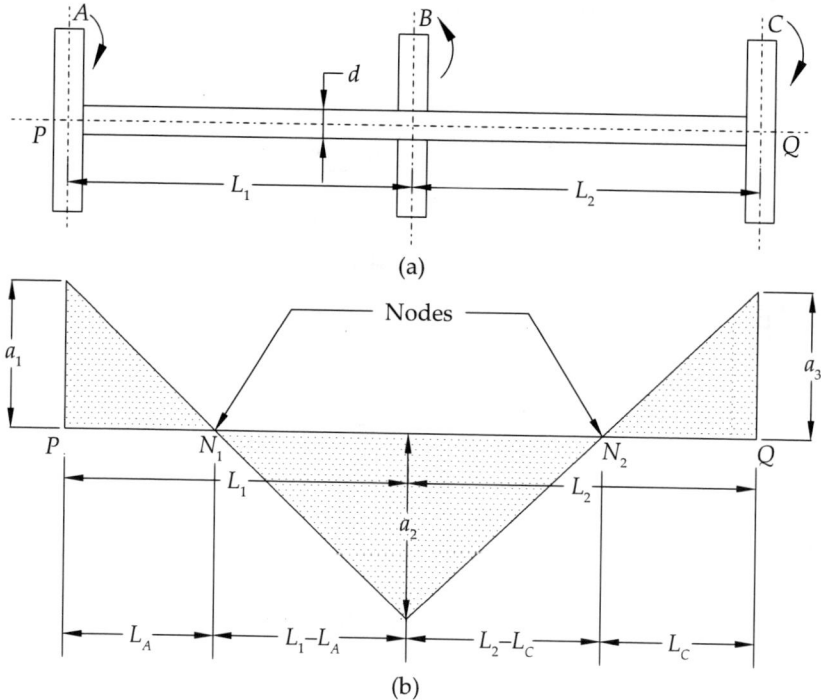
Fig. 16.4

Let N_1 and N_2 be the two nodes as shown. Let L_A and L_C be the distances of the node N_1 from rotor A and node N_2 from C as shown. Let q be torque required to twist rotor through one radian.

Figure 16.5(a) shows a three rotor system where the two adjacent rotors rotates in the opposite direction of the other rotor.

Natural frequency of torsional vibrations of Rotor A,

$$f_{nA} = \frac{1}{2\pi}\sqrt{\frac{q_A}{I_A}} = \frac{1}{2\pi}\sqrt{\frac{CJ}{I_A L_A}} \tag{1}$$

and rotor C,

$$f_{nC} = \frac{1}{2\pi}\sqrt{\frac{q_C}{I_C}} = \frac{1}{2\pi}\sqrt{\frac{CJ}{I_C L_C}} \tag{2}$$

and rotor B

$$f_{nB} = \frac{1}{2\pi}\sqrt{\frac{q_B}{I_B}} = \frac{1}{2\pi}\sqrt{\frac{CJ}{I_B L_B}} \tag{3}$$

where q_B is the torque required to twist B through one radian. When the shaft is regarded as fixed at the nodes N_1 and N_2, then q_B is the sum of the torques required to produce a twist of one radian in each of the lengths $L_1 - L_A$ and $L_2 - L_C$.

Therefore,

$$q_B = \frac{CJ}{(L_1 - L_A)} + \frac{CJ}{(L_2 - L_C)}$$

$$= CJ\left[\frac{1}{(L_1 - L_A)} + \frac{1}{(L_2 - L_C)}\right]$$

and substituting in equation (3) gives

$$f_{nB} = \frac{1}{2\pi}\sqrt{\frac{q_B}{I_B}}$$

$$= \frac{1}{2\pi}\sqrt{\frac{CJ}{I_B}\left[\frac{1}{(L_1 - L_A)} + \frac{1}{(L_2 - L_C)}\right]} \tag{4}$$

Since the torsional frequencies of all the three rotors must be equal, i.e. $f_{nA} = f_{nB} = f_{nC}$. Equating expressions (1) and (2) gives

$$L_A I_A = L_C I_C$$

$$L_A = \frac{L_C I_C}{I_A} = pL_C \tag{5}$$

where $p = \dfrac{I_C}{I_A}$

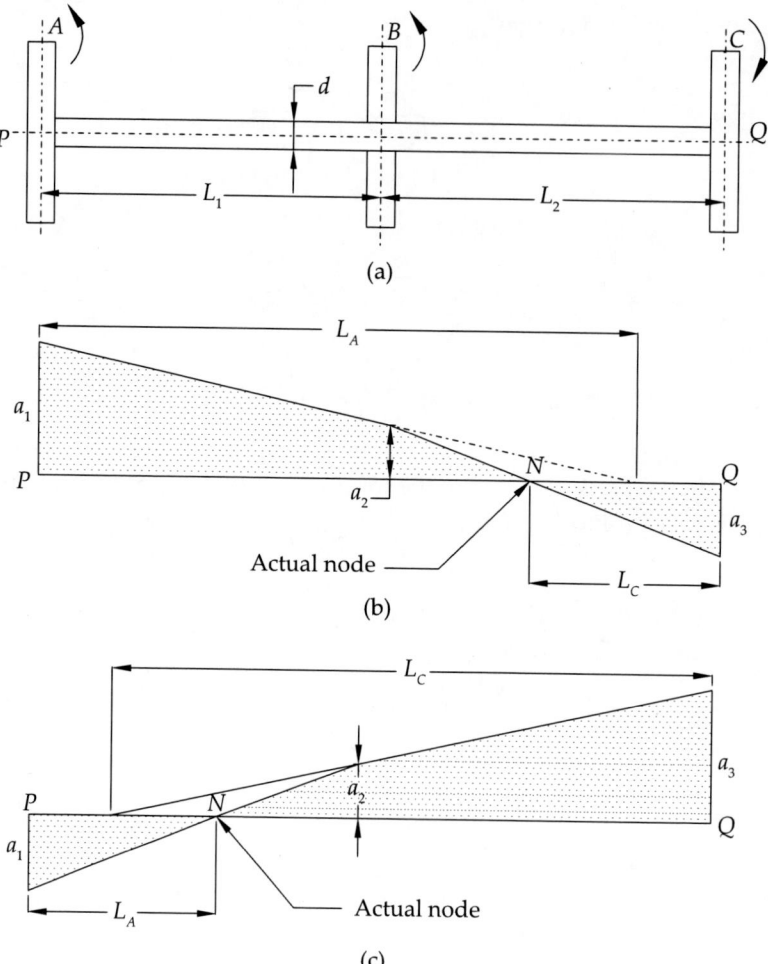

Fig. 16.5

Equating expressions (2) and (3) gives

$$\frac{1}{I_B}\left[\frac{1}{(L_1 - L_A)} + \frac{1}{(L_2 - L_C)}\right] = \frac{1}{L_C I_C}$$

$$L_C\left[\frac{1}{(L_1 - L_A)} + \frac{1}{(L_2 - L_C)}\right] = q \quad (6)$$

where $q = \dfrac{I_B}{I_C}$

After substituting for L_A from (5) in (6), and on simplification, a quadratic equation in L_C can be obtained in terms of L_1, L_2, p and q as follows.

$$(pq + q + 1)L_C^2 - (pqL_2 + qL_1 + L_2 + L_1)L_C + qL_1L_2 = 0 \quad (7)$$

The solution of it gives two values of L_C and corresponding two values of L_A.

One set of L_A and L_C values when used in either (1) or (2) gives a two-node frequency as shown in Fig. 16.4(b). Another set of L_A and L_C values when used gives a one-node (the second node beyond the limits) frequency as shown in Fig. 16.5(b,c).

The frequency corresponding to single node is called a fundamental frequency or a single node frequency. When the amplitude of vibration or twist a_1 for rotor A is known, then the amplitude of vibration or twist of rotor B, a_2 is given by $\left[\dfrac{(L_A - L_1)}{L_A}\right] a_1$ and for the rotor C, a_3 is given by $\left[\dfrac{L_C}{(L_C - L_2)}\right] a_2$.

As there are two sets of values for L_A and L_C, therefore, there will be two sets of values of amplitudes corresponding to one node and two node vibrations respectively.

W E 16.3: Three identical masses are equidistantly mounted on a uniform shaft. Show that the two natural frequencies of torsional oscillations of the shaft are in the ratio $1 : \sqrt{3}$.

Given:
As shown in Fig. 16.5(a), all the three rotors have the same moment of inertias, so $I_A = I_B = I_C$. Hence, the values of $p = 1$ and $q = 1$.

Since $f_{nA} = f_{nC}$

$$\frac{1}{2\pi}\sqrt{\frac{CJ}{I_A L_A}} = \frac{1}{2\pi}\sqrt{\frac{CJ}{I_C L_C}} \qquad (1)$$

then $L_A I_A = L_C I_C$. As $I_A = I_C$, hence

$$L_A = L_C \qquad (2)$$

Since $f_{nB} = f_{nC}$, then

$$\left[\frac{1}{(L_1 - L_A)} + \frac{1}{(L_2 - L_C)}\right] = \frac{1}{L_C}$$

Also given that $L_1 = L_2 = $ say 1

Substitute the values of p, q, L_1 and L_2 in the quadratic equation,

$$(pq + q + 1)L_C^2 - (pqL_2 + qL_1 + L_2 + L_1)L_C + qL_1 L_2 = 0$$

which gives $3L_C^2 - 4L_C + 1 = 0$.

The two roots are

$$\frac{\left[4 \pm \sqrt{(16 - 4 \times 3 \times 1)}\right]}{2 \times 3} = \frac{(4 \pm 2)}{6}$$

giving L_C equal to 1 and $\dfrac{1}{3}$.

Hence, substitute the two values of L_C in $\dfrac{1}{2\pi}\sqrt{\dfrac{CJ}{I_C L_C}}$ to get the two frequencies as $\dfrac{1}{2\pi}\sqrt{\dfrac{CJ}{I_C}}$ and

$\frac{1}{2\pi}\sqrt{\frac{3CJ}{I_C}}$. Hence, the torsional oscillations of the shaft are in the ratio $1 : \sqrt{3}$.

W E 16.4: Three rotors A, B and C having moment of inertia of 2000, 6000, and 3500 kg-m² respectively are carried on a uniform shaft of 0.35 m in diameter. The length of the shaft between the rotors A and B is 6 m and between B and C is 32 m. Find the two natural frequencies of the torsional vibrations. The modulus of rigidity for the shaft material is 80 GN/m².

Given:
Here the three rotors are unequal and their moment of inertias are

$I_A = 2000$ kg-m²; $I_B = 6000$ kg-m²; $I_C = 3500$ kg-m²; $C = 80 \times 10^9$ N/m²; $L_1 = 6$ m and $L_2 = 32$ m, Diameter $d = 0.35$ m, Polar moment of inertia $J = \dfrac{\pi(0.35)^4}{32} = 0.00148$ m⁴.

The ratio of the moment of inertias

$$p = \frac{I_C}{I_A} = \frac{3500}{2000} = 1.75$$

and

$$q = \frac{I_B}{I_C} = \frac{6000}{3500} = 1.74$$

Substituting the values of p, q, L_1 and L_2 in the quadratic equation

$$(pq + q + 1)L_C^2 - (pqL_2 + qL_1 + L_2 + L_1)L_C + qL_1L_2 = 0$$
$$(1.75 \times 1.74 + 1.74 + 1)L_C^2 - (1.75 \times 1.74 \times 32 + 1.74 \times 6 + 32 + 6)L_C + 1.74 \times 32 \times 6 = 0$$

which gives a quadratic equation $5.74L_C^2 - 144.44L_C + 334 = 0$. The two roots are

$$\frac{\left[144.44 \pm \sqrt{(144.44^2 - 4 \times 5.74 \times 334)}\right]}{2 \times 5.74} = \frac{(144.44 \pm 114.87)}{11.48}$$

giving the two values for L_C as 22.588 m and 2.576 m and the corresponding values of L_A are 39.53 m and 4.508 m with a single node and two-node systems.

Natural frequency of the system taking $L_C = 22.588$ m,

$$f_{n1} = \frac{1}{2\pi}\sqrt{\frac{CJ}{I_CL_C}}$$

$$= \frac{1}{2\pi}\sqrt{\frac{80 \times 10^9 \times \pi \times 0.35^2}{32 \times 3500 \times 22.588}}$$

$$= \frac{29.199}{\sqrt{22.588}} = 6.144 \text{ Hz Fundamental frequency}$$

Natural frequency of the system taking $L_C = 2.576$ m,

$$f_{n2} = \frac{1}{2\pi}\sqrt{\frac{CJ}{I_C L_C}}$$

$$= \frac{1}{2\pi}\sqrt{\frac{80 \times 10^9 \times \pi \times 0.35^2}{32 \times 3500 \times 2.576}}$$

$$= \frac{29.199}{\sqrt{2.576}} = 18.19 \text{ Hz}$$

Try: A set for operating a D.C. motor consists of an A.C. motor direct coupled to two D.C. generators in tandum. The three armatures in order along the shaft have moments of inertia of 1168, 1029, 584 kg cm^2 and the equivalent length of 19 mm diameter steel shaft between each unit is 152 mm. Determine the lower natural frequency of torsional oscillations of system. **[Ans. 2764 Hz]**

16.4 TORSIONALLY EQUIVALENT SHAFT

So far a shaft of uniform diameter was considered in all the above cases. But in actual practice, the shaft with variable diameter for different lengths will be used. In such a case, the shaft may be theoretically replaced by an equivalent shaft of uniform diameter.

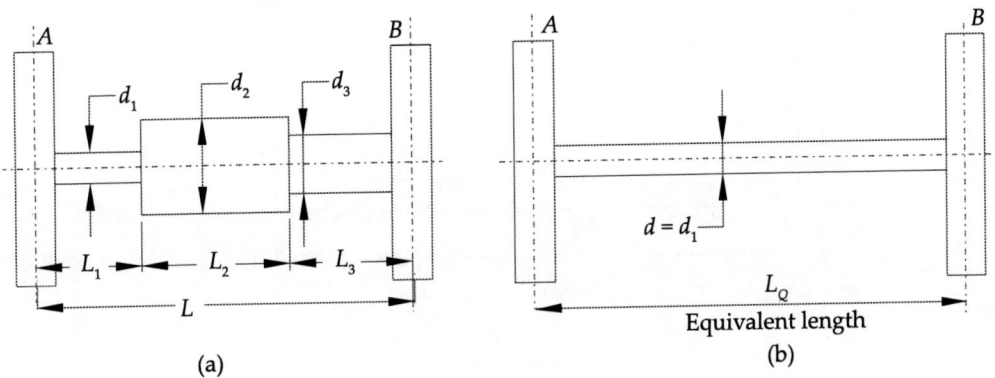

Fig. 16.6

Figure 16.6(a) shows a shaft having three steps with diameters d_1, d_2, d_3. Their lengths are L_1, L_2 and L_3 respectively. Let θ be the total angle of twist of total shaft = sum of the twists of the three individual steps. Hence, $\theta = \theta_1 + \theta_2 + \theta_3$. But $\theta = \dfrac{TL_Q}{CJ}$ where L_Q is the length of the equivalent shaft of diameter say d_1 and similarly for each step

$$\theta_1 = \frac{TL_1}{CJ_1} \quad \theta_2 = \frac{TL_2}{CJ_2} \quad \theta_3 = \frac{TL_3}{CJ_3}$$

Hence,
$$\frac{TL_Q}{CJ} = \frac{TL_1}{CJ_1} + \frac{TL_2}{CJ_2} + \frac{TL_3}{CJ_3};$$

Taking out $\frac{T}{C}$ commonly throughout gives

$$\frac{L_Q}{J} = \frac{L_1}{J_1} + \frac{L_2}{J_2} + \frac{L_3}{J_3} \qquad (1)$$

Since $J = \frac{\pi d^4}{32}, J_1 = \frac{\pi d_1^4}{32}, J_2 = \frac{\pi d_2^4}{32}, J_3 = \frac{\pi d_4^3}{32}$ and after replacing J and multiplying throughout with d_1^4 where the diameter of the equivalent shaft d is taken as d_1. Then the length of the equivalent shaft

$$L_Q = L_1 + L_2 \frac{d_1^4}{d_2^4} + L_3 \frac{d_1^4}{d_3^4}$$

Note: Whenever stepped shaft is given, first convert into an equivalent shaft of uniform diameter and then determine the frequencies as discussed above for two rotors or three rotors as the case may be.

W E 16.5: A steel shaft 1.5 m long is 95 mm in diameter for the first 0.6 m of its length, 60 mm in diameter for the next 0.5 m of the length and 50 mm in diameter for the remaining 0.4 m of its length.

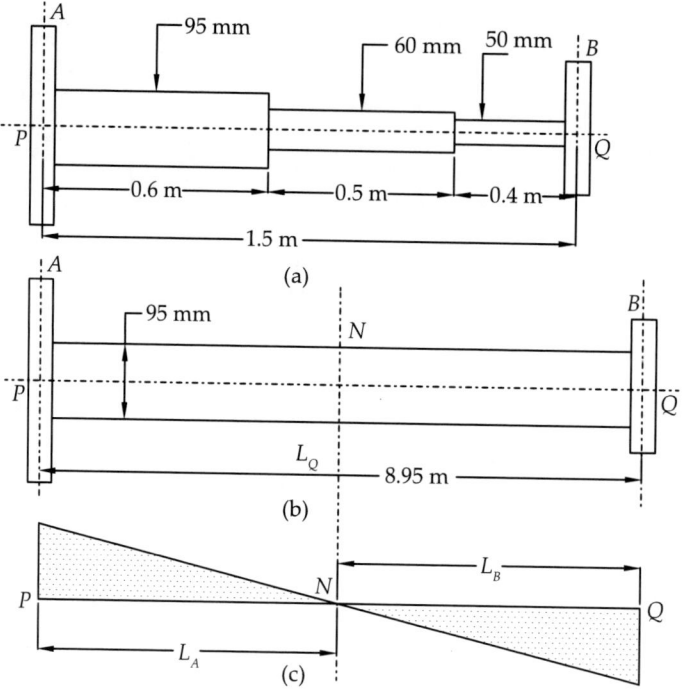

Fig. 16.7

The shaft carries two flywheels at two ends, the first having a mass of 1000 kg and 1.0 m radius of gyration located at the 95 mm diameter end and the second having a mass of 600 kg and 0.8 m radius of gyration located at the other end. Determine the location of the node and the natural frequency of free torsional vibration of the system. The modulus of rigidity of shaft material may be taken as 80 GN/m².

Given:
As shown in Fig. 16.7, the total length of the shaft $L = 1.5$ m $= L_1 + L_2 + L_3$; $d_1 = 95$ mm $= 0.095$ m,
$L_1 = 0.6$ m; $d_2 = 60$ mm $= 0.06$ m, $L_2 = 0.5$ m. $d_3 = 50$ mm $= 0.05$ m, $L_3 = 0.4$ m;
$C = 80$ GN/m² $= 80 \times 10^9$ N/m². $m_A = 1000$ kg; $k_A = 1.0$ m; $m_B = 600$ kg; $k_B = 0.8$ m;

The actual shaft and its equivalent shaft are shown in Fig. 16.7(a) and (b).

Let the length of the equivalent shaft be L_Q. Assuming the diameter of equivalent shaft as $d_Q = d_1 = 95$ mm $= 0.095$ m. Then

$$L_Q = L_1 + L_2 \frac{d_1^4}{d_2^4} + L_3 \frac{d_1^4}{d_3^4}$$

$$= 0.6 + 0.5 \frac{0.095^4}{0.06^4} + 0.4 \frac{0.095^4}{0.05^4}$$

$$= 0.6 + 3.14 + 5.21 = 8.95 \text{ m}$$

Location of the node: Let the node (N) of the equivalent shaft lie at a distance L_A from the flywheel A and at L_B from the flywheel B.

Hence, the M.I. of A is
$$I_A = m_A(k_A)^2$$
$$= [1000 \times (1.0)^2] = 1000 \text{ kgm}^2$$

and the M.I. of B is
$$I_B = m_B(k_B)^2$$
$$= [600 \times (0.8)^2] = 384 \text{ kgm}^2$$

Since $f_{nA} = f_{nB}$;
$$L_A = \frac{L_B I_B}{I_A} = \frac{L_B \times 384}{1000} = 0.384 L_B$$

But $L_Q = 8.95$ m $= L_A + L_B = 0.384 L_B + L_B = 1.384 L_B$.

Therefore, $L_B = \dfrac{8.95}{1.384} = 6.47$ m. So $L_A = 8.95 - 6.47 = 2.48$ m.

Hence, the node lies at 2.48 m from flywheel A or 6.47 m from flywheel B on the equivalent shaft.

The position of the node on the original shaft from flywheel, A

$$= L_1 + (L_A - L_1) \frac{d_1^4}{d_2^4}$$

$$= 0.6 + (2.48 - 0.6) \frac{0.06^4}{0.095^4} = 0.9 \text{ m}$$

The polar moment of inertia of the equivalent shaft

$$J_Q = \frac{\pi d_1^4}{32} = \frac{\pi 0.095^4}{32} = 8 \times 10^{-6} \text{ m}^4$$

Hence the frequency of the system

$$f_n = f_{nA} = \frac{1}{2\pi}\sqrt{\frac{CJ_Q}{I_A L_A}}$$

$$= \frac{1}{2\pi}\sqrt{\frac{80 \times 10^9 \times 8 \times 10^{-6}}{2.48 \times 1000}} = 2.56 \text{ Hz}$$

W E 16.6: A steel shaft PQ 1.5 m long has flywheel at its ends. The mass of the flywheel A is 600 kg and has a radius of gyration of 0.6 m. The mass of the flywheel D is 800 kg and has a radius of gyration 0.9 m. The connecting shaft has a diameter of 50 mm for the portion AB which is 0.4 m long: and has a diameter of 60 mm for the portion BC which is 0.5 m long: and has a diameter of d mm for the portion CD which is 0.6 m long. Determine: (1) The diameter d of the portion CD so that the node of the torsional vibration of the system will be at the centre of length BC and (2) the natural frequency of the torsional vibrations. The modulus of rigidity for the shaft material is 80 GN/m².

Given:
Total length $L = 1.5$ m $= L_1 + L_2 + L_3$; $m_A = 600$ kg; $k_A = 0.6$ m; $m_D = 800$ kg; $k_D = 0.9$ m; $d_1 = 50$ mm $= 0.05$ m; $L_1 = 0.4$ m; $d_2 = 60$ mm $= 0.06$ m; $L_2 = 0.5$ m; $d_3 = d$; $L_3 = 0.6$ m; $C = 80$ GN/m² $= 80 \times 10^9$ N/m².

The actual shaft is shown in Fig. 16.8(a). First let us find the length of equivalent (L_Q) shaft assuming its diameter as $d_Q = d_1 = 50$ mm as shown at (b).

$$L_Q = L_1 + L_2 \frac{d_1^4}{d_2^4} + L_3 \frac{d_1^4}{d_3^4}$$

$$= 0.4 + 0.5\frac{0.05^4}{0.06^4} + 0.6\frac{0.05^4}{d^4}$$

$$= 0.4 + 0.24 + \frac{3.75 \times 10^{-6}}{d^4}$$

$$= 0.64 + \frac{3.75 \times 10^{-6}}{d^4} \tag{1}$$

1. Diameter 'd' of the shaft CD

The node of the equivalent shaft lies say at N as shown in Fig. 16.8(c) and let L_A and L_D are the distances from flywheel A and D. The mass moment of inertia of the flywheel A and D be I_A and I_D. They are equal to $m_A(k_A)^2$ and $m_D(k_D)^2$. On substituting the values of m_A, k_A, m_D and k_D give 216 and 648 kg-m².

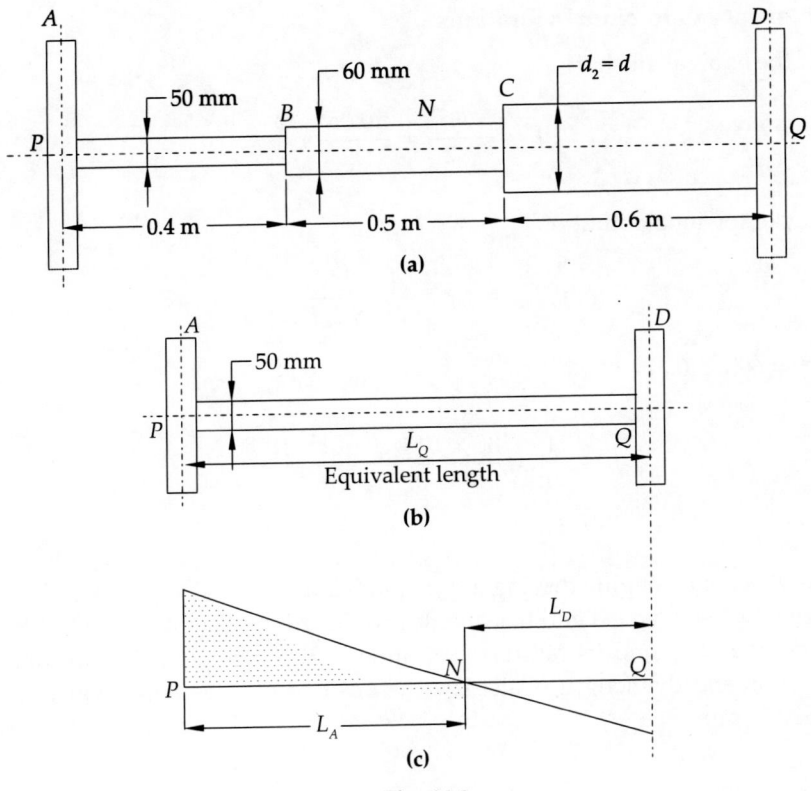

Fig. 16.8

Since $f_{nA} = f_{nD}$; $L_A I_A = L_D I_D$ or

$$L_D = \frac{L_A I_A}{I_D} = \frac{L_A 216}{648} = \frac{L_A}{3} \qquad (2)$$

In order to make the node lie at the centre of length BC in the given original system, its equivalent length from the rotor A, is

$$L_A = L_1 + \frac{L_2 d_1^4}{2 d_2^4} = 0.4 + \frac{0.5 \times 0.05^4}{2 \times 0.06^4} = 0.52 \text{ m}$$

Therefore,

$$L_D = \frac{L_A}{3} = \frac{0.52}{3} = 0.173 \text{ m}$$

From (1) and (2), as

$$L_Q = L_A + L_D;$$

$$= 0.64 + \frac{3.75 \times 10^{-6}}{d^4} = 0.52 + \frac{0.52}{3}$$

on simplification gives $d = 91.7$ mm.

2. Natural frequency of torsional vibrations

The polar M.I. of the equivalent shaft

$$J_Q = \frac{\pi d_Q^4}{32} = \frac{\pi 0.05^4}{32}$$

as $d_Q = d_1$.

Natural frequency of torsional vibrations

$$f_n = f_{nA} = f_{nB} = \frac{1}{2\pi}\sqrt{\frac{CJ}{I_A L_A}}$$

where $I_A = m_A K_A^2 = 600 \times (0.6)^2 = 216$

$$f_n = \frac{1}{2\pi}\sqrt{\frac{80 \times 10^9 \times 0.614 \times 10^{-6}}{0.52 \times 216}}$$

$$= 3.33 \text{ Hz}$$

W E 16.7: The flywheel of an engine driving a dynamo has a mass 180 kg and a radius of gyration of 30 mm. The shaft at the flywheel end has an effective length of 250 mm and is 50 mm diameter. The armature mass is 120 kg and its radius of gyration is 22.5 mm. The dynamo shaft is 43 mm diameter and 200 mm effective length. Calculate the position of node and frequency of torsional oscillation. $C = 83$ kN/mm².

Given:
Flywheel mass, $m_{fl} = 180$ kg; Radius of gyration $k_{fl} = 30$ mm $= 0.03$ m; $d_1 = 50$ mm $= 0.05$ m; $L_1 = 250$ mm $= 0.025$ m; Armature mass, $m_{ar} = 120$ kg; Radius of gyration, $k_{ar} = 22.5$ mm $= 0.0225$ m; $d_2 = 43$ mm; $L_2 = 200$ mm $= 0.2$ m; $C = 83$ kN/mm² $= 83 \times 10^9$ N/m².

Here there are two steps. Let L and d be the length and diameter of the equivalent shaft, their ratio is given in terms of L_1, d_1, L_2 and d_2 is

$$\frac{L}{d^4} = \frac{L_1}{d_1^4} + \frac{L_2}{d_2^4} \tag{1}$$

Taking the diameter of the equivalent shaft $d = d_1$ and multiplying with d_1^4 equation (1) gives

$$L = L_1 + L_2 \frac{d_1^4}{d_2^4} = 0.25 + 0.2\frac{50^4}{43^4} = 0.6156 \text{ m}$$

Mass moment of inertia of flywheel, $I_{fl} = 180 \times (0.03)^2 = 0.162$ kg m²;

Mass moment of inertia of armature, $I_{ar} = 120 \times (0.0225)^2 = 0.061$ kg m²

Since $I_{fl}.L_{fl} = I_{ar}.L_{ar}$, therefore,

$$L_{ar} = L_{fl} \times \frac{0.162}{0.61} = 2.656 L_{fl}$$

But $L_{fl} + L_{ar} = 0.6156$ m; so $L_{fl} + 2.656 L_{fl} = 0.6156$ m.

Hence, $L_{fl} = \dfrac{0.6156}{3.656} = 0.1684$ m; and $L_{ar} = L - 0.1684 = 0.4472$ m;

Polar moment of inertia

$$J = \dfrac{\pi 0.05^4}{32} = 6.13 \times 10^{-7} \text{ m}^4$$

Natural frequency of vibration

$$f_n = f_{nA} = f_{nB} = \dfrac{1}{2\pi}\sqrt{\dfrac{CJ}{I_{fl}L_{fl}}}$$

$$= \dfrac{1}{2\pi}\sqrt{\dfrac{83 \times 10^9 \times 6.13 \times 10^{-7}}{0.1684 \times 0.162}}$$

$$= 217 \text{ Hz}$$

The nodal distance from flywheel $= 0.25 + (0.1684 - 0.25)\dfrac{43^4}{50^4}$

$$= 0.2054 \text{ m}$$

16.5 FREE TORSIONAL VIBRATIONS OF A GEARED SYSTEM

Assumption: *Inertia of gears and shafts are negligible.*

Conditions to be satisfied by an equivalent system are

(a) Kinetic energy (KE) of the equivalent system must be same as the original system

From Fig. 16.9:

KE of shaft L_1 + KE of shaft L_3 = KE of shaft L_1 + KE of shaft L_2

Therefore, KE of shaft L_3 = KE of shaft L_2 or

$$\dfrac{I_{B'}(\omega_{B'})^2}{2} = \dfrac{I_B(\omega_B)^2}{2} \quad \text{since } \omega_{B'} = \omega_A$$

$$\dfrac{I_{B'}(\omega_A)^2}{2} = \dfrac{I_B(\omega_B)^2}{2} \qquad (1)$$

(b) Strain energy (SE) of the equivalent system must be same as the original system:

SE of shaft L_1 + SE of shaft L_3 = SE of shaft L_1 and SE of shaft L_2;

Therefore, SE of shaft L_3 = SE of shaft L_2

$$\dfrac{T_3 \theta_3}{2} = \dfrac{T_2 \theta_2}{2}$$

$$\dfrac{T_3}{T_2} = \dfrac{\theta_2}{\theta_3} \qquad (2)$$

Assuming the power transmitted through l_3 and l_2 as same,

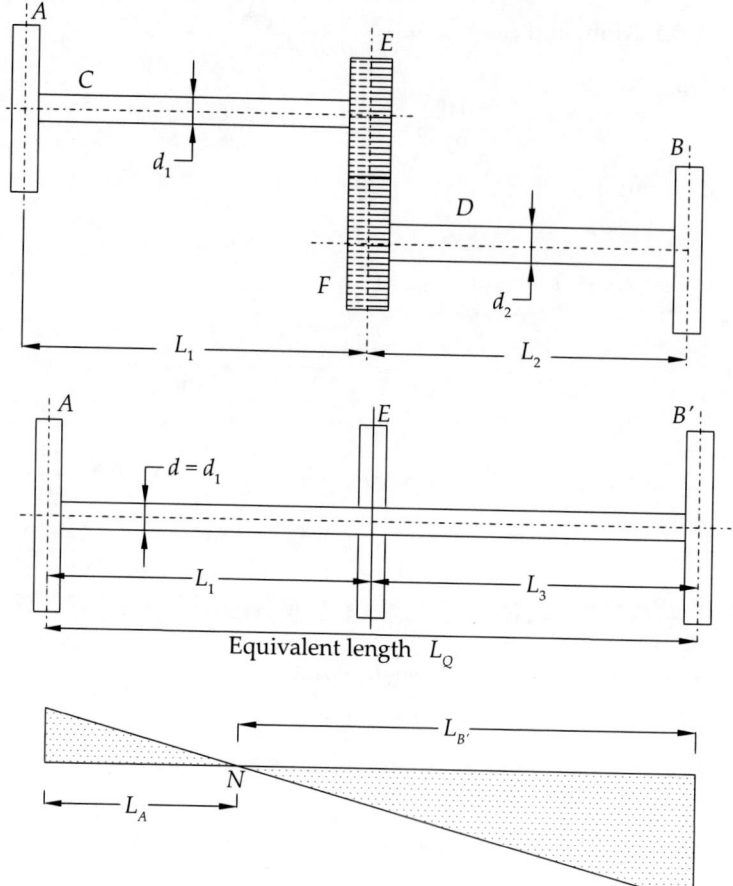

Fig. 16.9

Therefore,

$$T_3 \omega_A = T_2 \omega_B \quad \text{or} \quad \frac{T_3}{T_2} = \frac{\omega_B}{\omega_A} = \frac{1}{G} \tag{3}$$

Combining (2) and (3)

$$\frac{T_3}{T_2} = \frac{\theta_2}{\theta_3} = \frac{1}{G} \tag{4}$$

Torsional stiffness $q = \dfrac{T}{\theta} = \dfrac{CJ}{L}$

for L_3,

$$\frac{T_3}{\theta_3} = \frac{CJ_3}{L_3} \tag{5}$$

and for L_2,

$$\frac{T_2}{\theta_2} = \frac{CJ_2}{L_2} \tag{6}$$

(5)/(6)
$$\frac{T_3}{T_2} = \frac{J_3 \theta_3 L_2}{J_2 \theta_2 L_3}$$

from (4)
$$\frac{1}{G} = \frac{J_3 G L_2}{J_2 L_3} \quad \text{and} \quad L_3 = G^2 L_2 \left(\frac{d_1}{d_2}\right)^4$$

Assuming the diameter of equivalent shaft d as d_1; and $d_3 = d_1$

$$J_3 = \frac{\pi d_1^4}{32}; \quad J_2 = \frac{\pi d_2^4}{32}; \quad \frac{J_3}{J_2} = \left(\frac{d_1}{d_2}\right)^4$$

Hence,
$$L_3 = G^2 L_2 \left(\frac{d_1}{d_2}\right)^4 \tag{7}$$

Therefore, total length of the equivalent shaft $L_Q = L = L_1 + L_3 = L_1 + G^2 L_2 \left(\frac{d_1}{d_2}\right)^4$ as shown at Fig. 16.9(b).

The natural frequency of the torsional vibration of a geared system that was converted as a two rotor system can be determined as follows.

Let L_A and $L_{B'}$ be the distances from the node N, then the frequencies f_{nA} and $f_{nB'}$ can be calculated using $\frac{1}{2\pi}\sqrt{\frac{CJ}{I_A L_A}}$ and $\frac{1}{2\pi}\sqrt{\frac{CJ}{I_{B'} L_{B'}}}$. Since $I_A L_A = I_{B'} L_{B'}$ and $L_Q = L_A + L_{B'}$ as shown at Fig. 16.9(c). Find the lengths and then the natural frequencies just as two rotor system.

Assumption: Inertia of gears considered

When the mass moment of inertias of the gears E and F are considered, then an additional rotor showed on the equivalent system must be introduced at a distance L_1 from the rotor A. This rotor will have a mass moment of inertia $I'_E = I_E + \left(\frac{I_F}{G^2}\right)$, where I_E and I_F are the moment of inertia of the pinion and wheel respectively. The system then becomes a three rotor system and the frequency of such a system may be obtained as discussed earlier.

W E 16.8: An electric motor is to drive a centrifuge, running at four times the motor speed through a spur gear and pinion. The steel shaft from the motor to the gear wheel is 54 mm diameter and L metre long; the shaft from the pinion to the centrifuge is 45 mm diameter and 400 mm long. The masses and radii of gyration of motor and centrifuge are respectively 37.5 kg and 100 mm; 30 kg and 140 mm.

Neglecting the inertia effect of the gears and the shaft, find the value of L if the gears are to be at the nodes for torsional oscillation of the system and hence determine the frequency of torsional oscillation. Assume modulus of rigidity for material of shaft as 84 GN/m².

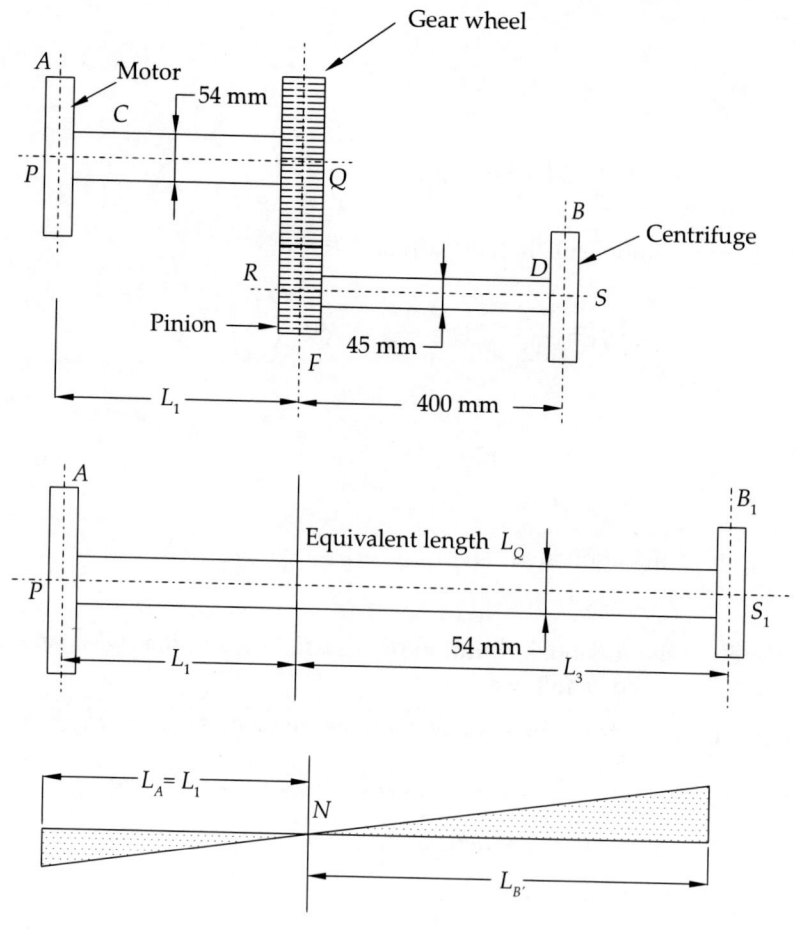

Fig. 16.10

Given:

Speed ratio $G = \dfrac{\text{Speed of motor }(N_M)}{\text{Speed of centrifuge }(N_C)} = \dfrac{N_A}{N_B} = \dfrac{1}{4} = 0.25$; $d_1 = 54$ mm $= 0.054$ m; $L_1 = L$ m;

$m_A = 37.5$ kg; $k_A = 100$ mm $= 0.1$ m; $d_2 = 45$ mm $= 0.045$ m; $L_2 = 400$ mm $= 0.4$ m;
$m_B = 30$ kg; $k_B = 140$ mm $= 0.14$ m; $C = 80$ GN/m² $= 80 \times 10^9$ N/m².

$$I_A = m_A k_A^2 = 37.5(0.1)^2 = 0.375 \text{ kg-m}^2$$

$$I_B = m_B k_B^2 = 30(0.14)^2 = 0.588 \text{ kg-m}^2$$

The given system and its equivalent neglecting the inertia of gears are shown in Fig. 16.10(a) and (b) respectively.

Mass moment of inertia of the equivalent rotor

$$B' = I_{B'} = \frac{I_B}{G^2} = \frac{0.588}{(0.25)^2} = 9.4 \text{ kgm}^2$$

Additional length of the equivalent rotor B shaft by assuming its diameter as d_1 is

$$L_3 = G^2 L_2 \frac{d_1^4}{d_2^4} = (0.25)^2 \times 0.4 \times \frac{0.054^4}{0.045^4} = 0.0518 \text{ m} = 51.8 \text{ mm}$$

Total length of the equivalent shaft

$$L_Q = L_1 + L_3 = 300 + 700 = 1000 \text{ mm} = 1.0 \text{ m}$$

The node N for torsional oscillations of the system is to lie at the gears as shown in Fig. 16.10(c). $L_A = L_1 = L$ and $L_{B'} = L_3 = 0.0518$ m; since $I_A L_A = I_{B'} L_{B'}$; or

$$L \times 0.375 = 0.0518 \times 9.4 = 0.487 \text{ or } L = 1.3 \text{ m}$$

The polar moment of inertia of the equivalent shaft

$$J_Q = \frac{\pi d_1^4}{32} = \frac{\pi 0.054^4}{32} = 8.35 \times 10^{-7} \text{ m}^4$$

Torsional frequency,

$$f_n = \frac{1}{2\pi} \sqrt{\frac{84 \times 10^9 \times 8.35 \times 10^{-7}}{1.3 \times 0.375}} = 60.4 \text{ Hz}.$$

16.6 EXERCISE

16.6.1 Short Answer Questions

1. Develop an equation for natural frequency of transverse vibration explaining each term.
2. Explain two rotor and three rotor vibrations.
3. Find the ratio of amplitudes of rotors of torsional vibrations of a two rotor system.
4. Describe a three rotor vibratory system and find the ratio of their amplitudes.
5. Discuss free torsional vibrations of geared system.
6. Explain 'torsionally equivalent shaft'.

16.6.2 Problems

Single rotor

1. Find from first principles, the whirling speed of steel shaft 2.5 cm diameter and 1.5 m long carrying a disc weighing 50 N at its centre. Assuming that the bearings give no constraint on the direction of the shaft and $E = 2.4 \times 10^5 \text{ N/cm}^2$.

2. A 1.2 m long shaft has a diameter of 45 mm for half the length and 60 mm for the remaining length. One end of the shaft is fixed and the other carries a rotor of 200 kg mass with a radius of gyration of 45 mm. Find the frequency of free torsional vibration neglecting the inertia of the shaft. Take $G = 84$ GN/m². [Ans. 3.88 Hz]

3. A steel shaft 75 mm diameter and 900 mm long fixed at one end carries a flywheel of 1 tonne weight and radius of gyration 5 cm at its free end. Find the frequency of the torsional vibration. $E = 2 \times 10^6$ kg/cm².

Two rotors

1. The moments of inertia of the left and right side rotors are 75 and 50 kg-m² respectively. The lengths l_1, l_2, l_3 and l_4 are 300 mm, 400 mm, 100 mm and 260 mm and the diameters d_1, d_2, d_3 and d_4 are 150 mm, 100 mm, 190 mm and 130 mm respectively. Determine the frequency of natural torsional oscillation of the system. $G = 85$ GN/m². [Ans. 35.5 Hz]

2. Two flywheels with moment of inertias of 40 N-m² and 60 N-m² and separated by a uniform shaft 75 cm long. The stiffness of the shaft is 27×10^5 N-m/rad. Compute the position of the node and the natural frequency of torsional vibrations.

Three rotors

1. A single cylinder oil engine drives directly a centrifugal pump. The rotating mass of the engine, flywheel and the pump with the shaft is equivalent to a three rotor system. The mass moment of inertia of the rotors is 0.15, 0.3 and 0.09 kg-m² respectively. Find the natural frequency of the torsional vibration. The modulus of rigidity for the shaft material is 84 kN/mm². The diameter of the uniform shaft is 70 mm and the total length of the shaft is 2.5 m and the distance between the rotors is 1.5 m and 1 m.

Geared system

1. A centrifugal pump rotating at 400 r.p.m. is driven by an electric motor at 1200 r.p.m. through a single stage reduction gearing. The moments of inertia of the pump impeller and the motor are 1500 kg-m² and 450 kg-m² respectively. The lengths of the pumps shaft and the motor shaft are 500 and 200 mm and their diameters are 100 and 50 mm respectively. Neglecting the inertia of the gears, find the frequency of torsional oscillations of the system. $G = 85$ GN/m². [Ans. 4.74 Hz]

16.6.3 Multiple Choice Questions

1. A torsional vibratory system having two rotors connected by a shaft has
   (a) one node   (b) two nodes   (c) three nodes   (d) no node [Ans. (a)]

2. A torsional vibratory system having three rotors connected by a shaft has
   (a) one node   (b) two nodes   (c) three nodes   (d) no node [Ans. (b)]

3. The natural frequency of free torsional vibration of a shaft is

   (a) $2\pi\sqrt{\dfrac{CJ}{IL}}$ (b) $\dfrac{1}{2\pi}\sqrt{\dfrac{CJ}{IL}}$ (c) $\dfrac{1}{2\pi}\sqrt{CJIL}$ (d) $\dfrac{1}{2\pi}\sqrt{CJ(I_A L_A)}$ [Ans. (b)]

4. At the nodal point the amplitude is

   (a) minimum   (b) maximum   (c) zero.   [Ans. (c)]

5. Two rotors connected by a shaft when subjected to torsional vibrations will have

   (a) two nodes   (b) three nodes   (c) one node   (d) no node   [Ans. (c)]

6. Three rotors connected by a shaft when subjected to torsional vibrations will have

   (a) two nodes   (b) three nodes   (c) one node   (d) no node   [Ans. (a)]

7. Rotating shafts tend to vibrate violently in transverse directions at certain speed. This speed is called

   (a) critical speed   (b) whirling speed   (c) whipping speed   (d) all of the above.   [Ans. (d)]

8. The critical speed of the shaft carrying a mass m at the centre of the span is given by

   (a) $\omega_C = \sqrt{\dfrac{s}{m}}$ (b) $\omega_C = \sqrt{\dfrac{g}{\delta}}$ (c) both a and b   (d) none of the above.   [Ans. (b)]

9. When heavy rotating masses are connected by a shaft and equal and opposite torques are applied to these masses (rotors).

   (a) the rotors vibrate torsionally in the same direction
   (b) the rotors vibrate torsionally in the opposite direction
   (c) there is one point on the axis of shaft which remains undisturbed by vibration
   (d) both (b) and (c).   [Ans. (d)]

10. Three n rotors mounted on the shaft and when subjected to torsional vibration there will be

    (a) n nodes   (b) $(n-1)$ nodes   (c) $(n+1)$ nodes   (d) any number of nodes.   [Ans. (b)]

Bibliography

[1] Bevan Thomson, *Theory of Machines*, CBS Publishers, 3rd Edition, 2004.

[2] Joseph Edward Shigley, *Theory of Machines and Mechanisms*, Tata McGraw Hill Education Pvt., Ltd., 2nd Edition.

[3] Henry T. Brown, *Five Hundred and Seven Mechanical Movements*, General Books.

[4] Rattan S.S., *Theory of Machines*, Tata McGraw Hill Education Pvt. Ltd., 3rd Edition.

[5] Ballaney P.L., *Theory of Machines and Mechanisms*, Khanna Publications.

[6] Khurmi R.S. and Gupta J.K., *Theory of Machines*.

[7] e-Library.

Index

A

Acceleration 51
    Coriolis component of 78
    in mechanisms 70
Acceleration analysis 71
Ackermann steering gear 18, 35
Angle of friction 151
Angular acceleration 70, 98
Angular displacement 52
Angular velocity 52, 53
Anti-friction bearings 178
Approximate straight line mechanism 31
Arc of contact 335

B

Band brake 170, 174
Base circle 289
Beam engine 18
Belt drives, types of 230
    velocity ratio of 232
Belt, slip of 234
    power transmitted by 241
    types of 227
Bevel gears 347
Block shoe brake 168
Block brake 174
Brakes, types of 168
Bull engine 14, 15

C

Cam 285
    types of 288
    with specified contours 312
Cam angle 290
Cam profile 289
    construction of 303
Centrifugal couple, effects of 270, 275
Centrifugal governors 189
Centrifugal tension, effect of 241
Centrodes 53
Chain drives 245
Chains, types of 246
Circular arc cam 312
Circular belt 229
Clutches 164
Coefficient of fluctuation of speed 131
Coefficient of insensitiveness 219
Completely constrained motion 6
Compound belt drive 231
    velocity ratio of 233
Compound epicyclic gear train 371
Compound gear train 358
Compound mechanisms 12
Cone clutch 166
Cone pulley 232
Connecting rod
    angular velocity of 98
    force along 105
    weight of 109

Constant velocity ratio 332
Crank and slotted lever quick-return mechanism 13
Critical speed 468
Crossed belt, length of 237
Cycloidal motion 297
Cycloidal teeth 345

D

D' Alembert's principle 94
Davis steering gear (exact) 34
Degrees of freedom 11
Differential band brake 170
Direct and reverse crank method 435
Direct crank 441
Displacement 51
Displacement diagrams
    construction of 299
    for cycloidal motion 302
    for uniformacceleration and retardation 301
Double Hooke's joint 41
Double slider crank chain 15
Dunkerley's method 464
Dynamic friction 150
Dynamic magnifier 478
Dynamically equivalent system 109, 110
    determination of 111
Dynamometers 176

E

Elements 3
    nature of contact between 5
    relative motion between 5
Ellipse trammels 15
Energy (Rayleigh's) method 464
Energy method 453
Energy, fluctuation of 129
Engine, reciprocating parts of 104
Epicyclic gear train 362
    torque in 369
    with bevel gears 375

Equilibrium of body on a rough inclined plane 153
Equivalent dynamical system 110

F

First equation using pairs 8
Five-cylinder radial engine 434
Flat belt 228
Flywheel 130, 209
    design of 132
    energy stored in 131
    in punching press 141
Follower
    analysis of motion of 291
    types of 286
Forced closed pair 10
Forced damped vibration, frequency of 476
Four-bar mechanisms 17
Four-stroke internal combustion engine 128
Four-wheeler, stability of 268
Free damped vibrations, frequency of 471
Free longitudinal vibrations, natural frequency of 451
Free torsional vibrations 493
Free torsional vibrations, natural frequency of 487
Free transverse vibrations 459
Friction
    between dry surfaces 151
    between rough surfaces 151
    cause of 149
    laws of 150
    self adjusting 151
    types of 178

G

Gear drive 355
Gear train 355
Gears, terminology used in 328
Governors 189, 209
    parts and terms used in 191
Grasshopper mechanism 32
Greasy friction 180

Gyroscope 257
Gyroscopic couple 258
    effect of 259, 264, 265, 266, 269, 274

H

Hammer blow 419
Hart mechanism 29
Hartnell governor 209
Hartung governor 213
Helical gears 346
Higher pairs 5
Hooke's joint 36

I

Incompletely constrained motion 7
Initial tension, effect of 243
Instantaneous centre method 53
Instantaneous centres
    application of 57
    in a mechanism 55
    location of 55
    properties of 54
    types of 55
Internal expanding shoe brake 176
Inversion 12
Inverted tooth 247
Involute gears, interference in 344
Involute tooth 346

K

Kennedy's theorem 56
Kinematic chain 7
    with three lower pairs 11
Kinematic link 3
Kinematic pair 5
Klein's acceleration diagram 101
Klein's velocity diagram 100

L

Law of gearing 332
Laws of friction 150
Lift 290
Linear and angular acceleration, vector form
    between 70

Linear velocity 53
Link 3
    classification of 4
    types of 5
Loading, types of 459
Locomotives, types of 420
Logarithmic decrement 474
Longitudinal vibrations 451
Lower pairs 6

M

Magnification factor 478
Maximum efficiency 155
Maximum power, condition for 242
Mechanical advantage 158
Mechanism 11
Motion 51
Motion down the plane 154
Motion up the plane 154
Multi-cylinder in-line engines 428
Multicylinder petrol engine 120
Multi-plate clutch 165

N

Natural frequency 459

O

Oldham coupling 16
Open belt, length of 235
Oscillating cylinder mechanism 12

P

Pantograph 27
Partial balancing 416
    in two-cylinder locomotives 417
Peaucellier mechanism 29
Pendulum engine 14, 15
Piston effort 105
Piston
    acceleration of 97
    displacement of 96
    velocity of 97
Pitch circle 291
Pitch curve 290

Pitch point 290
Pivot and collar friction 159
Porter governor 194
Pressure angle 290
Pressure, uniform intensity of 161
Prime circle 290
Proell governor 197
Prony brake 176

Q
Quick-return mechanisms 14

R
Radial engines 434
Radian 52
Ratio of tensions 239, 243
Rayleigh's method 454
Reciprocating engine 4
    inertia forces in 113
Reciprocating mechanism 12, 415
    analytical method for 95
    Klein's construction for 100
Relative velocity method 64
Retardation 296
Reverse crank method 441
Reverted gear train 359
Right-hand screw rule 53
Roberts mechanism 32, 33
Roller chain 246
Rolling friction 150, 152, 178
Rope 229
Rope brake dynamometer 177
Rope drive 243
Rotary I.C. engine 14
Rotary motion 51
Rotating element
    balancing of 392
    checking of 391

S
Scotch yoke mechanism 16
Scott-Russell mechanism 30
Screw friction 156

Screw jack, effort and weight lifted by 157
Second equation using joints 9
Self-closed pairs 10
Several unbalanced rotating masses 392, 393
    rotating in the same plane 399
    rotating in several planes 402
Shaft, whirling speed of 468
Ships, special terms used in 263
Silent chain 247
Simple band brake 172
Simple gear drive 355
Simple gear train 355
    power transmitted by 357
Simple harmonic motion 294
Simple harmonic motion, displacement diagram for 300
Simple mechanisms 3, 12
Simple Watt governor 191
Single cylinder engine 119
Single slider crank chain 12
Single unbalanced rotating mass 392
Single-cylinder double-acting steam engine 127
Single-plate clutch 165
Sliding friction 150
Speed 51
Speed ratio 232, 356
Speed value 356
Spiral gears 348
Square thread 156
Static friction 149
Steering gear mechanism 33
Stepped pulley 232
Straight line motions, mechanisms for 28
Stroke 290
Successfully constrained pairs 7
Swaying couple 419

T
Tangent cam 315
Tchebicheff straight line mechanism 32
Teeth, forms of 344

Three-centres-in-line theorem 56
Toothed gearing, classification of 325
Torsional vibrations 451
Torsional vibrations 488, 493
       of a geared system 505
Torsionally equivalent shaft 499
Trace point 289
Tractive force 418
Train value 356
Translatory motion 51
Transverse vibrations 451
Two-way brand brake 172
Two-wheeler, stability of 273

U

Uniform acceleration 296
Uniform velocity 292
Uniform velocity displacement, diagram for 300

Universal joint 36
Unknown instantaneous centres 57

V

V-belt 228
Velocity 51
Velocity ratio 356
    effect of creep on 235
V-engines 439
Vibrations, types of 450, 451
Vibratory system, basic elements of 449
Viscous damping 471
V-thread 158

W

Watt mechanism 31
Wear, uniform rate of 162
Wilson-Hartnell governor 219